ATOMIC and MOLECULAR PHYSICS

Third US/Mexico Symposium

ATOMIC and MOLECULAR PHYSICS

Third US / Mexico Symposium

Morelos, Mexico 13 – 16 March 1991

Editors:

C. Cisneros
I. Alvarez
IFUNAM

T. J. Morgan
Wesleyan University

World Scientific
Singapore • New Jersey • London • Hong Kong

Published by

World Scientific Publishing Co Pte Ltd
5 Toh Tuck Link, Singapore 596224
USA office: 27 Warren Street, Suite 401-402, Hackensack, NJ 07601
UK office: 57 Shelton Street, Covent Garden, London WC2H 9HE

British Library Cataloguing-in-Publication Data
A catalogue record for this book is available from the British Library.

ATOMIC AND MOLECULAR PHYSICS

ISBN-13 978-981-02-0843-1
ISBN-10 981-02-0843-X

FOREWORD

The Third US/Mexico Symposium on Atomic and Molecular Physics — "Few Body Problems, Electron Correlation Effects and Beyond" was held at Cocoyoc, Morelos, Mexico, March 13–16, 1991. This meeting continues the tradition of the early conferences in reviewing some aspects of atomic and molecular physics. The symposium brought together 45 leading investigators in the fields of recombination, molecular dissociation, ionization, electron impact collisions, simply heavy particle collision systems, excited states, correlation effects, photoionization and photodetachment, multiply-charged ion collisions, and synchrotron radiation. Both theory and experiment were well represented. The papers in this volume are based on participants' invited talks at the symposium and represent an important collection reviewing the most up-to-date work in these fields.

The diversity of research areas represented, coupled to a common focal point, revealed different perspectives on such questions as: "How do particles depend on one another?" Correlation provides a good starting point, but are there other useful conceptual approaches which will yield a new view of the phenomena and a more complete understanding?

The contributions in this volume provide different perspectives and bring together a collection of papers which serve as a basis for further progress in this field.

The Organizers

ORGANIZING COMMITTEE

Carmen Cisneros — Instituto de Fisica, Universidad Nacional Autonoma de Mexico, Cuernavaca, Mexico

Thomas J. Morgan — Physics Department, Wesleyan University, Middletown, CT USA

EDITORS

Ignacio Alvarez, Carmen Cisneros — Instituto de Fisica, Universidad Nacional Autonoma de Mexico, Cuernavaca, Mexico

Thomas J. Morgan — Physics Department, Wesleyan University, Middletown, CT USA

MEETING SPONSORED BY

— Consejo Nacional de Ciencia y Tecnologia, CONACyt, Mexico

— Instituto de Fisica, Universidad Nacional Autonoma de Mexico, IFUNAM, Cuernavaca, Mexico

— National Science Foundation, NSF, United States of America

CONTENTS

IV. SIMPLE SYSTEMS

V. ELECTRON IMPACT COLLISIONS

VI. EXCITED STATES

IX. HIGHLY CHARGED ION COLLISIONS

X. SYNCHROTRON RADIATION

I. RECOMBINATION

RECENT PROGRESS ON RECOMBINATION WITH HIGHLY CHARGED IONS

Lars H. Andersen
Institute of Physics
University of Aarhus, DK-8000 Aarhus C
DENMARK

Abstract

Radiative- and dielectronic-recombination measurements have been performed for a number of highly charged ions in the past few years. The status of the field will be recapitulated. Emphasis will be put on the impact of recent experimental progress. Radiative recombination for bare ions as well as for ions carrying electrons is considered. For dielectronic recombination, we will focus on selected systems which are still a challenge to theory. Finally, we discuss the possible influence on electromagnetic fields on the two recombination processes.

1. Introduction

Collisions between free electrons and positive ions take place in many places such as in man-made plasmas, the solar corona, in gaseous nebulae, and in the interstellar space. Much of the early concern about electron-ion recombination was indeed related to astrophysics. Since much of our information about distant parts of the Universe exists in the form of photons, which may be regarded as information carriers, we may learn about the history of the Universe by knowing the *physical* origin of photons as well as the interaction between matter and light in the media in which the photons travel.

The recombination of an electron with a positive ion normally results in the emission of one or more light quanta. The field of electron-ion recombination, however, is not only motivated by astrophysics and plasma physics. Since, as will be discussed below, doubly excited states are often involved, as well as high Rydberg states, precise knowledge about the *structure* of atoms and ions is required to accurately predict the rate of recombination between free electrons and positive ions and molecules. Some of the states involved are extremely sensitive to external fields and interactions between different electronic configuration. This is reflected in the experimental data, and thus experimental information on the recombination processes is essential to guide us to the right theoretical comprehension.

Electrons are present both in solids and gases, and the free-electron - ion interaction finds its equivalent in ion-atom and ion-solid collisions. The first step to understand such more complex collision processes is to understand the collisions between the constituent particles, electrons and positive ions. Among other processes like ionization and excitation, recombination is one such process which may take place when a free electron interacts with a positive ion.

As an electron approaches a highly charged ion, it is accelerated, and large kinetic energies may be obtained, especially in the vicinity of the nucleus. To become bound to the ion, the electron must exchange energy and momentum with a third 'particle'. In the case of radiative recombination (RR), this exchange process takes place due to the coupling between matter and the electromagnetic field. During the large acceleration near the nucleus, a photon may be created, and as a result, the electron has no longer sufficient energy to escape the Coulomb field of the positive ion. Predominantly, final states with low main quantum number n are populated. The radiative-recombination process may be written as:

$$X^{Z+} + e^- \rightarrow X^{(Z-1)+}(n) + h\nu , \qquad (1)$$

where X is the positive ion with total charge Z, $h\nu$ the emitted photon, the energy of which equals the kinetic energy of the free electron plus the binding energy of the state n. The RR process may take place at any initial energy of the free electron.

The recombination process, known as dielectronic recombination (DR), is a resonant process in which the incoming electron excites an electron which is present on the ion. It thereby loses energy and becomes bound to the ion, provided the initial kinetic energy is lower than the excitation energy. In this way, doubly excited ionic states are created. If these states decay radiatively, the electron cannot escape from the ion, and the DR process is completed. If the doubly excited states decay via autoionization, the process appears as a resonance in the elastic-scattering cross section. The DR process can be represented by

$$X^{q+} + e^- \rightarrow (X^{(q-1)+})^{**} \rightarrow (X^{(q-1)+})^{*} + h\nu . \qquad (2)$$

Unlike radiative recombination, dielectronic recombination occurs only when the ion X carries electrons. If the core-excitation energy is ΔE and the initial kinetic energy of the electron is E_r, then for capture to the state nl, energy conservation yields

$$E_r + E_{nl} = \Delta E , \qquad (3)$$

where E_{nl} is the binding energy of the state designated nl. In the case where ΔE is small compared to the total binding energy of the core electron, high Rydberg states (nl) are populated.

At this point it should be mentioned that another process, called ternary recombination (TR) or three-body recombination, exists. It is similar to dielectronic recombination except that the exchange of momentum and energy is between the incomming free electron and another free electron rather than a bound one. Obviously, TR becomes important when the density of free electrons is high. The cross section scales like $T^{-9/2}n_e^2$, where n_e is the electron density and T is the electron temperature in the rest frame of the ion. Except for the situation with ultracold electrons or extremely high electron densities, TR is less important than DR and RR. In Fig. 1 is shown the three recombination processes and their laser-induced variants, to be discussed later.

From an experimental point of view, recombination between a free electron and an ion, with well defined relative energy, is difficult to measure. The cross sections connected with recombination are normally quite small, and hence, a signal due to electron capture from atoms or ions present in the experimental setup may easily dominate the true recombination signal originating from the target of free electrons. If the vacuum conditions are modest ($\approx 10^{-10}$ torr), intense sources of monoenergetic electrons are needed. A very compressed electron beam as a target may, however, deteriorate the energy resolution in the experiment. In order to achieve good energy resolution, the electron target has to be relatively thin and as a consequence, the vacuum conditions good ($\approx 10^{-11}$-10^{-12} torr).

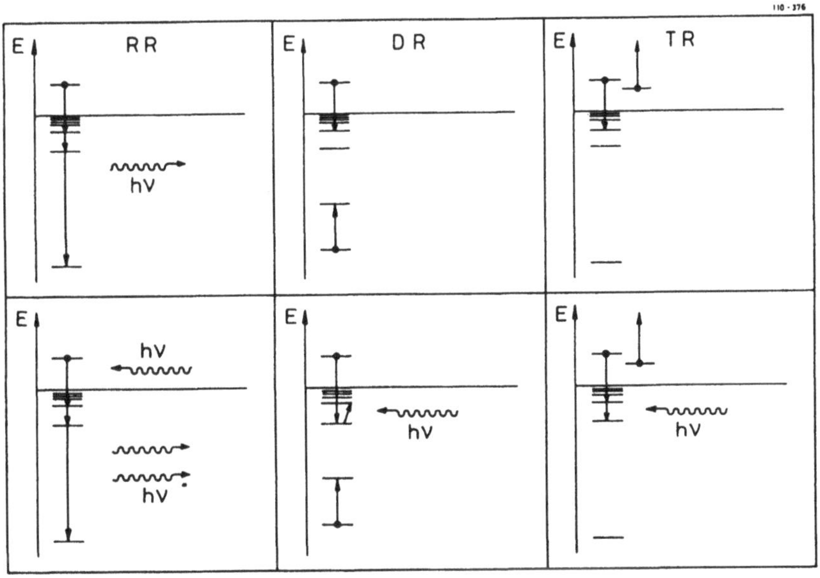

Fig. 1. Diagram of three different recombination mechanisms and their laser-stimulated variants (see later in text).

With the new developments of electron coolers for ion-storage rings, the experimental situation has indeed improved. With these devices, which normally have to operate under ultra-high vacuum conditions, it is possible to obtain good energy resolution and at the same time sufficiently high electron densities for merged-beam experiments to be performed. With this technique, the first results of RR have been obtained[1]. DR measurements with intrashell excitations ($\Delta n'=0$)[2,3] and with $\Delta n'>0$ have been performed[4]. It is now possible to test the various theoretical approaches on a high level. Due to a rather narrow energy distribution of the electron beam (0.15 eV), the magnitude of the cross section as well as spectroscopic information is available from the experiments.

At the University of Aarhus, an electron-cooler device was built to serve as a prototype for a future electron cooler at the heavy-ion storage ring ASTRID. The cooler also served as a target of electrons for studies of collisions between low-energy electrons and highly charged ions. A detailed discussion of the apparatus is found elsewhere[2]. In Fig. 2 is shown the experimental apparatus. In the remainder of this paper, we give examples of measurements with this cooler of radiative recombination for both bare ions and ions carrying electrons. Further, we consider dielectronic recombination associated with $2s$ to $2p$ excitations in various ions.

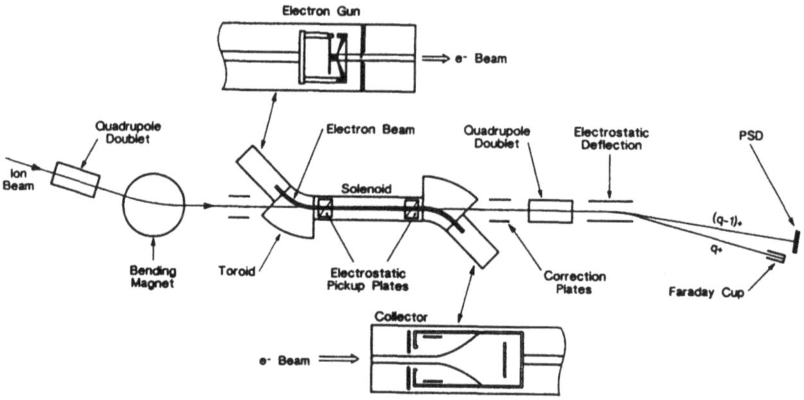

Fig. 2. The experimental setup used at the University of Aarhus.

2. Radiative Recombination

In the case, where the wave length of the photon is much larger than the extension of the bound state (dipole approximation), Bethe and Salpeter[5] obtained the following simple analytical expression for the cross section for radiative capture to a hydrogenic level with main quantum number n:

$$\sigma_n = 2.10x10^{-22}cm^2 \frac{Z^4 E_0^2}{nE_e(Z^2E_0 + n^2E_e)} , \qquad (4)$$

where E_e is the kinetic energy of the free electron, E_0 is the Rydberg energy, and Z is the charge of the fully stripped ion. The theory of Bethe and Salpeter was based on the assumption that the oscillator strength is continuous across the continuum limit. Formula (4) is thus expected to be most accurate for high values of n. Indeed, corrections must be applied for the lower states. For $n=1$, 2, and 3, where the largest cross sections are found, a more accurate calculation was given by Stobbe[6] in 1930. At low energies, these corrections are 20% for $n=1$, 12% for $n=2$, and 9% for $n=3$, the cross sections of Stobbe being smaller than those of Bethe and Salpeter.

Figure 3 presents the experimental results obtained with bare ions. The rate coefficient $<v\sigma>$ is shown as a function of kinetic energy of the initially free electron. Energies between 0 and 1 eV are being considered. The theoretical curve shown in Fig. 3 was obtained from

$$<v\sigma> = \int \sum_n \sigma_n v f(v)\, dv , \tag{5}$$

where the electron velocity distribution $f(v)$ is given as a product of two Maxwell distributions. The theoretical cross section is given by Bethe and Salpeter, formula (5), corrected according to Stobbe for $n=1, 2$, and 3. In the experiment, the ion-charge-state separation field removes the outer Rydberg electrons. In the evaluation of $<v\sigma>$, only values of n corresponding to the actual analyzer field were included. The uncertainty of the experimental rate coefficient is estimated to be 20%. This is mainly due to the uncertainty in the determination of the effective target thickness and the ion-detection efficiency. Evidently, for the bare ions, radiative recombination at low energy is theoretically well predicted.

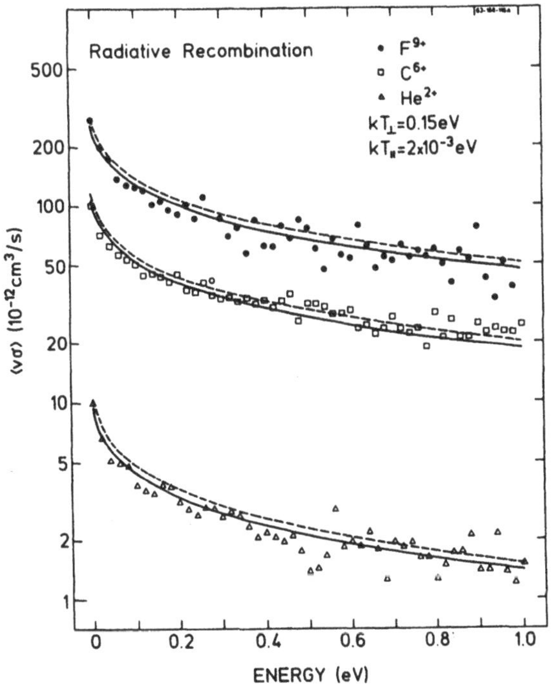

Fig. 3. The rate coefficient $<v\sigma>$ as a function of energy for RR with bare ions of He, C, and F. The solid curve is the calculation based on Eq. (4). The dashed curve is without the corrections by Stobbe.

Radiative recombination with non-bare ions is much more difficult to treat theoretically than radiative recombination with bare ions due to the much more complex wave functions involved. To gain knowledge about RR with non-bare ions, it was obvious to test how well the theory for bare ions could be used to predict the rate coefficient obtained with H-like ions. To this end, we measured the RR rate coefficient for C^{5+}, O^{7+}, and F^{8+} at relative energies between 0 and 1 eV[7]. The experimental results for C^{5+} are shown in Fig. 4. The curves shown in the figure result from formula (4), corrected according to Stobbe's theory for $n=1,2$, and 3. Further, the Pauli exclusive principle was taken into account by summing only over open shells in Eq. (5). We used an effective charge Z_{eff} in Eq. (4) which was chosen to be the asymptotic (total) ion charge. To compare with this, we performed a calculation with Z_{eff} being equal to the nuclear charge of the ion. The agreement between the calculated and experimental data is excellent when the total charge is used in the calculation.

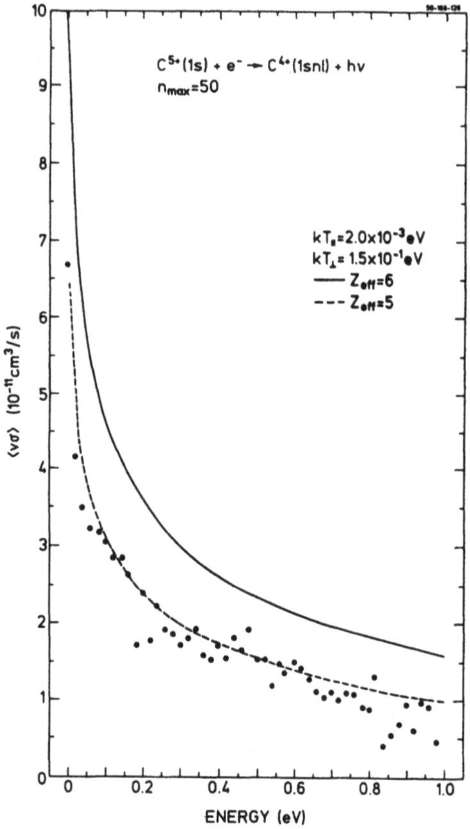

Fig. 4. The RR rate coefficient $<v\sigma>$ with H-like C^{5+}.
Solid curve: $Z_{eff}=Z_{nucl}$, dashed curve: $Z_{eff}=Z_{nucl}$.

To qualitatively understand the charge dependence of RR with non-bare ions, we note that at low energies ($E << Z^2E_0/n^2$), σ_n scales as $1/n$, meaning that a large part of the total cross section relates to the excited states. Further, the final states φ_{nl} mostly have angular momentum numbers l which are larger than zero[1]. Such states do not penetrate much into the core region, and consequently, the matrix element $<\varphi_{nl} | r | \psi>$ is determined in the region of space outside the core. Hence, the total cross section is expected to be characterized by the total ion charge, in agreement with suggestions by McLaughlin and Hahn[9]. They suggested that the effective charge Z_{eff} should be

$$Z_{eff}=\frac{Z_C+Z_I}{2}-\left(\frac{Z_C-Z_I}{2}\right)\left(\frac{\tilde{n}^2}{\tilde{n}^2+1}\right), \quad \tilde{n}\equiv\frac{Z_C}{p_C} \tag{6}$$

where Z_C and Z_I is the core charge and the total ion charge, respectively. Z_C is the momentum of the free electron. For $p_C \rightarrow 0$, it is seen that $Z_{eff} \rightarrow Z_I$, the total ion charge.

As a final example of RR with non-bare ions, the data obtained with O^{5+} are shown in Fig. 5. The solid curve is the calculation based on formula (4) (corrected for $n=1$, 2, and 3), with the effective charge being equal to the nuclear charge ($Z_{eff}=8$). The dashed curve is that obtained with $Z_{eff}=5$, the total charge. In both cases, the Pauli exclusive principle was taken into account by excluding the two 1*s* orbitals and the one 2*s* orbital. A very good agreement is found for relative energies lower than 0.4 eV when the total charge is used in the calculation. At higher energies, the experimental data exceed the theoretical prediction by as much as 50%. This difference has recently been accounted for by McLaughlin and Hahn[8]. They applied different effective charges for the different shells. The dot-and-dash curve in Fig. 5 is their calculation with $Z_{eff}=6.5$ for n=2 and $Z_{eff}=5.1$ for $n \geq 3$.

Radiative recombination is equivalent to the spontaneous electronic transition from one bound state to another. Like in bound-bound transitions, it is accomplished by the emission of light and may also be stimulated by intense electromagnetic radiation (lasers). Following the lines of an earlier paper[10], we obtain the following expression for the gain G_n, defined as the ratio between the rate of induced recombination (to a specific state n) and the total rate of recombination to all states:

$$G_n=\frac{1}{n}\frac{4.42x10^{-6}}{\sqrt{kT_\perp(eV)\,kT_\parallel(eV)}\,(h\nu(eV))^3}\,I(W/cm^2) \tag{7}$$

where $kT_\perp$ and $kT_\parallel$ are the two temperatures which charaterize the electron beam, $h\nu$ is the photon energy in the frame of the ion, and I is the intensity of the applied electromagnetic field. Equation (7) is valid for electrons with zero energy relative to the heavy ion. Obviously, low temperatures of the electron beam are desirable. This means that the laser communicates with a large fraction of the electrons which may all participate in the spontaneous process. The photon energy $h\nu$ is equal to the binding energy ($Z^2 13.6eV/n^2$) for a given ion. Thus, G_n scales as n^5, which means that capture to high n states is much favored. In other words, it is best to use a laser with small photon energy to obtain a large gain. As an numerical example, consider a Nd:YAG laser with $h\nu$ equal to 1.17 eV, and assume $kT_\perp=0.15$eV and $kT_\parallel=0.002$ eV. For a 1 Mev/amu He^{2+} beam, the photon energy will be Doppler-shifted to 1.11 eV, which is equal the binding energy of $n=7$. In order to obtain a gain G_7 of 1, the laser intensity must be in the order of 40 kW/cm^2. This can be realized by the pulsed-laser technique.

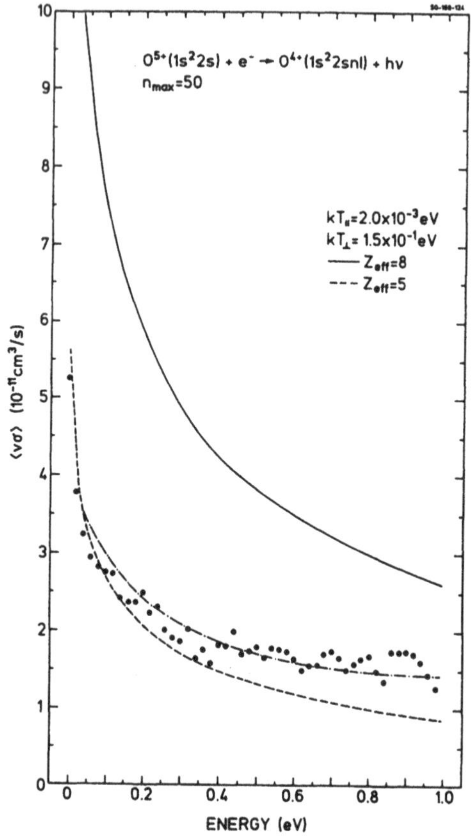

Fig. 5. The rate coefficient $\langle v\sigma \rangle$ for RR with Li-like O^{5+}. Solid curve: $Z_{eff}=Z_{nucl}$, dashed curve: $Z_{eff}=Z_{nucl}$, dot-and-dash curve: Calculation by McLaughlin and Hahn[8].

3. Dielectronic Recombination

The cross section for dielectronic recombination into an isolated resonance nl is given as

$$\sigma_{nl}=\sigma_0\ 2(2l+1)\frac{A_a(n,l)A_r(n,l)}{A_a(n,l)+A_r(n,l)}, \qquad (8)$$

where σ_0 is a proportionality constant and $2(2l+1)$ is the statistical weight of the final state. A_a and A_r are the autoionization decay rate and the radiative decay rate, respectively. Since the first step in the DR process is an inverse Auger process, it is proportional to A_a. The second step is the radiative stabilization which happens with a probability given by $A_r/(A_a+A_r)$. In a given situation, the Auger and the stabilizing radiative decay may happen via several channels, and the cross-section calculation must include a sum over partial decay rates. The radiative stabilization is normally due to the transition of the core when the final state is a high Rydberg state. A_r is then nearly independent of l, whereas A_a strongly decreases with increasing l. When, as is often the case for low values of l, $A_a >> A_r$, it is seen from Eq. (8) that $\sigma_{nl} \propto 2(2l+1)A_r(nl)$. For high values of l, $A_a << A_r$ and the cross section are negligibly small. Thus there exists a critical value l_c of l defined by $A_a(n,l_c) \approx A_r(n,l_c)$. All levels with $l < l_c$ contribute to the DR process while levels with $l > l_c$ may be neglected.

We shall here discuss DR associated with the $2s \rightarrow 2p$ excitation of different ions. Obviously, Li-like ions with the ground state $1s^2 2s$ are candidates for DR resonances related to this excitation. In Table I, we list the excitation energies of this transition for C,N,O, and F ions. As is seen, the excitation energy is low ($\approx$10eV) compared to the ionization potential of the respective Li-like ions. Thus, rather high Rydberg states are populated according to Eq. (3). DR data obtained with Li-like ions have previously been obtained by the group from Oak Ridge National Laboratory[11]. Unfortunately, the energy resolution in these early experiments was not good enough to resolve the individual DR resonances.

Table I
Excitation energy (eV) for $2s \rightarrow 2p$ in Li-like ions.

C	N	O	F
8.0	10.0	12.0	14.0

As an example of DR with Li-like ions, Fig. 6. shows a spectrum obtained with Li-like N^{4+}. The DR signal clearly exhibits resonance structure which is related to capture into different Rydberg states $\{nl\}$. The resonance energy E_{nl} is obtained from Eq. (3), which yields:

$$E_{nl} = \Delta E - \frac{z^2\, 13.6\text{eV}}{n^2} \qquad (9)$$

This is a good approximation when high Rydberg states are populated. The present energy resolution ($kT_\perp$) is 0.15 eV, and as is seen from Fig. 6, resonances associated with n=5-11 are resolved. Note that the resonance energies are observed midway down on the high-energy side of the experimental resonances as expected[2]. The dominating feature is the contribution from the many very high-n states. The dominance is due to the high density of states. The use of a strong electric field to separate signal (N^{3+}) from the incoming beam (N^{4+}) in the experiment sets an upper limit to the number of Rydberg states which actually contribute to the signal. Typically, n_{max} is around 50.

One of our primary concerns about the Li-like DR data is the influence of small electric DC fields in the interaction region on the size of the rate coefficient for the Rydberg contribution. A small electric field may perturb the Rydberg states and cause a mixing of the low-l states which are primarely populated via the inverse Auger process, with higher-l states which are stable against autoionization. As discussed by Harmin[12], this has two effects. It results in the formation of states which are more likely to stabilize radiatively, and in an increased l_c, which means that more l-states contribute to the cross section, as discussed above. The total effect of the field cannot, however, be obtained just by counting available states; one must consider interference effects from overlapping, interacting manifolds. The cross section is expected to increase with increasing electric fields up to some maximum field.

A small DC electric field of about 2-5V/cm is expected in the interaction region due to the space-charge potential created by the electron cloud. A comparison between a distorted-wave calculation[13] and previous DR measurements[2] with Li-like C and O ions showed that as one increases the atomic number of the ion, an increasing field strength had to be applied in the calculation in order to bring experiment and theory in agreement. This trend was also confirmed by data obtained with F^{6+}. The experimental rate coefficients from the Oak Ridge group with Li-like ions also exceeded the theory. Currently, experiments are being prepared with the application of a small external electric field perpendicular to the beam direction to study these effects in more detail for the Li-like ions.

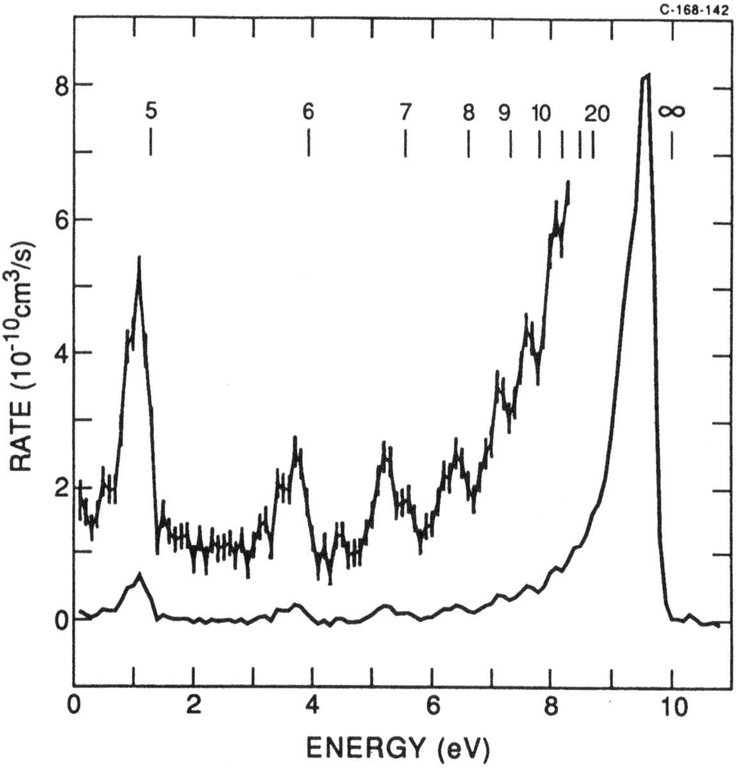

Fig. 6. The rate coefficient for DR with Li-like N4+.
The expected positions of the resonances are indicated.

Another very interesting case of $2s \rightarrow 2p$ excitation is that obtained with metastable He-like ions. With the use of highly charged He-like ions (C^{4+}, N^{5+}, O^{6+}, F^{7+}), a substantial amount of $1s2s(^1S)$ and $1s2s(^3S)$ states are present in the beams. The ^3S metastable fraction is typically $\approx 20\%$, while the ^1S metastable is in the order of 1%. The lifetime of the ^3S state is always long compared to the ion flight time, whereas the ^1S lifetime is sometimes much shorter than the ion flight time, and a notable reduction due to the exponential decay is encountered. With the ground-state $1s^2$ ions, no DR resonances are expected in the low-energy region of 0 to 20 eV considered here. The DR process in this case has been studied at the Test Storage Ring in Heidelberg[14]. With the metastable states, the following transitions may happen:

$$\begin{aligned} 1s2s(^1S)+e^- &\rightarrow 1s2p(^1P)nl \\ 1s2s(^3S)+e^- &\rightarrow 1s2s(^1S)nl \\ &\rightarrow 1s2p(^3P)nl \\ &\rightarrow 1s2p(^1P)nl \end{aligned} \tag{10}$$

The excitation energies are given in table 2. The strength of the individual series depends on the radiative rate of the doubly excited intermediate states. Clearly, the $1s2p(^1P)$ core state has a very large decay rate associated with the transition to the $1s^2(^1S)$ state. The transition energy is several hundreds of electron volts. Calculations of DR with He-like ions have been carried out by Ballantone and Hahn [15] and by Badnell et al.[16].

Table II
Excitation energies (eV) for metastable He-like ions[15].

	$^1S\rightarrow{}^1P$	$^3S\rightarrow{}^1S$	$^3S\rightarrow{}^3P$	$^3S\rightarrow{}^1P$
C^{4+}	3.50	5.41	5.42	8.91
N^{5+}	4.28	6.63	6.52	10.91
O^{6+}	5.06	7.83	7.61	12.89
F^{7+}	5.86	9.04	8.71	14.90

As an example of DR with metastable He-like ions, Fig. 7 shows a spectrum obtained with F^{7+}. In this case, the ^1S initial state has almost entirely decayed before the electron target. The main resonances are due to the formation of $(^1P)nl$ resonance states from the ^3S initial state. As noted above, the dominance of this transition is due to the strong stabilization of the ^1P core. The interesting thing to note from the spectrum is the ratio between the different n resonances of the $^3S\rightarrow{}^1P$ series. Normally, the cross section for members of the same series varies as $1/E$. This apparently is not the case for $n=7,8,9$ and 10. Here the cross section increases with energy. Similar peculiarities have been seen in other He-like systems[2].

It was early suggested by Taulbjerg and Macek[18] that such enhancement could be an effect of interaction between the ^1P configuration and the two configurations ^1S and ^3P. As is seen from Fig. 8, the $n=9$ and $n=10$ members of the $1s2p(^1P)nl$ series are very close in energy to $n\approx17$ and $n\approx30$-40 members of the $1s2s(^1S)nl$ and $1s2p(^3P)nl$ series, respectively. If there were sufficient interaction between the series, doubly excited resonance states like, for instance, $1s2s(^1S)30l$, would mix with the $1s2p(^1P)10l$ states. Due to the strong radiative transition $1s2p(^1P)\rightarrow 1s^2(^1S)$, this could result in a considerable increase in the DR cross section, as seen in the experimental data. It is possible that the mixing could be Stark-induced by the space-charge field in the interaction region. This has not yet been tested neither experimentally nor theoretically.

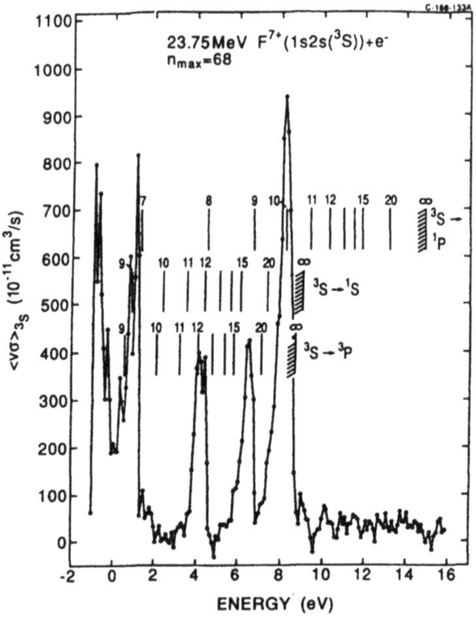

Fig. 7. The rate coefficient for DR with metastable He-like F^{7+} ions.

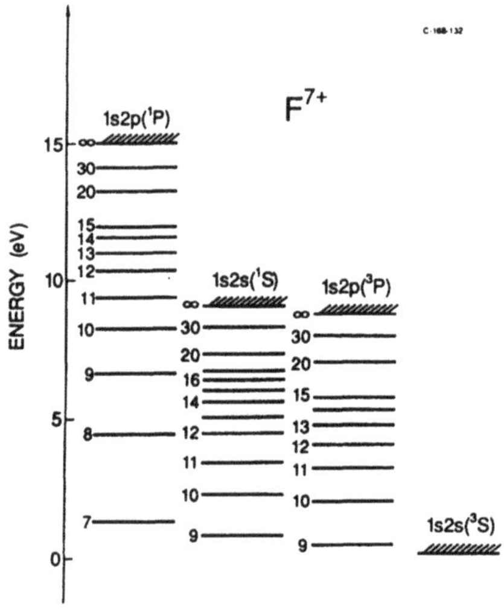

Fig. 8. Level diagram for F^{7+}.

Above the ^1S and ^3P continua, $1s2p(^1P)nl$ $(n > 10)$ states may **autoionize** to the these continua. This would result in a reduction of the DR cross section, as observed experimentally.

Since DR is a two-step process: first a formation of a resonance state and then a radiative decay of this state, the largest cross sections are found for those states which are easily formed (large A_a) and then decay by photon emission with a large probability (large A_r). When, as is often the case, A_a is much larger than A_r, an enhancement of the DR rate may occur when the resonance state (with low A_r) is coupled to other states with larger radiative decay rate. As discussed above, this coupling between states with different decay properties could be due to configuration interaction. We could call this the 'intrinsic' gain. It is also possible to introduce an 'extrinsic' gain by applying an electromagnetic (laser) field to the resonance state. In order to achieve a gain for the DR process, one would have to couple the resonance state φ_{nl} with low A_r to another state φ^*_{nl} which has a substantially larger radiative decay rate. The intensity of the applied field has to be strong enough to compete with the Auger decay rate of the resonance state φ_{nl}. The state would otherwise autoionize before it had time to couple to φ^*_{nl}. By recording the DR gain as a function of the laser frequency, one would be able to study the properties of such doubly excited resonance states. So far, no experimental results have been obtained with 'extrinsic' DR gain (laser-stimulated dielectronic recombination).

Acknowledgements

This work has been supported by the Carlsberg Foundation and the Ib Henriksen Foundation.

References

1. L.H. Andersen and J. Bolko, Phys.Rev A. **42** (1990) 1184
2. L.H. Andersen, J. Bolko, and P. Kvistgaard, Phys.Rev. A **41** (1990) 1293
3. L.H. Andersen, P. Hvelplund, H. Knudsen, and P. Kvistgaard, Phys. Rev.Lett. **3** (1989) 2656
4. G. Kilgus et al., Phys.Rev.Lett. **64** (1990) 737
5. H. Bethe and E. Salpeter, *'Quantum Mechanics of One- and Two-Electron Systems'*, Vol. **35** of *Handbuch der Physik* (Springer, Berlin, 1957).
6. M. Stobbe, Ann.Phys. **7** (1930) 661
7. L.H. Andersen and J. Bolko, J.Phys. B. **23** (1990) 3167
8. D.J. McLaughlin anf Y. Hahn, Phys. Rev. A **43** (1991) 1313
9. J. McLaughlin and Y. Hahn, private communication.
11. P.F. Dittner, S. Datz, P.D. Miller, P.L. Pepmiller, and C.M. Fou, Phys. Rev. A **35** (1987) 3668
12. D.A. Harmin, Phys. Rev. Letters **57** (1986) 1570
13. D. C. Griffin, M. S. Pindzola, and P. Krylstedt, Phys. Rev. A **40** (1989) 6699
14. G. Kilgus et al., private communication
15. R. Ballontone and and Y. Hahn, Physica Scripta **42** (1990) 650
16. N. R. Badnell, M.S. Pindzola, and D.C. Griffin, Phys. Rev. A **41** (1990) 2422
17. G.W. Drake, Can. J. Phys. **66** (1988) 586
18. K. Taulbjerg and J. Macek, Phys. Rev. Lett. **62** (1989) 2766

NEW EXPERIMENTS ON H_3^+ RECOMBINATION.

J.B.A. Mitchell
Dept. of Physics
University of Western Ontario
London, Ontario, Canada
N6A 3K7

and

B.R. Rowe
Dépt. de Physique Atomique et Moleculaire
Université de Rennes
35042 Rennes
France

Abstract.

New flowing afterglow experiments have yielded a room temperature rate for the dissociative recombination of H_3^+ which is much larger than that found by Adams et al[1]. A description of these measurements is given along with a discussion of the state of excitation of the ions in afterglow experiments.

Introduction.

H_3^+ has a central importance in interstellar chemistry. It is formed via the reaction scheme:

$$H_2 + \text{photon/cosmic Ray} \rightarrow H_2^+ + e \quad \ldots(1)$$
$$H_2^+ + H_2 \rightarrow H_3^+ + H \quad \ldots(2)$$

The latter reaction leaves the H_3^+ ions in a wide range of vibrational states[2], but in the low density environment of interstellar clouds, they have time to relax down to the ground state before undergoing further collisions. In dense interstellar clouds, H_3^+ is destroyed primarily by reaction with carbon monoxide:

$$H_3^+ + CO \rightarrow HCO^+ + H_2 \quad \ldots(3)$$

In diffuse clouds however, it was believed for many years that the destruction of H_3^+ would be dominated by dis dissociative recombination, (DR), thus:

$$H_3^+ + e \rightarrow H + H + H \quad ...(4a)$$
$$\rightarrow H_2 + H \quad ...(4b)$$

In 1983, Adams et al.[2] at the University of Birmingham in England, published the results of a Flowing Afterglow Langmuir Probe, (FALP) experiment which indicated that in fact reaction (4) had a small rate coefficient ($< 10^{-8}$ cm^3s^{-1}). This was more than an order of magnitude lower than previous measurements[3] had indicated and challenged astrophysicists to rethink their models of diffuse cloud interstellar chemistry. It also had important implications for Jovian chemistry since it means that the ionosphere of Jupiter would be dominated, not by H^+ but by H_3^+. Discussion of many of these issues are given in ref. (4).

The results of Adams et al. had theoretical justification since calculations[5] of the potential energy curves involved in the recombination of H_3^+ indicate that the direct recombination mechanism should proceed very slowly for ions in the v=0 level. This is because, the overlap between the ion state and the neutral dissociating state lies far from the minimum of the ion curve. (See figure 1).

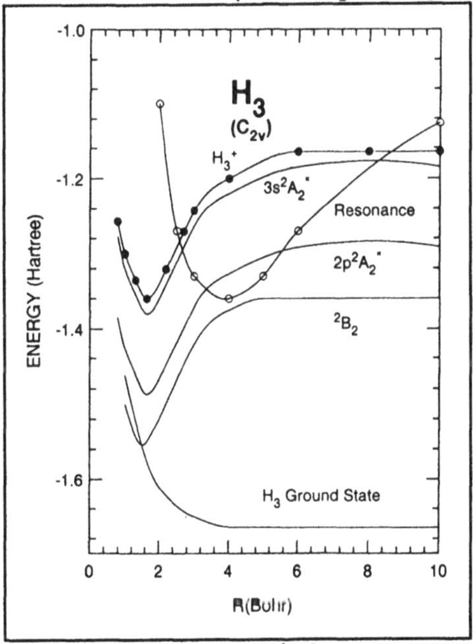

Figure 1 Potential energy curves for H_3^+ and H_3 in D_{3h} symmetry. (From ref. 5).

Hus et al.[6] remeasured the cross section for H_3^+ recombination using cold ions, produced in a radiofrequency trap ion source injected into a merged beams apparatus. They found that indeed the cross section for cold ions was more than a factor of 10 lower than previous merged beam measurements obtained using ions produced in a conventional rf source[7]. This is consistent with a rate coefficient of about 1-2 x 10^{-8} cm^3s^{-1} at 300K. Hus et al. used measurements of the threshold energies for the process of dissociative excitation:

$$H_3^+ + e \rightarrow H + H_2^+ + e \quad ..(5)$$

to determine the internal energy of the ions in their experiments.

At the time, it was not clear that they had been able to achieve complete relaxation of the ions because the excitation results were somewhat ambiguous. In particular, the threshold was found to lie below the energy predicted theoretically for v=0 and yet above that predicted for ions in the v=1 level. Subsequent re-examination of the dissociative excitation process[8] has shown that in fact the threshold region contains sharp resonant structures due to the temporary formation of neutral, autoionising/autodissociating states of H_{3**}. These studies have shown that in fact the ions in the Hus et al. experiment were in the ground state.

The situation with regard to afterglow experiments was however perplexing. Adams and Smith had recalibrated their apparatus[9] and set a new upper limit of the room temperature rate coefficient of 1×10^{-11} cm^3s^{-1}. Meanwhile, Amano[10], using an infra-red absorption technique which involved studying the decay of individual vibrational states of H_3^+, produced in a hydrogen discharge, reported a DR rate of about 2×10^{-7} cm^3s^{-1}! Now, three different laboratories, using three different techniques had succeeded in obtaining DR rates that differed by five orders of magnitude. Clearly there was a need for an independent examination of this problem but one which duplicated one of the methods already used. Such a measurement has been performed and is discussed below.

The Rennes Flowing Afterglow Experiments.

The Flowing Afterglow apparatus at the University of Rennes in France, shown in figure 2, has been used for the measurement of the dissociative recombination of H_3^+. The method used in this experiment is somewhat different from that used by Adams et al. The main difference is in the ion chemistry used in the production of H_3^+. In the Birmingham experiment, hydrogen gas is added to a helium afterglow and the following reactions occur:

$$He + He + He^+ \rightarrow He_2^+ + He \quad ...(6)$$
$$He_2^+ + H_2 \rightarrow He + He + H_2^+ \quad ..(7)$$
$$He^* + H_2 \rightarrow H_2^+ + e + He \quad ...(8a)$$
$$\rightarrow H^+ + H + e + He \quad ...(8b)$$

H_3^+ is then formed via reaction 2. Other reactions involving HeH^+ and He_2H^+ ions also lead to the production of H_3^+. Care is taken to ensure that He^+ is only a minor constituent of the plasma, following hydrogen injection, for this recombines very slowly and would maintain the ionization along the flow tube.

In the Rennes apparatus, argon gas is injected into the flow downstream from the microwave cavity so that helium metastable atoms, produced in the cavity, are removed via

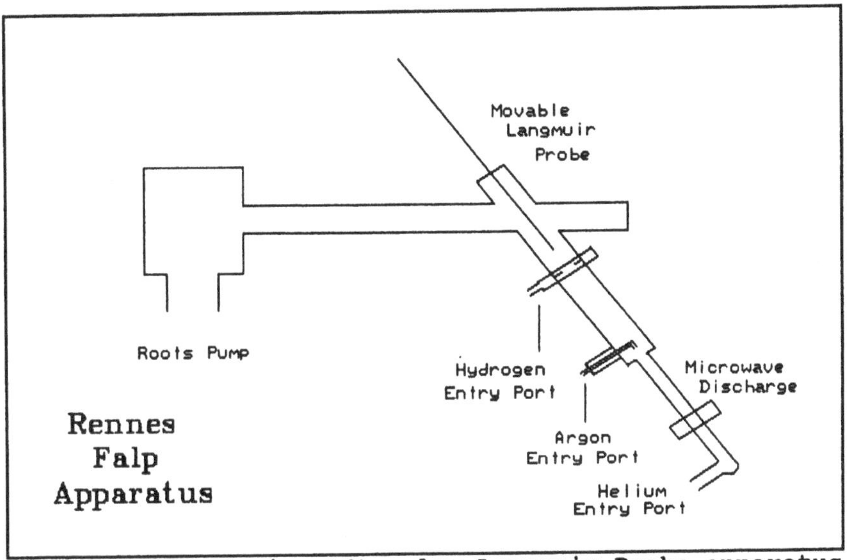

Figure 2 The Flowing Afterglow Langmuir Probe apparatus at Rennes.

the reaction:

$$He^* + Ar \rightarrow He + Ar^+ + e \quad (9)$$

Following hydrogen injection, H_3^+ is produced via the reaction scheme:

$$Ar^+ + H_2 \rightarrow ArH^+ + H \quad ...(10)$$
$$ArH^+ + H_2 \rightarrow Ar + H_3^+ \quad ...(11)$$

In both experiments, the decay of the electron density along the flow tube is measured using a langmuir probe.
The rate of change of H_3^+ with time is given by:

$$\frac{d[H_3^+]}{dt} = -\alpha [H_3^+][e]$$

$$- \text{(diffusion term)} ...(12)$$

where square brackets indicate concentrations and α is the recombination rate coefficient. If recombination is the major loss process, and assuming that $[H_3^+] = [e]$, then Eqn. 12 may be written.
Integration of Eqn. 13 yields:
Hence a plot of the reciprocal of the electron density

$$v\frac{d[e]}{dx}=-\alpha[e]^2 \ldots (13)$$

$$\frac{v}{[e]_x}=\frac{v}{[e]_0}+\alpha x \ldots (14)$$

vs distance along the flow tube yields α/v directly. v, the flow velocity can be determined by pulsing the discharge and monitoring the time delay in the resulting changes to the langmuir probe signal.

Mass spectrometric sampling downstream of the recombination region is performed to ensure that only one ion is present in the afterglow. The decay rates obtained in the Birmingham and Rennes experiments were found to differ. The former experiment showed an initial rapid fall, upon hydrogen injection, followed by a slow decrease characteristic of ion diffusion. The Rennes experiment[11] on the other hand displayed a consistent decrease in the electron density along the flow tube. The recombination rate deduced from this measurement is 1.5 x 10^{-7} cm^3s^{-1}!

A second experiment, using the apparatus shown in figure 3, has been performed at Rennes. In this apparatus, both the mass spectrometer and the langmuir probe can be independently displaced along the drift tube. The method of producing the ions in this experiment is the same as that described above but the determination of the rate coefficient is performed differently.

The problem with the first method is that it is very susceptible to impurity problems. Thus, if for example there is contamination of the gas lines leading into the apparatus, then a second ion may be present in the afterglow and the recombination might be dominated by that ion. This problem is more severe than might first appear for it is the density of the impurity ion, not the impurity itself which is important. If the production rate for the ion is higher than for example, the production rate of H_3^+, then the impurity ion can be present in concentrations approaching that of H_3^+ even though the impurity itself might only be present in trace amounts.

This second method successfully overcomes the impurity problem as can be seen in the following. One can write down the complete equation for the change of H_3^+ concentration along the flow tube as follows:

$$v\frac{d[H_3^+]}{dx}=\alpha[H_3^+][e]-k[H_3^+][I]$$

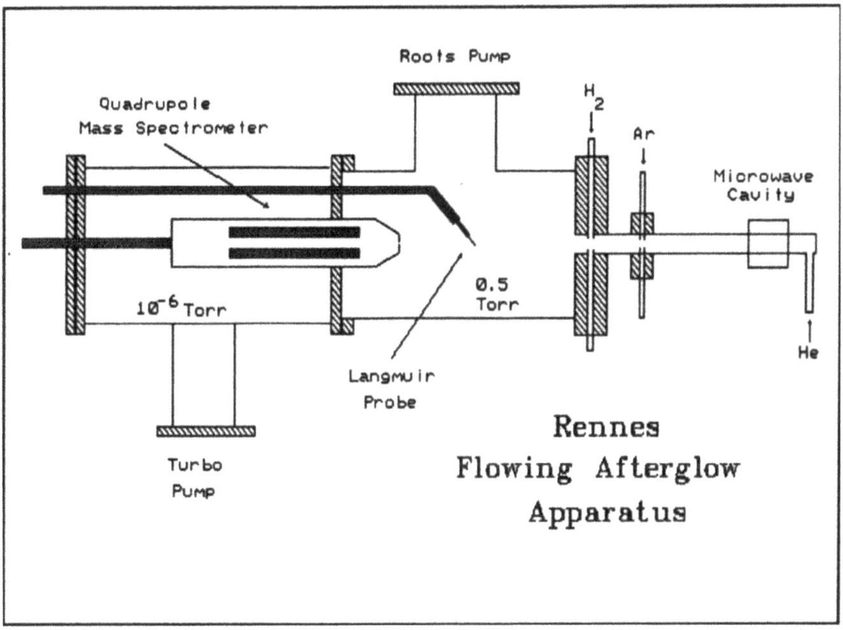

Figure 3 The new afterglow apparatus at Rennes with movable langmuir probe and movable mass spectrometer.

- (diffusion term) ...(15)

where $[H_3^+]_0$ is the initial H_3^+ concentration and $[H_3^+]_x$ the concentration at a distance, x, along the tube. k_I is the rate coefficient for the reaction of H_3^+ with the impurity, I.

Integration gives:

$$Log\frac{[H_3^+]_x}{[H_3^+]_0} = -\frac{\alpha}{v}\int [e]\,dx - \frac{k_I[I]\,\Delta x}{v}$$

-(integral of diffusion term) ...(16)

Thus, it can be seen that a plot of $\log([H_3^+]_x/[H_3^+]_0)$ vs $\int[e]dx$ should yield a straight line whose slope is α/v and which is independent of the impurity concentration. Hence by measuring $[H_3^+]_x$ and $[H_3^+]_0$ using the movable mass spectrometer and concurrently measuring [e] as a function of x, one can determine α without interference from impurity and diffusion effects. The ability to move the mass spectrometer means that the afterglow can be analyzed directly in the recombination region giving much better information on plasma purity. Furthermore, measurements can be made of individual rate coefficients using an afterglow that contains more than one ion. Measurement of the rate coefficient for H_3^+ using this apparatus again yield a rate coefficient of 1.5×10^{-7} cm^3s^{-1}.

Clearly these measurements are in strong disagreement with that of Adams et al. There is however an interesting aspect to this problem that has been overlooked and that concerns the state of excitation of the ions. Adams et al argued that, once formed, the H_3^+ ions would be rapidly de-excited in collision with H_2 via the proton transfer reaction:

$$H_3^+(v') + H_2 \rightarrow H_3^+ (v=0) + H_2 \quad ...(17)$$

The rate coefficient for this had been measured by Kim et al.[12] to be 2.7×10^{-10} cm^3s^{-1}. This rate is sufficiently high that the H_3^+ ions would decay to their ground state within a very short distance from their point of formation and indeed Adams et al. argued that the initial fast decrease in the electron density was caused by the recombination of vibrationally excited ions. Once de-excitation was complete, then the slow rate set in.

The problem with this argument is that Kim et al's. rate is too high. Subsequent studies of this reaction by Blakley et al.[13] showed that while such a high rate may apply for very excited ions, those in low vibrational states with $v \geq 2$ displayed a de-excitation rate of between 1×10^{-11} cm^3s^{-1} and $1 \times 10^{-12} cm^3s^{-1}$!. With these rates, the ions in the experiment of Adams et al. are likely to be excited. This then seriously degrades the theoretical justification of their results. The same argument applies to the Rennes experiment although in this case the H_3^+ ions are formed in a different manner and the initial population of vibrational states is unknown. The reverse of reaction (11) is endothermic but it can proceed for H_3^+ ions in the v=2 level. Hence this is an additional mechanism for the de-excitation of states with $v \geq 2$ and so the average internal energy of the ions in this experiment may well be less than in that of Adams et al.

Merged Beam Experiments.

With regard to the merged beam experiments, it was realized early on[14] that the ions, produced in a conventional rf source are likely to be excited because of this low de-excitation rate. The residence time (~35μs) is insufficient to allow for complete de-excitation. In the rf trap source however, the residence time is of the order of milliseconds and it is possible to produce $H_3^+(v=0)$. Both of these facts have been experimentally verified as discussed above[6,7]. If the rate of Kim et al. was correct, then the ions produced by the rf source should be vibrationally cold but this is not the case.

Recently[15], the DR cross section for $D_3^+(v=0)$ has been determined and these results are shown in figure 4. Also shown are DR results for D_3^+ and H_3^+ produced using a conventional rf source and for $H_3^+(v=0)$. It can be seen that the results for the vibrationally excited ions differ by almost a factor of 2 while the cross sections for $H_3^+(v=0)$ and $D_3^+(v=0)$ are about equal, at least at low energy. The reason for the difference in the former case is that the vibrational spacing of D_3^+, (0.26 eV) is much less than for H_3^+ (0.37 eV). This means that initially, the D_3^+ ions will be produced with a much wider range of vibrational states populated[2]. Since the de-excitation rate for highly excited states is much larger than for lower states, by the time they emerge from the ion source, the D_3^+ will actually have a lower average internal energy and so the measured cross section is lower.

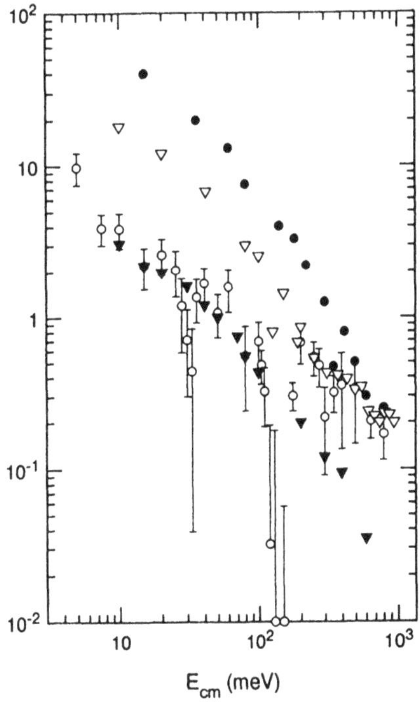

Figure 4 Dissociative recombination cross sections for ○ $D_3^+(v=0)$, ▼ $H_3^+(v=0)$, ● $H_3^+(v')$ and ▽ $D_3^+(v')$. (From ref. 8).

During the Rennes experiments, the rate for D_3^+ was also measured and the ratio of $\alpha(H_3^+)/\alpha(D_3^+)$ was found to agree with that found in the merged beam experiments using excited ions. This is further evidence that the afterglow experiments apply to vibrationally excited ions.

Dissociative Recombination Mechanisms.

The picture that is emerging is that the recombination rate for $H_3^+(v=0)$ is certainly greater than 1×10^{-11} cm^3s^{-1}. This means that some mechanism faster than radiative stabilization is operating for this ion. Clearly however, the direct mechanism cannot explain the observations. Michels and Hobbs[5] have argued that the indirect process should also be slow but their argument applied only to the conventional picture in which a transition from the ion to a neutral rydberg state occurs with a change of vibrational quantum number of +1. Hickman[16], in studies of H_2^+, has shown that in fact transitions involving a large change in v can play a significant role in DR.

Two possible mechanisms present themselves. The first involves a chain of transitions such as (See figure 1):

$$e + H_3^+(v=0) \rightarrow H_3(2p\ ^2A_2'')(v=10) \rightarrow H_3^*(^2A_1) \rightarrow H_3(^2B_2) \rightarrow H + H + H \quad ...(18)$$

The second involves capture into a high rydberg state of H_3^* which is subsequently predissociated by the ground state of H_3. Indeed such predissociations have been seen experimentally[16]. High vibrational levels of the npE' states are believed to act as intermediates in this process.

Conclusion.

There still remains a discrepancy between the value of the recombination rate for $H_3^+(v=0)$ as measured by Hus et al.[6] and Amano[10]. The Rennes data does show some curvature in the decay rate at late times when most of the ions have recombined and this may indicate that the rate for v=0 is less than 1.5×10^{-7} cm^3s^{-1} consistent with Hus et al's findings. It is hoped that further theoretical and experimental work will be forthcoming that will address these issues so that we might have a better understanding of this complex process of dissociative recombination

Acknowledgements.

The financial support of the CNRS, US AFOSR and the Canadian NSERC are gratefully acknowledged.

References.

1. N.G. Adams, D. Smith and E. Alge J. Chem. Phys. 81, 1778, 1984.
2. V.G. Anicich and J.H. Futrell Int. J. Mass. Spec. Ion Proc. 55, 189, 1983/4.

3. M.T. Leu, M.A. Biondi and R. Johnsen Phys. Rev. A 8, 413, 1973. J.A. MacDonald, M.A. Biondi and R. Johnsen Planet. Space. Sci 32, 651, 1984. B. Peart and K.T. Dolder J. Phys. B.7, 1948, 1974. D. Mathur, S.U. Khan and J.B. Hasted J. Phys. B. 11, 3615, 1978. Also refs. 7 and 13.
4. J.B.A. Mitchell and S. L. Guberman *Dissociative Recombination: Theory, Experiment and Applications*. World Scientific, Singapore, 1989.
5. H.H. Michels and R.H. Hobbs, Astrophys. J. 286, L27, 1984. K.C.Kulander and M.F. Guest J. Phys. B. 12, L501,1979.
6. H. Hus, F.B. Yousif, A. Sen and J.B.A. Mitchell Phys. Rev. A 38, 658, 1988.
7. J.B.A. Mitchell, C.T. Ng, L. Forand, R. Janssen and J.Wm. McGowan J. Phys. B. 17, L909, 1984.
8. F.B. Yousif, P. Van der Donk, M. Orakzai and J.B.A. Mitchell (In press, Phys Rev A. 1991).
9. N.G. Adams and D. Smith. Ref. 4, p. 124.
10. T. Amano Astrophys. J. 329, L121, 1988. J. Chem. Phys. 92, 6492, 1990.
11. A. Canosa, J.C. Gomet, B.R. Rowe, J.L. Queffelec and J.B.A. Mitchell (Submitted to J. Chem. Phys, 1991).
12. J.K. Kim, L.P. Theard and W.T. Huntress Int. J. Mass Spec. Ion. Proc. 15, 223, 1974.
13. C.R. Blakley, M.L. Vestal and J.H. Futrell J. Chem. Phys. 66, 2392, 1977.
14. D. Auerbach, R. Cacak, R. Caudano, T.D. Gaily, C.J. Keyser, J.Wm. McGowan, J.B.A. Mitchell and S.F.J. Wilk J. Phys. B. 10, 3797, 1977.
15. P. Van der Donk, F.B. Yousif and J.B.A. Mitchell (Submitted to Phys. Rev. A. 1991).
16. A.P. Hickman Ref. 4, p. 35.
17. C. Bordas and H. Helm Phys. Rev. A. (1 Apr 1991)

EXPERIMENTAL OBSERVATION OF LASER STIMULATED RADIATIVE RECOMBINATION BETWEEN PROTONS AND ELECTRONS

T. J. Morgan
Physics Dept. Wesleyan University
Middletown, CT 06459–0155
and
F. B. Yousif, P. Van der Donk, Z. Kucherovsky, J. Reis, E. Brannen,
J. B. A. Mitchell
Dept. of Physics, University of Western Ontario
London, Ontario, Canada N6A 3K7

ABSTRACT

Spontaneous radiative recombination of protons and electrons to produce hydrogen atoms with principle quantum numbers $8 \leq n \leq 19$ has been measured using a merged beam apparatus combined with electric field ionization detection. Infra–red radiation from a CO_2 laser has been used to induce stimulated radiative recombination to the n=11 and 12 states of hydrogen with subsequent field ionization providing specific detection of the recombination neutral products.

1. Introduction

When a low energy electron undergoes a close encounter with a free proton, during a very short interval, ($\sim 10^{-15}$ s), the system can have the configuration of a bound electron moving in an excited Bohr orbit. If during this interval, the quasi hydrogen atom undergoes a transition to a lower orbit, emitting a photon, the energy of the electron may be insufficient to escape, resulting in stabilization of the recombination process. This is known as radiative recombination. It is a very low probability process since the characteristic time for electromagnetic radiation emission is of the order of 10^{-9}s. Usually the electron will simply be scattered by the proton and continue on its way.

Bethe and Salpeter[1] derived the following expression for the radiative recombination cross section for capture into a given state with principal quantum number n:

$$\sigma = 2.1 \times 10^{-22}\text{cm}^2 \frac{E_0^2}{nE_e(E_0 + n^2 E_e)} \tag{1}$$

where $E_e = (1/2)mv_e^2$ is the kinetic energy of the electron and E_0 the ionization potential of the n = 1 level of atomic hydrogen. The frequency ν of the stabilizing photon is given by energy conservation:

$$h\nu = \frac{1}{2} mv_e^2 + \frac{E_0}{n^2} \tag{2}$$

An experimental verification of Eq (1) has recently been performed by Andersen et al.[2] for C^{6+} ions over the collision energy range near zero to 1.0 eV.

If an electron–proton collision takes place in an intense photon field, it is possible that the stabilizing transition can be induced by photon impact. That is, stimulated emission of a stabilizing photon is possible with the frequency of the stimulating and emitted photons being equal. This process, referred to as Laser Stimulated Radiative Recombination can be expressed as follows:

$$h\nu + e^- + H^+ \rightarrow H(n) + 2h\nu. \qquad (3)$$

The hydrogen atom is left in a particular n state determined by the laser photon frequency and the energy of the electron, as given by Eq. (2). Neumann et al.[3] have examined this process theoretically, and its possible application to the production of atomic antihydrogen. They have given an equation for the gain G given by the ratio of the stimulated recombination rate to the spontaneous rate:

$$G = \frac{\sigma^{stim}}{\sigma^{spon}} = \frac{Pc^2}{A\Delta\nu 8\pi h\nu^3} \qquad (4)$$

where P is the laser power is watts, A is the cross sectional area of the laser beam and ν is the laser frequency. (Equation 4 assumes that there is complete overlap between the laser beam and the electron–ion interaction region). In order to stimulate an electron to recombine with a proton, the photon frequency must match the frequency of the photon which would be emitted and this is given by the sum of the binding energy of the level into which the electron capture occurs and the kinetic energy of the electron, as given by Eq. 2.

In practice, the frequency spread of the laser is much smaller than the energy spread of the electrons in the center–of–mass frame of reference and so the $\Delta\nu$ term in Eq. 4 is dominated by the electron velocity spread, Δv_e:

$$\Delta\nu = \frac{mv_e}{h}\Delta v_e \qquad (5)$$

In the experiment reported here a CO_2 laser is used to induce the recombination of electrons and protons colliding at low energies in a merged beam apparatus. The center–of–mass collision energy is given by:

$$E_{cm} = E_+ + E_e - 2(E_+E_e)^{1/2}\cos\Theta \qquad (6)$$

where Θ is the intersection angle between the two beams, E_e is the electron beam energy and E_+ the reduced ion energy given by $E_+ = E_i\, m/m_i$, E_i and m_i being the ion energy and mass. For small Θ Eq. 5 can be written:

$$E_{cm} = (E_e^{1/2} - E_+^{1/2})^2 + (E_eE_+)^{1/2}\Theta^2 \qquad (7)$$

Differentiation of this equation with respect to E_+, E_e and Θ, enables one to derive an expression for the uncertainty in the center–of–mass collision energy:

$$\Delta E_{cm} = [\{[1-(\frac{E_+}{E_e})^{1/2}]\Delta E_e)^2+\{[1-(\frac{E_e}{E_+})^{1/2}]\Delta E_+\}^2+[2(E_eE_+)^{1/2}\Theta\Delta\Theta]^2]^{1/2} \quad (8)$$

When $E_e = E_+$, the first two terms go to zero and the uncertainty is determined mainly by Θ and $\Delta\Theta$. Using well collimated beams and careful alignment, these quantities can be made very small. In principle at very low c.m. collision energy the merged beam approach can produce very high energy resolution measurements even though the uncertainties in the electron and ion energies are quite large. Calibration of the center–of–mass energy was performed using a procedure described in ref. (4).

A discussion of the experimental method and its relationship to antihydrogen production has been given in previous reviews[5-7].

2. Experimental Approach

The merged electron–ion beam machine has been described in detail elsewhere[8]. A proton beam of 10^{-8}A is produced using a radio frequency ion source in the terminal of a 400 keV Van de Graaff accelerator. The ions are mass analyzed and injected into the collision chamber where they are electrostatically deflected to remove neutrals formed in the beam line. An indirectly heated barium oxide cathode emits a 100μA electron beam which is merged with the protons by a trochoidal analyzer. The two beams interact over a distance of 86 mm, the overlap being measured at two points along the interaction region using scanners which intersect the beams both horizontally and vertically[9]. After the interaction region the ions are electrostatically analyzed with the protons being deflected into a Faraday cup.

The neutral hydrogen atoms formed in the interaction region due to both electron–ion recombination and charge exchange with the background gas, pass undeflected into an electric field ionizer made of two highly polished aluminum surfaces separated by 1.0 mm, each with a circular aperture of radius 0.6 mm. The ionizer produces an electric field of up to 150 KV/cm along the axis of the beam which ionizes atoms in excited Rydberg states with $n \geq 7$. The resulting ions are then deflected onto a surface barrier detector. Attention has been given to elimination of neutrals not formed by electron–ion recombination. Charge exchange with background gas is minimized by maintaining the pressure in the chamber below 10^{-10} Torr and a high primary ion beam energy (330 KeV) is used so that the total charge exchange cross section is low, ($\sim 5 \times 10^{-20}$ cm^2). The background is further reduced since the charge exchange cross section into a particular n state scales as n^{-3}.

Multiple apertures in the chamber minimize the chance of scattered particles hitting the detector. Final discrimination between background and true electron–ion signals is achieved by modulating the electron beam and counting the detector signal in and out of phase with this modulation. The background count rate is very low (typically 5 counts/minute) and given the small cross sections being measured, counting times of up to 24 hours per point were necessary to reduce statistical errors to acceptable values.

A photon beam from a 20 Watt CO_2 laser, operated in CW mode (TEM_{00}) at a wavelength of 10.535 μm, is reflected into the collision chamber through a zinc selenide window. It is crossed at right angles with the merged ion–electron beam in the interaction region over a length of 1.4 mm and then exits through a second window. A computer controlled feedback system maintains the laser tuned on the desired wavelength. Measuring the laser intensity profile, with a two dimensional scanning pinhole, just before and after entry, allows the divergence of the beam to be determined. Losses due to transmission through the windows are corrected for.

3. Results and Discussion

The detector was operated with an on–axis field of 105 kV/cm so that hydrogen atoms with $n \geq 8$ would be ionized. The resulting proton signal was counted as described above. Prior to arrival at the field ionizer, the atoms had to pass through the charge state analyzer where the field was 2.4 kV/cm which is sufficient to field ionize atoms with $n \geq 19$. A length of 1.4 mm of the interaction region was illuminated by the 10.535 μm beam from the CO_2 laser and the absolute value of the recombination cross section was measured as a function of center–of–mass energy. The results are shown in figures 1a and 1b.

The maximum in the cross section in figure 1a at 5.33 meV is due to enhancement of recombination to n=11 level of atomic hydrogen, by stimulated emission of radiation induced by the laser beam. This enhancement only occurs when the resonance condition given by equation 2 is satisfied. The width of the peak, (0.5 meV) reflects the high energy resolution of the experiment. Figure 1b shows the same process for the n=12 level. In this case, the peak width is 0.7 meV. The ion energy used for these measurements was 330 keV. In order to achieve such a high energy resolution, Eq. 7 indicates that Θ and $\Delta\Theta$ must be less than one tenth of a degree. Off resonance, the measured cross sections are for spontaneous radiative recombination of protons and electrons to form $H(8 \leq n \leq 19)$. The peaks are superimposed upon this background. The two data points shown as squares indicate measurements that were taken with the laser beam off so that there was no stimulated contribution. Using Eq. 1, the theoretical values for σ_{spon} $(8 \leq n \leq 19)$ is 4.92×10^{-19} cm^2 and 9.41×10^{-20} cm^2 at 5.33 and 23.3 meV, respectively. The corresponding measured values are $4.2 \pm 1.4 \times 10^{-19}$ cm^2 and $1.4 \pm 0.7 \times 10^{-19}$. Considering the difficulty of measuring such a small cross section, the agreement can be be considered to be very good. The peak values of the measured effective cross sections, including the stimulated contributions, are $16.0 \pm 3.8 \times 10^{-19}$ cm^2 and $11.3 \pm 3.0 \times 10^{-19}$ cm^2. Subtracting the spontaneous recombination contribution to the cross section in each case yields an effective stimulated recombination cross sections of $11.7 \pm 4.0 \times 10^{-19}$ cm^2 for n=11 and $9.8 \pm 3.1 \times 10^{-19}$ cm^2 for n=12. It should be noted that the effective stimulated recombination cross section as defined here is laser power dependent. Eq. 1 shows that $\sigma_{spon}(n)/\sigma_{spon}(8 \leq n \leq 19)$ is 0.097 and 0.089 for n=11 and 12. Using these ratios and the measured values of σ_{spon} $(8 \leq n \leq 19)$ at 5.33 meV and 23.3 meV, one obtains corresponding values for σ_{spon} (n=11) and σ_{spon} (n=12) of $4.2 \pm 1.4 \times 10^{-20}$ cm^2 and $1.3 \pm 0.6 \times 10^{-20}$ cm^2. Taking the ratio $\sigma_{stim}/\sigma_{spon}$ gives 28 ± 14 for n=11 and 78 ± 46 for n=12. The laser beam, however, has a FWHM diameter of 1.4 mm and intersects the interaction region (length 86 mm) at right angles. The stimulation occurs only over a small overlap region. If the entire interaction region had been illuminated, the resulting gains would have been 1720 ± 860 and 4790 ± 2830. The large errors are due to statistical errors in the measurements of the cross sections.

The laser power used was 12.6 watts and 15.3 watts for the measurements in figures 1a and 1b, respectively. Using these powers, Eq. 3 predicts gains of 2388 and 1139 respectively. Given the uncertainties associated with matching the profiles of the three beams, the agreement between the predicted and measured gains is reasonable. Another process which could reduce the efficiency of stimulating the free electron to a bound state is photon excitation of the free electron to a higher continuum state (free–free transitions). Estimates of this effect for our case, using the formalism of Neumann et al., shows that the ratio of the free–free transition rate to the stimulated transition rate into n = 11 and into n = 12 levels are about .03 and .07, respectively. This process appears to have little effect on our present results. However, since the ratio scales as $n(h\nu)^{-1}$ this process can become quite significant at higher principle quantum numbers. For zero energy electrons the ratio equals 1 for n = 29 under the present conditions.

Neumann et al.[3] have discussed the consequence of using too much laser power. This would result in the reionization of the recombined atom by the impact of a second photon. This phenomenon becomes important if the photon flux $\phi \geq (\tau\sigma_{ph})^{-1}$ where τ is the exposure time (or the lifetime of the n state, whichever is shorter), σ_{ph} is the photoionization cross section and $\phi = P/Ah\nu$. In our experiment $\tau \sim 0.1$ ns and $\phi \sim 3 \times 10^{22}$ photons cm^{-2}. Since $\sigma_{ph} \sim 7.7 \times 10^{-17}$ cm^2 for 10.535 μm photons interacting with the n=11 level of atomic hydrogen[10], the maximum photon flux before reionization becomes problematic is $\sim 10^{24}$ photons cm^{-2}. Consequently, this phenomenon has little effect on our results.

Clearly there are many improvements that can be made to this experiment, not least of which being to illuminate the interaction region more uniformly. Modifications are being done to the apparatus to facilitate this. For now, the data in figures 1a and 1b clearly show that strong enhancement of the radiative recombination cross section using laser stimulation has been demonstrated in the fundamental electron–proton system.

Note Added: Stimulated radiative recombination to the n=2 level of atomic hydrogen has been observed recently by Schramm et al.[11] Further discussion of the present work can be found in Reference 12.

4. Acknowledgements

The support of the United States Air Force Office of Scientific Research (Grant No. AFOSR–89–0327) and the Canadian Natural Sciences and Engineering Research Council is gratefully acknowledged.

5. References

1. H. Bethe and E. Salpeter *Quantum Mechanics of One– and Two– Electron Atoms* Plenum Publishing Co., New York, 1977.
2. L.H. Andersen, P. Hvelplund, H, Knudsen and P. Kvigtgaard Phys. Rev. Lett. **62**, (1989) 2656.
3. R. Neumann, H. Poth, A. Winnacker and A. Wolf, Z. Phys. A. **313**, (1983) 253.
4. F.B. Yousif, P. Van der Donk, M. Orakzai and J.B. A. Mitchell (Submitted to Phys. Rev A. 1990).
5. J.B.A. Mitchell, *Antihydrogen Production Schemes* in Antiproton Science and Technology (Eds. B.W. Augenstein, B.E. Bonner, F.E. Mills and M.M. Nieto) World Scientific, Singapore, (1988) 359.
6. J.B.A. Mitchell, *Antihydrogen Production Schemes in* Intersections between Particle and Nuclear Physics, Rockport, ME, 1988 (ed. G.M. Bunce) AIP Conf. Proc. **176**, AIP New York, (1988) 447.

7. T.J. Morgan and J.B.A. Mitchell, *Laser Stimulated Radiative Recombination: A Field Ionization Detection Approach in Dissociative Recombination,: Theory, Experiment and Applications* (eds. J.B.A. Mitchell and S.L. Guberman) World Scientific, Singapore, (1989) 175.
8. D. Auerbach et al. Phys. B. 10, (1977) 3797.
9. C.J. Keyser, H.R. Froelich, J.B.A. Mitchell and J. Wm. McGowan, J. Phys. E. **12**, (1979) 316.
10. R.D. Cowan, The Theory of Atomic Structure and Spectra, Univ. of California Press, Berkeley, 1981.
11. U. Schramm et al., Phys. Rev. Lett 67, 22 (1991).
12. F.B. Yousif, P. Van der Donk, Z. Kuchevovsky, J. Reis, E. Brannen, J.B.A. Mitchell, and T.J. Morgan, Phys. Rev. Lett. 67, 26 (1991).

6. Figure Caption

1. Measured electron–proton recombination cross section with, laser on, laser off. (a) Peak at 5.33 meV corresponds to stimulated recombination to the n=11 level. (b) Peak at 23.3 meV indicates stimulated recombination to the n=12 level.

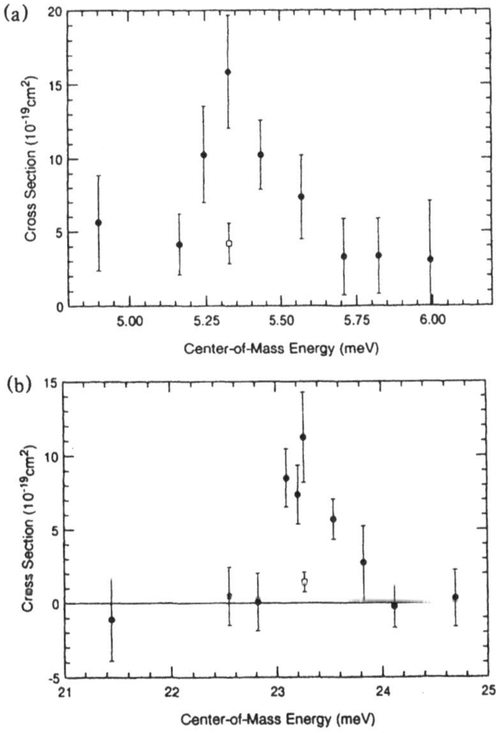

II. MOLECULAR DISSOCIATION

HYDROGEN MOLECULES IN INTENSE LASER FIELDS

H. HELM, M. J. DYER, H. BISSANTZ, D. L. HUESTIS

Molecular Physics Laboratory
SRI International
333 Ravenswood Avenue
Menlo Park, CA 94025

ABSTRACT

A high resolution photoelectron study of the response of molecular hydrogen to very intense laser fields (>10^{13} W/cm^2) at wavelengths between 309 and 330 nm shows that with increasing intensity direct multiphoton ionization of the molecule is surpassed by photodissociation of the neutral. This process leads to the formation of excited atomic hydrogen photofragments which are subsequently photoionized. Photoelectrons from this process form the dominant signature in the electron spectra at high field strength. The photoelectron spectra remain strongly dependent on wavelength at even the highest intensities investigated (8×10^{13} W/cm^2). By contrast, the total ionization rate does not change perceptibly in the wavelength range studied.

The molecular response is discussed in the framework of combined effects of AC-Stark shifting of molecular Rydberg states and predissociation of the molecular Rydberg states induced by the intense laser field.

Introduction

At intensities above 10 TW/cm^2 atoms exhibit above threshold ionization (ATI) phenomena, that is they absorb more photons than required for ionization.[1-3] The photoabsorption is generally followed by emission of photoelectrons, and the measurement of the photoelectron energy is a primary diagnostic tool for ATI. Molecules respond similarly as atoms do, however an additional degree of freedom appears in the form of their dissociation continua. Thus in multiphoton excitation the dissociation process can be a competitor to ionization, or cause the photodissociation of the molecular ion produced. Photodissociation of neutral H_2 has been observed in many stepwise photo-excitation experiments and has been interpreted as resulting from of sequential excitation, first to singly excited states, followed by excitation of the core electron.[4,5,6]

Here we report observations of intensity-dependent photodissociation of H_2 that competes with the direct ionization channel. This phenomenon is seen as a manifestation of (1) the deformation of molecular bonds that arises from bound-free absorption of the molecular ion core and (2) the AC-Stark shifting of molecular Rydberg states in the intense laser field.

Experimental

The experiment is carried out in a magnetic bottle photoelectron spectrometer using light from a tunable dye-laser system with frequency doubling. The tunable oscillator is a synchronously pumped dye laser operating with DCM dye. The 3-500 mW average output of the dye laser is fed into 7m of single mode fiber to chirp and a double pass grating stage to compress the pulses down to 400 fs leading to an average output of typically 50 mW on the compressed light. Ten pulses of this train are amplified per second by a four stage amplifier system consisting of three Bethune prism cells with increasing bore diameters of 1, 3, and 6 mm. The beam is then expanded to a diameter of 20 mm and it double passes an open axicon amplifier. The amplifiers are pumped by a 4-ns 450-mJ frequency-doubled YAG laser at 532 nm. The final output at the red wavelengths is between 2 and 4 mJ at pulse lengths between 400 fs and 2 ps. Frequency doubling in KDP leads to pulses up to 1 mJ in energy.

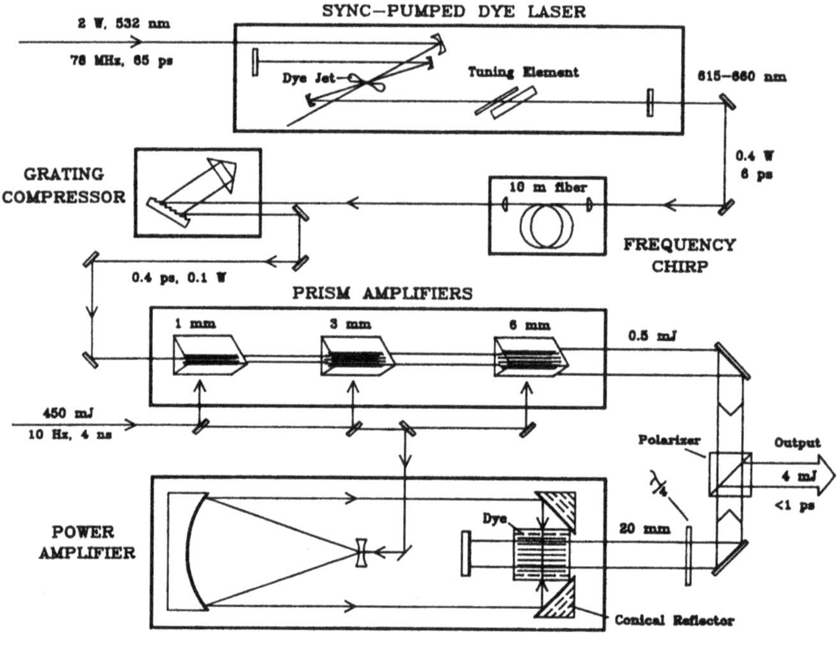

Figure 1. Schematic of dye laser system and amplifiers. The pulses from the synch pumped dye-laser are chirped and compressed and then amplified in three prism amplifiers. The final stage amplifier is a reflective axicon design for even amplification of the 20 mm diameter laser beam.

The photoelectron spectrometer is a magnetic bottle design[7] that uses permanent magnets and a 64 cm time-of-flight section. The photoelectron signal is monitored by a constant fraction discriminator whose output feeds a chain of 16 time-to-digital converters, each with 1 ns resolution. The pressure in the interaction region is adjusted such that several (typically less than 10) electrons are recorded per laser shot. In studies of resonant ionization of H_2 at low intensity we have demonstrated[8] a resolution of 50 meV at 1eV electron energy using this device. The laser energy is encoded with the arrival time of each electron detected and stored on hard disk of an IBM personal computer. Final data are restricted to laser energy bins that fall within ± 20% of the quoted value. Experiments are carried out using linear laser polarization unless otherwise specified.

The base pressure in the photoionization region of the spectrometer is around 1×10^{-8} Torr, and the hydrogen pressure added reached as high as 5×10^{-6} Torr at the lowest intensities studied. The laser beam is focused using a 15 cm lens. Based on the near Gaussian and near diffraction limited beam profile we estimate the focal diameter ($1/e^2$) to be around 14 μm although a direct measurement of the focal properties at full power have not yet been carried out. All of the hydrogen work described below was obtained with the dye laser output of ~2 ps, obtained from autocorrelation measurements and assuming a $sech^2$ pulse shape. The bandwidth of the laser falls typically in the range of 5 Å. Intensities quoted below are assumed lower limits, averaged over a conservatively chosen 20 μm spot diameter.

Experimental Results

The wavelength range chosen for this investigation is such that the three-photon energy lies below the $C^1\Pi_u$ state (3-photon threshold: 300 nm) while the four-photon energy lies near the ionization limit. Thus at the wavelengths employed here (309 -330 nm) three-photon resonances are only expected to involve the $B^1\Sigma_u^+$ state while four-photon resonances may occur with a variety of vibrational levels of the Rydberg manifolds belonging to n = 3 and higher. We observed that scanning the dye laser in 5 Å intervals over a portion of the wavelength range studied (328 to 318 nm) showed no noticeable wavelength dependence in the total ionization yield at intensities around 3×10^{13} W/cm^2. The yield was comparable at the other wavelengths investigated where photoelectron spectra were recorded at intervals of 25 Å. Despite the insensitivity of the total ionization rate to wavelength we observe that the intensity distribution in the photoelectron energy spectra is strongly wavelength dependent.

In order to introduce the photoelectron features we first discuss relevant potential energy curves of molecular hydrogen. In Figure 2 we show a simplified diagram of molecular excited states, together with the lowest states of the molecular ion. Also indicated in this figure are the three, four, and five photon energies for photoexcitation at 325 nm, together with the vibrational wavefunction of ground state H_2, v = 0.

It is evident that in this case the Franck Condon overlap into the $B^1\Sigma_u^+$ state is poor and that the first "good" resonant states occur at the four-photon energy. This leads us to assume that an important ionization channel may involve four-photon excited states of the molecule:

$$H_2 + 4\,h\nu \rightarrow H_2^{**}(v) \qquad (1)$$

that are subsequently ionized by a fifth photon forming molecular hydrogen ions in specific vibrational levels v^+:

$$H_2^{**}(v) + h\nu \rightarrow H_2^+(v^+) + e\,. \tag{2}$$

Alternatively a non-resonant five photon process might occur

$$H_2 + 5\,h\nu \rightarrow H_2^+(v^+) + e \tag{3}$$

which could also form molecular hydrogen ions in a variety of vibrational levels.

The observation is that indeed, at low intensities, the spectra show photoelectrons with energies that are consistent with the formation of H_2^+. As an example we give in Figure 3 spectra obtained at 3175 Å at various intensities.

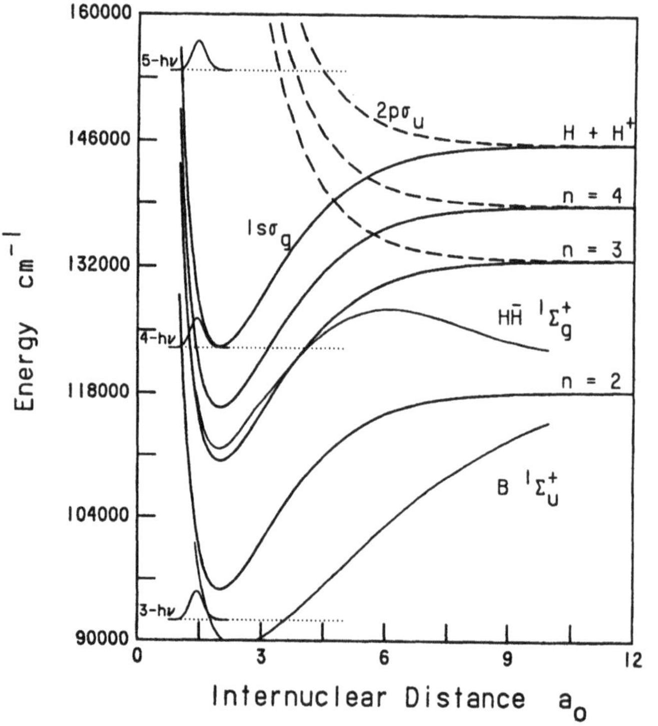

Figure 2. Selected potential energy curves of H_2 and the ground states of H_2^+. Also indicated are the resonant energies for 3, 4, and 5 photons at 325 nm together with the vibrational wave function of v = 0 of ground state H_2. The bound and repulsive Rydberg states are shown for zero quantum defect.

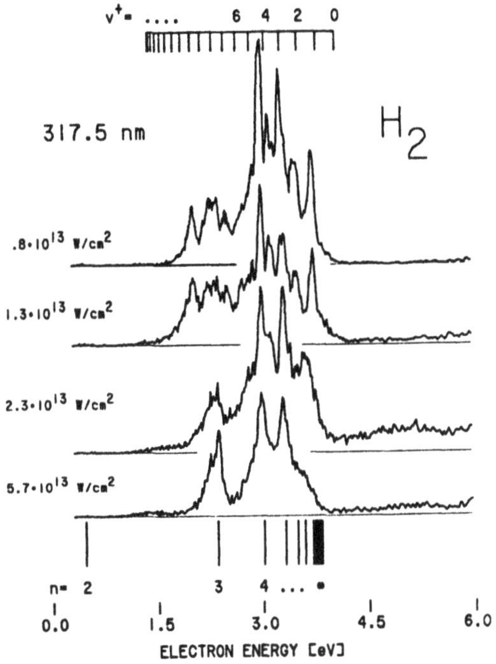

Figure 3. Photoelectron spectra of H_2 at 3175 Å as a function of intensity. The tickmarks at the top refer to photoelectron energies as they should appear from five-photon nonresonant ionization forming H_2^+. The tickmarks at the bottom refer to photoelectrons produced in one-photon ionization of excited atomic hydrogen fragments $H(n\ell)$.

The surprising observation in Figure 3 and in the spectra recorded at other wavelengths is that as the laser intensity is increased the photoelectron signal from the molecular hydrogen ion channel is apparently surpassed by the signal from one-photon ionization of excited atomic hydrogen

$$H(n\ell) + h\nu \rightarrow H^+ + e \tag{4}$$

where $n \geq 3$. This observation is interpreted as being due to a multiphoton dissociation process. Plausible channels for the formation of excited atomic hydrogen are either the photodissociation of molecular Rydberg states:

$$H_2^{**}(v) + h\nu \rightarrow H(1s) + H(n\ell) \tag{5}$$

or also dissociative ionization processes involving at least eight photons:

$$H_2 + nh\nu \rightarrow H^+ + H(n\ell) + e\ . \tag{6}$$

In Figure 3 we indicate at the top of the figure the electron energy peak location expected for non resonant five-photon ionization (process (3)) when the molecular ion is formed in the lowest rotational level. The one-photon ionization process of atomic hydrogen gives rise to a unique energy position that unambiguously identifies the excited atomic hydrogen fragment H(nℓ) and the energies expected from process (4) are marked at the bottom of the figure. The rather distinct and simple features in the photoelectron spectrum at high intensities clearly suggest that reaction (4) is the dominant channel there.

As mentioned above we find that the intensity distribution in the photoelectron spectra is strongly wavelength dependent. This dependence exists at low intensities where the presence of many individual peaks indicates that processes (1-3) lead to the formation of H_2^+ ions in a variety of vibrational levels but also at higher intensities where the spectrum simplifies and specific atomic excited states are the dominant final products.

As an example we give in Figure 4 photoelectron spectra obtained at 3250 Å for a range of intensities. The spectra appear quite different from those obtained at 3175 Å. An additional surprising feature in the spectra in Fig. 4 is that the dissociation process tends to favor lower dissociation limits H(nℓ) as the intensity grows. While dissociation into H(n = 3) is only a weak channel at the lowest intensity (see the lowest trace in Figure 4) this channel dominates the spectrum the highest intensity value. We also note in this figure the appearance of the first ATI peaks of the photoelectron spectra at the higher intensities. ATI peaks of similar magnitude are present at all other wavelengths studied.

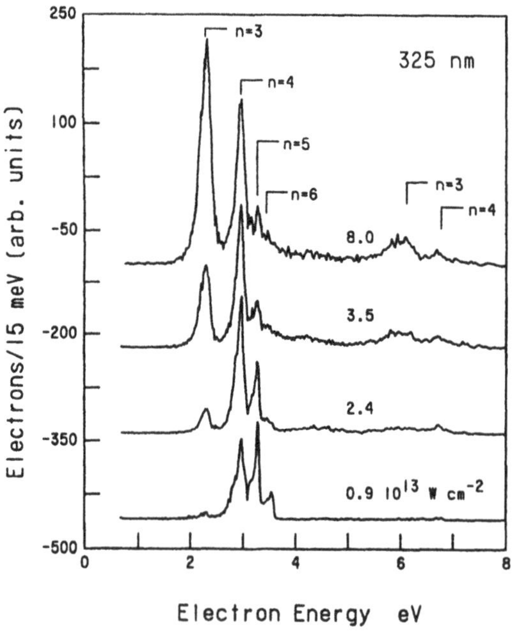

Figure 4. Photoelectron spectra of H_2 at 3250 Å as a function of intensity. An intensity dependent shifting of population into lower dissociation limits is apparent.

Discussion

Numerous papers have appeared in recent years on photoionization of molecular hydrogen in intense fields.[9-12] The sole common feature than can be extracted from reading this literature appears to be that the ionization process remains wavelength specific at even high intensities such that no single common rule has yet emerged. This fact is different from the atomic case where the dominant path appears when selected bound states are AC-Stark shifted into resonance - giving rise to specific ionization features that can be correlated with individual bound states.

The phenomenon of strong wavelength dependence is not surprising in view of the various electronic, vibrational and rotational degrees of freedom of the intermediate excited states that can participate in the multiphoton excitation process H_2. The data presented here add another dimension to the ongoing discussion on molecular hydrogen in strong fields and their interpretation must be viewed as applicable to the specific wavelength range investigated. We have carried out similar studies at wavelengths from 610 to 660 nm and find photoelectron spectra that are quite different from those reported here in that individual peak structures are lost at high intensity at the red wavelengths.

Before we discuss the mechanism of photoionization and photodissociation via four-photon excited intermediate states to explain the observed photoelectron features we first discuss specific properties of the excited states of molecular hydrogen as well as the effect of ponderomotive forces on the appearance of the photoelectron spectra.

Potential Energy Diagram

The potential energy diagram for molecular hydrogen given in Figure 2 is highly schematic in its representation of the bound and dissociative Rydberg states. The states shown are drawn assuming zero quantum defect. The full diagram known to date for the excited singlet manifold of states is considerably more dense owing to the fine structure arising from the non zero quantum defects of the various ℓ-components in each principal quantum number. Furthermore, the avoided crossings[13] between states of the same symmetry that are either the singly excited Rydbergs (orbital configuration $(1s\sigma_g)(n\ell\lambda)$ or the doubly excited states $(2p\sigma_u)(n\ell\lambda)$ modify the appearance of the adiabatic states at large $(R>4a_0)$ internuclear distances. In addition, the $^1\Sigma$ state potential energy curves are modified at large separation by the strongly attractive ion-pair character. This is apparent for the $B^1\Sigma_u^+$ and the $H\bar{H}^1\Sigma_g^+$ curves which are included in Figure 2.

Despite these simplifications the potential energy diagram in Figure 2 serves to indicate the strongest transition expected for reaching excited states at the wavelengths considered in this paper: At the internuclear distance of the vibrational wavefunction of the v = 0 level of the ground state of H_2, a vertical multiphoton excitation may find a first energy resonance with either the $B^1\Sigma_u^+$ state (3-photon transition) or else with the gerade states of the manifolds n = 3 or higher at the four photon energy. At the short internuclear separation where vertical excitation can be efficient the latter are nearly purely the singly excited states $(1s\sigma_g)(n\ell\lambda_g)$ with ℓ = s, d, f and h. At the wavelengths of this experiment the Franck-Condon factors to reach the bound B-state levels is always at least an order of magnitude smaller than the best overlaps with vibrational levels of the four-photon excited states. We therefore initially concentrate on ionization processes that involve four photon excitation of H_2 to the bound Rydberg states and only later discuss the role of three-photon resonances with the B-state.

Ponderomotive Effects

When photoelectron energy spectra are recorded with intense lasers and short pulse lengths it is paramount to consider the effect that the ponderomotive forces can have on the measured photoelectron energy. We note that the pulse lengths utilized here (~2ps) are sufficiently small, and the interaction volume sufficiently large (≥ 15 μm) such that the ponderomotive energy is not recovered by slow photoelectrons because they cannot exit the laser focus during the time of the laser pulse. To prove this point we used our laser in a photoionization experiment in xenon for which numerous previous data exist for comparison. We show a typical photoionization spectrum of xenon recorded at 620 nm in Figure 5. At this wavelength seven photons are required to reach above the zero-field ionization threshold. We have marked in the figure where photoelectrons were expected

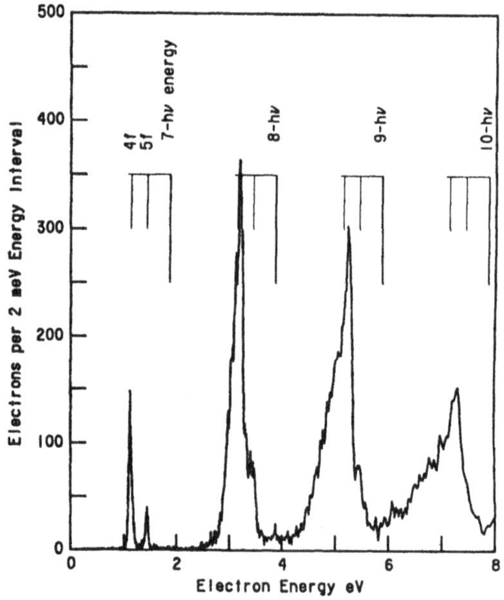

Figure 5. Photoelectron energy spectrum obtained in xenon at ~$3 \cdot 10^{13}$ W/cm^2. The pulse length is 2 ps and the wavelength is 6200 Å. The Photoelectron peak position expected from non-resonant 7-photon ionization to form $Xe^+(^2P_{3/2})$ is marked at the top of the figure. The peak energies in the 7-photon peaks as well as the higher ATI peaks are consistent with resonant excitation via the 4f and 5f resonance levels.

from non-resonant seven photon ionization to form $Xe^+(^2P_{3/2})$. We note that the observed electron energy peaks are shifted to lower energies by an amount that is consistent with the process[1-3] where a six-photon resonance of bound neutral xenon is shifted into resonance as the intensity of the field grows and is subsequently ionized by an additional photon. To the extent that the excited states that form intermediate resonances experience an AC-Stark shift that is similar to the ponderomotive shift of the free electron

$$\varepsilon_p = 750\, I\, \lambda^2\ (\mathrm{cm}^{-1}) \tag{7}$$

the photoelectron's kinetic energy in the field is reduced from that of the non-resonant process

$$E_e = 7\ h\nu - IP \tag{8}$$

by an amount equivalent to ε_p. In Eq. 7, I is the laser intensity in units of 10^{12} W/cm^2 and λ is the wavelength in µm. The ponderomotive energy of Eq. 7 is converted into kinetic energy of the electron if the laser is on over a time span sufficient for the electron to "surf" off the focal volume[15] thereby converting the gradient of the ponderomotive potential into kinetic energy. If on the other hand the laser pulse is short compared to the time required to exit from the focal region the ponderomotive energy remains with the electromagnetic field and the electron's energy is given by

$$E_e = 7\ h\nu - IP - \varepsilon_p \tag{9}$$

For an orientation to the magnitudes involved we note that an electron of 2.7 eV energy travels 1 µm in 1 ps. The energy positions in the photoelectron spectra in Figure 5 indicate that the intermediate states 4f and 5f are the dominant contributors to the ionization process in xenon at this wavelength and that the AC-Stark shift of these states is quite similar to that given by Eq. 7 for the free electron. We note that no contributions from intermediates of the 7p and 8p manifolds of xenon appear and that this shifted resonance structure is reproduced in the higher order ATI peaks although the experimental resolution diminishes somewhat there.

Therefore we expect that under the laser pulse and focal properties employed in our experiment on molecular hydrogen the ponderomotive shifts of the free electron are not recovered from the electromagnetic field.

Wavelength Dependence

The absence of any marked wavelength dependence in the total photoionization yield indicates that either a non-resonant ionization process dominates over one that utilizes resonant intermediates or else that the manifold of excited states that participate in resonant intermediate formation is sufficiently dense as well as sufficiently broadened by the presence of the intense laser field such that excited state ro-vibrational resonance structures merge into a quasi-continuum. At least three processes can be identified that contribute to this broadening, and most likely all three participate with varying degree in our experiment.

For one, an energy broadening of the excited state will arise from the short lifetime of the excited state against one-photon ionization. Typical values for these lifetimes can be calculated using the generalized cross section formula given by Huestis.[16] A value of

200 fs is obtained for n = 4 at 10^{13} W/cm^2. It is quite evident that one-photon ionization of excited states should be quite efficient.

A second process responsible for broadening is related to the intensity dependent lifetimes of the molecular Rydberg states against photodissociation. We note here that this process involves the core electron of the Rydberg and we will discuss this process in detail in the next section.

A third factor is that even if no resonance appears at the four-photon energy at zero field, the AC-Stark shift of the excited states can shift rovibrational excited states into resonance at sufficient intensity, giving the appearance of continuous, wavelength-independent ionization. While the exact magnitude of the AC Stark shifts are not known, an order of magnitude estimate may be obtained from the ponderomotive shift (Eq. 7). At our wavelengths ε_p is about 100 mV per 10^{13} W/cm^2. The process of shifting intermediates into resonance is known to dominate atomic photoionization[1-3] at high fields and the example of xenon given in Figure 5 falls under this category. It it tempting to suspect a similar process in the molecular case but we must also consider the possibility of purely non resonant ionization.

If non-resonant ionization were an important contribution at our wavelengths then we would expect that the wavelength dependence of the electron energies formed in process (1) follows a five-photon dependence. This is contrary to what we find in the electrons' spectra. We show in Figure 6 the location of the dominant electron energies observed at the various wavelengths. It appears from the guiding lines drawn at a slope of one that

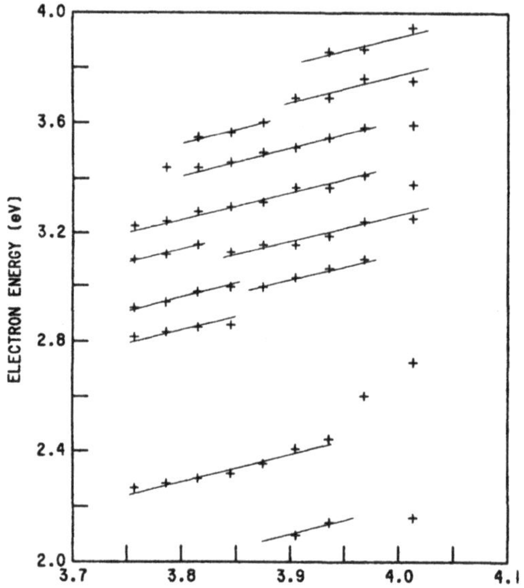

Figure 6. Energy of dominant features of the photoelectron spectra at wavelengths between 309 and 330 nm. Most peak positions follow short segments of slope of one, indicating that one-photon ionization of bound states that are AC-Stark shifted into resonance is the origin of these photoelectrons.

most peaks fall onto segments that vary with the one-photon energy. Also - if the five-photon process occurs in a purely non-resonant fashion at the intensities utilized in this work we would expect that the variation in the ponderomotive shift of the photoelectrons that are created at widely different intensities should lead to a distribution of electron energies for each final state v^+ generated.

We conclude that the most likely ionization channels involve four-photon excited intermediates that are shifted into resonance at a specific intensity. The photoionization of the shifted state will lead to a photoelectron energy that shifts linear with the one-photon energy over a wavelength range over which the Stark shift can be accomplished by the intensities employed.

Four-Photon Resonance Formation

The existence of four-photon resonant intermediates can be used to develop a picture that can explain the observed shifting from ionization of excited H_2 to dissociation into neutral fragments and it can also convey the origin for the switching to lower dissociation limits $H(n\ell)$ as the intensity grows.

One aspect of this picture involves the AC Stark shift of the Rydberg electron. At 50 TW/cm^2 the ponderomotive energy of a free electron amounts to 0.5 eV. If the highly excited states of H_2 experience AC-Stark shifts like that of a free electron, then the excited state manifold will shift upwards with respect to the four-photon dressed ground state as the intensity grows.

It would thus appear that bound states that lie for zero field strength at energies below the four-photon energy can be shifted into resonance as the intensity grows. Ionization may then be envisioned to occur whenever the four-photon dressed ground state encounters a resonance with good transition strength. This picture similar to that generally accepted in the atomic ionization case.

The open question is of course what is the actual Stark shift of the four-photon excited states at high field strength. At the wavelengths used here no bound transitions can be driven with one photon from Rydberg states with $n \geq 3$. A good first-order assumption may therefore be that the AC-stark shift of the higher Rydberg states is comparable to that of the free electron. However how will the core respond to the field? In the classical picture of a Rydberg state we may consider the ionic core as being separate from the Rydberg orbital. If we neglect electrostatic shielding of the core by the Rydberg orbital we should consider the response of the isolated core ion, H_2^+ to the field. This response has been the subject of several recent papers and we use the picture developed by Bandrauck et al.[17] and Gusti et al.[18] in the following discussion.

Dressed State Picture

We now combine the predicted response of the isolate core ion[18] with the picture of a Rydberg electron that experiences an AC-Stark shift as if it were a free electron. This leads to a potential energy diagram of adiabatic potential energy curves such as shown in Figure 7. The full curves give a simplified representation of the Rydberg states that originate from the dissociation limits H(1s) + H(n = 3) and H(1s) + H(n = 4), calculated for a field strength of 50 TW/cm^2. The unperturbed ground state of H_2^+ is indicated by the dashed line and the 4-photon energy at 325 nm is indicated at the left. The potential energy curves are the results of a Floquet calculation[18] that we performed for the diabatic singly and doubly excited states, $1s\sigma_g n\ell\lambda_g$ and $2p\sigma_u n\ell\lambda_g$, for 50 TW/cm^2. In this calculation the Rydberg states with n = 3 and n = 4 are considered separately, with no matrix element connecting them. In the dressed state picture the coupling between the singly excited gerade states and the doubly excited ungerade states leads to avoided crossings between the states {(n = 3)gerade + even photon number} and {(n = e)ungerade+odd photon number}, with different numbers of photons, N, and likewise for 4g and 4u. At molecular distances this coupling reduces the number of bound states, and remaining bound states are affected by predissociation through higher order couplings with the continua labelled 3,4g N = 2 and 3,4u N = 3.

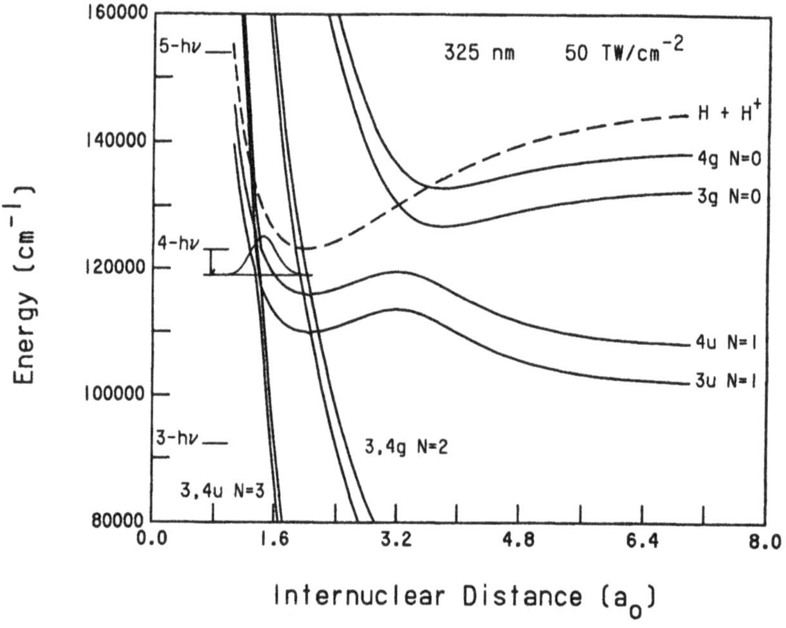

Figure 7. Results of a Floquet calculation for the n=3 and n=4 states of H_2 with the orbital configurations $1s\sigma_g n\ell\lambda_g$ and $2p\sigma_u n\ell\lambda_u$ at 50 TW/cm^2 and 325 nm. The ponderomotive shift for the location of the four photon energy relative to the unperturbed potential energy curve of H_2^+ (dashed curve) is also indicated.

As the intensity of the laser field increases a reduction of the number of bound levels occurs in the excited state manifolds and concurrent with this effect the four-photon resonant energy continually sweeps through the excited state manifold. In this fashion an intensity dependent competition between neutral dissociation and ionization occurs because with increasing intensity the higher vibrational levels in the Rydberg states will appear as continuum states that dissociate rapidly. The classical times for dissociation of molecular hydrogen is extremely short, it takes typically 10 to 20 fs to accelerate to distances of 10 Å. Consequently dissociation in competition with direct ionization of the neutral excited molecule is rather plausible in this adiabatic picture. Concurrent with opening the molecular bonds of the Rydberg the shifting of states in energy at short internuclear distance leads to a progressively better overlap with states from lower principal quantum number - thereby explaining the switching to lower dissociation limits as the intensity increases.

Conclusions

The wavelength and intensity dependence of photoelectron energies observed from molecular hydrogen at intensities above 10^{13} W/cm^2 and at wavelengths where five photons are required to reach the ionization continuum show that the most likely ionization channels involve four-photon excited intermediate states from the Rydberg manifolds with n = 3 - 6. While specific rovibrational assignment of these states cannot be given at this point (primarily because little experimental information exists about hydrogen states at these high energies) this identification allows to develop a framework of the behavior of molecular Rydberg states in intense fields. This framework should be generally applicable to molecules in strong fields.

In molecular Rydberg states the AC-Stark effects on both the Rydberg electron and on the core ion need to be taken into account. If the core ion has a transition that is strongly driven by the applied electromagnetic field then the core's response to the field must play a major role. In the framework of dressed states that are adiabatic in the electromagnetic interaction the core's response results in both a level shift (that is generally opposite to the direction of the ponderomotive shift of a free electron) and in a level broadening owing to predissociation and dissociation. The combination of this effect with the AC-Stark shifting of the Rydberg orbital provides a qualitative explanation of the observed switching from ionization to form H_2^+ to dissociation into neutral excited atomic hydrogen, as well as the observed switching in dissociation to progressively lower dissociation limits as the intensity grows.

This simple discussion of a Rydberg electron and a core ion that respond independently of each other to the electromagnetic field appears useful in rationalizing the experiments performed. There remain many quantitative questions that need to be investigated both experimentally and theoretically.

Acknowledgments

This research was supported by the National Science Foundation under grant number PHY-87-06332 and by the Air Force Office of Scientific Research under Contract No F49620-88-K-0006.

References

1. R. R. Freeman, P. H. Bucksbaum, H. Milchberg, S. Darack, D. Schumacher, and M. E. Geusic, Phys. Rev. Lett. **59**, (1987) 1092.
2. T. J. McIlrath, R. R. Freeman, W. E. Cooke, L. D. van Woerkom, and T. J. McIlrath, Phys. Rev. A **40**, (1989) 2770.
3. P. Agostini, P. Breger, A. L'Huillier, H. G. Muller, G. Petite, A. Antonetti, and A. Migus. Phys. Rev. Lett. **63**, (1989) 2208.

4. J.H.M. Bonnie, J.W.J. Verschur, H. J. Hopman, and H. B. van den Linden van den Heuvell, Chem. Phys. Lett. **130**, (1986) 43.
5. M. A. O'Halloran, S. T. Pratt, P. M.Dehmer, and J. L. Dehmer, J. Chem. Phys. **87**, (1987) 3288.
6. E. Y. Xu, T. Tsuboi, R. Kachru, and H. Helm, Phys. Rev. A **36**, (1987) 5645 .
7. T. Tsuboi, E. Y. Xu, Y. K. Bae, and K. T. Gillen, Rev. Sci. Instrum. **59**, (1988) 1357.
8. E. Y. Xu, A. P. Hickman, R. Kachru, T. Tsuboi, and H. Helm, Phys. Rev. A **40**, (1989) 7031.
9. T. S. Luk and C. K. Rhodes, Phys. Rev. A **38**, (1988) 6180.
10. J.W.J. Verschuur, L. D. Noordam, and H. B. van Linden van den Heuvell, Phys. Rev. A **40**, (1989) 4383.
11. A. Zariyev, P. H. Bucksbaum, H. G. Muller, and D. W. Schumacher, Phys. Rev. A **42**, (1990) 5500.
12. S. Allendorf, in press (1991).
13. L. J. Lembo, N. Bjerre, D. L. Huestis, and H. Helm, J. Chem. Phys. **92**, (1990) 2219.
14. P. Avan, C. Cohen-Tannoudji, J. Dupont-Roc, and C. Fabre, Le Journal de Physique **37**, (1976) 993.
15. P. H. Bucksbaum, M. Bashkansky, and T. J. McIlrath, Phys. Rev. Lett. **58**, (1987) 349.
16. D. C. Lorents, D. J. Eckstrom, and D. Huestis, "Excimer formation and decay processes in rare gases," Report No. MP 73-2, Stanford Research Institute, Menlo Park, CA (September 1973).
17. A. D. Bandrauk and N. Gélinas, J. Chem. Phys. **86**, (1987) 5257.
18. A. Guisti-Suzor, X. He, O. Atabek, and F. H. Mies, Phys. Rev. Lett. **64**, (1990) 515.

$H^+ + H^-$ Coincidence Studies Resulting From the Dissociation of Excited, Fast Moving H_3^+

D. H. JAECKS, O. YENEN, D. CALABRESE, AND L. WIESE
Behlen Laboratory of Physics
University of Nebraska
Lincoln, NE 68588-0111 U.S.A.

ABSTRACT

The three-particle decay mode of excited H_3^+ leading to $H^+ + H^- + H^+$ is of particular interest because at large separation the dominant interaction between all three particles is coulombic. This is the only known three-body system of massive particles with this characteristic that has been studied to date. All other studied systems involve two electrons and a massive nucleus. In order to determine the characteristics of the strongly correlated motion of these massive particles that will also test theory, we have designed and built an apparatus that will measure the six-fold laboratory differential cross section for the dissociation of excited H_3^+ into $H^+ + H^- + H^+$. There are three angles and three energies that are to be determined simultaneously by a coincidence experiment. We have most recently made coincidence measurements between H^+ and H^- from excited H_3^+, for H^+ and H^- emitted along the beam. The essential feature of these results is that there are no protons with near-zero energy in the H_3^+ c.m. that are found to be in coincidence with H^-. This result is consistent with our earlier interpretation that most of the available energy in the excited H_3^+ is given to the protons.

Three-Body Coulomb States of Massive Particles

The experimental determination of the characteristics of the decay mode of excited H_3^+ leading to the three particle-dissociative state, $H^+ + H^- + H^+$, is a problem of fundamental importance. This importance stems from the dominant long range Coulomb interactions at large particle-particle separations that, in turn, lead to highly correlated motion of the three particles. For internal energies of H_3^+ near and above the dissociation energy limit of $H^+ + H^- + H^+$, correlation effects are expected to dominate the final state motion.

A complete quantal description or wave function of such a highly correlated three-body system has not been found;[1] therefore such systems deserve further attention. Several theories based on Wannier's[2] classical description of threshold ionization have been developed. In all such theories, solutions are limited to regions about the Wannier Ridge and for energies near zero. Solutions are also limited to a hyperradius that lies outside the reaction zone. Feagin[1] discussed the history of this theoretical development starting with the semiclassical redevelopment of Wannier's classical theory by Peterkop[3,4] and Rau.[5,6] Klar[7] extended this development to include systems of arbitrary mass having total angular momentum $L = 0$. Feagin further developed the theory for arbitrary mass and for $L > 0$, within the WKB approximation.[1]

Mass-dependent effects of theory have not been tested. To date, experimental tests of the various theoretical descriptions of the three body problem have involved two electrons and a massive nucleus. These tests have been primarily concerned with the threshold behavior of the ionization cross sections of atoms by electrons, double photoionization of H^-,[8] and double photoionization of heavy atoms.[9] The emphasis of the experimental work has been the determination of the threshold power law. Some experiments have measured the energy sharing fraction of the two electrons.[8] Angular correlation measurements between the two outgoing electrons have also been carried out.[10]

Definitive experimental tests of the three-body description of three massive particles of equal or unequal mass have not been carried out. There also have been no tests for three light but equal mass particles such as $e^- + e^+ + e^-$. To test the full range of theory, it is imperative that measurements on massive particles be carried out. An ideal system would be slowly moving $\bar{p} + p + p$; however, the experimental difficulties for this system would be overwhelming. The three-body decay of excited H_3^+ into $H^+ + H^- + H^+$ appears to be the most viable candidate for such a study of massive particles.[11, 12, 13] Feagin[1] has calculated measurable physical parameters for the $H^+ + H^- + H^+$ system that can be tested. Other properties measured in threshold ionization, but not specifically addressed by Feagin, can also be carried out.

In such experiments the H_3^+ is excited in a keV collision with a gas target, most often He. It has been determined that the particular excited state of H_3^+ that is of interest is about 40 eV above the ground state. Nothing is known about the symmetry of the excited state, or its excitation mechanisms.

Alternate modes of excitation to that state have been looked into. Synchrotron radiation of sufficient intensity and energy is not yet available. However, such radiation would work if the transition to the excited state satisfies the yet unknown selection rules. Multiphoton excitation in an intense laser field might also work if the H^- can be made to survive the laser pulse and if the repetition rate can be made high enough. At this time, collisional excitation appears to be the only efficient and viable means of producing excited states of H_3^+ that decay to the three-body state of interest.

We have embarked on a program to determine the final state characteristics of the $H^+ + H^- + H^+$ system. To do so we have designed and built an apparatus that can measure the six-fold laboratory differential cross section for the dissociation of H_3^+ into $H^+ + H^- + H^+$. There are three angles and three energies that are to be determined simultaneously by a triple coincidence measurement. From the laboratory energies and angles of the three particles, one can determine all of the c.m. energies, angles between c.m. momenta, and the deflection angle of the c.m. The apparatus is designed to be sensitive to only c.m. velocity vectors that lie in the plane of the three particles.

The general measurement approach is shown in Fig. 1 where we show a side view of the electrostatic energy analyzer used to measure the lab energies and angles of the H^+. One H^+ will be detected on the beam axis, while a second wedge and strip position sensitive detector will detect H^+ with a variable lab energy and angle. The position-sensitive H^- detector will detect a range of energies and angles. The detector positions are moveable so that a range of angles and energies expected can be detected. Experimental problems are compounded by the fact that there are, in principle, an infinite number of ways the three particles can share any available internal energy. This in turn makes it difficult to know where (lab angle and energy) one should place the detectors. Added complications arise because the particles are of equal mass. This equality can cause all three particles to be deflected from the initial beam direction. This is not the case, for example, when massive atoms or ions emit light electrons in an ionization process. Thus, our approach has been deliberate and measured. The H^- energy distribution was first determined.[12,13] From the laboratory energy distribution, we have determined an approximate c.m. energy distribution of the H^-. This is shown in Fig. 2.[13] The c.m. energy distribution peaks at around 0.75 eV, with the distribution ranging from 0 to 2.5 eV. From the measured energy loss of 40 eV and the known energy of the final $H^+ + H^- + H^+$ state, relative to the initial H_3^+ state, we calculate an available energy of about 19 eV that is to be shared by the

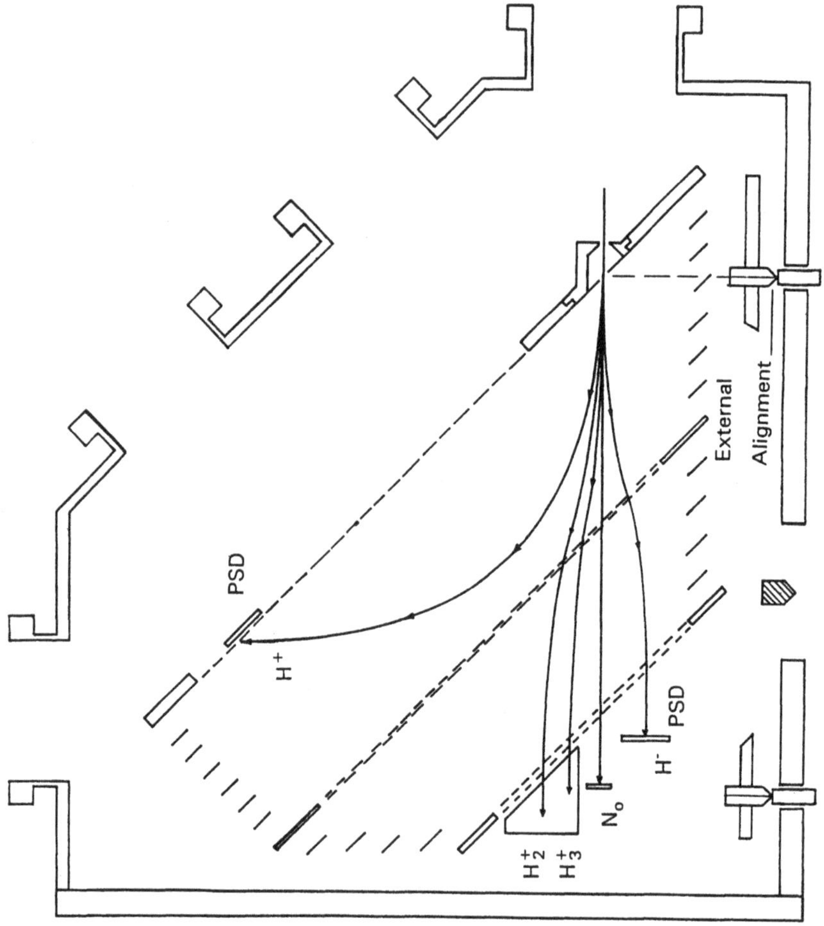

Figure 1

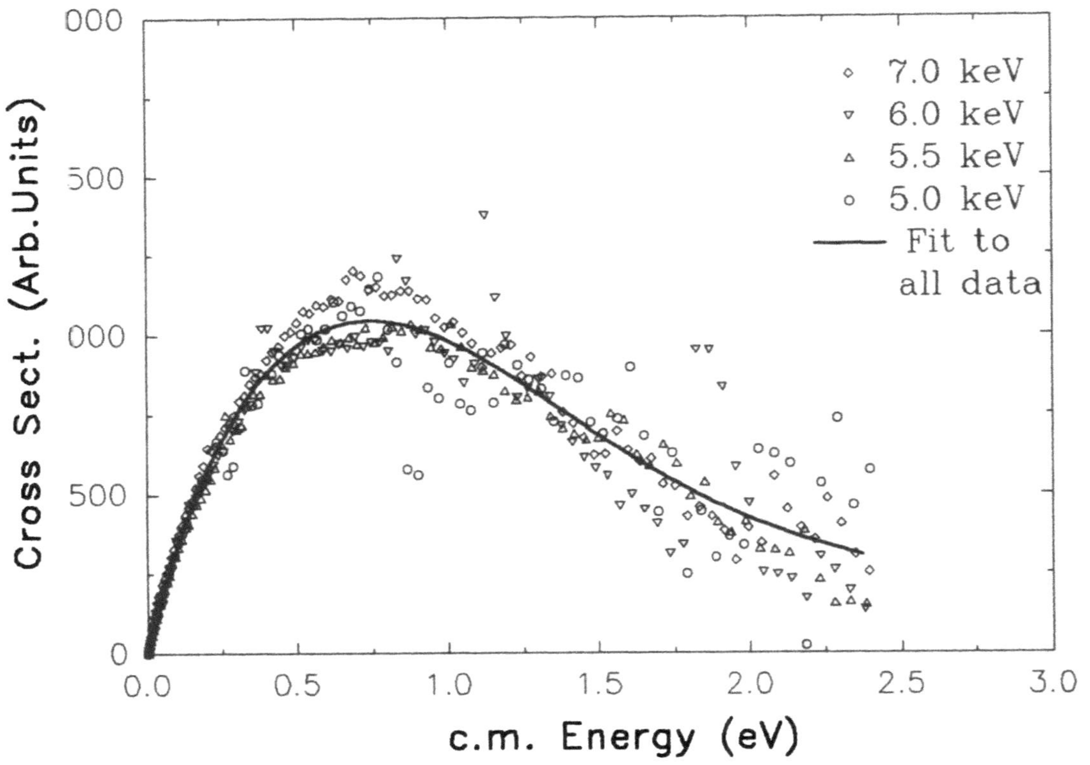
000
500
000
500
0
0.0
0.5
1.0
1.5
2.0
2.5
3.0
Cross Sect. (Arb.Units)
c.m. Energy (eV)
7.0 keV
6.0 keV
5.5 keV
5.0 keV
Fit to all data

Figure 2

three particles. This number is different than our original value of 39 eV which we published in reference 13. The error in our original published work resulted from the fact that the original calibration procedure for our parallel plate analyzer led to large uncertainties in the measured energies. This has been corrected. We have also taken into account the fact that the H_3^+ ion energy is shifted from the value of the potential placed on the source. We thank Dr. James Peterson of SRI for pointing out the discrepancies in our earlier work.

We have most recently made coincidence measurements between H^+ and H^- from excited H_3^+, for H^+ and H^- emitted along the beam in the laboratory. The laboratory acceptance angle for H^+ was $\pm$ 0.03° while the H^- acceptance angle was $\pm$ 0.25°. The limited range of acceptance of the H^- detector means that we are not seeing all possible coincidences. Our goal in these initial measurements was to determine whether *any* coincidences could be observed, as well as to determine the range of coincidence rates for various H^+ energies. We found no coincidences between H^+ and H^- with only the background gas. The background pressure was about 5×10^{-8} Torr.

Figure 3 shows examples of our measured coincidence start-stop time spectrum for two different voltages on the parallel plate analyzer. Figure 4 shows the total number of integrated coincidences for different analyzer settings. On the same figure is the total H^+ energy spectrum measured in the laboratory resulting from all processes. The central peak of the H^+ spectrum corresponds to particles having near-zero energy in the H_3^+ c.m. With negligible transverse velocities, the particles just move along with the beam and are collected with high efficiency. From the coincidence energy spectra, it is apparent that there are few protons of zero energy that are simultaneously formed with H^-, although such protons are collected with high efficiency. This is just what we would expect for the break-up of the three particles when the three particles lie along or near the beam axis with the H^- between the two protons. The symmetric position of the coincidence data occurs at an analyzer voltage that corresponds to an energy loss of about 40eV, a number that is consistent with the energy loss of our earlier total H^- energy spectra.[13] We emphasize that the present measurements are incomplete due to the limited range of angles accepted by the present H^- detector. However, the general viability of the coincidence technique has been illustrated with these initial two-particle coincidence measurements.

A complete determination of the final state characteristics await the three-particle coincidence measurements. From the three-particle coincidence measurements we expect to be able to determine the distribution in the correlation angle θ_{12} between the two protons,[1] for different ranges of available internal energy. The distribution of the sharing of the extra energy by the two protons can also be determined for each and every coincident evident, for different ranges of available internal energy. We believe that

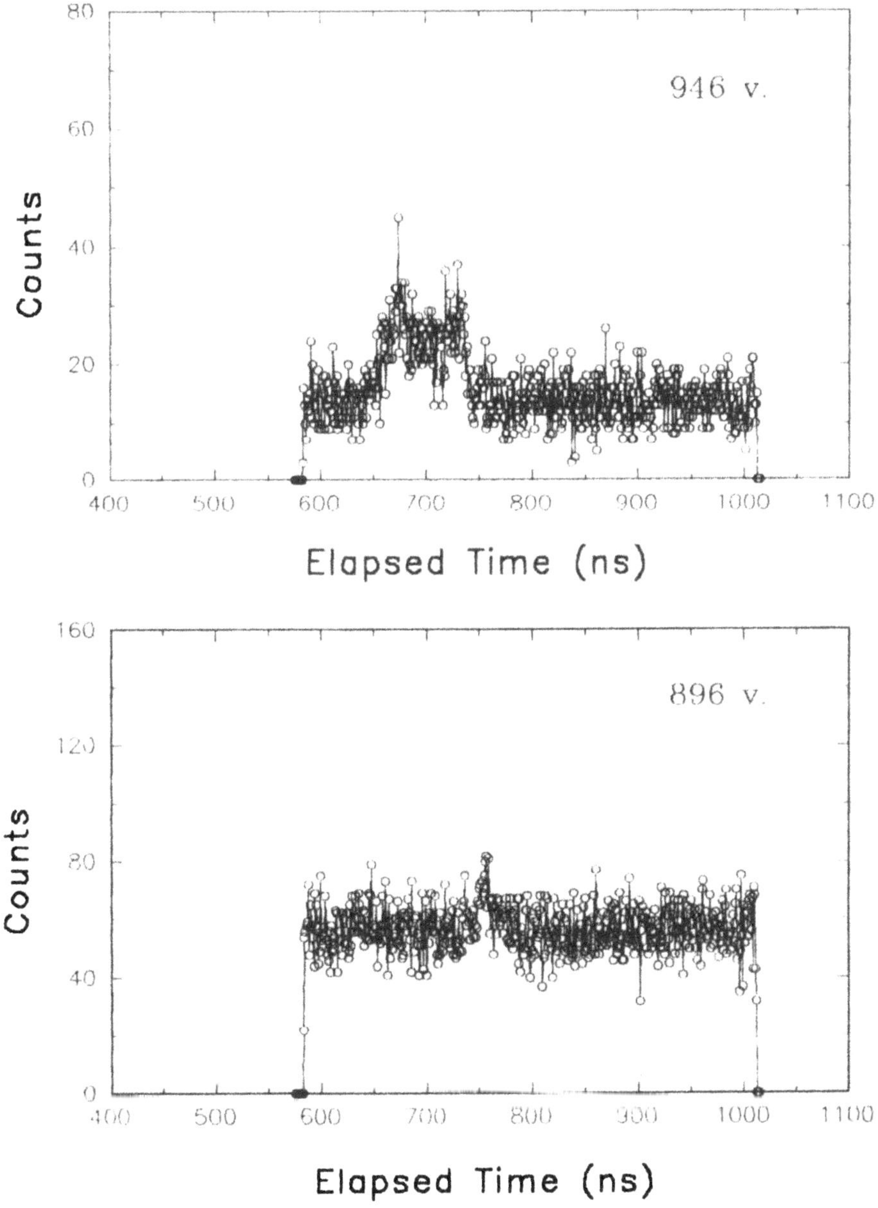

Figure 3

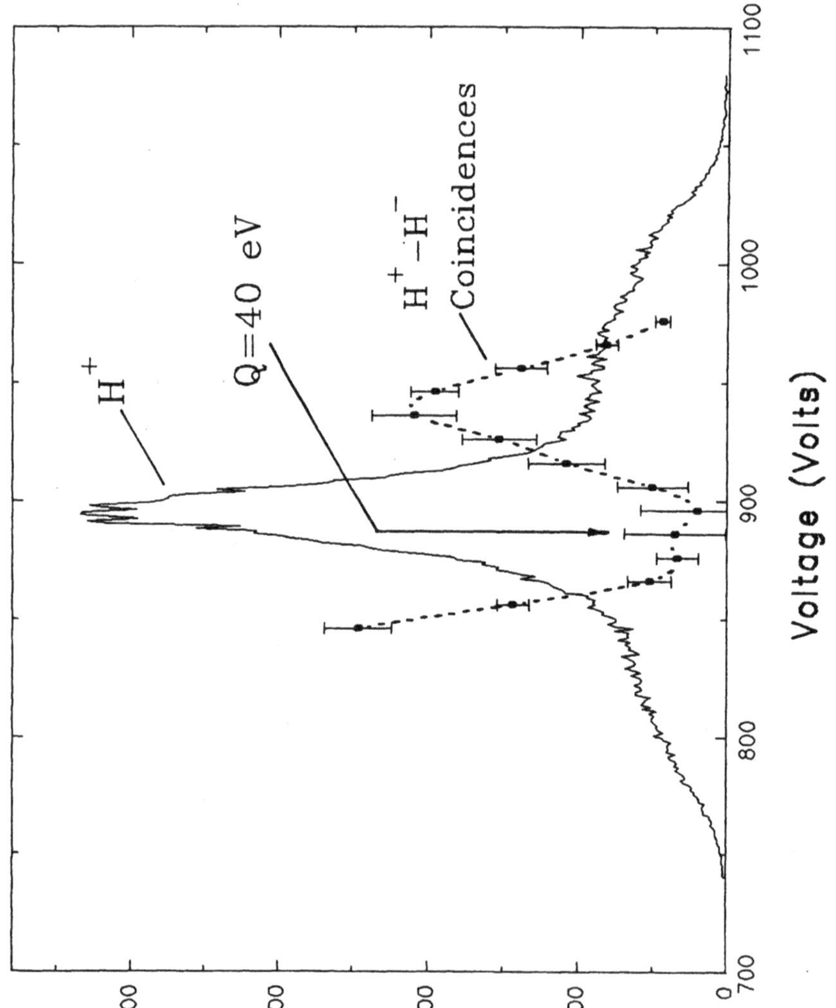

H^+
Q=40 eV
H^+–H^-
Coincidences
Intensity (Arb. Units)
0
5000
10000
15000
20000
700
800
900
1000
1100
Voltage (Volts)

Figure 4

three-particle measurements of this type hold great promise in the understanding of reactions and dissociations of three-particle systems.

Acknowledgements

The support of the National Science Foundation through Grant No. PHYS-8701905 is gratefully acknowledged.

References

1. J. M. Feagin, *J. Phys.* **B17** (1984) 2433.
2. G. H. Wannier, *Phys. Rev.* **90** (1953) 817.
3. R. Peterkop, *J. Phys.* **B4** (1971) 513.
4. R. Peterkop, *J. Phys.* **B9** (1983) L587.
5. A. R. P. Rau, *Phys. Rev.* **A4** (1971) 207.
6. A. R. P. Rau, *J. Phys.* **B9** (1976) L283.
7. H. Klar and W. Schlecht, *J. Phys.* **B9** (1976) 1699.
8. F. H. Read, in *Electron Impact Ionization*, eds. T. D. Mack and G. H. Dunn, (Springer Verlag: New York 1985).
9. H. Kossman, V. Schmidt, and T. Andersen, *Phys. Rev. Lett.* **60** (1988) 1266.
10. P. Selles, A. Huetz and J. Mazeau, *J. Phys.* **B20** (1987) 5195.
11. D. L. Montgomery and D. H. Jaecks, *Phys. Rev. Lett.* **51** (1983) 1862.
12. I. Alvarez, C. Cisneros, J. de Urquaijo, and T. J. Morgan, *Phys. Rev. Lett.* **53** (1984) 740.
13. O. Yenen, D. H. Jaecks, and L. M. Wiese, *Phys. Rev.* **A39** (1989) 1767.

PREDISSOCIATION OF H_3 n = 2 RYDBERG STATES: PRODUCT BRANCHING AND ISOTOPE EFFECTS

J. R. Peterson, P. Devynck, Ch. Hertzler, and W. G. Graham

Molecular Physics Laboratory
SRI International
333 Ravenswood Avenue
Menlo Park, CA 94025 USA

ABSTRACT

A translational spectroscopy technique is used to obtain predissociation kinetic energy release spectra from the lowest bound states of H_3, 2s A_1' and 2p A_2''. These states are formed in near-resonant electron capture by 3 keV H_3^+ in Cs vapor. Their ground rovibrational levels are energetically about 1 eV above the H + H + H dissociation limit, thus all levels can yield both H_2 + H and the 3-body products. The spectra contain both 3-body and 2-body components, which are deconvoluted to obtain the branching ratios. Data obtained from two different ion sources show that 3-body/2-body ratio increases with increased rovibrational energy in the H_3. The results are compared to recent theory, and with ratios obtained from dissociative recombination. The H- and D-atom spectra from the 2-body decay of D_2H show that the ejected H-atom is favored on a per-atom basis.

I. INTRODUCTION

The triatomic radical H_3 is purely dissociative in its ground electronic state, but bound states exist when one of the electrons is in an excited Rydberg orbital. Such "Rydberg" molecules have recently been reviewed by Herzberg,[1] who gave considerable attention to H_3. Optical spectra have shown that the lowest state of H_3, 2s $^2A_1'$, predissociates in about 0.2 ps.[2] The nearby 2p $^2A_2''$ state, however decays by rotational coupling with lifetimes of about 30 ps, for rotational level N = 1, decreasing to 5 ps for N = 4.[2] However its rotationless (N = K = 0) lower symmetric stretch and bending states are stable against rotational predissociation.[3,4] Their slow decay was originally considered to be purely radiative, to 2s $^2A_1'$, in about 87 μs.[3] However, very recent measurements have found lifetimes in the 600-800 ns range,[5] with the more rapid decay attributed to (a) radiation directly to the ground state accompanied by bending-mode excitation, and (b) possibly some spin-orbit coupled predissociation.

These two primarily predissociating H_3^* states lie about 1 eV above the H + H + H limit (see Figure 1), and so the predissociation of each can produce both H + H + H and H_2 + H products. Until this work, there have been no measurements of the relative yields of the 3-body and 2-body products. Peterson and Bae,[6] in a trial translational spectroscopy experiment, measured the D-atom energy distributions and yields from D_3^* formed by electron capture of D_3^+ in Cs, and noted branching into the 2-body and 3-body product channels. Cosby and Helm[7] have recently examined predissociation of nearly rovibrationless levels of the 3s A_2' and 3d E" states. By detecting the radial positions and time differences in the arrival of the 2-body decay products in a fast beam experiment, they were able to obtain detailed vibrational distributions in the H_2 products and found them to extend strongly into the continuum,[7] indicating the presence of 3-body products, but the detector does not sample 3-body distributions. Mitchell et al.[8,9] have measured 3-body/2-body branching ratios for dissociative recombination of H_3^+ with low energy electrons:

$$H_3^+ + e \rightarrow H_3^* \rightarrow \text{neutral products} \qquad (1)$$

This is also a predissociative process, but it originates through very high H_3^* Rydberg states near the ionization limit. Their results will be discussed below.

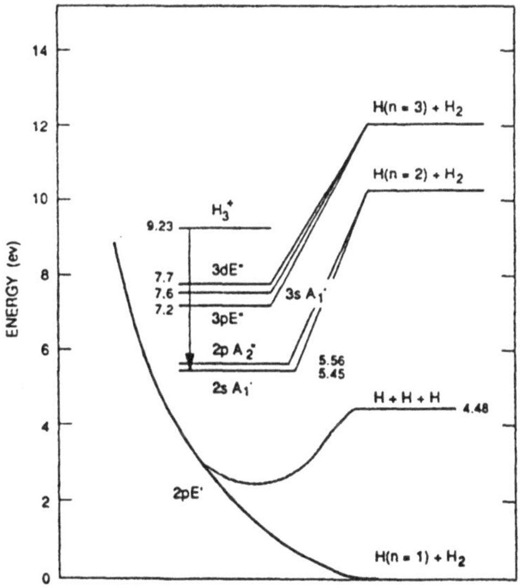

Figure 1. Schematic potential energy diagram of H_3 and H_3^+ The arrow indicates the electron capture process.

In this paper, we report measurements of the branching ratios, energy distributions, and isotopic dependences of the dissociation products of the n = 2 Rydberg states of H_3, D_3, and D_2H. These states can be formed very efficiently from an H_3^+ beam by electron capture from Cs (I.P. = 3.89 eV) in the reaction

$$H_3^+ + Cs \rightarrow H_3^* \,(n = 2) \; + Cs^+ \tag{2}$$

because the reaction is nearly "resonant" (little change in internal energy) and because the H_3^+ - H_3^* Franck-Condon factors are very large (~1 for $\Delta v = 0$). The reaction is endothermic by only 0.11 eV for the 2s $^2A_1'$ product and 0.22 eV for 2p $^2A_2''$. The near-resonant characteristic allows transitions at large impact parameters, and hence large cross sections. The measured total cross sections for Reaction (2) exceed 150 Å^2 at ion beam energies below 3 keV.[10]

The essentially diagonal Franck-Condon factors have another very important role in these experiments. Because of them, and the "softness" of the near-resonant collisions, the rovibrational distribution in the initial H_3^+ beam is strongly preserved in the H_3^* products of Reaction (2), as has been demonstrated in H_2^+ + Cs reactions.[11] The use of different ion sources allows us to vary the internal energy in the H_3^*.

Near-resonant capture is a strong factor in the final electronic state distribution primarily at lower collisional velocities (< ~5 × 10^7 cm/s); relative yields of higher n Rydberg molecules will increase at higher beam energies above 2 keV/amu, where the total cross section decreases.[10] Bound-bound radiation from n = 3 states of D_3 produced by electron capture of 5 keV D_3^+ beams in K vapor (endothermicities ≥ 2 eV) was observed by Figger et al.,[12] with an estimated partial cross section of ~10^{-19} cm^2, which is negligible compared to the total cross section of about 1.6 × 10^{-14} cm^2 under the present conditions.[10] The results support our belief that essentially all H_3^* formed in our experiment are in n = 2 Rydberg states.

II. EXPERIMENTAL TECHNIQUE

The technique used to probe the nature of the dissociation products following electron capture in Cs has been described elsewhere.[6,13] Briefly, we use a translational spectroscopy technique to obtain the distribution of total center-of-mass (c.m.) kinetic energies released to the dissociation fragments. This quantity, W, is the difference between the total internal energy (electronic and rovibrational) in the excited H_3^* before pre-dissociation, and the in the final products. We utilize the fact that following the predissociation:

$$H_3^* \rightarrow H_2 + H, \tag{3a}$$

or

$$H_3^* \rightarrow H + H + H, \tag{3b}$$

a small fraction of the fast H products, produced from ~3 keV H_3^+, can capture an electron in a subsequent collision within the cesium oven, to form a negative ion:

$$H + Cs \rightarrow H^- + Cs^+ \tag{4}$$

We then measure the laboratory kinetic energies of the H^- ions formed at 0° in the direction of the beam, which represent H atoms from (3) ejected from H_3^* at 0° and 180° c.m. angles. These data are then manipulated to obtain the c.m. kinetic energy release spectrum P(W).

A. Apparatus

A partial diagram of the apparatus is shown in Figure 2. In these experiments, the H_3^+ ions were formed in an ion source (described below) from H_2, D_2, or mixtures of them, extracted and accelerated to the final beam energy, focused, mass selected by a magnetic field, and directed into a Cs vapor oven, where they undergo Reactions (1) to (3). H^- ions emerging from the oven along the beam axis pass into the analysis chamber enter an electrostatic quadrupole deflector, Q1, where they are separated from the positive and neutral components and are deflected 90° to the entrance of an energy analyzer. Neutrals leaving the oven pass through Q1 and into a Faraday cup, FC1, where they are monitored by their secondary electron current. The voltages on the deflecting quadrupole and the energy analyzer are coordinated to pass the same energy. The ions that pass through the energy analyzer enter a microchannel-plate detector (MCP) whose output pulses are directed to an on-line computer.

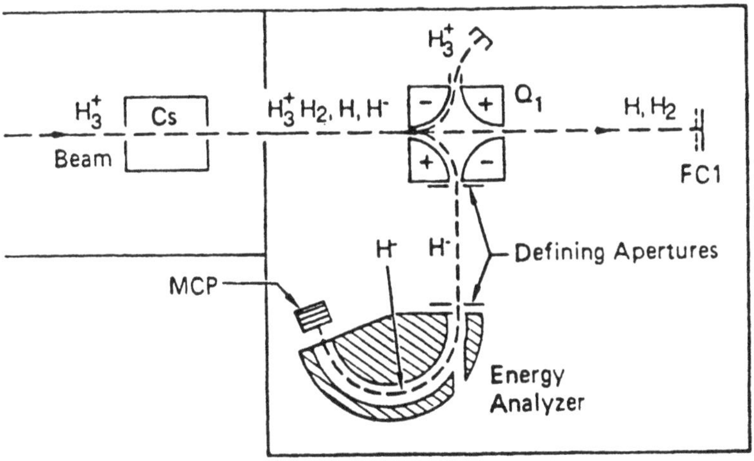

Figure 2. Diagram of the apparatus.

To examine the effects of internal rovibrational excitation in the H_3^* on the 3-body/ 2-body branching ratios, we have used two different ion sources. First, a Colutron-type sustained discharge source was used. It produces rovibrationally hot H_3^+ ions. Recently, we have used a cooled hollow-cathode source,[14] operating at about 1-3 Torr, which yields relatively cool ions. As mentioned above, the long-range nature of Reaction (2) and the diagonal nature of the ion-Rydberg Franck-Condon factors preserves the ions internal energy distribution in the product Rydberg states.

An example of the laboratory kinetic energy spectrum obtained from a 3.2 keV D_3^+ beam using the hollow-cathode source is shown in Figure 3.

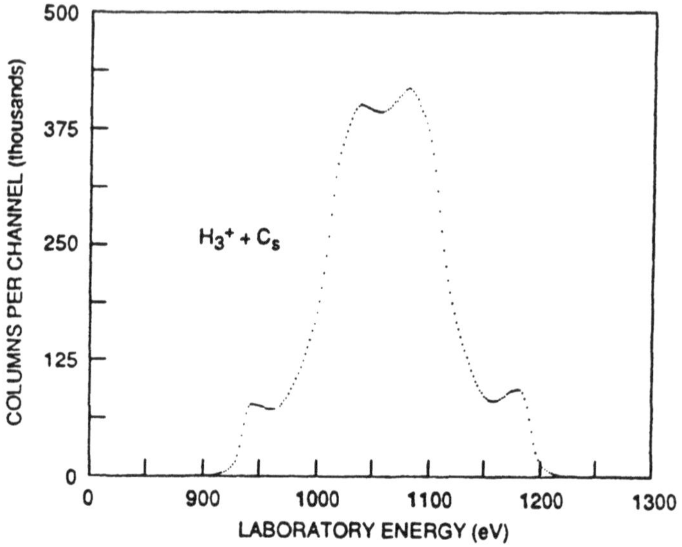

Figure 3. Laboratory energies of H^- formed at 0° by electron capture of H atoms from 3.2 keV H_3^+ in Cs.

B. Data Analysis

The data are first divided by the cross section for forming H^-, Reaction (4), using smoothed data from Miethe et al.,[15] then divided by the laboratory energy because the relative energy analyzer width ΔE/E is constant, and then transformed,[13] including the laboratory-c.m. solid angle conversion, to obtain the c.m. kinetic energy release spectrum. The energy scale for W used in the transformation is based on the 2-body geometry. The transformed data from Figure 3 are shown in Figure 4. The solid line links the data from the higher energy part of the laboratory spectrum in Figure 3, representing 0° c.m. fragments, and the and the dots are the data from the lower energy (180° c.m.) part. The close agreement between the two data sets validates the data treatment procedure. The spectrum has two components; the higher energy peak is logically attributed to the two-body products and the lower energy one to the 3-body component.

1. Deconvolution of the c.m. spectrum.

To obtain the branching ratios, the total 2-body and 3-body spectrum must be deconvoluted in some reasonable way without an accurate knowledge of the relative shapes of the two individual distributions. This process is simpler for the spectra obtained from the cooled hollow cathode ion source, and is demonstrated using the 0° and 180° data from Figure 4, which have been combined and smoothed in Figure 5. We first recall that the 2s and 2p states are 0.97 and 1.08 eV above the H_2 dissociation limit, thus the lower limit to the 2-body spectrum is about 1.0 eV. Also, we expect the ejected H_2 molecules to have a vibrational population extending into the continuum, since the 3-body products are observed. In Figure 5, we note that the full spectrum has a rather narrow low energy peak,

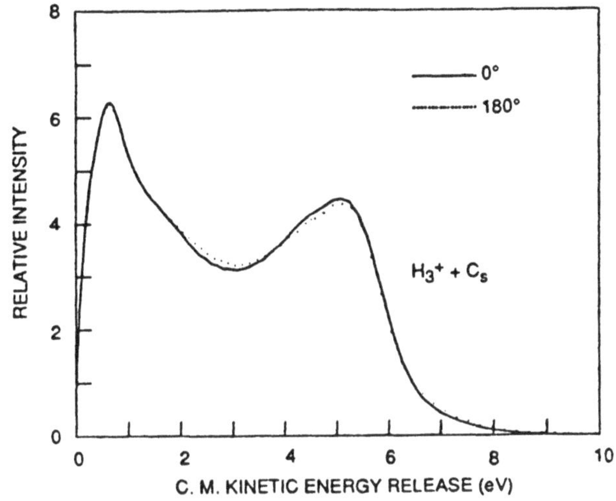

Figure 4. Transformed data from Figure. 4.

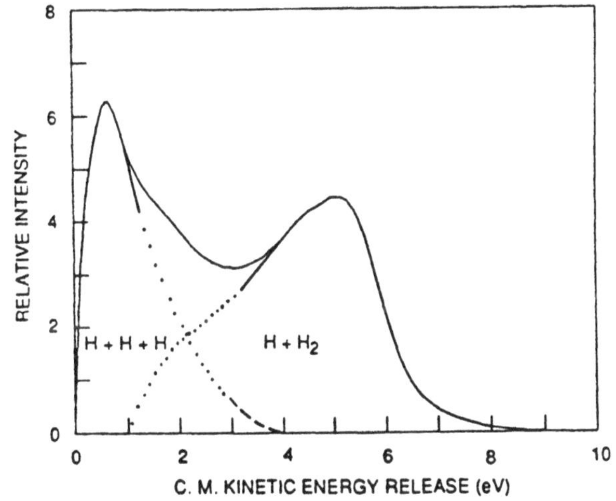

Figure 5. Deconvoluted distributions from Fig. 4. The initial H_3^+ ions were from the cooled hollow-cathode source.

and then begins to diverge toward the higher peak at 1 eV, which we attribute to the onset of contributions from the low-energy 2-body tail. We then draw a short straight line tangent to the total envelope at the point of divergence. Next, we note that the high-energy peak deviates toward the low-energy peak at about 4.2 eV. Another straight line is drawn here tangent to the full curve, and subtracted from it to yield the high-energy tail of the 3-body spectrum, shown in dashes. With the expectation that the 3-body spectrum should not show any structure, a smooth curve is then drawn monotonically from the dashed line, to join the lower tangent line just above 1 eV. This curve is then subtracted from the total to yield the 2-body spectrum. After several tries, a "reasonable" deconvolution is obtained, which does not contain any sharp substructure in either distribution. The different variations produce only minor changes in the areas.

It is also worth noting that the upper limit to the 2-body distribution for robvibrationless H_3^* is 5.5 eV. Higher energies in the W spectrum result primarily from internally excited H_3^* and from the bandwidth ΔE of the energy analyzer, which results in $\Delta W \propto W^2$ in the c.m. energies. This information was not used in the deconvolution, however, because of the unknown internal energy distributions.

In contrast to the hollow cathode source data, deconvolution of the spectra obtained using the hot discharge source was much less obvious. Figure 6 shows the c.m.-transformed data from H_3^+ produced in the colutron ion source. The low-energy peak is considerably wider and relatively larger in area than in the hollow cathode case. Both aspects result from increased internal energy in the H_3^+ ions, and consequently in the

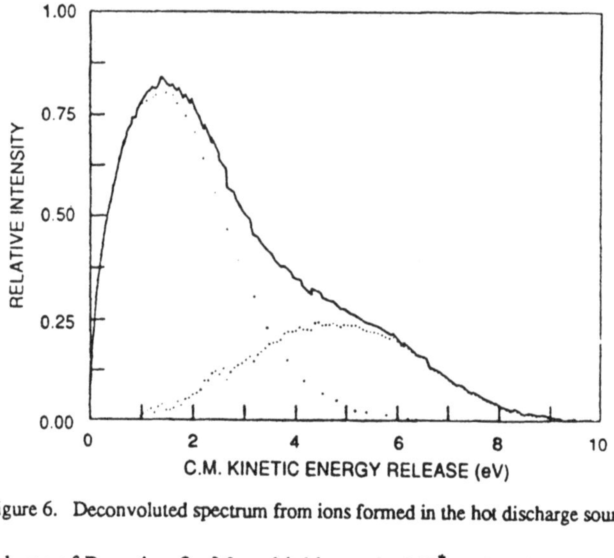

Figure 6. Deconvoluted spectrum from ions formed in the hot discharge source.

Rydberg products of Reaction 3. More highly excited H_3^* molecules clearly yield a more dominant and broader 3-body spectrum. In addition, the two-body spectrum is pushed to higher energies. As the internal energy in H_3^* increases, both the upper and lower limits of the 2-body spectrum, respectively corresponding to the H_2 in $v = 0$ and at its dissociation limit, are increased in energy. The deconvolution shown by the dotted lines in Figure 6 has a 2-body spectrum resembling one taken very recently using the position-sensitive detector of Helm and Cosby,[16] which can determine accurate 2-body kinetic energy distributions. Another, with a slightly more prominent low-energy 2-body tail resembling that from the hollow-cathode source was also used, and the results were averaged in obtaining the ratio and internal energy, as explained below.

2. Determination of the 3-body/2-body branching ratio.

Although the energy scale is important in determining the W values in the two distributions, it does not affect the determination of the branching ratios, which depend only on the areas under the two distributions. The reason for this independence of scale is that initially the spectra are obtained in the laboratory energies, and only one particle is detected per predissociation event; the other fragments have different energies and/or directions, and are not detected. The probability of an ion passing through the energy analyzer depends only on (a) its laboratory energy, because of the analyzer bandwidth, and (b) its laboratory velocity relative to the velocity of the parent molecule, because of the laboratory-c.m. solid angle transformation. Thus the detection probability is independent of the masses or velocities of the other fragments, which are considered in determining W.

Although the branching ratios can be determined from the areas under the two distributions using the 2-body scale, we must account for the multiplicity of the detected atoms among the fragments. Since all three H atoms in H_3 are ejected in a 3-body event, and only one in a 2-body event, only 1/3 of the area under the 3-body peak corresponds to individual events, compared to the area under the two-body distribution. If we measure D^- from D_2H, we use 2/3 of the area under the 3-body peak to compare with the full area under the 2-body peak, and if we detect H^- from D_2H, we use the full 3-body area.

With the 3-body peak area properly weighted by the inverse of the multiplicity of the detected H or D atom (the relative probability of detecting a single event) the resulting areas under the two distributions are proportional to the number of dissociation events in each channel.

3. Determination of energy scale for W.

The energy scale of Figures 4-6 is exact for the 2-body case. The 3-body scale can only be determined for fixed geometries. As earlier,[6] we use three simple cases as examples in the c.m. system. The first is identical to the 2-body case, except that two unbound atoms leave with the same velocities, with the third in the opposite direction. In the second, one atom is at rest and the others have opposing momenta. The energy scale for this case is 4/3 times the 2-body scale. In the third case, the three atoms depart symmetrically, at 120° to each other, with equal momenta. The scale for this case is 2.0 times the 2-body scale. The actual distribution of geometries will embrace all of these prototypes. The average geometry is unknown. In recent wave-packet studies of predissociation in H_3 Rydberg states, Orel and Kulander[17] find that the 3-body products tend to favor a collinear geometry, with one atom betwen two widely separated ones. With this finding in mind, we have weighted the first two cases equally, and the third with 1/2 their individual probability, in the order 2:2:1, yielding the average 3-body scale of 1.33 times the 2-body scale. This scale is used in determining the average energy of the three-body distribution as described below.

4. Estimate of the internal energy in the H_3^*.

To estimate the internal energies in the H_3^* molecules that yielded the measured kinetic energy distributions, we assumed that the average energy in the 3-body distributions was equal to the average total internal energy in the H_3^* (n = 2) distributions, relative to the H + H + H asymptotic limit of 4.48 eV above the H_2 + H ground state. This energy consists of 1.0 eV of electronic energy plus the average rovibrational energy. The average energies in the 3-body distributions were obtained using the 1.33 scaling factor.

Although this method of estimating the internal energy in the Rydberg states is somewhat arbitrary, it is based on logical assumptions with some theoretical guidance, it is the best that we could conceive, and is probably accurate with the ample acknowledged uncertainties. There is no possibility of an accurate experimental determination because the states are so rapidly predissociated (except for the negligible metastable rotationless 2p states). For the present work, we felt that it was appropriate to use the average W in the 3-body distribution.

III. RESULTS

A. Branching Ratios

The 3-body/2-body branching ratios were found to increase with the internal energy in the H_3^* Rydberg states, as determined from the average c.m. kinetic energies in the 3-body distributions. These ratios are plotted in Figure 7.

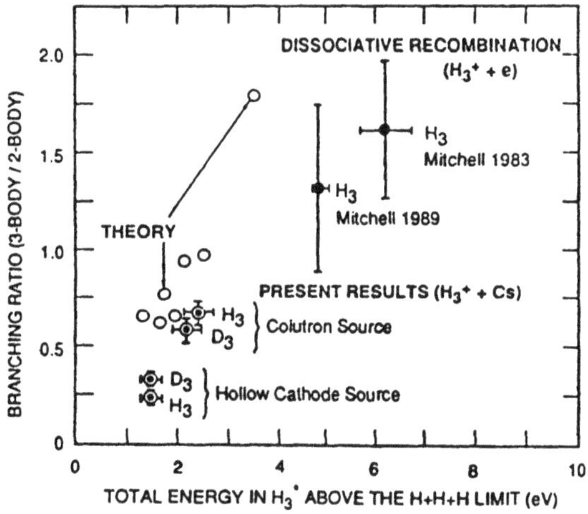

Figure 7. Present results, compared to theory and dissociative recombination experiments.

We found that the average energies in the 3-body distributions from the cooled hollow cathode source were about 1.48 ± 0.15 eV for H_3 and 1.52 ± 0.15 eV for D_3, and the 3-body/2-body ratios were 0.23 ± 0.02 and 0.35 ± 0.02, respectively. Evidently the internal energies in the D_3^+ ions were only slightly hotter than the H_3^+ ions emerging from the ion source, yet the branching ratio is noticeably higher from D_3. This result is a little surprising, if it represents a real effect. It could result from a higher proportion of vibrational (compared to rotational) energy in the D_3^*.

The results from the hot discharge source gave average internal energies of 2.42 ± 0.2 eV for H_3 and 2.16 ± 0.2 eV for D_3, and 3-body/2-body ratios of 0.65 ± 0.06 and 0.57 ± 0.06, respectively. In this case the H_3^+ ions were apparently somewhat hotter than the D_3^+ ions. These average energies imply rovibrational energies of 1.4 and 1.2 eV. They are in rough agreement with the calculations of Anicich and Futrell,[18] who found average vibrational energies of 1.3 and 1.5 eV, respectively. Since the ions are formed in the reaction

$$H_2^+ + H_2 \rightarrow H_3^+ + H, \qquad (5)$$

which is 1.83 eV exothermic, and the initial H_2^+ reactants are formed vibrationally excited (averaging 0.8 eV[18]) by electron impact, the initial H_3^+ ions are expected to be formed both rotationally and vibrationally excited. The distributions that emerge from the ion source depend on the gas pressures and the residence times that allow relaxation.

The rovibrational energies, about 1 eV less than the total internal energies (above the H + H + H asymptotic limit cited above), depend mostly on the pressures and residence times in the ion sources, which were only approximately similar in each source. The one clear result here is that the 3-body/2-body branching ratios increase strongly with the internal energy in the predissociating Rydberg molecules. This result is not surprising, and is predicted in the calculations of Orel and Kulander,[17] but these are the first experimental data to show the energy dependence.

Also shown in Figure 7 are the theoretical results of Orel and Kulander.[17] Their time-dependent wave-packet studies of H_3 2s $^2A_1'$ predissociation were carried out in a two-dimensional approximation, which considered the symmetric and bending modes in C_{2v} symmetry, with no rotation. We have plotted the R_3/R_2 ratios from their Table I against the total energy minus the 4.48 eV dissociation energy of H_2, to compare with our results. The energy dependence is similar, but our ratios are lower, probably in part because our total energy is likely to contain a considerable amount of rotation, resulting from the ion-molecule Reaction (5), and the 3-body channel is probably inhibited by a rotational barrier.

Also shown in Figure 7 are ratios obtained by Mitchell et al.,[10,11] for the 3-body/2-body product branching of the dissociative recombination of H_3^+ for near-zero electron energies. This reaction is primarily a predissociation reaction involving very high Rydberg states near the ionization limit, thus most of the internal energy is electronic. They used two ion sources: an r.f. discharge source, which produced hot ions, and an r.f. trap ion source, which produced ions primarily in v = 0. We have given the discharge source data the same rovibrational energy deduced in our Colutron source data, and have set their "cold" ion rovibrational energy at 0.07 eV. Note that H_3^+ (v = 0) is 4.75 eV above the H + H + H limit. The comparison of the dissociative recombination data with those resulting from predissociation of lower Rydbergs is interesting, indicating that the branching ratio is a steadily increasing function of total internal energy above the 3-body limit, and does not depend too much on whether the energy is electronic or vibrational. Comparison of our data to the theory of Orel and Kulander indicate that energy stored in rotation is not effective in producing 3-body products, compared to energy in vibrational modes, which is not surprising.

B. *Isotope Effects*

We performed only a few experiments on the isotopomers of H_3, and primarily on D_2H, since H_2D is indistinguishable from D_2 in our low resolution apparatus. Using D_2H ions produced in the hot discharge source, we measured the laboratory energies of both D and H atoms. From the two transformed spectra we were able to measure the 3-body/2-body ratios in each. After after equating the two 3-body spectra, normalized to the number of measured atoms in each, we found that among the 2-body products, the H atom has 2.7 times the probability of each D atom to emerge, on a per-atom basis! Of course, since there are two D atoms, the two-body products yield H atoms only 1.4 times as frequently as D atoms. The higher probability, per atom, of an H ejection indicates a more rapid predissociation rate, which can be ascribed to a better overlap of the H atom bound-free radial differential Franck-Condon factors between the bound and continuum states, which strongly affect radially-coupled predissociation rates.[19]

On the other hand, the 3-body/2-body branching ratio was 0.58 ± 0.1sd which is very similar to the rates obtained for both D_3 (0.58 ± 0.1) and H_3 (0.68 ± 0.1) using the hot discharge source. Thus the H atom is preferred at the expense of the D atoms in the 2-body case, but the 2-body/3-body ratio is unchanged.

SUMMARY

We have made the first determinations of the 3-body/2-body product branching in the predissociation of the n = 2 Rydberg states of H_3. The ratio is found to increase steadily with the internal energy in the molecules. The data are consistent with higher ratios obtained from the dissociative recombination of H_3^+, when plotted against the total internal energy above the H + H + H limit, indicative of a similar predissociation channel. The data show a similar energy dependence as ratios obtained from a two-dimensional calculation by Orel and Kulander,[17] and are consistent if the 3-body predissociation is inhibited by rotation.

A brief examination of the D_2H isotope indicated that the H atom has 1.6 times the D probability, per atom, of being ejected in the 2-body case, indicating that the H atom predissociation rate is higher, as found in other studies.

ACKNOWLEDGEMENTS

This work was supported by the Air Force Office of Scientific Research under Contract No. F-49620-89-K-0002 and by National Science Foundation Grant PHY-87 13309. CFH was supported by a grant from the Deutsche Forschungsgemeinschaft.

REFERENCES

1. G. Herzberg, *Ann. Rev. Phys. Chem.* **38** (1987) 27.
2. I. Dabrowski and G. Herzberg, *Can. J. Phys.* **58** (1980) 1238.
3. G. I. Gellene and R. F. Porter, *J. Chem. Phys.* **79** (1983) 5975.
4. H. Helm, *Phys. Rev. Lett.* **56** (1986) 42.
5. M. C. Bordas, P. C. Cosby, and H. Helm, submitted to *J. Chem. Phys.* (1990).
6. J. R. Peterson and Y. K. Bae, *Phys. Rev. A.* **30** (1984) 2807.

7. P. Cosby and H. Helm, *Phys. Rev. Lett.* **61** (1988) 298.
8. J.B.A. Mitchell, J. L. Forand, C. T. Ng, D. P. Levac, R. E. Mitchell, D. W. Mul, W. Claeys, A. Sen, and J. Wm. McGowan, *Phys. Rev. Lett.* **51** (1983) 885.
9. H. Hus, F. Youssif, A. Sen, and J.B.A. Mitchell, *Phys. Rev. A* **38** (1988) 658.
10. Y. K. Bae, M. J. Coggiola, and J. R. Peterson, *Phys. Rev. A* **31** (1985) 3627.
11. D. P. de Bruijn, J. N. Neuteboom, and J. Los, *Chem. Phys.* **85** (1984) 233.
12. H. Figger, M. N. Dixit, R. Maier, W. Schrepp, H. Walther, I. R. Peterkin, and J.K.G. Watson, *Phys. Rev. Lett.* **52** (1984) 906.
13. J. R. Peterson and Y. K. Bae, *Phys. Rev. A.* **34** (1986) 3517.
14. H. Helm, *Phys. Rev. A.* **38** (1988) 3425.
15. K. Miethe, T. Dreiseidler, and E. Salzborn, *J. Phys. B* **15** (1982) 3069.
16. H. Helm and P. C. Cosby, *J. Chem. Phys.* **86** (1987) 6813.
17. A. E. Orel and K. C. Kulander, *J. Chem. Phys.* **91** (1989) 6086.
18. V. G. Anicich and J. H. Futrell, *Int. J. Mass. Spect. Ion Proc.* **55** (1984) 189.
19. A. L. Roche and J. Tellinghuisen, *Mol. Phys.* **38** (1979) 129.

Angular and Velocity Distribution of the Diatomic Ions coming from the Dissociation of H_3^+, D_3^+ and HD_2^+.

I. Alvarez, H. Martínez, C. Cisneros and J. de Urquijo
Instituto de Física, UNAM, P.O.Box 139-B, Cuernavaca Mor., 62191. México

ABSTRACT

The angular and velocity distributions of the diatomic ion fragments resulting from collision-induced dissociation of the H_3^+, D_3^+ and HD_2^+ molecular ions incident on He, have been measured within the energy range 1 to 5 keV. These distributions were used to determine the binding energy of parent ions and to identified some of the main transitions involved in the dissociation process for the production of the fragments.

1. Introduction

The H_3^+ molecular ion and its isotopic variants HD_2^+, H_2D^+ and D_3^+ are the simplest polyatomic ions known regarding electronic structure. As examples of the great relevance for their study one can mention 1) the chemistry of dense molecular clouds in interstellar space.[1,2] 2) the heating or cooling of the clouds, 3) in Astrophysics, a knowledge of their vibrational and rotational spectra is very important [3,4], and 4) in ionized plasmas, a knowledge of their dissociative recombination is fundamental.[5]

The purpose of this paper is to present our experimental data on the production of the diatomic ionic fragments produced by the collision-induced dissociation of H_3^+, D_3^+ and HD_2^+ in He, in the energy range from 1 to 5 keV. Over this energy range we report the relevant angular distributions, and the energy distributions followed by a discussion of the results.

2 Experimental apparatus

The experimental apparatus has been described elsewhere.[6] The HD_2^+ molecular ions were formed in a Colutron-type ion source. Mixing equal quantities of H_2 and D_2 in the ion source gas feed, provided an adequate beam of HD_2^+; H_3^+ and D_3^+ were obtained from high purity reservoirs of hydrogen and deuterium. In all cases small amount of Ar (less than 10%) was allow to enter in the source to enhance the discharge and the ion current. The molecular ion beam was electrostatically accelerated at energies of 1 to 5 keV, velocity analyzed by a Wien filter and passed through a series of collimators before it entered the gas target cell, which was a cylinder 2.5 cm in length and 2.5 cm diam with a 1-mm entrance aperture and 2-mm-wide, 6-mm-long exit aperture. All apertures and slits had knife edges. Path lengths and apertures gave an overall angular resolution of the system of 0.1 deg.

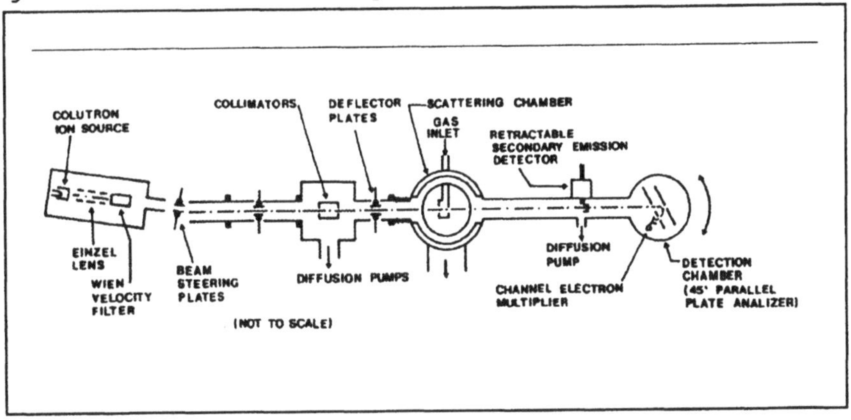

FIG.1. Schematic diagram of the apparatus.

The target cell was mounted in a computer controlled vacuum chamber such that the detector assembly rotated about the center of the gas cell. Located 47 cm away from the gas cell was the detector assembly, a Harrower-type parallel-plate analyzer[7] with a 0.36 mm entrance aperture and a channel electron multiplier (CEM) attached to its exit end. The beam entering the uniform field at an angle of 45 deg. By replacing the collimator located in the front of the CEM, this analyzer was used to measure both angular and energy distributions of the products. To measured the angular distributions, the collimator sited at the front of the CEM was an orifice of 1-cm diameter, while a slit, 1-cm-long, 0.01-mm-wide was placed in front of the CEM for energy distribution measurements. The energy distributions were obtained by varying the voltage on the analyzer; its calculated energy resolution was $\Delta\epsilon/\epsilon = 4.5 \times 10^{-3}$.

3 Measurement techniques

The measured quantities were I_0, the number of projectile triatomic ions incident per unit area per second; n, the number of helium atoms per unit volume; l, the effective length of the scattering chamber; and $I_f(\theta,\phi)$, the number of fragments of type f(i.e.,H_2^+, D_2^+ or HD^+) per unit solid angle per second detected at angles (θ,ϕ) with respect to the incident beam direction. With these measurements, the quantity

$$\frac{d\sigma_f}{d\varpi}=\frac{I_f(\theta,\phi)}{I_0 n l} \tag{1}$$

was evaluated. The angular distributions are displayed in Fig. 2. Several runs were made at different gas target pressures and "$d\sigma_f/d\varpi$" was determined for each run. These were compared in order to estimate the reproducibility of the experimental results as well to determine the limits of the "single collision regime" since the differential measurements reported are absolute. Although $d\sigma_f/d\varpi$ formally looks like a differential cross section, it is not a "differential scattering" cross section. The fragment detected at angle θ has not been "scattered". The substantial angle at which a fragment reaches the detector is due to the transverse component of the velocity acquired in the dissociation process.

The determination of both, the energy distribution of the fragments in the center of mass frame of reference and of their angular spread in the laboratory frame (LF), which are correlated, allows estimation of the binding energy of the molecular projectile, and also the association of some of the prominent structures to the related transitions conducting the excited molecular ion to the dissociation pathways. In doing this, we used for the three hydrogenic ions the available potential curves for H_3^+ assuming the D_{3h} symmetry.[8] The data were analyzed by using the relations[9,10]

$$(M+m)U=M(V-W-E)+mW\pm 2[MmW(V-W-E)]^{1/2} \tag{2}$$

$$\theta_{max}=\left[\frac{mW_{max}}{M(V-W_{max}-E)}\right]^{1/2} \tag{3}$$

Equation (2) refers to energy distributions, whereas Eq. (3) refers to angular distributions. In Eq.(2), (M+m) represents the mass of the molecular projectiles; M, the mass of the detected fragment (H_2^+, D_2^+, HD^+ or D_2^+); m, the mass of the undetected fragment ($H^{0,+}$ or $D^{0,+}$); U is the measured energy of the fragment of mass M detected at an angle of zero degrees in LF; V, the initial energy of the molecular projectile; E, the internal energy increase of the system and W, the kinetic energy released as a result of the dissociation process. The ± signs correspond to the forward or backward ejection of the fragment of mass M.[9,10]

Equation (3) represents the relation between the maximum kinetic energy W_{max} of the fragments and their maximum angular spread θ_{max}. This is the maximum angle between the direction of the primary beam and the direction of the observed fragment (M). As an example, Fig. 2 exhibits the angular distributions of the H_2^+ molecular fragment ions from H_3^+ + He at different energies. The same figure displays the position of the θ_{max} following a careful subtraction of the background so, its value can be read directly from the display.

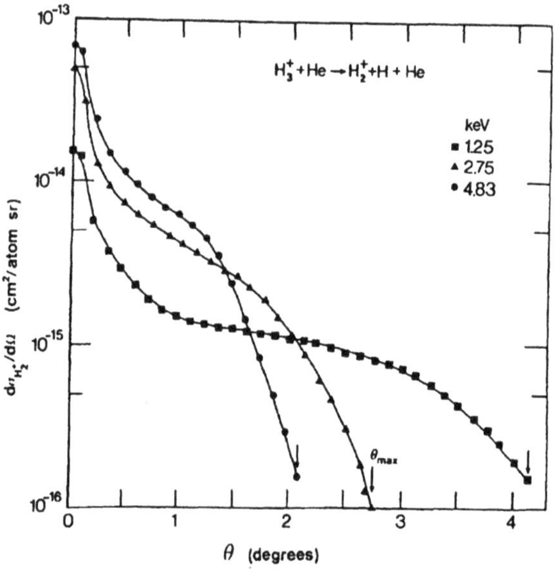

Fig.2. Angular distributions of the H_2^+ fragment, showing θ_{max}.

From Eqs. (2) and (3) we can relate the results from both angular and energy distribution measurements. The angular distribution of the dissociation products is due entirely to the transverse component of the dissociation velocity of the fragment. Thus, with the observed θ_{max} in the angular distributions, and provided that $V \gg W_{max} + E$, then W_{max} is evaluated from Eq.(3), and this value must be the same as that derived from Eq.(2) by taking the energy values at which the intensity reaches a minimum in the energy spectra. As it was pointed out[11] the maximum E + W ensures that the excitation originates in the lowest vibrational level of the triatomic ion and the maximum in W ensures the minimum residual internal energy.

The spectra of dissociation energies in collision-induced dissociation rarely consists of sharp lines, and the excitation cross sections depend strongly on the instantaneous separation R_{mM} of the molecular constituents M and m at the time of collision. Very often, the spectrum of dissociation energies consists of broad structures or shoulders due to the distribution of vibrational energies in which the projectile enters the collision. A similar explanation holds for angular distributions. The data were treated through an iteration method by taking the energies corresponding to the minimum intensity limits on the energy spectra as testing values, comparing the corresponding W and E for the two branches of the spectra, and then comparing this W value with the W_{max} as derived from the angular distributions. Finally the best E and W values compatible with both angular and energy distributions were obtained, with proper regard to the angular and energy resolutions of the experimental apparatus. The analysis of the energy spectrum was performed using Eq. 2, because of the lack of structure in the angular distributions does not allow any correlation between the energy and angular distributions. Moreover the experimental method used is not accurate enough to resolve the multiple contributions of the different states to the spectra, and hence it does not permit a definite assignment to the transitions.

4 Results and discussion

The H_2^+ fragment coming from H_3^+ + He.

In the energy spectrum of Fig. 3a the values of U_1 = 2124 eV and U_2 = 2525 eV at which the intensity goes to a minimum were chosen and for the angular distribution (Fig. 2) the corresponding θ_{max} was selected, and then, the E and W values calculated from this method were W = 13.0 ± 1 eV and 6.3 ± 0.5 eV. Also in Fig.3a, a broad structure can be seen around U_1 = 2214 eV and U_2 = 2443 eV. With these energy limits, values of W = 4.2 ± 0.32 eV and E = 5.0 ± 0.4 eV were obtained. With slightly different values of U, as U_1 = 2209 eV and U_2 = 2431 eV wc obtain W = 4.0 ± 0.3 eV and E = 16.5 ± 1.4 eV.

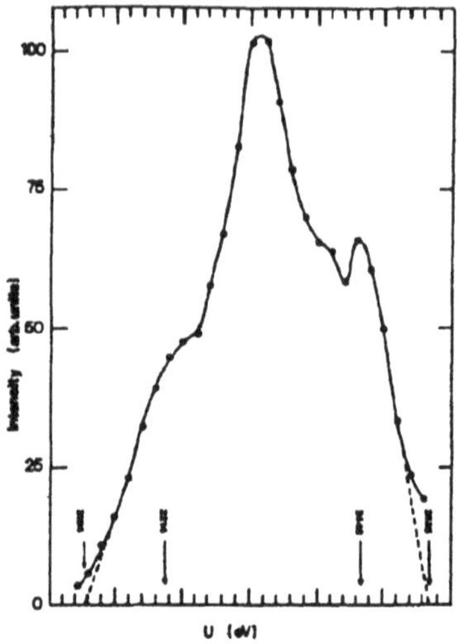

Fig. 3a

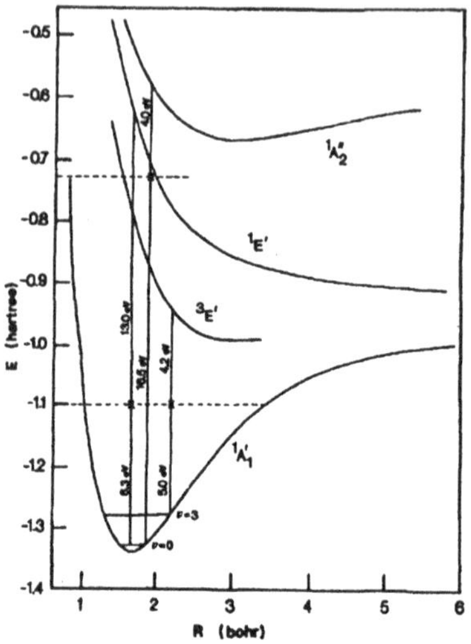

Fig. 3b

Fig. 3a display the energy spectrum of the H_2^+ fragment at 3.5 keV of H_3^+.

Fig. 3b show the relevant H_3^+ potential-energy curves in D_{3h} symmetry (equilateral structure). R is the length of the side of the triangle. Dotted lines show the dissociation limits.

The proposed transitions consistent with these values are:

1) $H_3^+(^1A')+He \dashrightarrow (H_3^+)^*(^1E')+He \dashrightarrow H_2^+(^2\Sigma_g^+)+H(2p)+He$

2) $H_3^+(^1A')+He \dashrightarrow (H_3^+)^*(^3E')+He$

to $(H_3^+)(^3A_1)+He \dashrightarrow H_2^+(^2\Sigma_g^+)+H(1s)+He$
or $(H_3^+)(^3B_2)+He \dashrightarrow H_2^+(^2\Sigma_g^+)+H(1s)+He$

and

3) $H_3^+(^1A')+He \dashrightarrow (H_3^+)^*(^1A''_2)+He \dashrightarrow H_2^+(^2\Sigma_g^+)+H(2p)+He$

This transitions are displayed in Fig. 3b.

The D_2^+ fragment coming from D_3^+ + He.

The angular distributions and the LF energy spectra of the D_2^+ fragment are show in Figs. 4a and 4b respectively

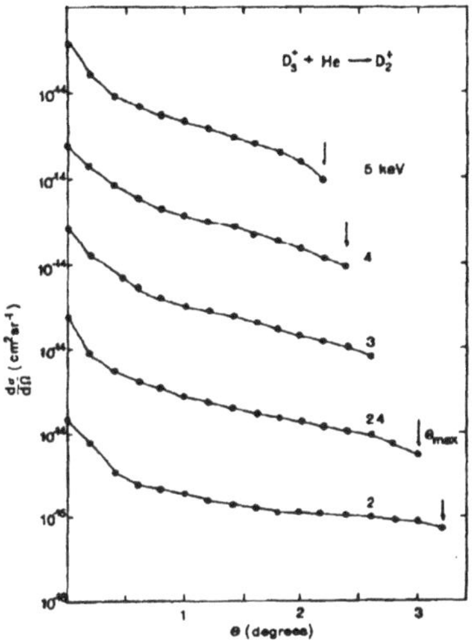

Fig. 4a

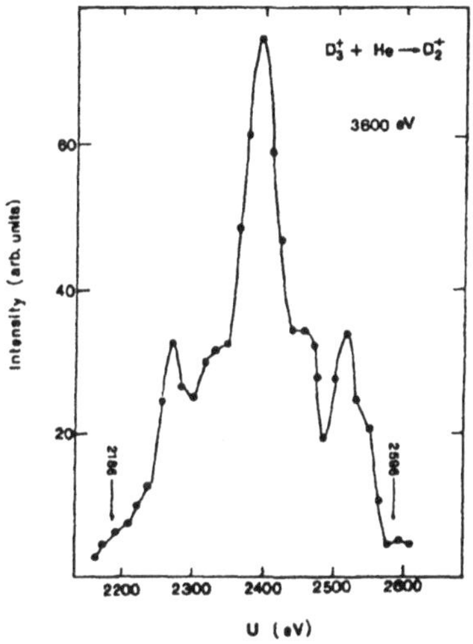

Fig. 4b

FIG. 4a. Angular distributions of the D_2^+ fragment, showing θ_{max}.
FIG. 4b. LF energy spectrum of the D_2^+ fragment at 3.8 keV.

Table 1 present the U values chosen, the W and E values obtained with the iteration method already described.

Table 1. Values for the D_2^+ fragment.

U_1 eV)	U_2 (eV)	W (eV)	E (eV)	Transition
1) 2186	2596	13.17±1	6.54±0.5	$^1A'_1 \rightarrow {}^1E'_1$
2) 2267	2522	5.08±0.42	5.64±4	$^1A'_1 \rightarrow {}^3E'_1$
3) 2312	2470	1.97±0.25	13±1.3	$^1A'_1 \rightarrow {}^1A''_2$
2226	**2560**	**8.75±0.79**	**6.42±0.5**	$A'_1 \rightarrow {}^3E'_1$
2227	**2561**	**8.75±0.85**	**4.94±0.51**	$^1A'_1 \rightarrow {}^1E'_1$

Also, in the table 1 are listed some of the proposed transitions congruent with the chosen U values, but is necessary to take into account that the experimental method used is not accurate enough to resolve the multiple contributions of the different states to the spectra. The lack of structures in the angular distributions does not allow any correlation between these and the energy spectra in addition to its minimum intensity points and the θ_{max} value, so the analysis of the energy spectra was performed using Eq. 3. Notice that the W and E values of the three first rows are practically the same as those obtained with the H_3^+ molecular ion, whereas the shaded ones appears now due to the richness of structures present in the D_2^+ energy spectra.

The HD^+ and D_2^+ fragments coming from HD_2^+ incident in He.

The angular distributions of these fragments are very similar to those of D_2^+ shown in Fig. 3a whereas the energy distributions are display in Figs. 5a and 5b respectively.

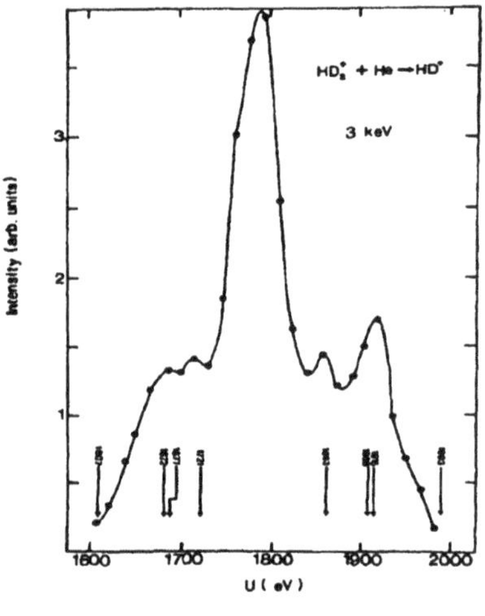

Fig. 5a

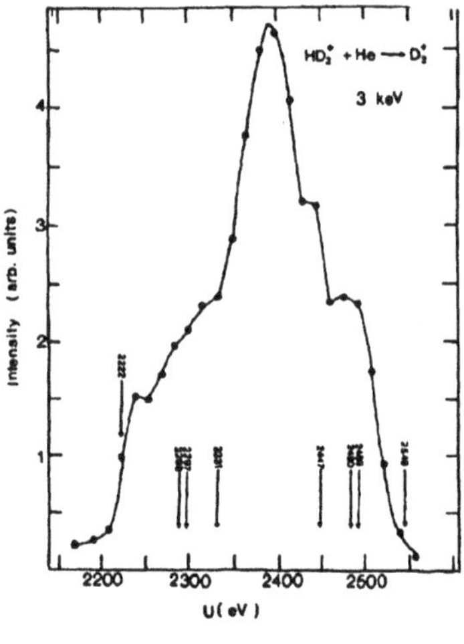

Fig. 5b

FIGS. 5a and 5b. LF energy spectra of the HD^+ and D_2^+ fragments from the HD_2^+ molecular ion colliding with He.

Table 2a and 2b present the U values chosen, the W and E values obtained with the iteration method already described for the HD^+ and D_2^+ fragments and the transitions congruent with these values.

Table 2a. Values for the HD^+ fragment.

U_1 (eV)	U_2 (eV)	W (eV)	E (eV)	Transition
1) 1607	1933	14±1.08	6.4±0.51	$^1A'_1 \rightarrow ^1E'_1$
2) 1677	1915	4.92±0.39	5.43±0.44	$^1A'_1 \rightarrow ^3E'_1$
3) 1672	1906	4.8±0.38	16.76±1.34	$^1A'_1 \rightarrow ^1A''$
1721	1863	1.76±0.14	12.59±1.01	$^1A'_1 \rightarrow ^1A''$

Table 2b. Values for the D_2^+ fragment.

U_1 (eV)	U_2 (eV)	W (eV)	E (eV)	Transition
1) 2222	2549	14±1.2	6.4±0.52	$^1A'_1 \rightarrow ^1E'_1$
2) 2297	2489	4.82±0.39	5.36±0.43	$^1A'_1 \rightarrow ^3E'$
3) 2288	2480	4.81±0.39	16.85±1.35	$^1A'_1 \rightarrow ^1A''$
2331	2447	1.76±0.14	12.67±1.01	$^1A'_1 \rightarrow ^1A''$

The analysis of both values, maximum kinetic energy W_{max} from the energy spectra of the HD^+ and D_2^+ and the θ_{max} from the angular distributions reveal the same binding energy value for HD_2^+, as those of the H_3^+ and D_3^+ molecular ions. From the inner structures in the energy spectra appears also the same important transitions involved in the molecular dissociation.

The results of the present work can be summarized as follows: a) We measured the differential cross sections and the energy distributions of the diatomic positive fragments as a consequence of the collision of H_3^+, D_3^+ and HD_2^+ on He.
b) As a result, the binding energy of the parent ions was measured and result to be the same under the experimental uncertainty for the three triatomic ions. The average of the determined (W + E), (transitions number 1) is basically the

state and in agreement with theory for the equilateral triangle configuration. c) From the structure in the energy distributions, other possible dissociation channels could be identified and were found to be essentially the same (transitions number 2 and 3), other possible transitions were identified by the presence of more structures in the distributions.d) Different shapes of the distributions result from different intensities due to the mass effect.

5 Acknowledgements

The authors wish to thank Mr. S. Pérez for his technical assistance. This work was supported by DGAPA grant number IN-104889.

6 References

1. E. Herbst and W. Klemperer, Astrophys. J. **185**,505 (1973).
2. W. D. Watson, Rev. Mod. Phys. **48**,513 (1976).
3. E. Herbst and W. Klemperer, Phys. Today **20**,23 (1976).
4. D. Smith, N. G. Adams and E. Alge, Astrophys. J. **263**,123 (1982).
5. J. B. Mitchell, T. C. Ng, L. Forand, R. Janssen and J. Wm. McGowan, J. Phys. B **17**,L909 (1984).
6. H. Martínez, C. Cisneros, J. de Urquijo and I. Alvarez, Phys. Rev. **A38**,5914 (1988).
7. G. A. Harrower, Rev. Sci. Instrum. **26**,850 (1955).
8. B. A. Colling, J. E. Polany, M. A. Smith, A. Stolow and A. W. Tarr, Phys. Rev. Lett. **59**,2551 (1987).
9. J. Los and T. R. Govers in *Collision Spectroscopy* edited by R. G. Cook (Plenum, New York, 1978) p. 289.
10. H. Martínez, I. Alvarez, J. de Urquijo, C. Cisneros and A. Amaya-Tapia, Phys. Rev. **A36**,5425 (1987).
11. S. C. Goh and J. B. Swan, Phys. Rev. **A24**,1624 (1981).

III. IONIZATION

DOUBLE IONIZATION OF HELIUM BY INTERMEDIATE TO HIGH-VELOCITY He^{+}, $O^{6,7+}$, AND $S^{6,13+}$ IONS

J. A. Tanis, S. M. Ferguson, J. L. Forest,
R. R. Haar, and K. Lifrieri
Department of Physics
Western Michigan University, Kalamazoo, MI 49008 USA

and

V. L. Plano
Physics Department
Kalamazoo College, Kalamazoo, MI 49008 USA

ABSTRACT

Single and double ionization of helium have been investigated in collisions with 0.125-3.0 MeV/u He^{+}, 0.5-2.5 MeV/u $O^{6,7+}$, and 0.16-1.8 MeV/u $S^{6,13+}$ ions. In order to investigate target ionization for different physical conditions and for different ranges of impact parameters, target ionization was associated with projectile ions undergoing capture, loss, or no charge change, respectively. Ratios of double-to-single target ionization were determined for each outgoing projectile charge state and the measured ratios are compared to previous studies and to theoretical predictions. Double ionization is considered in terms of two mechanisms: (1) <u>direct</u>, in which both target electrons are removed in separate interactions with the projectile, and (2) <u>shakeoff</u>, in which the first electron is removed in a direct interaction with the projectile while the second is ejected when the resulting ion "relaxes" to a continuum state.

1. Introduction

The double ionization of helium by fast ions, as well as by electrons and photons, is a few-body problem of fundamental interest[1-3] which has received considerable attention in recent studies.[4-11] It has been proposed[3,12] that double ionization by ions (or electrons) with greater than ~ MeV/u velocities can be understood in terms of two simple mechanisms: (1) a two-step process, second order in q/v (q is the charge and v is the velocity), in which both target electrons are removed in separate direct interactions with the projectile, and (2) a process, first order in q/v, in which the first electron is removed in a direct interaction with the projectile while the second electron is ejected when the resulting ion relaxes to a continuum state ("shakeoff"). The former process involves independent-particle interactions, and the latter is due to electron correlation in the initial state. Hence, the double-ionization cross section σ_{2I} can be written as the square of a sum of amplitudes

$$\sigma_{2I} = |a_{SO}(q/v) + a_{TS}(q/v)^2|^2 \qquad (1a)$$
$$= \sigma_{SO} + \sigma_{int} + \sigma_{TS} \qquad (1b)$$

where the notation SO refers to "shakeoff," TS refers to "two-step," and σ_{int} is the interference cross section resulting from the q^3 cross term. In the asymptotic limit of high projectile velocities ($q/v \ll 1$ with v in atomic units), shakeoff is expected to dominate since q/v is small corresponding to a small collision time, and thus the probability of two separate interactions is small.[3,12]

For bombardment by ions, double ionization can occur via (1) pure ionization in which the final charge of the projectile is left unchanged, (2) associated capture, a process called transfer ionization (TI), and (3) associated electron loss from the projectile. In the work discussed here, double ionization of helium is studied under each of these three conditions. Noteworthy is the fact that for TI one of the outgoing electrons, because it is captured from the target, leaves the interaction region with the speed of the projectile. In the case of target ionization accompanied by projectile loss, if the projectile is considerably heavier than helium and if it is highly stripped, then the study of double ionization of the helium target can be restricted to a limited range of small impact parameters.

In the work discussed here we are concerned with collisions in an energy range which includes both of the underlying (direct and shakeoff) mechanisms leading to double ionization. Helium is used as the target gas as opposed to heavier targets since double ionization can be unambiguously identified. In this work we will concentrate on the ratio R of double-to-single ionization, i.e., $R = \sigma_{2I}/\sigma_{1I}$ since this parameter contains the essential information on the mechanisms leading to double ionization.

Ionization is investigated for 0.125-3.0 MeV/u He^{+}, 0.5-2.5 MeV/u $O^{6,7+}$, and 0.16-1.8 MeV/u $S^{6,13+}$ ions colliding with helium targets. Ratios of double-to-single target ionization were determined for outgoing projectiles undergoing capture, loss, or no charge change, and the measured ratios are compared to previous studies and to theoretical predictions. The data presented here for $O^{6,7+}$ have been published[13,14] elsewhere, while some of the He^{+} data have been reported previously.[15]

2. Experimental Procedure

This work was done at Western Michigan University using the EN tandem Van de Graaff. Incident He^{+}, $O^{6,7+}$, or $S^{6,13+}$ ions were passed through a differentially pumped gas cell containing the helium target gas, and singly- and doubly-ionized helium ions formed in the collision region were identified using time-of-flight techniques. Recoiling helium ions were extracted by a transverse electric field, accelerated, and then detected with a microchannel plate. Projectile ions emerging from the gas cell were magnetically analyzed to separate the various charge-state components of the beam and detected with solid-state detectors. Coincidences between the recoiling helium target ions and projectiles capturing or losing an electron were observed with a time-to-amplitude converter using standard coincidence techniques. The main beam component (no charge change) was collected in a Faraday cup for normalization. For the pure ionization measurements, recoiling ions were detected in coincidence with projectile ions experiencing no charge change by reducing the incident beam intensity by factors of 100-1000 and detecting these ions with a solid-state detector.

The pressure in the gas cell was monitored with a capacitance manometer. All measurements were made for pressures $\leq$ 0.5 mTorr to ensure single-collision conditions as determined from the linear pressure dependence of the measured yields.

A typical time spectrum is shown in Fig. 1 for He^{+} + He collisions. From the measured yields of the singly- and doubly-ionized helium ions detected, the ratio of double-to-single ionization is determined directly. Absolute cross sections for each of the various processes studied were also obtained but will not be considered here.

3. Results

Of primary interest in this work is the double ionization of helium. As mentioned above, we will focus on the ratio R of double-to-single ionization, i.e., $R = \sigma_{2I}/\sigma_{1I}$, since this parameter contains the essential information on the mechanisms leading to double ionization. Fig. 2 shows the measured ratios R for ions undergoing capture, loss, or no charge change, respectively. The O^{6+} data and the S^{6+} data are not significantly different from those for O^{7+} and, hence, are not shown here.

In the case of pure ionization (no projectile charge change), the ratios are smaller than 5%, 10%, and 20%, respectively for He^{+}, O^{7+}, and S^{13+} over the energy range studied and the ratios monotonically decrease with

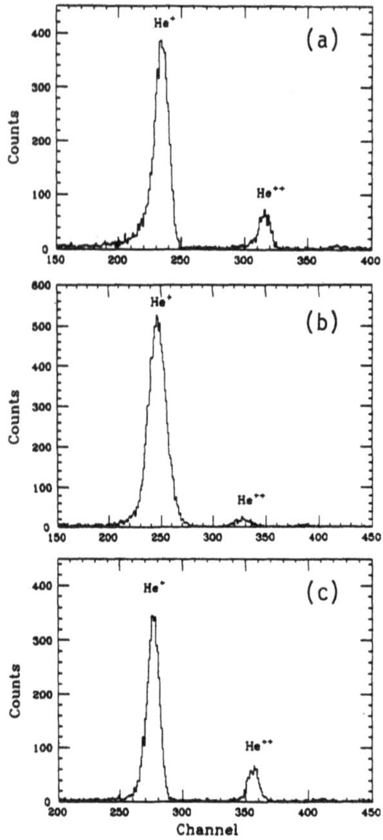

FIG. 1. Time-of-flight spectra for 0.25 MeV/u He^{+} ions colliding with He at a pressure of 0.3 mTorr. Peaks due to He^{+} and He^{++} ions are indicated and the ratio of double-to-single ionization is determined directly from the relative intensity of these peaks. Spectra (a), (b), and (c) are for recoiling target ions coincident with electron capture, no charge change, and electron loss, respectively.

increasing energy. For target ionization accompanied by projectile capture (TI), R reaches values in excess of 1.0 (for O^{7+} and S^{13+}), while for target ionization accompanied by projectile loss, R reaches values in excess of about 75% (again for the O^{7+} and S^{13+} data). For O^{7+} and S^{13+} at the highest energies, the R values for ionization accompanied by capture and loss are nearly identical. The present R values for He^{+} agree well with similar measurements for this same collision system by Wood et al.[16] in the region of overlap (see Fig. 3a). The smooth curves show the predicted ratio for pure ionization obtained from the empirical scaling of Knudsen et al.[4] which will be discussed below. Because of the capabilities of the accelerator facility, only the He^{+} data extend to velocities approaching the high-velocity limit, i.e., $q/v < 0.1$.

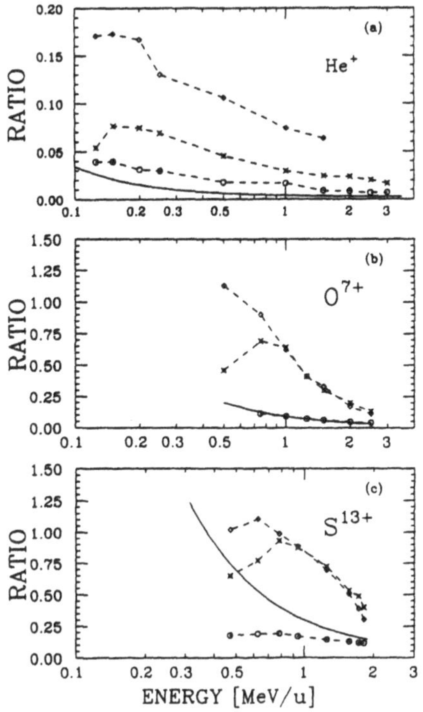

FIG. 2. Measured ratios of double-to-single ionization as a function of projectile energy (MeV/u) for He^+, O^{7+}, and S^{13+} ions striking He targets. Symbols represent target ionization ratios associated with projectiles undergoing capture (◇), loss (⌘), and no charge change (○), respectively. The dashed lines are drawn to guide the eye. The solid curves are the predicted ratios obtained from the empirical scaling of Knudsen et al. (Ref. 4) (see Eq. 4 below) for target ionization associated with no projectile charge change.

4. Discussion

Single ionization by fast, highly-charged ions striking light targets is well understood theoretically[17] and experimentally,[18] and the cross sections for double ionization can be written in terms of the cross sections for single ionization as follows:

$$\sigma_{SO} = (\text{const}) \times (q/v)^2 \ln v \quad (2a)$$
$$\sigma_{int} = (\text{const}) \times (q/v)^3 \quad (2b)$$
$$\sigma_{TS} = (\text{const}) \times (q/v)^4 \quad (2c)$$

where v is the projectile velocity. The ln v term comes from integration over the tail of the long-range Coulomb potential of the projectile.[17] Hence, the ratio $R = \sigma_{2I}/\sigma_{1I}$ can be written as

$$R = C_{SO} + C_{int} \frac{(q/v)}{\ln v} + C_{TS} \frac{(q/v)^2}{\ln v}. \quad (3)$$

Again, the first term in Eq. (3) is due to the SO mechanism, the last term results from the TS mechanism, and the middle term comes from the interference between the SO and TS mechanisms. At intermediate energies both the direct (second order in q/v) and the "shakeoff" (first order in q/v) mechanisms are expected to contribute significantly to the double ionization of helium, and, furthermore, there may be contributions from the interference between these two processes.

If, however, the interference term is neglected, Knudsen et al.[4] have found that

$$R = 2.20 \times 10^{-3} + 4.55 \times 10^{-3}\ q^2/E\ln(13.12\sqrt{E}) \quad (4)$$

where E is measured in MeV/u. Hence, this latter equation predicts that R should be a linear function of $q^2/[E\ln(13.12\sqrt{E})]$. This expression was obtained for projectiles with atomic numbers ranging from 1-8 and energies from 0.13 to 15 MeV/u incident on helium targets. More recently, Reading and Ford[7] have presented ab initio calculations which are consistent with Eq. 4. Even if interference is important, the second term of Eq. 4 should still be valid in the region where the two-step (TS) mechanism dominates.

In Fig. 3 we show the values of R obtained in the present work compared with the predictions of Eq. 4. Only the pure ionization ratios for O^{7+} agree well with the straight line behavior predicted by Eq. 4, while the measured S^{13+} values fall below the predicted curve. This latter discrepancy can be attributed to the fact that the velocities investigated for S^{13+} fall below the region where the perturbation theory used to derive the two-step direct mechanism[4] is valid. For He^+, ionization with no projectile charge change is seen to exceed everywhere the predictions of Eq. 4 although the $q^2/E\ \ln(13.12\sqrt{E})$ dependence is similar. The reason for this behavior is not clear.

For target ionization accompanied by projectile electron capture or loss it is seen that: (1) the R values are enhanced significantly in all cases when compared to the corresponding values without accompanying projectile capture or loss, and (2) the R values deviate significantly from the energy and charge-state behavior observed for pure ionization, i.e., they exhibit a different slope when plotted as a function of $q^2/E\ln(13.12\sqrt{E})$.

Since the TS and SO mechanisms given by Eqs. (2) are operative for the large impact-parameter dominated pure ionization collisions, they are also expected to be operative for small impact parameters, i.e., for target

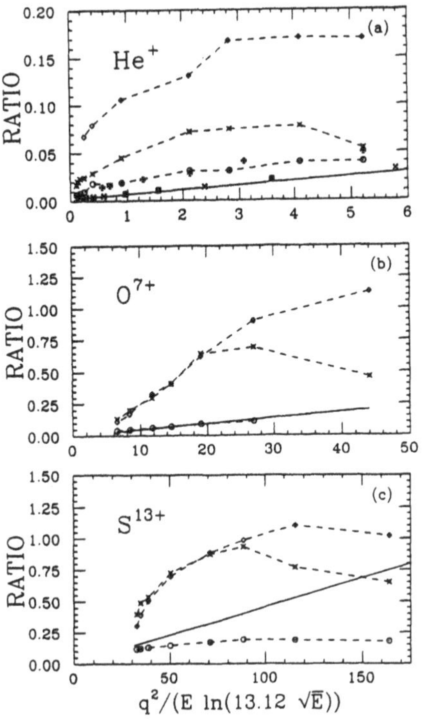

FIG. 3. Measured ratios of double-to-single ionization plotted as a function of the scaling parameter $q^2/E\ln(13.12\sqrt{E})$ given by Eq. 4 (Ref. 4). Symbols represent target ionization ratios associated with He^+ projectiles undergoing capture (◇), loss (⌧), and no charge change (O - present, + - Ref. 16). Data for He^{2+} projectiles undergoing no charge change are indicated by: □ - present, x - Ref. 4. Dashed lines are drawn to guide the eye. The solid curves are the predicted ratios obtained from the empirical scaling of Eq. 4 (Ref. 4) for target ionization associated with no projectile charge change.

ionization accompanied by projectile capture or loss events. It is emphasized again, however, that for the present range of energies and charge states, the TS mechanism is expected to dominate. A partial explanation for the large enhancement in R is that projectile capture or loss plus target ionization events occur at smaller impact parameters than pure target ionization events, thereby leading to a higher probability for double ionization of the target in the former cases. The observed deviation of the energy and charge-state behavior of R seen in Fig. 3 for these small impact parameter collisions compared to the case of pure ionization is more difficult to understand, however. Qualitatively, even though the magnitude of R is enhanced for small impact parameters, it might be expected that R should exhibit the same dependence on q and E, i.e., $q^2/[E\ln(13.12\sqrt{E})]$, with a consequent higher asymptotic SO value (i.e., the intercept with the ordinate in Fig. 3). This does not appear to be the case, however. Perhaps the interference term in Eq. 3 is significant thereby giving rise to the observed deviations.

An alternative way of looking at the data, one which reveals more explicitly the q and v dependence, is obtained by plotting the double-to-single ionization ratios as a function of $(v/q)^2$ as shown in Fig. 4. From Eq. 3, the ratio R should be a universal function of q/v (or v/q) and, thus, universality should be exhibited in this figure. For the pure ionization (no charge change) data, there does indeed appear to be a universality for the three ions and charge states shown although the data do not agree with the scaling predictions of Eq. 4. It should also be noted that the O^{6+} and S^{6+} data (not shown) obtained in the present work follow the same dependence on q and v. For target ionization accompanied by projectile electron capture or loss, however, there appears to be no common scaling. Thus, these latter data indicate that double ionization accompanied by projectile capture or loss are not simply described by Eq. 3. It should be noted that Eq. 3 is

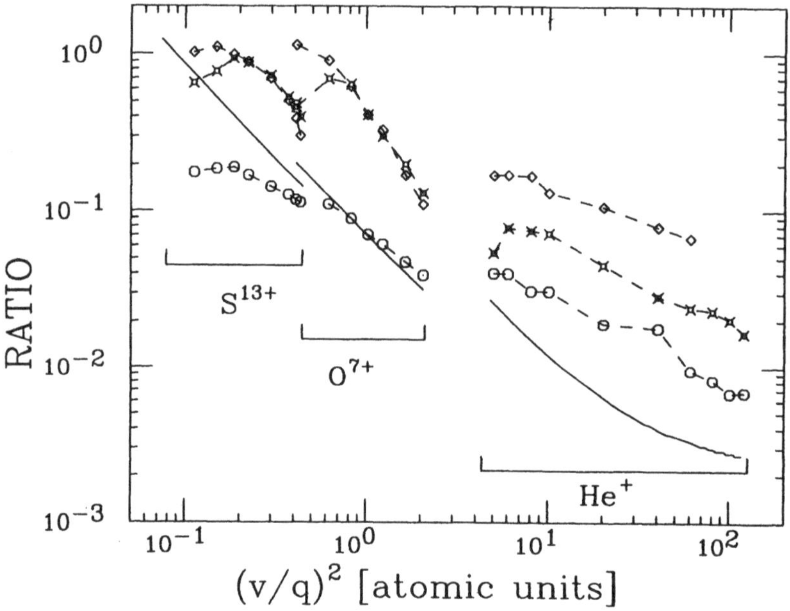

FIG. 4. Measured ratios of double-to-single ionization plotted as a function of v^2/q^2. Symbols represent target ionization associated with outgoing projectile charges as in Figs. 2 and 3, and the solid curves are obtained from the empirical scaling of Knudsen et al. (Ref. 4).

expected to be valid only for $v/q > 1$ due to the fact that perturbation theory is used in the derivation of this expression.

In the asymptotic limit ($q/v \ll 1$), shakeoff should be the primary mechanism by which double ionization occurs.[3] For pure ionization and velocities up to about 15 MeV/u Knudsen et al.[4] have found that R approaches a value of about 0.0022 (see Eq. 4) and thus $C_{SO} \simeq 0.22\%$ in Eq. 3. In the case of ionization by high-energy photons the shakeoff limit is found[19] to be about 3.5%, a factor of ten higher. The difference between the SO limit of 0.22% found for charged-particle ionization and the SO limit of 3.5% observed for photoionization has been attributed[3,18] to the fact that the first electron is not ejected at high speed in the former case, i.e., ionization by charged particles is dominated by large impact parameter collisions. Hence, the high-velocity charged-particle ratio R_{ch} of double-to-single ionization found in Ref. 4 does not strictly correspond to the photon value R_{ph}. McGuire has suggested[18] that R_{cp} can be related to R_{ph} by integrating the latter weighted by the density distribution of ejected electron energies for charged-particle ionization.

For double ionization associated with capture (TI), however, the first electron leaves with the speed of the projectile. Then, in the high-velocity limit, McGuire et al.[20] argue that the ratio of TI to pure single capture should go to a constant value independent of projectile velocity and charge. This follows from the original suggestion of Horsdal Pedersen and Larsen[21] that, at high velocity, TI should resemble photoionization followed by "shakeoff" since the first electron is removed from the target quickly in either case. Knudsen et al.,[22] however, have suggested that second Born capture mechanisms may dominate the TI process at high velocity thereby giving a value dependent of the projectile charge.

Since the measurement of target ionization accompanied by projectile-electron capture or loss limits the range of impact parameters investigated to relatively small values, it might be expected that R should approach the higher SO limit of 3.5% for these close collisions or at least a value between the two extremes. And, in the case of TI, the captured electron leaves the collision with the speed of the projectile, so for very high-velocity collisions one would expect R to approach the value found for photo double ionization. Extrapolation to zero of the ionization plus capture and loss data of Fig. 3b for O^{7+} does not indicate a value significantly different from that found for pure ionization, however, while the S^{13+} results shown in Fig. 3c indicate that the linear region for ionization accompanied by projectile capture or loss is not reached for the velocities studied.

Only for He^+ + He do the data approach the expected asymptotic limit, i.e., $q/v < 0.1$ (or $v^2/q^2 > 100$), for the highest velocities studied (see Fig. 4). Thus, for incident He^+ a more quantitative analysis of the asymptotic behavior is possible as shown in Fig. 5. Here, only the data for $q/v \leq 0.22$ ($v^2/q^2 > 20$) are plotted and it is seen that R is nearly linear for all three cases, i.e., capture, loss, and no charge change.

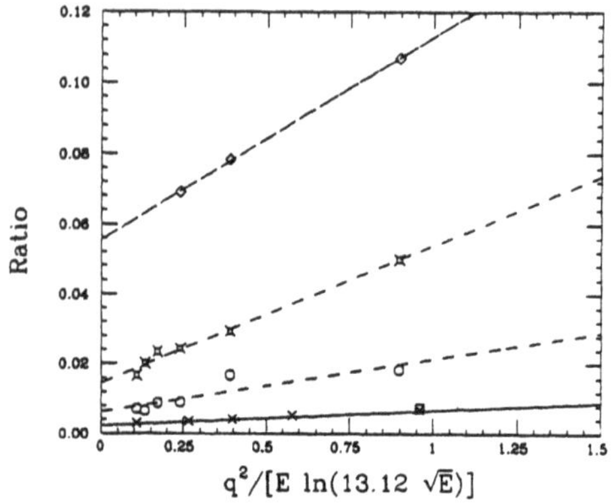

FIG. 5. Measured ratios of double-to-single ionization for He^+ projectiles for $q/v \leq 0.22$ ($v^2/q^2 \geq 20$) (see Fig. 4). The ratios are plotted as a function of the scaling parameter of Fig. 3. Symbols are as in Figs. 2, 3, and 4. The dashed curves are linear least squares fits to the data and the slopes and intercepts are given in Table I. The solid line is the empirical scaling given by Eq. 4 (Ref. 4).

From the indicated fits to these data, the shakeoff limit and the strength of the q/v dependence can be determined by fitting a (straight line) function of the form of Eq. 4 to the data. The results are listed in Table I. It is seen that the asymptotic limit (C_{so}) for target ionization with no projectile charge change, ionization accompanied by projectile loss, and ionization accompanied by capture are factors of 3, 7, and 25 larger, respectively, than pure ionization by He^{2+}. It is interesting to note that the C_{so} value obtained for target ionization accompanied by capture is very close to the value obtained for photo double ionization.

As mentioned earlier, for target ionization associated with projectile capture, the first electron leaves with the speed of the projectile, and, if this velocity is high enough ($q/v < 0.1$), the mechanism for double ionization is expected to be qualitatively and quantitatively similar to that for photo double ionization and this does indeed appear to be the case. For target ionization accompanied by projectile loss, the average velocity of the first electron ejected is undoubtedly higher than for the larger impact parameter pure ionization collisions, and therefore it might be expected that the asymptotic value of C_{so} is higher for the former case as it is indeed observed to be. For target ionization by He^+ with no charge change, it is not clear why the ratio is three times larger than for He^{2+}. One possible

TABLE I. Values of C_{SO} and C_{TS} obtained from fitting the function $R = C_{SO} + C_{TS}\, q^2/E \ln(13.12\sqrt{E})$ to the data shown in Fig. 5. The value of C_{SO} gives the expected ratio of double-to-single ionization in the asymptotic limit ($q/v \ll 1$) while the value of C_{TS} gives the strength of the q/v (or q^2/E) dependence in the two-step regime.

	He^{2+}		He^{+}	
	C_{SO}[1]	C_{TS}[1]	C_{SO}	C_{TS}
no charge change	2.2×10^{-3}	4.55×10^{-3}	6.3×10^{-3}	15×10^{-3}
electron loss	---	---	15×10^{-3}	39×10^{-3}
electron capture	---	---	56×10^{-3}	57×10^{-3}

[1]From Ref. 4 (see Eq. 4 of present text).

explanation is that the electron attached to the He^+ projectile acts independently in contributing to ionization of the target. This latter point needs further investigation.

5. Conclusions

Double ionization of helium has been investigated for a wide range of energies and charges states ($q/v = 0.1$-3.0) thereby covering regimes where both the direct (two-step) and shakeoff (one-step) mechanisms are expected to contribute significantly to double ionization. Ionization is strongly dependent on both q and v and, hence, the results can be used to test the validity of theoretical predictions. Since perturbation theory is used to formulate the double ionization processes, deviations from the theory may be expected for $q/v > 1.0$.

It is found that double ionization is strongly dependent on the outgoing charge of the projectile, an effect which apparently reflects the different impact parameter regimes selected by the different outgoing charge states. For pure double ionization (no projectile charge change) it is found that the ratio of double-to-single ionization scales as $(q/v)^2$ for values of $q/v < 1.0$ as expected. Furthermore, double ionization associated with capture and loss by the incident projectile is substantially enhanced and the q/v dependence is altered significantly from that observed for pure ionization. The large enhancement is qualitatively understood from the fact that target ionization accompanied by projectile capture or loss occurs at smaller impact parameters, thereby giving a higher probability for double ionization. The origin of the observed deviation of the energy and charge-state behavior of double ionization compared to pure ionization is not clear, however.

For He^{+} + He, the measurements approach the asymptotic limit ($q/v < 0.1$) and thus the results can be compared with the empirical predictions of Knudsen et al. (Ref. 4). It is found that double ionization accompanied by no projectile charge change, loss, and capture, respectively, is 3, 7, and 25 times larger than for pure ionization by He^{2+}. For target ionization accompanied by capture, the ratio of double-to-single ionization in the asymptotic limit approaches that for photo double ionization (~ 5%). This result can be understood from the fact that the captured electron leaves the interaction region with the speed of the projectile, and, hence, for high velocities the physical situation is essentially the same as that for ionization by a high-energy photon.

In summary, although several features of double ionization of helium are qualitatively understood, it is clear that much of the data are not explained within the framework of existing theories. Hence, additional theoretical and experimental work is indicated in order to understand this important fundamental problem in atomic physics.

6. Acknowledgements

This work was supported in part by the U. S. Department of Energy, Office of Basic Energy Sciences, Division of Chemical Sciences.

7. References

1. M. H. Mittleman, Phys. Rev. Lett. 16 (1966) 498.
2. F. W. Byron, Jr. and C. J. Joachain, Phys. Rev. Lett. 16 (1966) 1139.
3. J. H. McGuire, Phys. Rev. Lett. 49 (1982) 1153; and Proceedings of the Second US-Mexico Symposium on Atomic and Molecular Physics: Two-Electron Phenomena, Cocoyoc, Mexico, 1986, published in Notas de Fisica, I. Alvarez, C. Cisneros, J. de Urquijo, and T. J. Morgan eds., (Instituto de Fisica, UNAM, Laboratorio Cuernavaca, Mor., Mexico, 1987), Vol. 10, No. 2, pp. 129-142.

4. H. Knudsen, L. H. Andersen, P. Hvelplund, G. Astner, H. Cederquist, H. Danared, L. Liljeby, and K.-G. Rensfelt, J. Phys. **B17** (1984) 3545.
5. L. H. Andersen, P. Hvelplund, H. Knudsen, S. P. Møller, K. Elsener, K.-G. Rensfelt, and E. Uggerhøj, Phys. Rev. Lett. **57** (1986) 2147.
6. J. F. Reading and A. L. Ford, Phys. Rev. Lett. **58** (1987) 543.
7. J. F. Reading and A. L. Ford, J. Phys. **B20** (1987) 3747; and J. Phys. **B21** (1988) L685.
8. J. P. Giese and E. Horsdal, Phys. Rev. Lett. **60** (1988) 2018.
9. E. Y. Kamber, C. L. Cocke, S. Cheng, and S. L. Varghese, Phys. Rev. Lett. **60** (1988) 2026.
10. J. F. Reading, A. L. Ford, and X. Fang, Phys. Rev. Lett. **62** (1989) 245.
11. O. Heber, B. B. Bandong, G. Sampoll, and R. L. Watson, Phys. Rev. Lett. **64** (1990) 851.
12. J. H. McGuire, J. Phys. **B17** (1984) L779.
13. J. A. Tanis, M. W. Clark, R. Price, and R. E. Olson, Phys. Rev. **A36** (1987) 1952.
14. J. A. Tanis, E. M. Bernstein, M. W. Clark, S. M. Ferguson, and R. N. Price, Phys. Rev. **A43** (1991), in press.
15. J. A. Tanis, E. M. Bernstein, M. W. Clark, S. M. Ferguson, R. N. Price, and W. Woodland, Nucl. Instrum. Meth. Phys. Res. **B42** (1989) 523.
16. R. M. Wood, A. K. Edwards, and R. L. Ezell, Phys. Rev. **A34** (1986) 4415.
17. M. Inokuti, Rev. Mod. Phys. **43** (1971) 297; Rev. Mod. Phys. **50** (1978) 23.
18. H. K. Haugen, L. H. Andersen, P. Hvelplund, and H. Knudsen, Phys. Rev. **A26** (1982) 1950.
19. V. Schmidt, N. Sandner, H. Kuntzemüller, P. Dhez, F. Wuilleumier, and E. Källne, Phys. Rev. **A13** (1976) 1748.
20. J. H. McGuire, E. Salzborn, and A. Müller, Phys. Rev. **A35** (1987) 3265 .
21. E. Horsdal Pedersen and L. Larsen, J. Phys. **B12** (1979) 4085.
22. H. Knudsen, L. H. Andersen, P. Hvelplund, J. Sørensen, and D. Ciric, J. Phys. **B20** (1987) L253.

Ionization in Few Electron Atom - Atom Collisions*

R.D. DuBois

Pacific Northwest Laboratory, P.O. Box 999, Richland WA, 99352 USA

Abstract

Understanding ionizing collisions involving projectiles which initially possess loosely bound electrons is one of the next major steps in advancing our knowledge of atomic collisions. For this purpose, studies of simple, few-electron systems are desirable. This paper discusses experimental data obtained for two such systems, namely hydrogen and helium atom impact on helium. Experimental studies employing projectile ion-electron coincidence methods combined with projectile ion-recoil ion coincidence methods and standard non-coincidence electron emission methods are providing new insights into how one or two loosely bound projectile electrons influence the ionization process. These data provide information about 1) events where both the projectile and the target are ionized in a single collision, 2) how the screening of the nuclear charges by the bound electrons influence the ionization probabilities and 3) where current theoretical models of these collisions need to be improved. The He atom impact data also provide information about differential electron emission resulting from double ionization of the projectile and about singly differential cross sections for electron loss from the projectile. Brief examples and discussions of these data are provided.

Introduction

Information about atomic structure and how individual atoms interact with other atoms or molecules can be obtained from studies of the differential electron emission occurring as the result of ionizing collisions. Past studies have demonstrated that the first Born approximation can successfully model single ionization for fast proton impact on atomic targets.[1] However, adding even a single loosely bound electron to the projectile severely complicates the picture. Now the theory must account for the possibilities of ejecting an electron(s) from either the target or from the projectile and whether the collision partner is excited or ionized as well. In addition, theory must properly include screening of the nuclear charges by the bound electron clouds. For projectiles possessing several loosely bound electrons, the problem becomes even more complicated.

Thus, modeling these collisions requires that theory calculate probabilities for each process individually and then combine them before comparing with electron emission measurements obtained using standard non-coincidence techniques. When agreement is not achieved, it is not always clear where the fault lies. In recent years, however, using Projectile Ion-Recoil Ion COincidence (PIRICO) and Projectile Ion-Electron COincidence (PIECO) techniques, differential and total cross sections for the individual components, not just the overall sum, of these ionization cross sections are being obtained. These data permit more detailed tests of theory to be made.

Experiments of these types have been performed at the University of Aarhus (Denmark), the University of Frankfurt/M (Germany) and the Pacific Northwest Laboratory (USA) to study ionization resulting from fast hydrogen and helium atom impact on a helium target. These simple, few electron systems offer the first real opportunity to systematically study the influence of loosely bound electrons on the ionization process and to test existing theoretical models on a level heretofore unavailable.

Results and Discussion

Using the PIRICO technique, total cross sections for target and projectile ionization resulting from fast H-He collisions have been investigated.[2,3] Similar data have been obtained for fast He-He collisions at the Pacific Northwest Laboratory. Some of these data for the electron loss (projectile ionization) channel are shown in Fig. 1. The open symbols connected with the solid curves represent total cross sections for electron loss from the projectiles. For He impact, both the single and double electron loss cross sections are shown. Fig. 1 also contains cross sections for ionizing both the projectile and the target during the collision (the filled symbols). These "double ionization" probabilities clearly cannot be calculated using first order single electron transition theories but, as seen in Fig.1, are not negligible.

Investigating these "simultaneous" projectile-target ionization events in more detail can be done by using the PIECO method to study the differential electron emission. For the purposes of this paper, "simultaneous" means "in a single collision." It is not meant to imply, nor to exclude, that the emission is a correlated phenomenon since, at this point, there have been no experimental or theoretical studies designed to test this possibility.

Doubly differential cross sections for electron emission observed at 30° in 0.5 MeV H-He collisions are shown in Fig. 2. These data were recently obtained at the Univ. of Frankfurt.[4] The coincidence cross sections (Δ) represent events where projectile ionization (events resulting in a free electron and an ionized projectile) has occurred <u>and</u> events where simultaneous ionization (events resulting in a free target electron <u>and</u> a free projectile electron <u>and</u> an ionized projectile) have occurred.

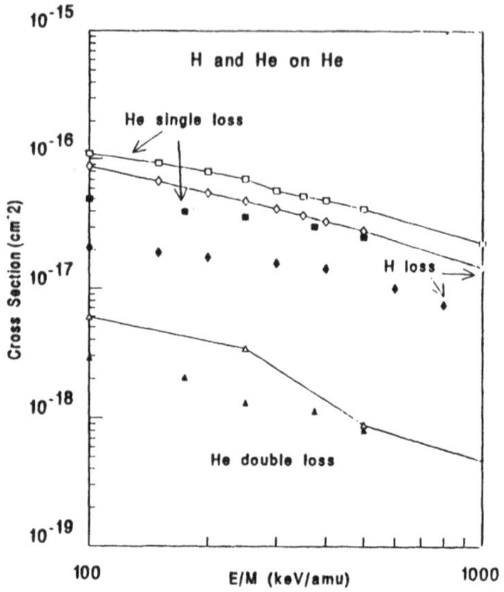

Fig. 1. Total cross sections for electron loss in H-He and He-He collisions. The data are from ref. 2,3 and from unpublished measurements obtained at the Pacific Northwest Laboratory. The open symbols are total electron loss cross sections while the solid symbols represent collisions where both the projectile and the target are ionized. The curves through the data serve merely to guide the eye.

Like target ionization, projectile ionization results in the vast majority of the emitted electrons having very small energies. Therefore, the kinematic transformation from the projectile to the laboratory frame shifts these electron to a laboratory velocity equal to the projectile velocity. Thus we observe an "electron loss" peak centered at the projectile velocity (near 250 eV in the example shown). This peak has a width indicative of the bound electron velocity distribution and the transformation process. Hence information about projectile ionization can be derived from this peak.

One method of providing more detailed information about the electron loss mechanism is to measure ionized projectile ion-emitted electron coincidences. These data, shown in Fig. 2 demonstrate that the electron loss peak sits on a background continuum which extends to larger and smaller electron energies. Kinematically, electrons possessing these laboratory energies cannot originate from the projectile; they are target electrons which originate in events where both the projectile and the target are ionized. That both collision partners are ionized is known because of the coincidence conditions imposed and the fact that it is extremely unlikely that projectile electrons have these laboratory energies. Hence these data provide information about the differential cross section for simultaneous ionization events that were discussed for Fig. 1.

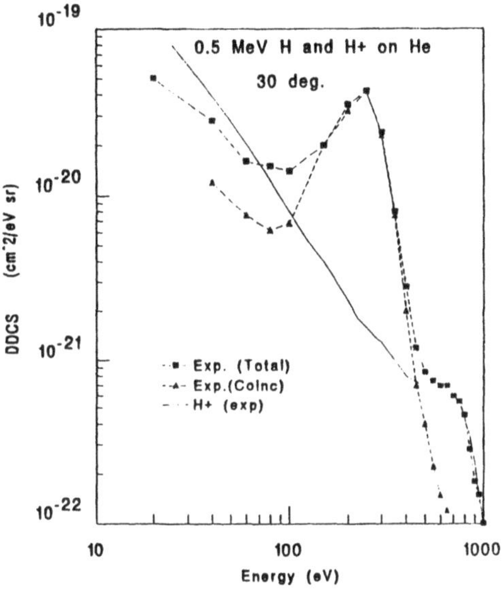

Fig. 2. Doubly differential cross sections for electron emission at 30° in 0.5 MeV H- and H^+ - He collisions. The data are from the University of Frankfurt.[4] ■, total (non-coincidence) electron emission; Δ, projectile ion-electron coincidence cross sections; ——, proton impact. The dashed curves are only to guide the eye.

Fig. 2 also displays the doubly differential cross sections (DDCS) for the total (non-coincidence) electron emission at 30° for H and H^+ impact. Comparing these data with each other provides information about screening of the nuclear charge by the bound 1s electron on the hydrogen atom. From a simple physical picture it is obvious that when the hydrogen atom is far from the target, i.e. for large impact parameters, this screening is larger than for small impact parameters where the H and He electron clouds penetrate each other. Thus at small impact parameters (which generally lead to the emission of energetic electrons), the screening is ineffective and the H and H^+ cross sections should be the same. This is seen in the data between 600 and 1000 eV. At large impact parameter collisions which result in the emission of low energy electrons, the screening is more effective and the cross sections for hydrogen atom impact are smaller than for proton impact since the coulomb force is smaller.

One of the primary purposes of these DDCS data are for testing and improving our theoretical understanding of collisions involving non-bare projectile impact. Recently Steve Manson has performed first Born calculations for the H-He system[5] using procedures previously outlined for He^+ impact.[6,7] In his theory he has included processes where one collision partner is ionized and the other remains in the ground state or is excited to a discrete or a continuum state.

In Fig. 3 theoretical cross sections for the total electron emission (obtained by summing all the components of the theory), ——, and cross sections which should be an upper limit for the experimentally measured coincidence cross sections, - - -, are shown. Note that the cross sections were calculated for specific electron energies and straight lines were used to connect these points.

Qualitatively the calculated cross sections demonstrate the same features as the experimental data shown in Fig. 2; however quantitatively they tend to underestimate the cross sections for both projectile and for target ionization. This is attributed to improper treatment of the long range interactions since a) the poorest agreement for target ionization is at low electron energies and b) the intensity of the electron loss peak is primarily due to low energy electron emission. In the theoretical model, long range interactions were not included since both collision partners were neutral. The theory also severely underestimates the amount of simultaneous ionization at low electron energies and overestimates it at large electron energies. This indicates that more sophisticated calculations are needed.

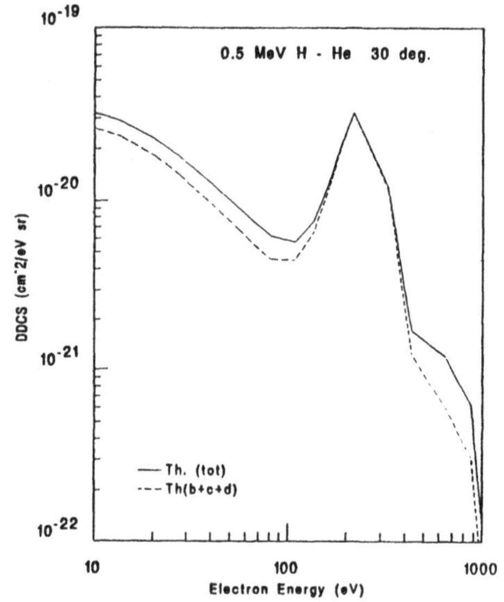

Fig. 3. Calculated DDCS for electron emission at 30° in 0.5 MeV H-He collisions obtained using the Born approximation.[5] Solid curve, total electron emission cross sections; dashed curve, an upper limit calculated for the simultaneous projectile-target ionization process.

The next step in understanding clothed particle impact is the influence of a second loosely bound projectile electron. Thus data for 0.5 MeV/amu He impact were also measured at the University of Frankfurt.[8] An example of these data are shown in Fig. 4. Features similar to those already discussed for H impact are seen. Note that the data shown imply that the cross section is decreasing at the lowest energy point, unlike the behavior observed in Fig. 2 for H impact. However, because of the poor statistics for these low energy data in Fig. 3, this downward trend is considered to be an experimental artifact.

Since both single and double ionization of the helium projectile can occur, both single and double loss from the projectile were investigated. As seen in Fig. 4, the double loss probabilities are approximately an order of magnitude smaller than the single loss probabilities but exhibit the same general features. From the double loss data, information about the differential electron emission cross section for double ionization can be obtained.

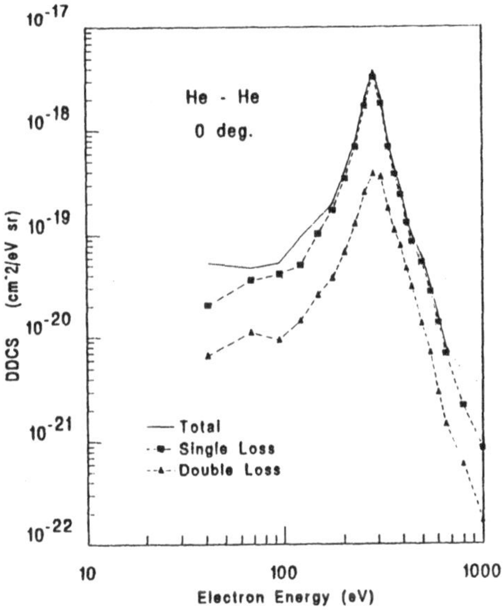

Fig. 4. DDCS for electron emission at 0° in 0.5 MeV He - He collisions. The data were obtained at the University of Frankfurt.[8] ——, total electron emission; projectile ion-electron coincidence cross sections, ■ , single loss channel, Δ , double loss channel.

When we compare the single and double loss data, we find that the width of the electron loss peak is greater for double loss than for single loss. This is because the width of the peak is indicative of the bound electron momentum, i.e. it is larger for more tightly bound electrons. Thus the broader peak for double loss than for single loss. For the same reason the width of the single loss peak for He impact is greater than for H impact.

Information about projectile ionization can also be obtained by studying the intensity of the electron loss peak as a function of laboratory emission angle. To obtain the singly differential cross section for electron loss, the doubly differential cross sections (background subtracted) were integrated over the peak. These singly differential cross sections are similar to those obtained by measuring the differential elastic scattering cross section on helium. Comparing the integrated peak intensities for He and H atom impact shows that their relative importances varies as a function of laboratory emission angle (see Fig. 5). For example, at 0° it is approximately 50% more probable to ionize H than He, at 10° the probabilities are nearly equal, but at 30° and 50° it is approximately 60% more probable to ionize He than H.

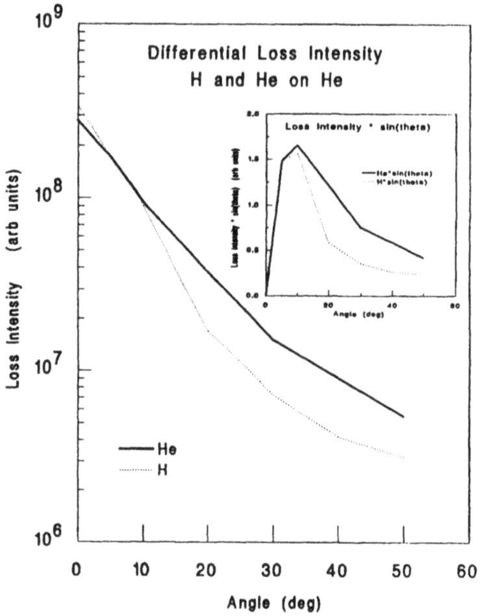

Fig. 5. Intensity of the single electron loss signal in 0.5 MeV/amu H-He and He-He collisions obtained by integrating the doubly differential cross sections over the background subtracted electron loss peak. Solid curve, He impact; dotted curve, H impact. The insert shows these differential intensities times sinθ with the areas under the curves being proportional to the total single electron loss cross sections.

The total single loss cross sections for He impact on He are known[9] to be approximately 60% larger than the similar cross sections for H impact.[10] See Fig. 1. Since the total cross sections are obtained by integrating the singly differential cross sections shown in Fig. 5 over angle and the integration is done by taking the singly differential cross section times sinΘ and plotting versus Θ, the inset in Fig. 5 shows that most of the area contributing to the total cross section lies in an angular range above 10°. This is exactly where the differential loss data for He impact are larger than for H impact. Hence the important message to be derived from Fig. 5 is that although one measures the largest electron loss signal when the electron spectrometer is positioned at 0°, only a small portion of the total solid angle contributes to this signal. For larger Θ angles, a smaller signal is measured for a particular Θ, Φ combination, but summing over the Φ angles leads to a larger contribution to the total cross section. Thus, conclusions about the total cross section derived from measurements made only at 0° may be erroneous.

Conclusions

Cross sections for total and doubly differential electron emission resulting from projectiles having one and two loosely bound electrons, namely H and He, impacting on helium have been discussed. These data demonstrate the importance of processes where both the projectile and the target are ionized in a single collision. Comparing the H and He impact data with those obtained for a fully stripped projectile, H^+, demonstrated the influence of screening of the projectile nuclear charge by one or two loosely bound projectile electrons. Comparing the H impact data for differential electron emission with calculations made using first Born theory indicated that the long range interactions are important even though the collision partners are neutral atoms. The comparison also implies that future theoretical treatments of these collisions must include processes where both particles are excited to discrete or continuum states. Lastly it was observed that the total single electron loss cross sections are larger for He impact than for H impact because the differential single loss cross sections for He impact are larger than those for H impact for larger laboratory emission angles which contribute most to the total cross section. At 0° the cross sections are largest for H impact.

Acknowledgements

* This work was supported by the Office of Health and Environmental Research, U.S. Dept. of Energy under contract No. DE AC06-76RLO 1830.

References:

References:

1. S. T. Manson, L. H. Toburen, D. H. Madison and N. Stolterfoht, Phys Rev A 12, 60 (1975).

2. R. D. DuBois and A. Kövér, Phys Rev A 40, 3605 (1989).

3. L. H. Andersen, L. B. Nielsen and J. Sørensen, J Phys B 21, 1587 (1988).

4. O. Heil, R. Maier, M. Kuzel, K-O. Groeneveld and R. D. DuBois, submitted to XVII ICPEAC, Brisbane, Australia (1991).

5. S. T. Manson, R. D. DuBois, O. Heil, R. Maier, M. Kuzel and K-O. Groeneveld, submitted to XVII ICPEAC, Brisbane, Australia, (1991).

6. R. D. DuBois and S. T. Manson, Phys Rev Lett 57, 1130 (1986).

7. R. D. DuBois and S. T. Manson, Phys Rev A 42, 1222 (1990).

8. O. Heil, R. Maier, R. DuBois, M. Kuzel and K-O. Groeneveld, Vth Int. Conf. on the Physics of Highly-Charged Ions, Giessen, FRG, (Sept. 1990).

9. E. Horsdal-Pedersen and P. Hvelplund, J Phys B 7, 132 (1974).

10. E. Horsdal-Pedersen and L. Larsen, J Phys B 12, 4099 (1979).

ELECTRON CORRELATION IN HIGH-ENERGY ION-ATOM COLLISIONS

A. L. FORD and J.F. READING
Physics Department, Texas A&M University
College Station, Texas 77843, USA

Abstract

We use two-electron processes in helium to illustrate correlation effects in energetic ion-atom collisions. For helium double ionzation we discuss what is known and unknown, and use the Born series as a tool for describing the different physical processes that enter. We briefly review our own forced-impulse method results and comment on other types of calculations. Differential cross sections are discussed, and we briefly compare double ionization to other two-electron processes.

1. Introduction

We will use two-electron processes in helium excited by energetic ion projectiles as the specific system to use to illustrate what we hope are general concepts. We will comment on what we know about these processes, and especially about what is still unknown or unresolved.

We will start with helium double ionization by bare ions in the 100 keV/u to 40 MeV/u energy region. The projectiles considered are protons, antiprotons, and bare ions of positive charge Z_p, where $Z_p = 2$ to 8. The double ionization results are presented in terms of the ratio R of double to single ionization. It is useful to analyze the data in terms of the Born series for the transition probability. (We use the language of the semiclassical impact parameter method where the projectile moves on a classical trajectory.) The Born series gives the transition amplitude as an expansion in powers of the projectile charge Z_p,

$$a = Z_p a_1 + Z_p^2 a_2 + Z_p^3 a_3 + \dots \quad (1)$$

Thus the transition probabiltiy is given as

$$P = |a|^2 = Z_p^2[|a_1|^2] + Z_p^3[2Re(a_1 a_2^*)] + Z_p^4[|a_2|^2 + 2Re(a_1 a_3^*)]. \quad (2)$$

2. Double Ionization of Helium

2.1 First Born at High Collision Velocities

For double ionization (or any other two-electron process) first Born without correlation gives zero. The experimental results for high collision velocity do obey the Born Z_p^2 scaling. (As can be seen, for example, by comparing p^+ and He^{2+} projectile results.[1]) The experimental high-energy limit for R seen for e^-, e^+, p^+, p^- and He^{2+} projectiles of about 0.3% is in excellent agreement with our first Born calculations performed with configuration-interaction (correlated) wavefunctions.[2] Note that R is quite small, less than 3% over the entire energy region above 100 keV.[2] One question is whether R at high energies is a constant, independent of energy. That is, does the first Born (with correlation) for double ionization have the same collision velocity dependence as the first Born for single ionization (where correlation is not needed to get a nonzero result)? For double ionization in the first Born one electron is ionized by the projectile and then the second electron is removed by the electron-electron interaction, either by a shake-off type process due to a change in screening or by an electron-electron collision. It would seem that the distribution of ejection energies of the first electron would depend on the projectile speed and that this would in turn affect the probability of the second electron being removed, and hence give an R that depended on the collision velocity. Our first Born calculations give an R that is only very weakly dependent on collision velocity above 1 MeV/u, and the experiments also suggest a constant high-energy limit. A very recent analysis by Sidorovich[3] gives a double ionization cross section with energy dependence $(C + \ln E)/E$ where $C = -5.54$ when E is in keV. This shows that the double ionization has the same $\ln E/E$ dependence as single ionization asympotically, but that deviations due to the C/E term will persist to high energies.

2.2 Proton/Antiproton Difference

For collision energies between 0.5 MeV and 5 MeV/u antiprotons give about twice as much double ionization as do protons.[2] In this same energy region the two projectiles give virtually identical single ionization,[4] and the single ionization is accurately described by first Born. From the point of view of the Born series, the proton/antiproton difference must come from terms in the expansion for the transition probability that involve odd powers of Z_p, and all such terms arise from cross-terms (interference) between different orders in the amplitude. For example, the Z_p^3 term in Eq. 2 comes form the interference between first and second Born amplitudes. Note that if only the lowest two terms in the Born expansion are kept,

$$P \simeq Z_p^2[|a_1|^2] + Z_p^3[2Re(a_1 a_2^*)]. \tag{3}$$

But if this expression is accurate then p^- and p^+ would give cross sections symmetrically placed on either side of the first Born, and this is observed to be the case (both in the experiments and in our coupled-channels calculations to be described later) only at the higher energies.[5] Higher Born terms must also play an important role. It is also interesting to examine the probability $P(b)$ (from our calculations) as a function of impact parameter b for p^+ and p^- compared to first Born; there is qualitatively different behavior for different b and for different E. For example, at 20 MeV/u the p^-, p^+ results are approximately symmetrically placed on either side of the first Born. For 2.3 MeV/u p^- is above first Born and p^+ is below for $b < a_0/2$, but both are above first Born for larger b. And at 0.30 MeV/u both p^- and p^+ results are above the Born at all b.[6]

2.3 Independent Particle Model (IPM)

In the independent particle model[7] for ionization of two electrons of opposite spin, such as being considered here, the probability P_d of double ionization is the square of the probability P_s of single ionization, $P_d = (P_s)^2$. Since at the energies we are considering the single ionization is given accurately by first Born, $P_s \approx Z_p^2$ so $P_d \approx Z_p^4$ and $R \approx Z_p^2$. This IPM prediction disagrees with the observed high energy behavior, where R is independent of Z_p, and also predicts no p^+/p^- difference, in strong disagreement with what is seen experimentally at lower energies. The IPM is part of the second Born term. There are two contributions to second Born, where the projectile interacts twice with the target system. (1) The projectile interacts twice with one electron. Then correlation is required for double ionization. (2) The projectile interacts once with each electron. This includes the IPM contribution. Correlation is not necessary but may still play a role. Presumably this second contribution dominates, since correlation is not required.

2.4 Role of Non-Dipole Transitions

Single ionization at these energies is over 90% dominated by dipole-allowed transitions (to ^1P states): $(1s)^2\,{}^1S \rightarrow (sp)\,{}^1P$. But if only dipole transitions are induced by the projectile, then since the electron-electron interaction conserves total orbital angular momentum L the final states reached in second Born: $(1s)^2\,{}^1S \rightarrow (pp)\,{}^1S$ or ${}^1D$ are different from those reached in first Born: $(1s)^2\,{}^1S \rightarrow (sp)\,{}^1P$. Different final states means no 1st and 2nd Born interference, which means no Z_p^3 term, which means no p^+/p^- difference. Thus the interference between different Born amplitudes that produces the p^+/p^- difference requires non-dipole transitions.

2.5 Born Terms Extracted from Calculations and Experiment

In our coupled-channels calculations we do not calculate individual terms in the Born series. Rather, we calculate a cross section $\sigma(Z_p, E/\text{u})$ for a specified projectile charge Z_p and collision energy E. We can however fit $\sigma(Z_p, E/\text{u})$ for each E/u to a power series in Z_p and extract the terms in the Born series at each collision velocity. This is useful in discussing different mechanisms.[5] We have also found that we can very accurately extract p^- results from the power series obtained by fitting to our calculations for positive Z_p. That is, the p^- double ionization cross section obtained from the fit is in close agreement to what we calculate directly with $Z_p = -1$ in our collision code. The Born series, at least as represented in our calculations, can be analytically continued to negative Z_p. We suggest that the same thing be attempted with experimental cross sections taken at the same collision velocity for different positive bare ions, as a consistency check on the results from the very difficult p^- experiments. At lower energies where capture becomes important this analytic continuation to $Z_p = -1$ may no longer be possible.

2.6 Our Forced-Impulse Method (FIM) Calculations

The FIM[5] uses an impulse type approximation applied with respect to the residual interaction (which in our calculations done so far has always been the full $1/r_{12}$ interaction), but the collision is chopped into short time steps to force the impulse approximation to converge. The qualitative idea is to start in a correlated state, let the system time-evolve in the IPM for a short time step, then project back onto correlated states, and repeat the procedure to step out in time until the collision is over. The FIM exploits the fact that the IPM is a rather good description for the most probable processes. It avoids the huge computational expense of doing a direct coupled-states expansion in a basis of two-electron states. At these high collision velocities the projectile has sufficient energy and momentum to do almost anything to the target electrons. There are an infinity of open channels including numerous ionization-excitation channels, and a large basis is needed. For example, we have 423 correlated pseudostates in our *sp*-basis calculations. This number rises to over 1800 when *d*-orbitals are included as well.

In addition to developing the FIM to efficiently include the electron correlation effects in the collision dynamics, to calculate double ionization we had to solve what we have called the interpretation problem.[5] In all our coupled-channels calculations we have used a discrete pseudostate representation of the continuum. For one-electron systems the pseudostates representing the bound states are identified by their energy and then by unitarity the rest must represent the ionization continuum. But for a two-electron system the double ionization continuum is overlapped by single ionization continua. Pseudostates from diagonalizing the helium hamiltonian that have energies above the double ionization threshold are a mixture of these continua and we must project out the double ionization part. This we did by constructing channel states, that satisfy single or double ionization boundary conditions. Since the double ionization cross section is much less than the single ionization (about 1% of it), this separation out of the double ionization part of each pseudostate must be done very accurately.

Our FIM calculations give a good account of the sign, magnitude and energy dependence of the p^+/p^- double ionization difference that is observed experimentally.[5] We also agree well[5] with the He^{2+} + He double ionization results of Knudsen *et al.*[8] and of Shah and Gilbody,[9] and with the results for projectiles of higher charge,[5] up through O^{8+} at selected energies (2.3 MeV/u, for example) from Knudsen *et al.*[8] But there are still several problems with the calculations as we have done them thus far: (1) Our p^- results are somewhat below experiment; the p^+ results are in much better agreement. (2) We know that the FIM calculations we have done so far, that are restricted to only s and p orbitals, are not converged in l. We have done calculations with d-orbitals only in the first Born approximation, and even there more l's are needed to demonstrate convergence. (3) We have in fact not used small time steps in the FIM but large ones. Our calculations have been done in a three-stage approximation where the projection onto correlated states is done initially before the collision, at the distance of closest approach, and again after the collision is over. It is in fact surprising that our results with such large time steps contain so much of what is seen in the experiments. Apparently even though the results are completely wrong if correlation is omitted entirely, it need be included only in lowest order to obtain fairly accurate results.[10] (4) Another defect of our calculations is that we have not yet included capture channels, that will be important at lower collision energies. (5) Finally, even though our calculations may yield accurate numerical cross sections it is difficult to infer a physical mechanism from them.

2.7 Other Calculations of Helium Double Ionization by Protons and Antiprotons

The single and double ionization of helium by protons and antiprotons has been calculated in the CTMC method by Olson,[11] using an atomic model that includes the electron-electron interaction. His calculations show a p^+/p^- difference roughly in agreement with experiment, but he reports results only up to a collision energy of 1 MeV/u. The relation between his classical calculations and our quantum ones, where the p^+/p^- difference comes from interference between different Born amplitudes, is unclear. It would be very interesting to see what the CTMC calculation gives above 1 MeV, since the p^+/p^- difference persists to nearly 10 MeV. But the double ionization cross section becomes quite small as the energy is increased and it it is very difficult to get sufficient statistical accuracy in the Monte Carlo method. Olson also found a greater difference between p^+ and p^- single ionization than we obtain in our calculations and or is seen experimently.[4]

Model calculations of helium double ionization have been carried out by Végh.[12] His calculations are based on a specific mechanism for how correlation gives more ionization by antiprotons than by protons. Végh considered the effect of radial correlation between the electrons during the collision in a model in which both electrons are removed by the projectile. Simply put, the antiproton pushes out the electron closest to it to ionize the electron. The radial correlation that keeps the electrons on opposite sides of the nucleus then causes the second electron to swing around and become closer to the projectile, thereby decreasing the effective impact parameter for the second ionization and increasing the probability that the second electron will also be ionized. The reverse happens for the proton projectile. The physical basis of the mechanism is compelling, but since the process is 2nd Born plus correlation it is unclear how much it contributes to the total double ionization.

2.8 Approach to the First Born Limit for Projectiles with Larger Z_p

Apart from $\ln(v)$ terms the first Born amplitude has a $1/v$ collision velocity dependence and the second Born a $1/v^2$ dependence. Single ionization is given accurately by first Born so can be written as

$$\sigma_{\text{sing}} = \frac{Z_p^2}{v^2}\sigma_{1\text{B}}^{s}. \tag{4}$$

The three lowest order terms in the double ionization are

$$\sigma_{\text{doub}} = \frac{Z_p^2}{v^2}\sigma_{1\text{B}}^{d} + \frac{Z_p^3}{v^3}\sigma_{\text{interf}}^{d} + \frac{Z_p^4}{v^4}\sigma_{2\text{B}}^{d}. \tag{5}$$

The double to single ionization ratio R then is

$$R = R_{1\text{B}} + \frac{Z_p}{v}R_{\text{interf}} + \frac{Z_p^2}{v^2}R_{2B}. \tag{6}$$

The collision velocity required so that only the first term contributes and the common high velocity limit is reached increases as Z_p is increased. That is, Z_p/v is the relevant expansion parameter and must be small to get to the first Born limit. We have fit our FIM calculations for up to $Z_p = +8$ and for E between 2 and 30 MeV/u to the energy dependence suggested by the semiempirical model of Andersen *et al.*,[13] and find

$$\begin{aligned} R = \left[1.7\times 10^{-3}\frac{\ln(18(v/v_0))}{\ln(3(v/v_0))}\right] - \left[\frac{0.016}{(v/v_0)\ln(3(v/v_0))}\right] Z_p \\ + \left[\frac{0.25}{(v/v_0)^2\ln(3(v/v_0))}\right] Z_p^2, \end{aligned} \tag{7}$$

where v_0 is the Bohr velocity. If the first Born limit is reached at 10 MeV/u (where $v/v_0 = 20$) for protons, the energy must be increased a factor of 7^2, to 490 MeV/u, to reach the first Born limit for N^{7+} projectiles. Recent experiments by Heber *et al.*[14] confirm this. For 10 to 30 MeV/u the R they measure for N^{7+} is about a factor of 4 above the high energy 0.27% limit for protons.

There is a problem raised, however, by the Heber *et al.*[14] results. As noted above, our FIM calculations agree very well with the results of Knudsen *et al.*[8] for projectiles up through O^{8+} at 2.3 MeV/u. However, our calculated values of R for N^{7+} are about a factor of two below the Heber *et al.* results at 20 and 30 MeV/u. Our calculations agree that at these energies for N^{7+} the first Born limit is not reached, but there is this discrepancy about the magnitude of the difference. We also note that the semiempirical model of Andersen *et al.*,[13] based in part on a fit to p^+, p^-, and He^{2+} data up to 10 MeV/u is in excellent agreement with the N^{7+} data.[15] But terms in the model relatively unimportant for small Z_p dominate for $Z_p = +7$ so there might be doubt about the validity of the extrapolation of their model to higher Z_p. The situation needs to be resolved by further experimental work, and by our pushing our own calculations to convergence.

2.9 Impact-Parameter Dependent Probabilities

Our FIM calculations treat the projectile motion classically, so automatically yield impact-parameter dependent probabilities $P(b)$. We have previously noted[6] that the ratio of double to single ionization probabilities at small b, that is $r(0) = P_d(0)/P_s(0)$, is much larger than the corresponding ratio of cross sections, $R = \sigma_d/\sigma_s$. And the p^+/p^- difference is larger in $r(0)$ than in R. There is great interest in experimental tests of these predictions, either by measurement of the cross section $d\sigma/d\Omega$ differential in projectile scattering angle or by experiments that somehow select out small impact parameters. One example of the latter is experiments with large Z_p ions carrying one electron into the collision. Measurement of electron loss from the projectile in coincidence with target double ionization selects out small impact parameters, of the order of the K-shell radius of the projectile ion. Preliminary results from such experiments are reported in these proceedings by John Tanis.

The ratio of differential cross sections $d\sigma/d\Omega$ for double and single ionization has been measured.[16,17] But we cannot simply convert our $P(b)$ to $d\sigma/d\Omega$ by the classical relation between b and Ω because in the collision appreciable amounts of projectile transverse momentum is given to the target electrons;[18] that is, the projectile can scatter off the target electrons in a hard binary collision as well as off the target nucleus. The angle of 0.55 mrad is the maximum angle through which a free electron at rest classically can scatter a proton. If this happens the electron gains a large amount of momentum and is ejected at high energy. The high-energy ejection of one electron enhances the removal of the second electron by the electron-electron interaction and the double to single ratio is much larger than in soft collisions where the projectile is scattered mostly by the target nucleus and the primary electron is ionized into a low-energy continuum state. But since a single binary scattering off an electron gives at most a scattering of 0.55 mrad, at large angles (≥ 2 mrad) the projectile is scattered mostly by the nucleus (in a small b collision) and our $P(b)$ can be converted to $d\sigma/d\Omega$ and compared to experiment. Our FIM are in reasonable agreement (about 25 to 35%) too low) with the experiments of Giese and Horsdal;[16] the remaining discrepancy may be due to the truncation to s and p states of the basis for our calculation. Further experimental work has been reported recently by Kristensen and Horsdal-Pedersen.[19] It would be very interesting to have similar experiments, even just at large angles, for p^- and/or for projectiles with larger positive Z_p.

The Giese and Horsdal experiments also show a peak in the double to single ratio at about 1 mrad, for $E = 0.300$ to 1.0 MeV. The shape and location of the peak at 1.0 MeV is well reproduced by CTMC calculations of Olson *et al.*,[20] but the absolute magnitude of the calculated double to single ratio is about a factor of three too large. The CTMC calculations were done in an IPM, so correlation was not included. But we know that correlation is essential in describing the cross section ratios at this energy. In the calculations double ionization is produced by the projectile making successive independent scatterings off each electron. We showed[21] that a peak was also obtained, but at a somewhat smaller angle, if the process in which the projectile makes a hard binary collision with one electron and the other is shaken off is added (incoherently) to our FIM result. The former process produces one fast and one slow electron whereas in our FIM calculations both electrons are ionized into low energy continuum orbitals. Hence the final states for the processes are different and do not interfer. Végh[22] has also proposed a triple collision mechanism, similar to the Thomas mechanism for charge capture. Better quantal calculations are needed, that include in a single consistent calculation the production of both fast and slow electrons and that account for all momentum transfers among the projectile, the target nucleus, and the target electrons. Also, comparison of theory and experiment needs to be made at all energies for which there is data, not just at 1 MeV/u.

3. Other Two-Electron Processes in Helium

It is now well established that for 1 to 6 MeV antiprotons produce about twice as much double ionization as do protons. Another related process is excitation-ionization, and a similar situation obtains: the cross section is about a factor of two larger for antiprotons.[23] To date there have been no theoretical calculations reported for this process. Presumably the mechanism for the p^+/p^- difference in excitation-ionization is the same as in double ionization.

The situation appears to be qualitatively different for double excitation, to autoionizing resonances such as $(2s2p)\,{}^1P$ and $(2p)^2\,{}^1D$. Two experiments done with protons and equivelocity electrons give no difference within experimental error.[24] The one published calculation[25] differs in several respects from the experimental results. It is important to note that these experiments measure the yield of electrons in the autoionization peaks rather than the excitation cross sections. And in the experiments, where the electron yield differential in ejection angle and energy is measured, there is detail to which no theoretical comparison has been made.

4. Comments on Other Recent Work

We will conclude with comments on three recent related developments in the literature. One is the very recent publication by Ermolaev[26] of the p^-/p^+ ionization ratio for atomic hydrogen that he calculates in an asymmetric two-centered expansion. The ratio $R_{\text{ion}} = \sigma_{\text{ion}}(p^+)/\sigma_{\text{ion}}(p^-)$ is much larger than unity for E below about 5 keV; this is the well known binding/antibinding effect of the Brandt perturbed stationary states model, widely used in analysis of low to medium energy K-shell vacancy production experiments for multi-electron targets. The ratio R_{ion} drops to 0.9 between 100 to 300 keV; this is presumably the polarization/antipolarization effect, also well known in the Brandt model. But the ratio that Ermolaev calculates rises above unity above 500 keV, reaching 1.07 for 1 to 1.5 MeV. The physical origin of the p^-/p^+ difference at such high energies is unknown. This result is very interesting and needs to be understood. Even though this is a one-electron system and correlation is not present, the mechanism producing the p^-/p^+ difference here might also play a role in the helium case. We need to find out if the Ermolaev result is pointing to something that is missing from our coupled-channels calculations; we have yet to see a similar effect.

The second recent development we wish to discuss is the recent publication by Skogvall and Schiwietz[27] of highly differential measurements of single and double ionization in p + He collisions at 300 keV. For a projectile scattering angle of 0.5° (which corresponds to an impact parameter of 0.02 a_0) they measured the ejected electron energy distribution at various ejection angles. They found that the IPM applied to the single ionization measurements to obtain $P(b)$ for double ionization overpredicts their double ionization measurements by a factor of 6. They also found that the angular distribution of ejected electrons at emission angles below 40° was markedly different from IPM predictions, with the IPM predictions being much too large. They concluded that correlation plays an important role and suggested two correlation effects. One is the initial state correlation, that produces a reduction of the probability density for finding one electron near the other. The other is a mean field effect; the second electron is harder to ionize than the first because it moves in a different mean field due to the change in screening that occurs when the first electron is removed. These differential measurements for double ionization pose a stringent challenge to theory.

Finally, we comment on two other recent calculations of helium double ionization. One is a paper of Shingal and Lin.[28] They calculated single- and two-electron processes for He^{2+} through F^{9+} projectiles on He. They included capture as well as ionization. The two-electron processes they calculated were transfer ionization and double ionization, and for these processes they used an IPM. They obtained a generally good fit to experiment for all processes, for example for He^{2+} + He for 50 to 300 keV. However, they found that they had to use one model potential (that was fit to single-electron excitation energies) for single-electron processes and another (with $V = -Z_{\text{eff}}/r$ where Z_{eff} is chosen so each electron is bound with one-half the double ionization threshold energy) for two-electron processes. They concluded that "It is not possible to find a single independent electron model for both one and two-electron processes simultaneously." We also note that for the lower energies and large Z_p they considered the second Born term should dominate over the first so interference (Z_p^3) effects not present in their IPM might not be very important. Deb and Crothers[29] calculated double ionization in He^{2+} + He collisions and got good agreement with the experiments of Shah and Gilbody.[9] They used the CDW method for the ionization calculation and used a pseudo-IPM. That is, they took the double ionization probability to be $P_d = P_1 P_2$ where P_1 and P_2 are single-electron ionization probabilities but calculated in different atomic potentials. (P_1 is calculated for He, with correlation included in the initial state, and P_2 for He^+. This allows for the change in the mean field after one electron is removed.) They give results only up to 1.4 MeV/u; presumably their calculation will fail at higher energies since it scales as Z_p^4 (2nd Born) and at high energies first Born plus correlation (Z_p^2) dominates. It would be interesting to see what their approach gives for p^+ and p^-.

4. Acknowledgements

This work was supported by the National Science Foundation, under grant PHY-9009717.

5. References

1. A.L. Ford and J.F. Reading, *J. Phys. B* **20** (1987) 3747-3769.
2. L.H. Andersen, P. Hvelplund, J. Knudsen, S.P. Møller, J.O.P. Pedersen, S. Tang-Petersen, K. Elsener, and E. Morenzoni, *Phys. Rev. A* **40** (1989) 7366-7368.
3. V.A. Sidorovich, *Phys. Lett. A* **152** (1991) 53-58.
4. L.H. Andersen, P.Hvelplund, H. Knudsen, S.P. Møller, J.O.P. Pederesen, S. Tang-Petersen, E. Uggerhøj, K. Elsener, and E. Morenzoni, *Phys. Rev. A* **41** (1990) 6536-6539.
5. J.F. Reading and A.L. Ford, *J. Phys. B* **20** (1987) 3747-3769.
6. A.L. Ford and J.F. Reading, *J. Phys. B* **23** (1990) 2567-2578.
7. J. F. Reading and A. L. Ford, *Phys. Rev. A* **21** (1980) 124-131.
8. H. Knudsen, L.H. Andersen, P. Hvelplund, G. Astner, H. Cederquist, H. Danared, L. Liljeby, and K.-G. Rensfelt, *J. Phys. B* **17** (1984) 3545-3564.
9. M.B. Shah and H.B. Gilbody, *J. Phys. B* **23** (1990) 3131-3137.
10. A.L. Ford and J.F. Reading, *J. Phys. B* **23** (1990) 3131-3137.
11. R. E. Olson, *Phys. Rev. A* **36** (1987) 1519-1521.
12. L. Végh, *Phys. Rev. A* **37** (1988) 992-994.
13. L.H. Andersen, P.Hvelplund, H. Knudsen, S.P. Møller, A.H. Sørensen, K.-G. Rensfelt, and E. Uggerhøj *Phys. Rev. A* **36** (1990) 3619-3629.
14. O. Heber, B.B. Bandong, G. Sampoll, and R.L. Watson, *Phys. Rev. Lett.* **64** (1990) 851-853.
15. L.H. Andersen, P.Hvelplund, H. Knudsen, S.P. Møller, J.O.P. Pederesen, A.H. Sørensen, E. Uggerhøj, K. Elsener, and E. Morenzoni, *Phys. Rev. Lett.* **65** (1990) 1687.

16. J.P. Giese and E. Horsdal, *Phys. Rev. Lett.* **60** (1988) 2018-2021.
17. E.Y. Kamber, C. L. Cocke, S. Cheng, and S.L. Varghese, *Phys. Rev. Lett.* **60** (1988) 2026-2029.
18. R. Dörner, J. Ullrich, H. Schmidt-Böcking, and R.E. Olson, *Phys. Rev. Lett.* **63** (1989) 147-150.
19. F.G. Kristensen and E. Horsdal-Pedersen, *J. Phys. B* **23** (1990) 4129-4149.
20. R.E. Olson, J. Ullrich, R. Dörner, and H. Schmidt-Böcking, *Phys. Rev. A* **40** (1989) 2843-2846.
21. J.F. Reading, A.L. Ford, and X. Fang, *Phys. Rev. Lett.* **62** (1989) 245-248; X. Fang and J.F. Reading, *Nucl. Instr. Meth.* **B53** (1991) 453.
22. L. Végh, *J. Phys. B* **22** (1989) L35-L38.
23. J.O.P. Pedersen and F. Folkmann, *J. Phys. B* **23** (1990) 441; K.-H. Schartner, B. Lommel, and D. Detleffsen, *J. Phys. B* **24** (1991) L13-L17; J.O.P. Pedersen, *Phys. Scr.* **42** (1990) 180-186.
24. J.O.P. Pedersen and P. Hvelplund, *Phys. Rev. Lett.* **62** (1989) 2373; J.P. Giese, M. Schulz, J.K. Swenson, H. Schöne, M. Benhenni, S.L. Varghese, C.R. Vane, P.F. Dittner, S.M. Shafroth, and S. Datz, *Phys. Rev. A* **42** (1990) 1231-1244.
25. W. Fritsch and C.D. Lin, *Phys. Rev. A* **41** (1990) 4776-4782.
26. A.M. Ermolaev, *Phys. Lett. A* **149** (1990) 151-154.
27. B. Skogvall and G. Schiwietz, *Phys. Rev. Lett.* **65** (1990) 3265-3268.
28. R. Shingal and C.D. Lin, *J. Phys. B* **24** (1991) 251-264.
29. N.C. Deb and D.S.F. Crothers, *J. Phys. B* **23** (1990) L799-L803.

IV. SIMPLE SYSTEMS

TWO-ELECTRON TRANSITIONS IN SIMPLE ATOMIC COLLISION SYSTEMS

ERHARD SALZBORN

Institut für Kernphysik, Universität Giessen
6300 Giessen, Germany

ABSTRACT

Two-electron transitions are studied in three crossed-beams experiments with H^- ions colliding with ions H^+, Ar^{q+} (q $\leq$ 8), and H^- at keV energies. Absolute cross sections have been measured for transfer ionization, two-electron removal and mutual ionization, respectively. Calculations based on (rather) simple models are in satisfactory agreement with the experimental data.

1. Introduction

Charge-changing collisions involving negative ions have been the subject of intense research activities during the past decades. However, in virtually all atomic collision experiments neutral atoms - either free or bound in solids - have been used as targets for incident negative projectile ions.[1]

With the advent of experimental set-ups employing crossed-beams or merged-beams techniques, investigations of charge-changing collisions between negative and singly-charged positive ion became available.[2] In recent years, this field of research has extended to cover collisions of negative ions with multiply-charged positive ions.[3] Only recently, the very first collision experiment with two negative ions has been performed.[4]

The principal difficulties of ion-ion crossed-beams experiments arise from the tenuous thickness provided by the ionic target. As a result, comparatively low ion-ion count rates are obtained which are completely masked by background counts due to ion interactions with residual gas particles. Especially for negative ions, which are easily stripped, background events dominate by a few orders of magnitude. Therefore, efficient techniques for signal recovery have to be employed in the measurements.

The present paper reports on three crossed-beams experiments with H^- ions. In collisions with the ions H^+, Ar^{q+} (q $\leq$ 8), and H^-, absolute cross sections have been measured in the keV energy range for transfer ionization, two-electron removal and mutual ionization, respectively. All processes studied involve two-electron transitions in rather simple atomic collisions systems.

2. Experimental technique

The experimental data discussed here have been obtained by means of the intersecting-beams technique. The general principles of this method have been described in detail previously [5] and only an outline of the experimental approach will be given here.

The method, in principle, appears to be straightforward: Two well-collimated ion beams of variable energies are arranged to intersect at some angle in an ultra-high vacuum region. The collision products formed in both beams are separated downstream from the interaction region from their parent ion beams and analyzed with respect to their charge states.

Experimentally, inherent difficulties however arise from the tenuous densities of the ion beams which have to be limited in order to avoid space-charge effects. Even at ultra-high vacuum (10^{-10} mbar) in the interaction region the residual gas density exceeds the ion densities within the beams by typically one or two orders of magnitude. These conditions not only result in comparatively low signal count rates (typically between $10^{-2}s^{-1}$ and $10^{2}s^{-1}$), but also in poor signal-to-background ratios (typically 10^{-2} to 10^{-4}). Therefore, signal-recovery methods have to be employed in order to separate required signal from background events which arise from the interaction of both beams with residual gas particles. Especially in the present experiments with negative ions, the loosely bound outer electron of the H^- ion is easily subject to stripping processes which result in high background count rates even at a very low vacuum pressure. Therefore, the present measurements were performed in an ultra-high vacuum of typically a few 10^{-11} mbar with both beams "on".

Fig.1 illustrates the essential features of the crossed-beams apparatus used in Giessen. A "slow" beam of ions Ar^{q+}, H^+ or H^- (energies: q (5-15)keV; q $\leq$ 8; ion currents: (2-200)nA) is arranged to intersect a "fast" H^- ion beam ((5-300) keV; (5-100)nA) at an angle $\theta = 45\,^\circ$. In order to reduce background, both ion beams (which are collimated to about 2 mm x 2 mm) are cleaned, shortly before intersection, by electrostatic deflectors E1, E2 from particles in other charge states which result from charge-changing collisions in the residual gas. The charge states of the collision products formed in both beams are analyzed a short distance down-stream from the interaction region by deflectors E3 and E4. The parent ion beams are recorded by biased Faradaycups F1 and F2, whilst the reaction products are counted by channeltron-based single-particle detectors D1 and D2. In order to avoid degradation of the ultra-high vacuum in the intersection region, the beam dumps F1, F2 are installed in a separately pumped vacuum region. Details of the experimental arrangement[6] and the signal-recovery technique[6,7] used have been described previously.

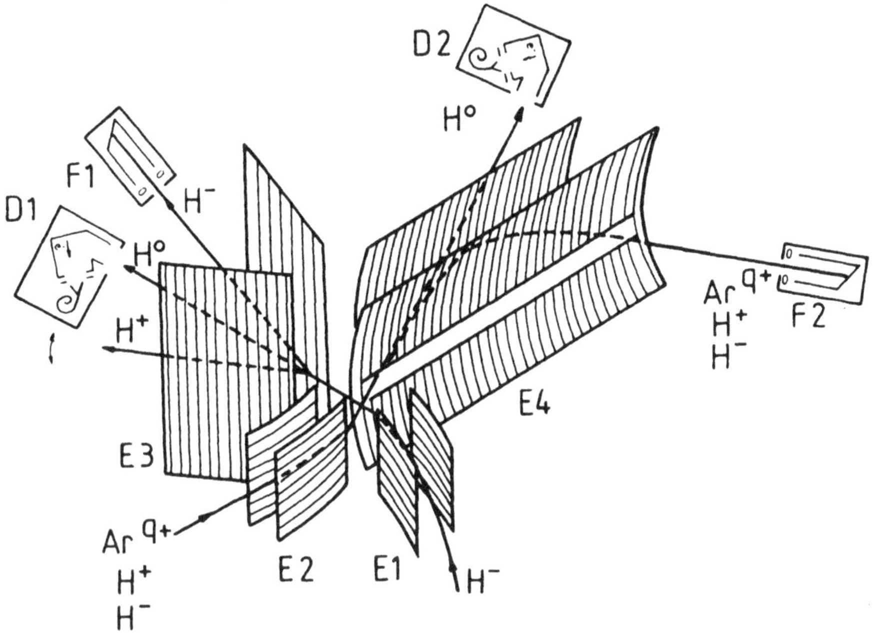

Fig.1: Schematic diagram of the crossed-beams arrangement used in Giessen

3. Transfer ionization in $H^- + H^+$ collisions [8]

Transfer ionization is a two-electron process in which one of the target electrons is captured into the projectile and the other is ejected. In recent years, this process has been intensively studied in collisions between multiply-charged ions and atoms.[9] At least at low energy, its mechanisms are rather well understood in terms of different Auger relaxation channels of a highly-excited quasimolecular state.[10]

Collisions between H^- and H^+ ions provide the simplest system to investigate transfer ionization in ion-ion collisions. With only two electrons involved, the $H^- + H^+$ system is simple enough hopefully to be treated theoretically without major uncertainties regarding the electronic structure, so that attention can be focused on the dynamical aspects. On the other hand, it is complex enough to offer intricate rearrangement possibilities.

Fig.2 shows absolute cross sections for the transfer-ionization reaction

$$H_a^- + H_b^+ \longrightarrow H_a^+ + H_b^\circ + e \qquad (1)$$

at center-of-mass energies ranging from 50 eV up to 40 keV. The data from 880 eV to 40 keV has been measured in Giessen employing the crossed-beams method outlined above, whereas the cross sections below 700 eV have been measured by Prof.Brouillard's group in Louvain-la-Neuve with a merged-beams set-up. In both experiments[8] a coincidence technique in the detection of the reaction products H_a^+ and H_b° has been used for signal recovery.

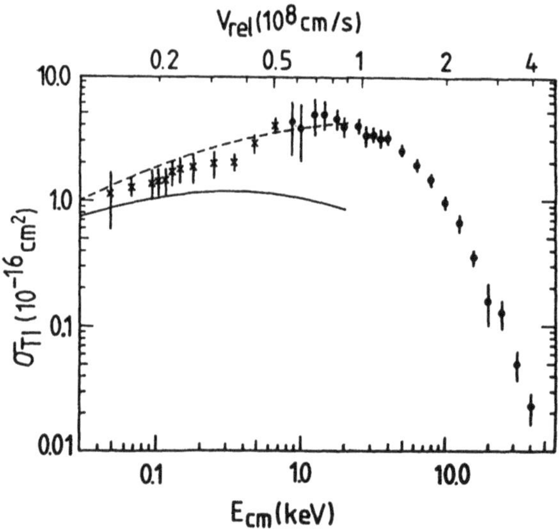

Fig.2: Cross sections for the transfer-ionization reaction
$H_a^- + H_b^+ \longrightarrow H_a^+ + H_b^\circ + e$
Experiment: • Giessen
x Louvain-la-Neuve
Theory: ——— f_ν^A calculated (see text)
- - - - f_ν^A set to 1/2

A full theoretical treatment of the transfer-ionization process is an extremely difficult task since many discrete states and a whole continuum are strongly coupled. However, in the low-energy range up to a few keV, the cross section can be estimated on the basis of a rather simple model. In that model[8], the transfer-ionization process in $H^- + H^+$ collisions is due to the evolution of the colliding system along several antibonding molecular states and its decay when these states enter the H_2^+ continuum. There are three such states: one *gerade* (state I, Fig.3a) and two *ungerade* (states II and III, Fig. 3b). Therefore, the first step in the calculation deals with the mutual neutralization into H(1s) + H(n=2) or H(1s) + H(n=3), at large internuclear distances. The next step is to calculate, for each of the states ν = I, II and III, the probability that it actually autoionizes in the collision. This probability is the product of two factors: the first one gives the survival probability across the Rydberg series below the continuum, and the second, f_ν^A, gives the probability of autoionization over the time spent in the continuum. The result of this model calculation (for details see Ref.8) is shown by the solid curve in Fig. 2. At low energy, there is satisfactory agreement with the experiment although the data are systematically underestimated. This might be an indication that some autoionization is produced by coupling to other doubly-excited configurations. An upper limit to the effect of such couplings is readily obtained by setting all f_ν^A equal to 1/2. This results in the dashed curve in Fig. 2. In summary, the low-energy behavior of transfer ionization in $H^- + H^+$ collisions can be rather well understood in terms of a two-step mechanism, where mutual neutralization is followed by autoionization.

At energies above 2 keV this model is not expected to be adequate since it neglects electron momentum translational effects. Furthermore, it also ignores the direct detachment of H^- which, at high energy, can supersede the mutual neutralization as the primary mechanism.

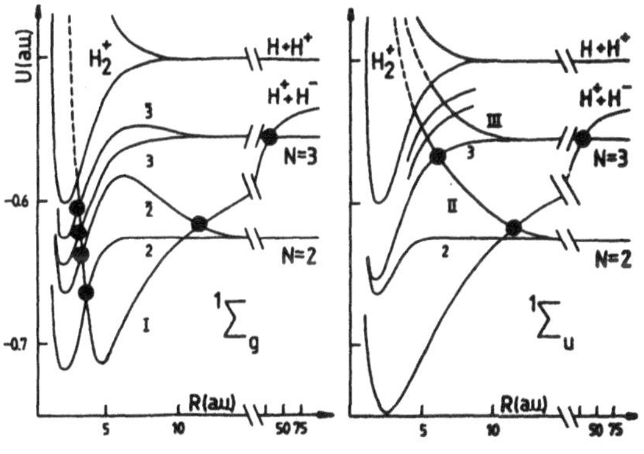

Fig. 3: Potential energy curves with associated crossings.[8]

4. Two-electron removal in $H^- + Ar^{q+} (q \leq 8)$ collisions

Transfer ionization is also one of the dominating processes contributing to the two-electron removal in H^- collisions with multiply-charged ions, apart from double ionization and two-electron capture processes.

Presently, there is much interest in cross sections for single- and double-electron removal from H^- in energetic collisions with positive ions. These cross sections are urgently needed for design and modelling studies for plasma neutralizers, which offer a considerably higher neutralization efficiency than gas targets (≈ 85 % vs 60 %) for conversion of energetic, multi-megawatt H^- beams into neutral H° beams needed for auxiliary heating of next generation fusion plasmas. Further, the use of a plasma neutralizer based on multiply-charged ions is expected[11] to benefit from a substantially reduced optimum line density due to the larger electron removal cross sections. Experimental data for the two-electron removal reaction is especially important since this channel directly degrades the efficiency of a plasma neutralizer.

In order to meet these data needs, absolute cross sections have been measured[12] for the single-electron removal

$$\sigma_{-\circ}: \qquad H^- + Ar^{q+} \longrightarrow H^\circ + \ldots \tag{2}$$

and the double-electron removal reactions

$$\sigma_{+\circ}: \qquad H^- + Ar^{q+} \longrightarrow H^+ + \ldots \tag{3}$$

A small 5 GHz-ECR ion source has been utilized for the production of intense and stable beams of argon ions in charge states $q \leq 8$. In the measurements of reactions (2) and (3) a beam modulation technique[7] had to be employed for signal recovery since the signal of 4 to 200 counts/s was masked by a background which was a factor 2000 to 50 higher, respectively. Typical measurement times were up to 4 hours for one cross section.

In Fig. 4 the cross sections $\sigma_{-\circ}$ and σ_{-+} for reactions (2) and (3), respectively, are displayed at a center-of-mass energy E_{cm} = 50 keV (v = 3.1 $\cdot 10^6$ m/s) and compared to three-body classical trajectory Monte-Carlo (CTMC) calculations [12] shown by solid lines. The dot-dashed line is the Bethe-Born calculation of Ref.13. The dashed line represents the electron removal cross section for neutral H° (Ref.14)

For the single-electron removal ($\sigma_{-\circ}$) , both the Bethe-Born and the CTMC methods reproduce the q = 1 experimental result. However, as the charge state of the ion increases, the q scaling predicted by first-order Born approximation becomes increasingly invalid even though the collision energy of 50 keV/u is well above the projectile-electron velocity-matching region of about 1.4 keV/u.

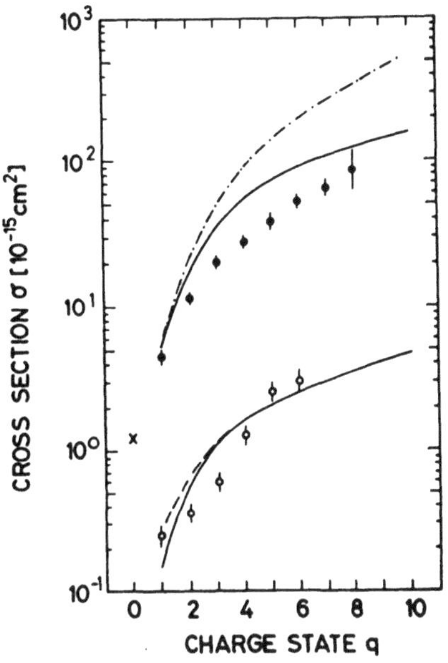

Fig.4: Experimental single (•) and double (∘) electron removal cross sections for $H^- + Ar^{q+}$ (q≤8) collisions at E_{cm} = 50 keV. An experimental data point (x) for single-electron removal in $H^- + Ar^\circ$ collisions (Ref.15) is shown for comparison. The solid lines are CTMC calculations for the respective processes. The dot- dashed line is the Bethe-Born calculation of Ref.13. The dashed line is the electron removal cross-section for neutral H° (Ref.14).

For the two-electron removal (σ_{-+}), there is qualitative agreement between experiment and the CTMC results. The calculations are based on a particularly simple model for the mechanism of two-electron removal. It is simply related to the product of the probability of removing the outer, loosely bound electron , P_o, times the probability of removing the more tightly bound inner electron, P_i :

$$P_{2i} = P_o \cdot P_i \tag{4}$$

The cross section is given by

$$\sigma_{-+} = 2\pi \int P_o \cdot P_i \, b \, db \, . \tag{5}$$

However, since the limits of integration of (5) are restricted to small impact parameters by P_i, where P_o is approximately constant (Fig. 5), we can rewrite (5) as

$$\sigma_{-+} \;=\; 2\pi P_{\circ} \int P_{i}\, b\, db\ . \tag{6}$$

One can immediately see that eq. (6) is simply

$$\sigma_{-+} \;=\; P_{\circ} \cdot \sigma_{\circ +}\ , \tag{7}$$

where $\sigma_{\circ +}$ is the well known single-electron removal cross section from neutral H° :

$$H^{\circ} + Ar^{q+} \longrightarrow H^{+} + \ \ldots \tag{8}$$

In Fig. 5 the transition probabilities calculated by the CTMC-method for electron removal (ionization + capture) for the loosely bound outer electron (dashed lines) and the tightly bound inner electron (solid lines) at E_{cm} = 50 keV for ion charge states q=1,2,5,10 versus impact parameter b are displayed. It is easily seen that the overlap in transition probabilities for removing both the inner and outer electron is concentrated at small impact parameters where the outer electron is ionized with an ≈ 100 % probability. This leads to the expectation that the two-electron removal cross section from H^{-}, σ_{-+}, is approximately equal to one-electron removal from neutral H°. The dashed line in Fig. 4 displays the parameterized cross sections $\sigma_{\circ +}$ for neutral H°.

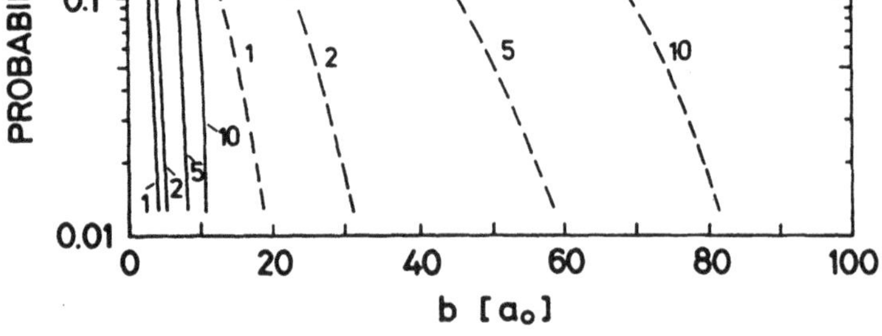

Fig. 5.: Transition probabilities calculated[12] by the CTMC-method for electron removal (ionization + capture) for the loosely bound outer electron (dashed lines) and the tightly bound inner electron (solid lines) at E_{cm} = 50 keV for ion charge states q = 1, 2, 5, 10 vs. impact parameter.

The CTMC calculations indicate that the two-electron removal cross sections shown in Fig. 4 are primarily composed of both a double ionization component

$$H^- + Ar^{q+} \longrightarrow H^+ + Ar^{q+} + 2e \qquad (9)$$

and a transfer-ionization component

$$H^- + Ar^{q+} \longrightarrow H^+ + Ar^{(q-1)+} + e\ . \qquad (10)$$

For singly-charged argon ions, the above two reactions contribute approximately equally to the two-electron removal. However, the transfer-ionization component becomes increasingly important with higher charge states and is calculated to be approximately 70 %, 90 % and 97 % of the two-electron removal cross section for charge states q=2, 5 and 10, respectively.

In summary, the two-electron removal from H^- in collisions with multiply-charged ions can be well understood in a simple model predicting cross sections which are approximately equal to one-electron removal from neutral $H°$, i.e.,
$\sigma_{-+}\ (H^-) \approx \sigma_{o+}\ (H°)$.

5. Mutual ionization in $H^- + H^-$ collisions

Experience from the above measurements and the performance of our crossed-beams arrangement suggested to tackle the very first collision experiment between two negative ions. Representing the most basic system in this category, we have chosen to investigate mutual ionization in $H^- + H^-$ collisions.[4]

Absolute cross sections for the reactions

$$H^- + H^- \longrightarrow H° + H° + 2e \qquad (11)$$

(double ionization, σ_{oo}) and

$$H^- + H^- \longrightarrow H^+ + H° + 3e \qquad (12)$$

(triple ionization, σ_{o+})
have been measured in the center-of-mass energy range (1.5 – 90) keV and (4 – 40) keV, respectively. For signal recovery, the reaction products $H°$ and $H°$ (or H^+) have been detected in coincidence. True signal count rates ranged between 0.05 s^{-1} and 3 s^{-1} and have been masked by background events which typically dominated by three to five orders of magnitude. Fig. 6 shows a coincidence spectrum from reaction (11).

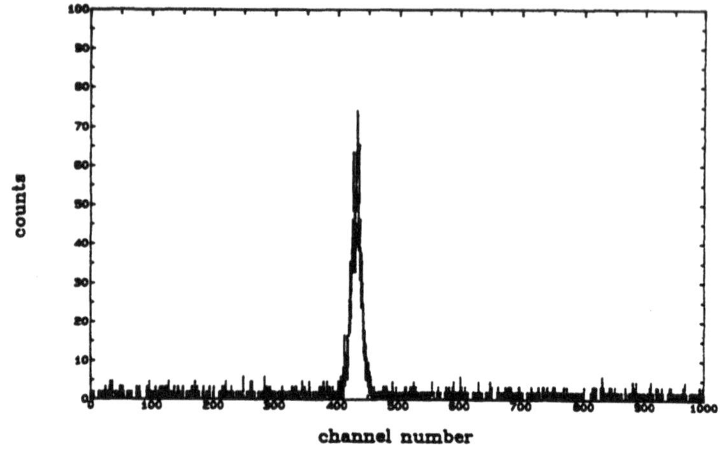

Fig. 6: Coincidence spectrum of H° atoms from the mutual ionization reaction $H^- + H^- \longrightarrow H^\circ + H^\circ + 2e$ measured [4] at $E_{cm} = 40.1$ keV. (One channel corresponds to a flight time difference of approximately 0.4 ns)

In Fig.7 the measured cross sections are compared with results of CTMC calculations[4] obtained with two different model potentials for the interaction between the inner electron and the H° core. In the calculations, the $H^- + H^-$ system has been modelled with a four-active-electron treatment in which the outer (inner) electron of the target interacts explicitly with the outer (inner) electron of the projectile through the Coulomb potential $1/r_{ij}$ as well as with the target and projectile cores (nuclei). The inner electron is assumed to interact with the unscreened nucleus of H^- through a Coulomb potential $-1/r$. For the interaction between the outer electron and the H° core (i.e. nucleus and inner electron) the following model potentials have been chosen:

$$V_{eff} = -\alpha/r \ ; \qquad \alpha = 0.2354 \tag{13}$$

$$V_{el} = -\tfrac{0.405}{r}[(1+\alpha r)e^{-2\alpha r} + (1+r)e^{-2r} + 0.94(3+\alpha) + e^{-(\alpha+1)r}] \tag{14}$$

The latter potential describes the electrostatic field associated with the one-electron charge distribution of H^-.

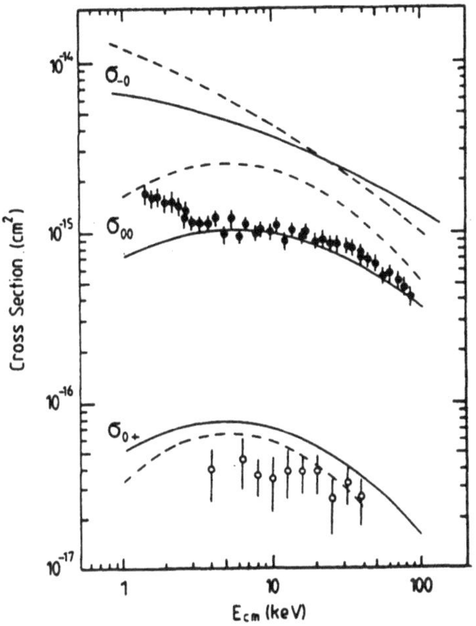

Fig.7 Cross sections for double ionization (σ_{oo}, full circles) $H^- + H^+ \longrightarrow H^\circ + H^\circ + 2e$ and for triple ionization (σ_{o+}, open circles) $H^- + H^- \longrightarrow H^\circ + H^+ + 3e$. The full and broken curves represent CMTC results obtained with the model potentials V_{el} and V_{eff}, respectively. Theoretical estimates obtained with these potentials for the single ionization (σ_{-o}) $H^- + H^- \longrightarrow H^- + H^\circ + e$ are also shown.

Comparing the experimental results with the CMTC calculations using V_{el}, both sets of data show nice agreement in the high energy range, whereas at low energies, the discrepancy amounts to a factor of two. The results obtained with V_{eff} are systematically above the ones obtained with V_{el}, in poorer agreement with experiment. According to both CMTC calculations, the cross section has a (flat) maximum at about 5 keV, a prediction that is not supported by the experiment. A possible reason for this discrepancy could be that the interchange of target and projectile electrons, which becomes increasingly important for decreasing impact energies, is not modelled accurately in the theoretical treatment.

The CTMC calculations for triple ionization (reaction 12) systematically overestimate the measured cross sections. Although the relative error is comparatively large because of poor counting statistics, this cross section for simultaneous production of an H° atom and an H^{+} ion seems to be almost constant in the energy range investigated and its magnitude is about a factor of 25 smaller than the cross section for double ionization.

Note that the reaction products, H° and H^{+}, are detected only in the "slow" and "fast" H^{-} beam line, respectively. Since an identical rate of reaction products is present (although not detected) in the other beam branch, the measured cross sections have been multiplied by a factor of two in order to compare them with the theoretical results for triple ionization.

Additionally, calculated cross sections for single ionization,$\sigma_{-\circ}$,

$$H^{-} + H^{-} \longrightarrow H^{-} + H^{\circ} + e \tag{15}$$

are shown in Fig. 7. These values await rigorous experimental test, and should be considered to have an accuracy of no better than a factor of two or three. The cross sections $\sigma_{-\circ}$ are needed, e.g., to determine confinement times for intense H^{-} beams in storage rings.

In summery, for the first time charge-changing reactions in collisions between two negative ions have been investigated experimentally by means of the crossed-beams technique. The measured absolute cross sections for double and triple ionization in $H^{-} + H^{-}$ collisions are partly well reproduced by CTMC calculations.

6. Conclusions

The three experiments discussed involve very simple atomic collision systems, which partly represent the most basic systems in their respective categories. However, in none of the cases a rigorous theoretical treatment is available, which is inherently intricate due to the two-electron transitions involved in the reaction processes. Nevertheless, it could be shown that calculations based on (rather) simple models are found to be in satisfactory agreement with the experimental data.

This work has been supported by BMFT under contract No 06GI307. The author wishes to acknowledge his current Giessen coworkers, especially Dr.Frank Melchert, as well as the collaboration with Prof. Brouillard's group at Louvain-la-Neuve (Belgium) and Prof. Olson's group at Rolla (USA).

References

1. see e.g., J.S.Risley, in *Electronic and Atomic Collisions* , eds. N.Oda and K.Takayanagi (North Holland, Amsterdam 1980) p. 619
2. see e.g., S.Szücs, M.Karemera, M.Terao and F.Brouillard, J.Phys.B **17** (1984) 1613; B. Peart and S.J.Foster, J.Phys.B **20** (1987) L 691
3. see e.g., M.Terao, S.Szücs, M.Cherkani, F.Brouillard and J.R.Allan, Europhys. Lett. **1** (1986) 123; B.Peart and M.A.Bennett, J.Phys. B **19** (1986) L 321; B.Peart and P.M.Wilkins, J.Phys. B **19** (1986) L 515
4. R.Schulze, F.Melchert, M.Hagmann, S.Krüdener, J.Krüger, E.Salzborn, C.O.Reinhold and R.E.Olsen, J.Phys. B **24** (1991) L 7
5. K.T.Dolder, in *Case Studies in Atomic Collision Physics*, Vol 1, eds. E.W.Mc Daniel and M.R.C. Mc Dowell (North Holland, Amsterdam 1969) p. 249
6. K.Rinn, F.Melchert and E.Salzborn, J.Phys.B **18** (1985) 3783
7. K.Rinn, F.Melchert, K.Rink and E.Salzborn, J.Phys. B **19** (1986) 3717
8. W.Schön, S.Krüdener, F.Melchert, K.Rinn, M.Wagner, E.Salzborn, M.Karemera, S.Szüs, D.Fussen, J.Janev, X.Urbain and F.Brouillard, Phys.Rev.Lett. **59** (1987) 1565
9. see e.g., E.Salzborn and A.Müller in *Atomic Processes in Electron-Ion and Ion-Ion Collisions* , ed. F.Brouillard, NATO Advanced Study Institutes Series B: Physics, Vol. 145 (Plenum Press, New York 1986) p. 357
10. R.K.Janev, L.P.Presnyakov and V.P.Shevelko, *Physics of Highly-Charged Ions* (Springer Verlag, Berlin 1985)
11. A.S.Schlachter, K.M.Leung, J.W.Stearns, R.E.Olson, in *Production and Neutralization of Negative Ions and Beams* , ed. J.G.Alessi, AIP Conf. Proc. No. 158 (1987) p. 631
12. F.Melchert, W.Debus, M.Liehr, R.E.Olson, and E.Salzborn, Europhys.Lett. **9** (1989) 433
13. Y.-K.Kim and M.Inokuti, Phys.Rev. A **3** (1971) 665
14. R.E.Olson, K.H.Berkner, W.G.Graham, R.V.Pyle, A.S.Schlachter, J.W.Stearns, Phys.Rev.Lett. **41** (1978) 163
15. C.J.Anderson, R.J.Girnius, A.M.Howald and L.W.Anderson, Phys.Rev. A **22** (1980) 822

Coulomb Focusing and "Path" Interference of Autoionizing Electrons Produced in 10 keV He^+ + He Collisions

J. K. Swenson
Lawrence Livermore National Laboratory, Livermore, CA 94550 USA

J. Burgdörfer
University of Tennessee, Knoxville TN 37996 USA

F. W. Meyer, C. C. Havener, D. C. Gregory, and N Stolterfoht‡
Oak Ridge National Laboratory, Oak Ridge, TN 37831-6372 USA

ABSTRACT

Autoionizing electrons emitted following low energy ion-atom collisions may scatter significantly from the receding spectator ion's attractive Coulomb field. In such cases the observed electron intensity is "focused" in the direction of the scattering ion as a result of the effective compression of the emission solid angle. In addition, interference may occur between trajectories, correspondiong to electrons scattering around opposite sides of the ion, which lead to the same final laboratory electron energy and emission angle. This Coulomb "path" interference mechanism manifests itself in the uncharacteristically rapid angular dependence of the He target $2s^2\,{}^1S$ autoionizing state measured near 0° following low energy He^+ + He collisions. A classical trajectory model for Coulomb focusing is presented and a semi-classical approximation is used to model the Coulomb "path" interference mechanism. In this description we account for the evolution of the phase of the autoionizing state until its decay and the *path dependence* of the amplitude of the emitted electron following decay of the autoionizing state. Calculated model lineshapes, which include contributions from adjacent overlapping resonances, reproduce quite well the angular dependence observed in the data near 0°.

1. Introduction

In low energy ion-atom collisions, post-collision interaction (PCI) of ejected autoionization electrons with the attractive Coulomb field of the slowly receding spectator ion leads to spectral lineshapes which are broadened and shifted towards lower electron energies. In the classical description of this effect, as first proposed by Barker and Berry[1], the spectral intensity as a function of electron energy may be written

$$I_{B.B.}(\Delta E) = \frac{\lambda_n}{\Delta E^2}\exp\left(\frac{-\lambda_n}{\Delta E}\right) \tag{1}$$

where the parameter $\Delta E = \varepsilon_n - E = q/R$ is the shift in the electron's kinetic energy as measured in the laboratory frame which results when an autoionizing state with a resonance energy ε_n decays at a distance R from the spectator ion of charge q. The term $\lambda_n = q\Gamma_n/v_p$, is a measure of the strength of PCI on a given state with decay width Γ_n, where v_p is the velocity of the ion. Eq. 1 represents a asymmetrically broadened lineshape with a shifted peak intensity $E_{PEAK} = \varepsilon_n - \lambda_n/2$. In this model the ejected electron velocity u_n is much greater than that of the receding ion, $u_n >> v_p$, and the electrons are assumed to follow straight line trajectories as they emerge from a time-independent potential. Therefore, in this basic description, only the laboratory energy of the detected electron is affected by the presence of the ion.

For the more general case where the electron's velocity is on the order of the ion velocity, $u_n \approx v_p$, the previous approximation is not valid. Depending upon the direction of emission and the proximity of the spectator ion, significant deflection of the electron's classical trajectory may result as it emerges from the interaction region. Such an influence is illustrated in Fig. 1. In low energy collisions, an autoionizing state with resonance energy ε_n (He^{**}) decays exponentially with a lifetime $\tau_n = 1/\Gamma_n$, emitting an electron with a velocity $u_n = \sqrt{2\varepsilon_n}$ (atomic units) at a time t, when the collision partners are separated by a distance $R(t) = v_p t + \delta$. The "unknown" quantity δ is the separation at t = 0 at which the autoionizing state is populated. The parameters u ,υ and ϑ, θ are the electron velocity and emission angle before and after scattering, respectively.

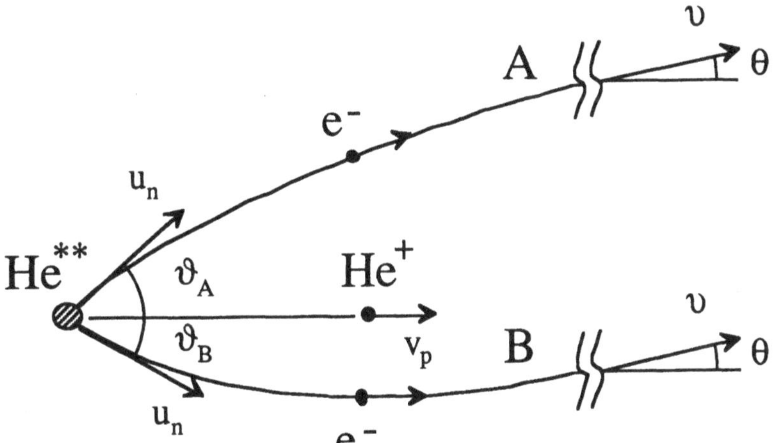

Figure 1. Effect of spectator ion's Coulomb field on the classical trajectories of ejected autoionizing electrons following low energy He^+ + He collisions. Paths A and B represent scattering around the near and far side of the ion, respectively such that the electron emerges with the same final energy $E = \upsilon^2/2$ and emission angle θ.

In this picture the electron interacts with a *time-dependent* Coulomb field. The laboratory-frame energy shift experienced by the electron is now $\Delta E = q/R + \vec{v}_p \cdot \Delta \vec{u}$, where $\Delta \vec{u} = \vec{u}_n - \vec{\upsilon}$ is the total change in the electron's velocity due to the Coulomb deflection. The shift now depends upon the ion velocity and the angle through which the electron is scattered[2]. For velocity changes $\Delta \vec{u}$ in the direction of the moving ion (i.e. for forward electron emission) the shift towards lower electron energy is increased since the effect of the ion motion is to extend the time the electron spends under the influence of the Coulomb field. For electron emission away from that of the ion the magnitude of the energy shift is decreased due to the reduced interaction time. For electron velocities less than that of the ion the electron's energy may actually be shifted towards higher energies since the ion leads the electron, "dragging" it forward.

In addition to the angle-dependent energy shift, there are other notable aspects of Coulomb deflection which manifest themselves in the spectra of autoionizing electrons emitted in low energy ion-atom collisions. The most apparent of these is the redistribution of the ejected electron's intensity which results in a strong enhancement or "focusing" of intensity in the direction of the spectator ion. This effect was shown to be important in the autoionizing spectra produced in low energy He^+ + He collisions measured near the collision axis[3]. A further interesting aspect arises if we consider that it is possible for the electron to scatter around either "side" of the projectile-ion and emerge with a given energy $E = \frac{1}{2}\upsilon^2$ and angle θ (see Fig. 1). Electrons following these trajectories are *indistinguishable* and therefore, the spectral intensity is computed as a coherent superposition of the amplitude for scattering along each possible path which leads to the same final state. For each laboratory energy E in the continuous spectrum, there are, in general, two indistinguishable paths which lead to a given final state. Interference between these "paths" has been recently shown to produce anomalous structure in the post-collision broadened autoionizing lineshape which has an uncharacteristically strong dependence upon emission angle near zero degrees[4].

In this paper we will present the classical trajectory model for Coulomb focusing and the semi-classical approximation used to model the Coulomb "path" interference mechanism. We use as a basis for the semi-classical approximation, Morgenstern's PCI-induced "overlapping" resonances model, which was first used to explain the complicated interference structure observed in quasimolecular electron emission in 1400 eV He+ + He collisions[5]. We make a simple extension of this model by accounting for the deflection of the electron in the Coulomb field and the path dependence of the amplitude of the emitted electron following decay of the autoionizing state. Calculated model lineshapes, which are coherent sums of contributions from adjacent PCI-broadened "overlapping" resonances for both near-side and far-side scattered trajectories, are found to reproduce the angular dependence observed for the He target $2s^2\,{}^1S$ state produced in 10 keV He^+ + He collisions near 0°. Finally we calculate the lineshape for the *isolated* $2s^2\,{}^1S$ state, in the absence of contributions from adjacent PCI-broadened states. We observe a similar interference structure which is solely attributed to the path interference mechanism.

2. Experimental He^+ + He Autoionization Spectra

The dominant mechanism through which autoionizing states are formed in low energy collisions ($v_p < 0.5$ a.u.) is electron promotion. For symmetric systems such as He^+ + He, this mechanism populates singlet autoionizing states with equal probability in both target and projectile[6]. Indeed, Bordenave-Montesquieu et al.[7] have shown, for collision energies ranging from 10 to 30 keV and observation angles from 20° to 160°, that electron emission from target and projectile autoionizing states is characterized by identical differential cross sections which are symmetric about 90° in the center-of-mass frame. In contrast, we have measured, at 0° and 180° emission angles, a strong enhancement in the intensity of autoionizing electrons produced in low energy (2-15 keV) He^+ + He collisions which we have attributed to deflection and *focusing* of the ejected electrons in the Coulomb field of the charged collision partner[3]. More recently, interference effects resulting from the diffraction of the electron by the Coulomb center have also been reported[4].

The experiments were performed at the Oak Ridge National Laboratory ECR ion-source facility using a target chamber and electron spectrometer temporarily transported from the Hahn-Meitner Institut, Berlin. Details of the experimental technique have been described previously[4,8]. Fig. 2 shows the background subtracted He target autoionizing spectra, acquired in non-coincidence or "singles" mode, measured at 0°, 5°, and 10° relative to the incident beam direction for 10 keV He^+ + He collisions. The spectrum was also measured at 30° and is similar in character to the 10° spectrum but is not shown here.

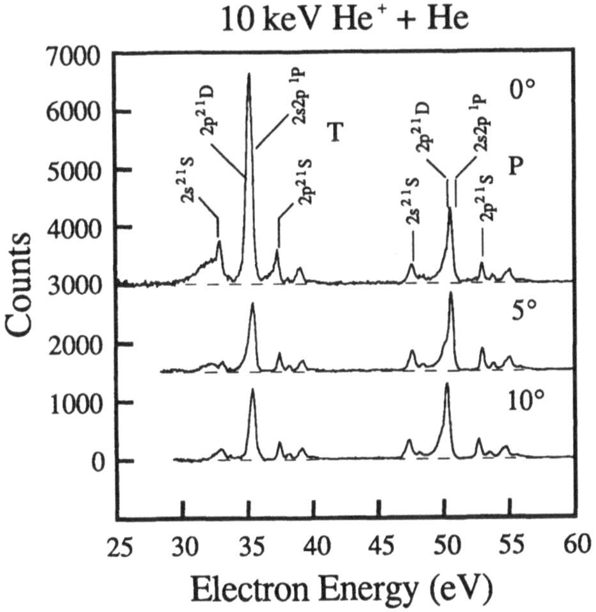

Figure 2. Experimental autoionizing electron spectra from 10 keV He^+ + He collisions, measured at 0°, 5°, and 10° with respect to the collision axis. Dominant target autoionizing transitions (T) are identified and are to be compared with the corresponding transitions in the projectile (P). The strong enhancement due to Coulomb focusing is evident in the 0° spectrum

The autoionizing states which are predominantly populated in both target (T) and projectile (P) in these low energy collisions are the $2s^2\,{}^1S$, $2p^2\,{}^1D$, $2s2p\,{}^1P$, the $2p^2\,{}^1S$ transitions. The effect of Coulomb focusing as seen in Fig. 2, represents a factor of ~4 enhancement in intensity of the He target autoionizing transitions compared to the projectile transitions at 0°. This enhancement is dependent upon many factors including collision energy, electron energy, and lifetime of the state in question. For the fastest decaying transition, the $2s^2\,{}^1S$ state, which exhibits the most pronounced PCI broadening, the target to projectile intensity ratio is ~8. This strong enhancement disappears for angles larger than about 5° where our results are in agreement with the electron promotion model and earlier measurements. At lowest collision energies where the enhancement is expected to be the greatest, the angular scattering of the heavy particles, which is ~10 keV•deg (1° @10 keV) for He^+ + He collisions which populate these doubly-excited states[9], washes out the Coulomb focusing effect[3].

Upon closer examination of the He target spectra, one notices a pronounced shoulder in the $2s^2\,{}^1S$ lineshape which becomes equal in intensity to the main part of the peak at 5°, and has essentially disappeared in the 10° spectrum. This uncharacteristically rapid angular variation in the ${}^1S$ lineshape is direct evidence for Coulomb "path" interference[4]. This mechanism is also active in the other nearby coherently excited $2p^2\,{}^1D$, and $2s2p\,{}^1P$ transitions. These two states are separated by ~0.25 eV and are unresolved in the spectra of Fig. 2. Post-collisional broadening of the spectral lineshapes cause all three resonances to overlap (see Fig. 3). The resulting composite lineshape is therefore calculated as a coherent sum of amplitudes for emission from each of these transitions[5]. In the low energy portion of the PCI-broadened ${}^1S$ state, constructive interference makes the angular variation in the lineshape readily observable. However, due to the destructive interference which exists in the low energy portion of the ${}^1D + {}^1P$ composite lineshape, the observed intensity is reduced, thus making any variations in this spectral feature more difficult to observe.

3. Time-Dependent Semi-Classical Model for PCI

3a. Coulomb Focused Autoionizing Lineshape

For target electron emission, the electrons are emitted and detected in the laboratory frame. However, the Coulomb deflection occurs in the moving projectile frame. We therefore write the doubly-differential electron emission cross section in the laboratory frame as a time integral over of the exponentially decaying autoionizing state of width Γ_n and energy ε_n, approximated by a delta function natural lineshape[7], as

$$\frac{d^2\sigma}{d\Omega dE} = \frac{d\sigma}{d\Delta}\int_0^\infty \Gamma_n \exp[-\Gamma_n t]\,\delta(\varepsilon-\varepsilon_n)\left(\frac{\varepsilon' E}{\varepsilon E'}\right)^{1/2}\frac{d\varepsilon'}{dE'}\frac{d\Delta'}{d\Omega'}dt \tag{1}$$

where we use a prime (') to indicate those parameters evaluated in the moving reference frame of the ion. The parameters, Δ and Ω, are the laboratory-frame emission solid angles before and after scattering respectively. The terms in the square-root are the solid angle transformations to and from the projectile frame[10]. We now make a change integration variable from *dt* to *dε* utilizing the projectile-frame energy conservation relation $\Delta E' = \varepsilon' - E' = q/v_p t$, where, for simplicity, we have neglected the small distance δ. We have,

$$dt = \frac{q}{v_p\,\Delta E'^2}\,\frac{dE'}{d\varepsilon'}\,\frac{d\varepsilon'}{d\varepsilon}\,d\varepsilon \tag{2a}$$

where the term,

$$\frac{d\varepsilon'}{d\varepsilon} = \frac{\varepsilon + \varepsilon' - \varepsilon_p}{2\varepsilon} \tag{2b}$$

derives from the kinematic relationships between the electron energy in the lab and projectile frames as a function of emission angle where the parameter $\varepsilon_p = \frac{1}{2}v_p^2$ is the "reduced" ion energy[10]. Integration over the delta function now picks out the resonance energy $\varepsilon = \varepsilon_n$ in each of the functions in the integrand and we are arrive at the target autoionizing lineshape,

$$\frac{d^2\sigma}{d\Omega dE} = \frac{d\sigma}{d\Delta}\,\frac{\lambda_n}{\Delta E'^2}\exp\left(\frac{-\lambda_n}{\Delta E'}\right)\left(\frac{\varepsilon_n + \varepsilon_n' - \varepsilon_p}{2\varepsilon_n}\right)\left(\frac{\varepsilon_n' E}{\varepsilon_n E'}\right)^{1/2}\frac{d\Delta'}{d\Omega'}\ . \tag{3}$$

This result consists of a Barker-Berry lineshape modified by kinematuc transformation terms and a term, $d\Delta'/d\Omega' = d\cos\vartheta'/d\cos\theta'$, which describes the scattering of the ejected electron in the ion-frame. In this way, the scattering once again, takes place in a time-*independent* Coulomb field. The kinematic terms are typically $\lesssim 1$ for low energy collisions and approaches unity as the projectile velocity goes to zero. The main modification to the PCI lineshape, therefore, derives from the scattering function which has its greatest effect for target electron emission in the forward direction near $\theta' = 0°$ where this function is singular. A similar relationship may be derived for projectile electron emission where the focusing is found to be peaked at 180° in the direction of the ionized target atom.

The explicit form for the scattering function depends upon the physical approximations used to model the post-collision deflection of the ejected electrons. In our time-dependent description, the electrons are deflected in the single-center Coulomb field of the receding spectator ion and therefore follow classical hyperbolic trajectories[11]. In this simplified two-body "half collision" the relationship between the angle of emission before and after scattering is written as a function of the dimensionless parameter $K' \equiv \Delta E'/\varepsilon_n'$, as

$$K' = \frac{2\sin\vartheta'\sin(\vartheta' \mp \theta')}{1 + \cos\theta'}\ , \tag{4}$$

where the $-\{+\}$ sign corresponds to scattering around the near{far}-side of the ion. For electron emission into $\theta' = 0°$, the above relation simplifies to $K' = \sin^2\vartheta'$. Differentiating

Eq. 4, remembering that the electron energy ε_n' is a function of the projectile frame emission angle ϑ', the scattering function takes the form,

$$\frac{d\Delta'}{d\Omega'} = \frac{-\frac{K'}{2} \pm \frac{\sin\vartheta'\cos(\vartheta' \mp \theta')}{\sin\theta'}}{\frac{K'\,(1+\cos\theta')}{\sqrt{\varepsilon_n'/\varepsilon_p + \cos\vartheta'}} + \frac{\sin(2\vartheta' \mp \theta')}{\sin\vartheta'}} \tag{5}$$

where the upper{lower} sign again corresponds to the near{far}-side scattering. We see that as $\theta' \to 0$, $d\Delta'/d\Omega' \to 1/\sin\theta'$. Therefore, a singularity exists in the classical differential cross section at $\theta' = 0$. However, when the differential cross section is integrated over the acceptance solid angle of the electron spectrometer the observed cross section is finite. In addition, other effects tend to wash out the observed intensity enhancement. These include the inherent divergence present in the ion beam and the heavy particle scattering which distributes the orientation of the internuclear axis along which the enhancement occurs.

Using the above results, we may now calculate the Coulomb focused lineshape for the $2s^2\,{}^1S$, $2p^2\,{}^1D$ and $2s2p\,{}^1P$ He autoionizing states, with resonance energies $\varepsilon_n = 33.28$ eV, 35.29 eV, 35.54 eV, and widths $\Gamma_n = 0.138$ eV, 0.072 eV, 0.038 eV respectively[12]. The $2p^2\,{}^1S$ state, which is not included in this calculation, is much longer lived and exhibits no significant PCI broadening or enhancement due to Coulomb focusing. Figure 3 shows the Coulomb focused lineshapes for the three states (solid curves) calculated for the *near-side* scattering (A) evaluated for an emission angle of $\theta = 1°$. We assume equal populations for each state and an isotropic initial angular distribution. In this way the resulting lineshapes reflect purely the effect of the redistribution of intensity caused by the Coulomb deflection. We find that trajectories which make a significant contribution to the Coulomb focused lineshape originate from pre-scattered emission angles ranging up to $\vartheta \approx 30°$. The dot-dashed curve in the $2s^2\,{}^1S$ lineshape is the calculation for *far-side* scattering (B). Here the calculated intensity from the far-side scattered trajectories is suppressed by nearly 30%, owing to the fact that the probability for scattering around the far-side of the ion is decreasing rapidly with emission angle. The dashed curves represents the Barker-Berry lineshapes given in Eq. 1. From this figure it can be seen that the largest enhancement is found for $2s^2\,{}^1S$ state which is the shortest lived of the three transitions and which exhibits the most significant PCI-broadening. In addition to the overall enhancement in intensity observed very near 0° we find an increased broadening the spectral lineshape. On the other hand, for observation angles out beyond 10° where the Coulomb focusing enhancement factor become less than unity (see Fig. 4), broadening in the spectral lineshape is actually reduced. Examining the detailed lineshaps here would provide an interesting comparison with this model.

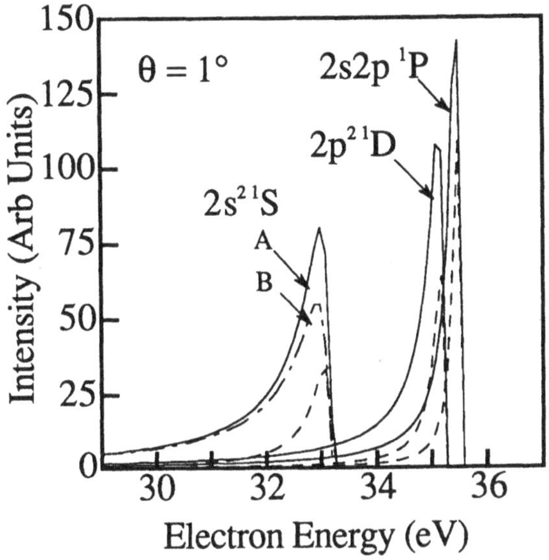

Figure 3. Calculated lineshapes for three He autoionizing transitions evaluated at θ = 1° for 10 keV collision energy. The *solid* curves are the Coulomb focused lineshapes given by Eq. 3 for near-side scattering (path A), The *dot-dashed* curve is the far-side (path B) lineshape and is shown only for the ^1S state. The *dashed* curves are the Barker-Berry lineshapes evaluated the absence of Coulomb focusing.

We now calculate the angular dependence of the intensity of target autoionizing electrons, emitted from the $2s^2\ ^1S$ autoionizing state in He for 10 keV collisions. The intensity is found by numerically integrating the Coulomb focused lineshape, as calculated above, adding the results for near-side and far-side scattering to obtain the total intensity at each observation angle. A polar plot of the $\log_{10}$ of the total integrated intensity, $\log(d\sigma_\epsilon/d\Omega)$, is shown as the solid curve in Figure 4. The most notable feature of the "tear-drop" shaped distribution is the strong enhancement in the direction of the projectile ion (**P**) at $\theta = 0°$. The dot-dashed curve represents the integrated lineshape where no deflection or focusing takes place, obtained by setting the scattering function $d\Delta'/d\Omega' = 1$. Comparing with the essentially isotropic non-focused result, we see that there is a corresponding *de-enhancement* in the electron intensity, as indicated by a narrowing of the calculated lineshape, for angles greater than ~10°. The result of the Coulomb deflection is simply a redistribution of electron intensity where the total cross section for electron emission, in this classical approximation, is conserved.

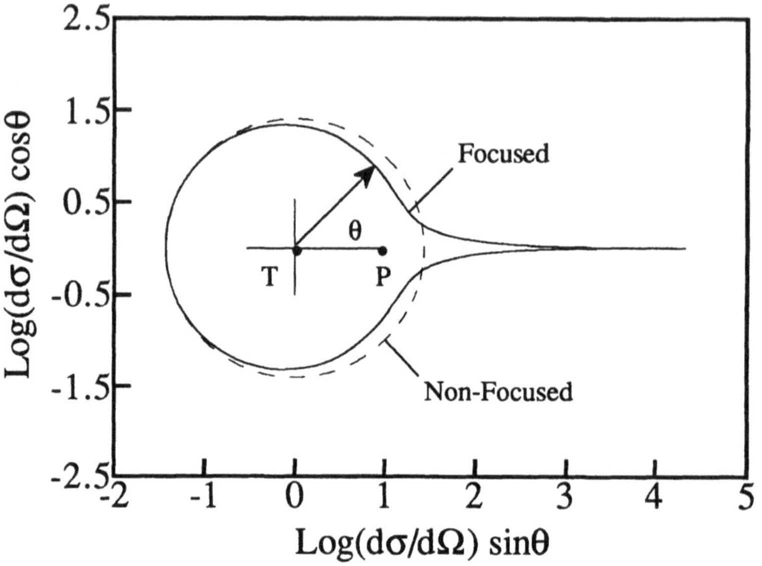

Figure 4. Polar plot of the logarithm of the integrated intensity as calculated from Eq. 3, i.e. Log(dσ/dΩ), as a function of emission angle θ for focused (solid) and non-focused results (see text).

3b. Model for Coulomb "Path" Interference

In the above analysis of He target autoionizing electrons emitted near the forward direction, we computed the angle-dependent electron intensity as a sum of the probabilities for scattering around the near and far-side of the ion. As we have pointed out, trajectories which lead to the same final electronic state are indistinguishable. Thus the intensity at a given energy and observation angle may be computed as a coherent sum of the amplitudes for emission along each of these paths[4]. In addition, the spectral intensity in the region of PCI-induced overlap between adjacent transitions can be expressed as a coherent sum of amplitudes for emission from the respective states[5]. Due to the restriction to the extreme forward angles, the enhancement due to Coulomb focusing is not significantly influenced by the weaker angular variation in intensity, resulting from constructive or destructive interference between adjacent overlapping resonances, which might be present. Nevertheless, to explain the detailed lineshape we observe for the He target $2s^2\ ^1S$ state and its unusually strong dependence upon observation angle near 0°, such interference effects must be properly included in the time-dependent PCI model.

Analysis of the measured lineshapes was performed using, as a starting point, the "overlapping" resonances model (OR) of Morgenstern et al.[5]. As we have shown, due to the induced PCI broadening of the spectral lineshapes, the $2s^2\ ^1S$, $2p^2\ ^1D$, and $2s2p\ ^1P$ states overlap. As a result of the indistinguishability of the detected electrons in the region of overlap, the composite lineshape is calculated as a coherent sum of amplitudes for emission from each of the three autoionizing resonances. Our extension of this semi-classical model derives from the observation that the OR model only accounts for the phase evolution of the autoionizing state from the time it crosses into the continuum to the time of its decay. The emitted electron is then assumed to "instantaneously" recede to infinity thus accumulating no additional phase. However, for He target autoionizing electrons produced in low energy He^+ + He collisions, where the electron velocity is "finite" and comparable to that of the projectile ion ($u_n/v_p \approx 5$) this approximation is no longer valid. A correct description of the near-forward autoionization spectrum must therefore include the scattering of the emitted electron by the spectator ion.

In our semi-classical approximation we write the angle-dependent spectral intensity which is a sum over the n autoionizing states (^1S, ^1D, and ^1P) each having two separate indistinguishable paths which contribute to the observed intensity at E and θ, as

$$I(E,\theta) = \left| \sum_{\text{paths}} \sum_{n} W_n(\vartheta,\varphi)\cdot C(E,\theta)\cdot \exp\left\{-i\left[\alpha(\Delta E') + \beta(p',l')\right]\right\} \right|^2 \quad (5)$$

where the effects of Coulomb scattering have now been properly included. This expression differs from the original result given by Morgenstern[5] in that the Barker-Berry amplitude has been replaced by the Coulomb focused amplitude, C(E,θ), which we take as the square root of the differential cross section given above in Eq. 3. In addition, we include the semiclassical scattering phases $\beta(p',l')$ and the sum over the paths (A and B) which contribute to the observed intensity. The phase accumulated by the autoionizing state prior to decay[5] $\alpha(\Delta E')$, is now evaluated in the ion frame, and is written

$$\alpha(\Delta E') = \frac{q}{v_p}\left[1 - \frac{\Delta E'}{q}\delta + \log\left(\frac{\Delta E'}{q}\right)\right]. \quad (6)$$

The term $W_n(\vartheta,\varphi)$ is the angle dependent autoionization transition amplitude, written as

$$W_n(\vartheta,\varphi) = \sum_{m_l} a_n^{m_l}(0)\cdot\omega_n^{m_l}(\vartheta,\varphi)\cdot\exp[-i\chi_n^{m_l}(0)] \quad (7)$$

which consists of contributions from different magnetic sublevels m_l, where $a_n^{m_l}(0)$ and $\chi_n^{m_l}(0)$ are the initial level populations and phases, and where (ϑ,φ) is the angular coordinate of the outgoing electron prior to Coulomb scattering. These initial amplitudes and phases are treated as free fitting parameters. We emphasize that by including the effects of Coulomb scattering of the ejected electron into this semi-classical model, we introduce no additional degrees of freedom which may be used to fit our experimental results.

The path dependent scattering phase $\beta(p',l')$ is derived from a classical action integral in the central field WKB approximation[13]. The accumulated phase from the classical turning point at $r = r_o$ to some distance r is given by

$$\beta(p,r,l) = \int_{r_o}^{r} p_r\,dr - r\cdot p_r + l\cdot\cos^{-1}\left(\frac{l}{p\cdot r}\right) \tag{8}$$

where $p_r = \sqrt{p^2 - l^2/r^2}$, is the radial component of the electron's linear momentum p, and $l = u_n R\sin\vartheta$ is its angular momentum. For convenience we have left off the primes on the various momenta and the radial coordinate r since these are *all* ion-frame quantities. Upon performing the integration we arrive at the result

$$\beta(p,r,l) = \frac{q}{p_o}\ln\left[2r\left(1+\frac{p_r}{p_o}\right)+\frac{2q}{p_o^2}\right] - l\cdot\sin^{-1}\left(\frac{q - l^2/r}{\sqrt{q^2 + l^2 p_o^2}}\right) + l\cdot\cos^{-1}\left(\frac{l}{p\cdot r}\right) \tag{9}$$

where $p_o = \upsilon$ is the electron's ion-frame linear momentum at $r = \infty$. The paths that the ejected electrons follows subsequent to emission are broken up into two parts which originate at the classical turning point of the trajectory. For the *outgoing* portion of the trajectory, the upper integration limit is chosen to be some large distance $r = r_{max}$, where the potential is negligible and any further accumulation of phase is path independent. For the *incoming* portion of the trajectory the upper limit is simply the internuclear separation, $R = R_A, R_B$, at the time of emission for paths A and B, respectively. The total scattering phase is then the sum of these two contributions as given by

$$\beta(p,l) = \beta(p,r,l)\Big|_{r_o}^{R_{AB}} + \beta(p,r,l)\Big|_{r_o}^{r_{max}}. \tag{10}$$

It is important to note that, although the total phase $\alpha(\Delta E') + \beta(p',l')$ is evaluated in the projectile frame, the phase differences which result from squaring the amplitudes in Eq. 5 are frame independent.

Figure 5 shows the results of the evaluation of the phase $\alpha(\Delta E')$ and the scattering phase $\beta(p',l')$ over the range of electron energies corresponding to the $2s^2\,{}^1S$ lineshape, for an emission angle $\theta = 5°$. A value of $r_{max} = 200$ a.u. was used in evaluating the scattering phase. We point out that choosing a different value for r_{max} will produce a different numerical result for β_A and β_B shown in the figure. However, the difference in phase which comes from evaluating the coherent lineshape, given by equation 5, is insensitive to the value chosen for r_{max} since out beyond ~100 a.u. the Coulomb field becomes weak and the scattering phase becomes path independent. For near-side scattering, β_A decreases monotonically as the electron energy $E \rightarrow \varepsilon_n$, the resonance energy for the ${}^1S$ state. While for far-side scattering, β_B diverges rapidly near the resonance due to the fact that the emission angle $\vartheta_B \rightarrow 0$ here, which corresponds to the electron passing through the origin at $r = 0$.

The rapidly changing phase in the vicinity of the resonance is the source of the strong angular variation in the interference structure observed in the $2s^2\,{}^1S$ lineshape near the forward direction. The slight differences in the magnitude of the phases α_A and α_B are due to the different emission angles ϑ_A and ϑ_B, thus, ion-frame electron energies which contribute to the spectral lineshape for the near and far-side scattered electrons.

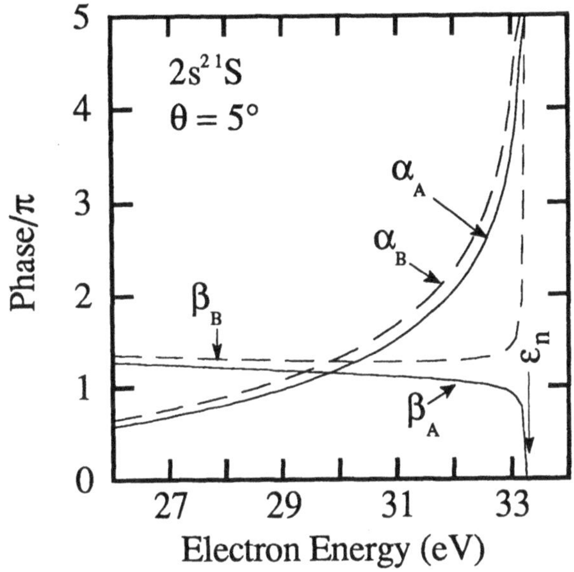

Figure 5. Calculated phases $\alpha(\Delta E')$ and $\beta(p,l)$ for $2s^2\,{}^1S$ state in He following 10 keV He^+ + He collisions. Solid(dashed) curve corresponds to near(far)-side scattered trajectories.

4. Angular Dependence of He Target $2s^2\,{}^1S$ Lineshape

Figure 6a shows the relevant portion of the He target autoionizing spectra of Figure 2 normalized to constant maximum intensity to remove the effects of Coulomb focusing and allow easy comparison of the three emission angles. Figure 6b shows the result of our calculations evaluated at 0.1°, 5°, and 10°, convoluted with a Gaussian spectrometer resolution function with a FWHM = 0.3 eV, also normalized to a constant maximum intensity at each angle. Calculations are done at 0.1° rather than at 0° due to the singularity which exists in the Coulomb focusing differential cross section there. For simplicity we assume each state consists of only a single magnetic sublevel which decays isotropically, as is the case for an ${}^1S$ state. We then have only four parameters, consisting of the initial ${}^1S$ and ${}^1D$ level populations and phase angles relative to the ${}^1P$ amplitude with which we fit the experimental spectrum at 0°. Holding their values fixed, we then simply vary the emission angle to obtain the calculated lineshapes at 5° and 10°. The solid curves in Figure 6b are the results of this fit ($\alpha_{1_S}(0) = 0.3$, $\alpha_{1_D}(0) = 0.4$, $\chi_{1_S}(0) = 0.7\pi$, and $\chi_{1_D}(0) = 1.0\pi$, $\alpha_{1_P}(0) = 1$, $\chi_{1_P}(0) = 0$), which include Coulomb path interference. The dashed curves are these same calculations, where we have now excluded the "far-side" or path B amplitude, thus suppressing the path interference mechanism. Although the path-dependent phase is still included in this "near-side only" calculation, the angular variation in the lineshape is essentially negligible over this range of emission angles. Clearly, only by including *both* "near-side" and "far-side" paths in the calculation do we reproduce the rapid angular dependence of the interference structure observed in the data. For comparison we show the lineshape obtained from the OR model alone which we obtain by suppressing both the path dependence of the amplitude and Coulomb focusing in Eq. 1. This result is shown in Figure 6b at 0°, as the dot-dashed curve. A comparison of the OR lineshape with the full Coulomb path interference calculation at 0° reveals that incorporating the Coulomb-focused amplitude eliminates the necessity to artificially decrease the lifetimes of the autoionizing states in order to increase the width of the lineshape thus improving the fit to the experimental spectra, as has been done in the past[5].

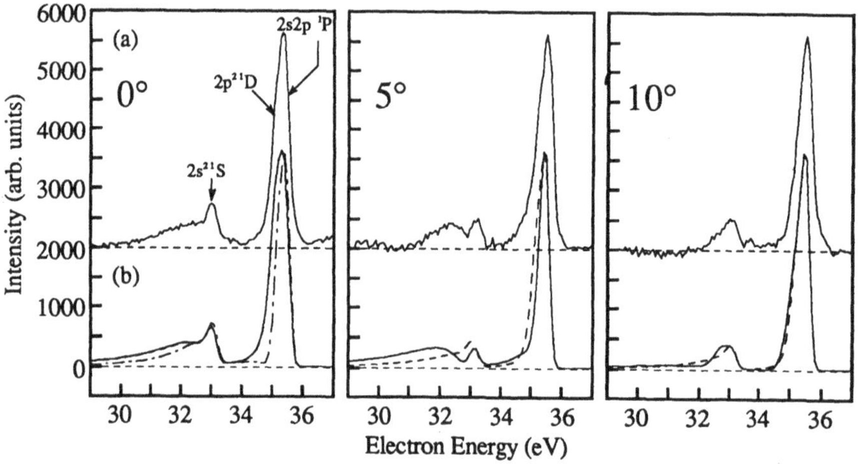

Figure 6. (a) Experimental He target autoionizing spectra for emission angles of 0°, 5°, and 10°.(b) Calculated lineshapes using Eq. 7 which includes "path" interference (solid) and where near-side (path A) *only* is accounted for (dashed). Dot-dashed curve in 0° calculation is best fit using the Morgenstern model alone which does not change significantly for the range of angles shown.

Finally, as an illustration, we calculate the path interference lineshape for the case of a single isolated resonance, keeping only the term in sum in Eq. 5 corresponding to the $2s^2$ ^1S state. The coherent lineshape is shown in Figure 7 for a range of emission angles near 0°. Here we have normalized the result at each angle to the maximum intensity in the incoherent lineshape to again remove the strong Coulomb focusing enhancement and to also show the change from constructive interference near 0° to destructive interference at larger emission angles which severely distorts the calculated lineshape from its monotonic incoherent form. As can be seen from the figure, we obtain an angular dependent lineshape similar to that which is observed in the previous calculation which includes contributions from the ^1P and ^1D states. We note that there is qualitative agreement with the calculated lineshapes of Barrachina and Macek[14] obtained using a CDW approximation to the final state in the autoionizing matrix. Observation of such isolated resonance lineshapes, which is not possible for the He^+ + He system, could provide a definitive test of our semi-classical model for post-collision interaction effects in low energy ion-atom collisions.

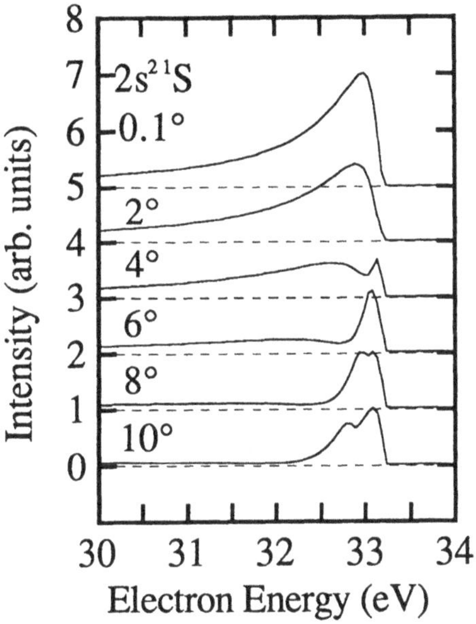

Figure 7. Isolated $2s^2\ ^1S$ lineshape calculated from Eq. 7 showing effect of Coulomb path interference in the absence of contributions from adjacent PCI-broadened states.

The authors would like to acknowledge J. Macek for illuminating discussions. This work was supported in part under the auspices of the U.S. Department of Energy by Lawrence Livermore National Laboratory under contract No. W-7405-Eng-48, the Office of Fusion Energy, under Contract No. DE-AC05-84OR21400 with Martin Marietta Energy Systems Inc, and the National Science Foundation. Support for NS was provided through Nato Collaboration Grant No. 0194/87.

‡Permanent Address: *Hahn-Meitner Institut für Kernforschung GmbH, Berlin, Germany.*

REFERENCES

1. R. B. Barker and H. W. Berry, *Phys. Rev.* **151**, 14 (1966).
2. P. van der Straten, R. Morgenstern, *J. Phys. B: Atom. Mol. Phys.*, **19** (1986) 1361; and P. van der Straten, *Ph.D. dissertation* (1987).
3. J. K. Swenson, C. C. Havener, N. Stolterfoht, K. Sommer, and F. W. Meyer, *Phys. Rev. Lett.* **63** (1989) 35.
4. J. K. Swenson, J. Burgdörfer, F. W. Meyer, C. C. Havener, D. C. Gregory, and N. Stolterfoht, *Phys. Rev. Lett.* **66** (1991) 417.
5. R. Morgenstern, A. Niehaus, and U. Thielmann, *Phys. Rev. Lett.* **37** (1976) 199; and R. Morgenstern, A. Niehaus, and U. Thielmann, *J. Phys. B: Atom. Mol. Phys.*, **10** (1977) 1039.
6. M. Barat and W. Lichten, Phys. Rev. **A 6**, (1972) 211.
7. A. Bordenave-Montesquieu, and A Gleizes, and P. Benoit Cattin, *Phys. Rev.* **A 25** (1982) 245.
8. F. W. Meyer, C. C. Havener, R. A. Phaneuf, J. K. Swenson, S. M. Shafroth and N. Stolterfoht, *Nucl. Instrum. Methods* **B 24/25** (1987) 106.
9. D. Bordenave-Montesquieu, and R. Dagnac, *J. Phys.* **B 12** (1979) 1233.
10. N. Stolterfoht, Phys. Rep. **146** (1987) 317.
11. See for example, H. Goldstein, *Classical Mechanics* (Addison-Wesley, Reading Mass, 1950).
12. W. Shearer-Izumi, At. Data Nucl. Data Tables **20** (1977) 531.
13. W. H. Miller, *Adv. Chem. Phys.* **30** (1975) 77.
14. R. Barrachina and J. Macek, *J. Phys. B: At. Mol. Opt. Phys.* **22** (1989) 2151.

THE ROLE OF THE VALENCE ELECTRON IN BINARY ATOMIC PROCESSES

C. E. BURKHARDT, M. CIOCCA AND J. J. LEVENTHAL
Department of Physics, University of Missouri - St. Louis, St. Louis, MO 63121 USA

and

J. D. KELLEY
McDonnell Douglas Research Laboratories, St. Louis, MO 63166 USA

ABSTRACT

The mechanism of reactive processes in binary atomic encounters may be investigated by employing a Rydberg atom as one of the reactants. In this way the excited valence electron is isolated so that its role in effecting the heavy body process can be assessed. In one series of experiments Penning ionization by a Rydberg atom was shown to have cross section significantly higher than might be inferred solely on the basis of overlap of the atomic wavefunctions of the ionic Rydberg core and the collision partner.

A model has been proposed to describe the dynamics of these reactive processes. In this model the nearly isolated Rydberg electron polarizes, and is therefore attracted by, the neutral ground state reactant. This attraction occurs at internuclear separation greater than that at which the ionic core of the Rydberg atom can substantially polarize the neutral. The polarization by the Rydberg electron leads to a transitory negative ion and concomitant Coulomb attraction to the ionic core. This Coulomb attraction can thus result in cross sections higher than calculated on the basis of atomic overlap.

1. Introduction

The extreme properties of Rydberg atoms, and their sensitivity to external perturbations, allow observation and analysis of effects that are difficult to see with less highly excited atoms. Rydberg atoms also produce effects different than those seen in ground-state or low-lying excited-state species. This is particularly true in collisions involving Rydbergs and perturbers, in which the large size, close level spacing, and high total energy of the Rydberg atom lead to a variety of inelastic processes, some with very large cross sections.

One of the inelastic collision processes that we have studied is Penning ionization

$$X^{**} + Y \rightarrow X + Y^{+} + e \tag{1}$$

The ionization potential of the perturber Y must be equal to or lower than the excitation energy of the Rydberg X^{**} for Penning ionization to occur without conversion of heavy-body collision energy to electronic energy. Ionization can in fact occur below this threshold, but the required energy conversion is inefficient, and the inefficiency increases rapidly with the energy defect. When the X^{**} energy exceeds the ionization potential of Y, the situation is not so straightforward because the mechanism by which ionization proceeds must be explicitly considered to assess the dependence on X^{**} energy, or equivalently, n and J, the principal quantum number and total angular momentum.

2. Discussion

We have measured Penning ionization cross sections for excited Na colliding with K atoms, and for excited Ba colliding with Na atoms. For Rydberg states just above the threshold energy, the cross sections are measured to be about 2500 Å^2. This value implies that the Penning process occurs when the collision partners approach within $R \approx 25 - 30$ Å, a distance far larger than the overlap range of the core orbitals, but much smaller than the radii of the n = 15 - 20 Rydberg atoms. Two general mechanisms suggest themselves: first, the ionization process could occur by simultaneous deexcitation of the Rydberg electron to the ground state and excitation of the perturber ground-state electron to the continuum, a resonance energy exchange process. After consideration of the poor overlap of the ground state wave functions with Rydberg or continuum wave functions and the lack of a strong interaction to mediate the process, it is apparent that this picture is inadequate. The process undoubtedly proceeds via a transfer of the perturber valence electron to the ion core of the Rydberg atom with the distant Rydberg electron departing with sufficient energy to conserve total electronic energy. This is simply resonant charge transfer with resonance enforced by the three-body final state.

The Rydberg and perturber must however approach within a few Å for charge exchange to occur, i. e. the initial and final core wave functions must overlap. Straight-line heavy-body trajectories are a clearly inadequate model. A long-range interaction exists between the ion core and perturber, namely the charge-induced dipole attraction which varies a α/R^4 where α is the perturber polarizability. Calculations using this interaction yield large cross sections for the Penning process, but not large enough - a factor of roughly four too small.

We have resolved this problem by including an additional polarization interaction, the interaction between the Rydberg electron and the perturber. This attractive interaction results in a delay of the Rydberg electron when it comes near the perturber, and is equivalent to an accumulation of negative charge around the perturber, thus producing a transitory negative ion. As a result, a Coulomb term with fractional, R-dependent charge is added to the ion core-perturber interaction. This additional long-range attraction allows collision pairs that approach within ~25 Å at thermal energies to spiral into one another, permitting charge transfer to occur and producing the relatively large observed cross sections. The mechanism is illustrated in Figure 1.

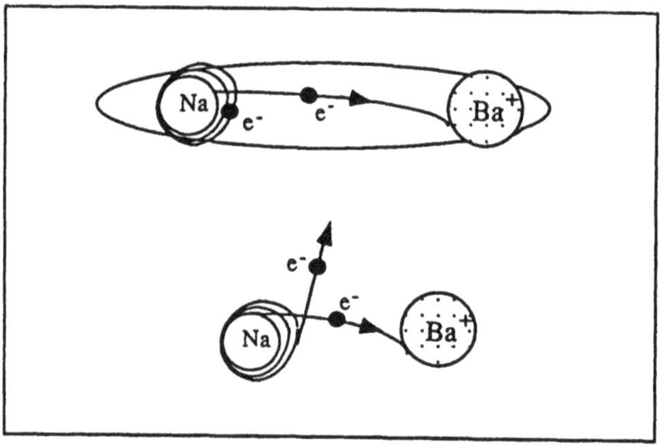

Figure 1. Schematic representation of the collision process described by the model.

This mechanism explains other features of our data as well. The sharp decrease in Penning cross section with increasing n above threshold results from destructive interference between the bound Rydberg electron wave function and the outgoing continuum wave function as the latter increases in energy. The observed cross sections should increase with perturber polarizability, and comparison of Na and K on Ba** shows this effect.

3. Acknowledgements

This work was supported by National Science Foundation Grants No. PHY-8907698 and No. INT-8719649 the McDonnell Douglas Independent Research and Development Program, and the University of Missouri Weldon Spring Fund.

ALGEBRAIC METHODS FOR AN ATOM-ANHARMONIC OSCILLATOR COLLISION

José Récamier A.

Laboratorio de Cuernavaca, Instituto de Física, UNAM.
Apartado Postal 139-B, 62191 Cuernavaca, Morelos, México.

ABSTRACT

In this work we use Lie algebraic methods to study the translation-vibration energy transfer in a collinear collision between an atom and a diatomic molecule within the semiclassical approximation. As an interaction between the atom and the diatomic we use a repulsive exponential function which is expanded in a Taylor series, as a model for the molecule we use an harmonic oscillator and also an anharmonic oscillator whose anharmonicity constants are obtained from a Taylor series expansion of a Morse potential retaining terms of up to fourth order. Numerical results are presented for different systems and different transitions as a function of the total collision energy and are compared with exact quantum results.

INTRODUCTION

The Landau-Teller model has been used frequently to study the vibrational energy transfer in a collision between an atom A and a diatomic BC.[1] Here, an harmonic oscillator is used as a first approximation to model the diatomic molecule and an exponential potential $De^{-r/L}$ accounts for the atom-molecule interaction. Though this is not a realistic model for the system, it is simple enough and several different approximations can be tested against exact quantum mechanical calculations.[2,3] It has been shown that, in some cases the dynamics of the system can be simplified using the semiclassical approximation.[4] In this approximation, the coordinate between the projectile and the first atom of the diatom r is expressed in terms of the coordinate of relative motion between the projectile and the center of mass of the molecule R and the internal coordinate of the molecule x. Solving the classical equations of motion for the coordinate R one obtains a time-dependent interaction potential which depends only on the internal coordinate of the molecule

$$\mathcal{V}_{sc}(x,t) = F(t)e^{\lambda x}, \tag{1}$$

and where:[5]

$$F(t) = E sech^2[(E/2\mu)^{1/2}t/L]$$

$$\lambda = m_C/(m_B + m_C)L, \quad E = De^{-\lambda x_0}$$

x_0 being the equilibrium position of the oscillator, the function F(t) is such that it goes to zero as the time goes to $\pm\infty$.

THEORY

Consider then, the collinear collision between an atom A and a diatomic molecule BC initially in a vibrational state ν and such that after the collision it gets to a final vibrational state ν', that is;

$$A + BC(\nu) \rightarrow A + BC(\nu') \qquad (1)$$

The Hamiltonian in semiclassical approximation is

$$\mathcal{H}_{sc}(x,t) = \mathcal{H}_0 + \mathcal{V}(R(t), \chi) \qquad (2)$$

where $\mathcal{H}_0$ is the time independent Hamiltonian corresponding to the free oscillator and $\mathcal{V}(R(t),x)$ is the interaction between the atom and the diatomic molecule.

To obtain the transition probabilities $P_{\nu,\nu'}$ between the vibrational states of the diatomic one has to evaluate the matrix element $S_{\nu,\nu'}$ of the scattering operator between eigenstates ν, ν' of the unperturbed oscillator.

The scattering operator S is defined as the long time limit of the time–evolution operator in the interaction picture. $S = \mathcal{U}(t \rightarrow \infty, t_0 \rightarrow -\infty)$ where

$$\mathcal{U}_I(t, t_0) = \mathcal{U}_0^{+}(t, t_0)\ \mathcal{U}(t, t_0) \qquad (3)$$

$\mathcal{U}_0 = e^{-i\mathcal{H}_0(t-t_0)/h}$ satisfies the differential equation

$$ih\partial t\ \mathcal{U}_I = |\ \mathcal{U}_0\ \mathcal{V}(R(t),\chi)\ \mathcal{U}_0|\ \mathcal{U}_I - \mathcal{H}_I\,\mathcal{U}_I \qquad (4)$$

with the initial condition $\mathcal{U}_I(t - t_0) = I$.

It is known that when the Hamiltonian can be expressed as a linear combination of operators which form a finite Lie algebra; the corresponding time evolution operator can be written as a product of exponentials,[7] and the solution is global when the algebra is solvable. This property has been exploited in problems dealing with atom–diatom[8] and diatom–diatom[9] energy transfer and also in the general case of scattering theory.[10]

THE HARMONIC CASE

When an harmonic oscillator model is used for the diatomic molecule, the unperturbed Hamiltonian is $\mathcal{H}_0 = \hbar\omega(a^\dagger a + 1/2)$ and the transformation given by Eq. 4. applied to the creation and anhilation operators $a^\dagger$, a gives

$$\mathcal{U}_0^\dagger a \mathcal{U}_0 = a e^{-i\omega t} \tag{5a}$$

$$\mathcal{U}_0^\dagger a^\dagger \mathcal{U}_0 = a^\dagger e^{i\omega t}, \tag{5b}$$

so that, the interaction picture Hamiltonian can be obtained easily. In this work, we propose

$$\mathcal{V}_{sc}(x,t) = \sum_{n=1}^{4} F_n(t) x^n \tag{6}$$

where we have made a Taylor series expansion of the interaction given by Eq. 1 and retained up to quartic terms in the displacement coordinate x.

Expressing the displacement coordinate x in terms of the boson operators a, $a^\dagger$,

$$x = \sqrt{\frac{\hbar}{2m\omega}}(a + a^\dagger)$$

we obtain

$$\mathcal{V}_{sc}(a, a^\dagger, t) = \sum_{m=1}^{15} \Psi_m(t) \mathcal{X}_m \tag{7}$$

with known functions $\Psi_m(t)$ and where, for convenience, we have written the operators $\mathcal{X}_m$ in normal order.[11]

Using the transformations given by Eqs. 5a. and 5b. to transform the operators appearing in Eq. 7 we obtain the Hamiltonian in the interaction picture as:

$$\mathcal{H}_I(t) = \sum_{m=1}^{15} \Phi_m(t) \mathcal{X}_m. \tag{8}$$

From the operators appearing in the above expression, the set $\{a, a^\dagger, a^\dagger a, a^2, a^{\dagger^2}, \mathcal{I}\}$ forms a finite Lie algebra and its time-evolution operator can be written in terms of a product of exponentials. Correspondingly, the time-evolution operator in the interaction picture is written as

$$\mathcal{U}_I = \mathcal{U}_l^{(1)} \mathcal{U}_{nl} \tag{9}$$

where

$$i\hbar\partial_t \mathcal{U}_l^{(1)} = \mathcal{H}_1 \mathcal{U}_l^{(1)} \tag{10}$$

and

$$i\hbar\partial_t \mathcal{U}_{nl} = [\mathcal{U}_l^{(1)\dagger}(H_I - H_1)\mathcal{U}_l^{(1)}]\mathcal{U}_{nl}. \tag{11}$$

We propose

$$\mathcal{U}_l^{(1)} = \prod_{n=1}^{6} e^{-\alpha_n \mathcal{X}_n} \tag{12}$$

with unknown, complex, time-dependent functions α_n which satisfy the initial condition $\alpha_n(t = -\infty) = 0$. Substitution into Eq. 10 yield a set of six, coupled, first order, differential equations for the α_n.

The time-evolution operator $\mathcal{U}_l^{(1)}$ is such that:

$$\begin{aligned} \mathcal{U}_l^{(1)\dagger} a \mathcal{U}_l^{(1)} &= M_{11}a + M_{12}a^\dagger + M_{13}I, \\ U_l^{(1)\dagger} a^\dagger \mathcal{U}_l^{(1)} &= M21a + M_{22}a^\dagger + M_{23}I \end{aligned}$$

where the M_{ij} are functions of the time and of the α_n and contain information about the solution of the time evolution operator corresponding to $\mathcal{H}_l^{(1)}$; the interaction obtained after the transformation given by Eq. 11 has the same set of 15 operators that we found in Eq. 8. It is clear that we can proceed as before and choose a subset which forms a finite Lie algebra and in so doing get a new interaction potential with again the same set of operators. Then, the time evolution operator in the interaction picture can be written as:

$$\mathcal{U}_I = \mathcal{U}_l^{(1)}\mathcal{U}_l^{(2)}\mathcal{U}l^{(3)}\ldots\mathcal{U}_{nl} \tag{13}$$

where the upper label indicates how many transformations have been applied and $\mathcal{U}_{nl}$ is the time-evolution operator corresponding to that part of the interaction which does not form a finite Lie algebra and which has to be obtained with approximate methods like perturbation theory, in this work we will simply neglect it's effects.

THE ANHARMONIC CASE

When we use an anharmonic oscillator to model the diatomic, the transformation given by Eqs. 5a. and 5b. to obtain the Hamiltonian in the interaction picture can not be carried out. To see that consider the BCH formula[12]

$$e^{\beta \mathcal{H}_0} a e^{-\beta \mathcal{H}_0} = a + \beta[H_0, a] + \frac{1}{2!}\beta^2[\mathcal{H}_0, [\mathcal{H}_0, a]] + \ldots,$$

when $\mathcal{H}_0$ contains higher than quadratic terms in the displacement coordinate x, each one of the commutators given above give as a result a different operator and then the set of operators in the interaction picture Hamiltonian is not finite. In order to obtain a finite set of operators for $\mathcal{H}_I$ we will make use of an approximate unperturbed Hamiltonian.

Consider the Morse Hamiltonian

$$\mathcal{H}_0 = \mathcal{T} + \mathcal{V}_{Morse}$$

with

$$\mathcal{V}_{Morse} = D\left[e^{-2bx} - 2e^{-bx}\right]$$

D being the depth of the potential and b an inverse scale parameter. Making a Taylor series expansion of the potential and retaining terms of up to fourth order in the displacement coordinate we get:

$$\mathcal{H}_0 = \mathcal{T} + \sum_{n=1}^{4} \phi_n x^n \tag{14}$$

now we write the Hamiltonian in terms of the boson operators a, $a^\dagger$ and make the decomposition

$$\mathcal{H}_0 = \mathcal{H}_0^{(1)} + \mathcal{H}_0^{(2)}$$

where $\mathcal{H}_0^{(1)}$ contains that part of $\mathcal{H}_0$ which forms a finite Lie algebra[13] and $\mathcal{H}_0^{(2)}$ contains the remainder. Notice however that $\mathcal{H}_0^{(1)}$ contains information from the anharmonic part of the potential since when expressed in terms of the boson operators the cubic and quartic terms contribute with terms proportional to a, $a^\dagger$, $a^\dagger a$, $a^{\dagger 2}$, a^2 and the identity operator $\mathcal{I}$.

The time-evolution operator corresponding to the unperturbed Hamiltonian $\mathcal{H}_0$ is then given as a product

$$\mathcal{U}_0 = \mathcal{U}_0^{(1)} \mathcal{U}_0^{(2)}$$

with

$$i\hbar \partial_t \mathcal{U}_0^{(1)} = \mathcal{H}_0^{(1)} \mathcal{U}_0^{(1)}$$

$$i\hbar \partial_t \mathcal{U}_0^{(2)} = [\mathcal{U}_0^{(1)\dagger} \mathcal{H}_0^{(2)} \mathcal{U}_0^{(1)}] \mathcal{U}_0^{(2)}.$$

Since we expect that the most important part of the anharmonicity has been taken into account with the linear and bilinear terms in the boson operators coming from the cubic and quartic terms, we will neglect the effects of $\mathcal{H}_0^{(2)}$. Comparison with exact numerical results will indicate if this approximation is correct or not.

Then, the time evolution operator for the unperturbed anharmonic oscillator is written as:

$$\mathcal{U}_0 \approx \mathcal{U}_0^{(1)} = \prod_{n=1}^{6} e^{-\beta_n \mathcal{X}_n} \tag{15}$$

where the complex time dependent functions β_n fulfill a set of six coupled, first order differential equations which are solved numerically.

With this form for $\mathcal{U}_0$ we can now obtain the Hamiltonian in the interaction picture through the transformation

$$\mathcal{H}_I = \mathcal{U}_0^{\dagger} \mathcal{V}_{sc} \mathcal{U}_0 \tag{16}$$

with $\mathcal{V}_{sc}$ given as in Eq. 7, the result is:

$$\mathcal{H}_I = \sum_{n=1}^{15} \Psi_n \mathcal{X}_n \tag{17}$$

where the set of operators appearing in $\mathcal{H}_I$ is the same as the one which appeared in the harmonic case, the difference being in the coefficients of the operators which now contain information from the anharmonicity of the oscillator. The construction of the time-evolution operator in the interaction picture is done in analogy with the harmonic case. The result is, neglecting the higher than bilinear terms in the boson operators:

$$\mathcal{U}_I = \mathcal{U}_I^{(1)} \mathcal{U}_I^{(2)} \ldots \tag{18}$$

RESULTS AND DISCUSSION

As mentioned in the introduction, for the numerical evaluation of transition probabilities we used a repulsive exponential function as interaction between the incident atom and the first atom of the diatomic because such a model has been used extensively and there are exact quantum results available for the harmonic[2] and for the Morse oscillator cases.[3] Notice that we are using a semiclassical approximation and the differences between our results and the exact quantum results may be due to this approximation, to the fact that we are replacing an exponential function with a fourth order polynomial and also to the approximate form in which we are treating the anharmonic oscillator (in the case of a Morse oscillator). In order to test the reliability of the algebraic treatment, we made a basis set expansion for the harmonic case in the semiclassical approximation, for an atom-diatom interaction written as a fourth order polynomial in the displacement coordinate. The results so obtained may be considered as the 'exact' semiclassical results for this polynomial interaction.

Table 1 shows the transition probabilities from the ground to the first excited state of an H_2 molecule colliding with an Hellium atom for several collision energies in the approximations $P^{(1)}_{Ho}$, $P^{(2)}_{Ho}$, P_{BS} and P_Q where the upper index indicates how many transformations were applied before neglecting the remaining interaction; that is $\mathcal{U}_I = \mathcal{U}_I^{(1)}$ or $\mathcal{U}_I = \mathcal{U}_I^{(1)}\mathcal{U}_I^{(2)}$; P_{BS} corresponds to the basis set expansion mentioned above and P_Q to the fully quantum calculation of Secrest and Johnson.[2]

Table 1. Transition probability $(0-1)$ as a function of the total collision energy in approximations $P^{(1)}_{Ho}$, $P^{(2)}_{Ho}$, P_{BS} and the exact quantum results[2] P_Q.

System $H_2 - H_e$

E/E_0	$P^{(1)}_{Ho}$	$P^{(2)}_{Ho}$	P_{BS}	P_Q[2]
4	9.20(-4)	9.73(-4)	9.84(-4)	7.20(-4)
6	3.50(-2)	3.76(-2)	3.89(-2)	2.95(-2)
8	1.38(-1)	1.47(-1)	1.55(-1)	1.33(-1)
10	2.61(-1)	2.73(-1)	2.87(-1)	2.92(-1)
12	3.45(-1)	3.49(-1)	3.59(-1)	4.28(-1)
16	3.44(-1)	3.29(-1)	2.96(-1)	4.07(-1)

As a result of the sequence of transformations, the transition probabilities approach the 'exact' semiclassical results in all the energy range considered and the differences between the 'exact' and any of the two approximations considered here are of the order of a few percent at most, for those cases where the collision energy is already rather large.

Notice also that the approximate results do not converge to the fully quantum results, in fact, the transition probabilities obtained after a single transformation are in better agreement with the quantum results than those obtained after two transformations.

Figure 1 shows the logarithm of the transition probability $1-2$ for the system $N_2 + (N_2)$, in approximation $P_1^{(1)}$ (full line), and P_Q (squares). For this system the differences between the results in the approximations $P_1^{(1)}$ and $P_1^{(2)}$ are so small that they can't be noticed in the graph. We see that the approximate semiclassical results display the correct behaviour as a function of the collision energy in all the energy range considered.

Figure 1. Logarithm of the transition probability $(1-2)$ as a function of the total collision energy measured in units of $\hbar\omega/2$ for the system $(N_2) + N_2$ in approximations $P_1^{(1)}$ (full line) and $P_Q{}^3$(squares).

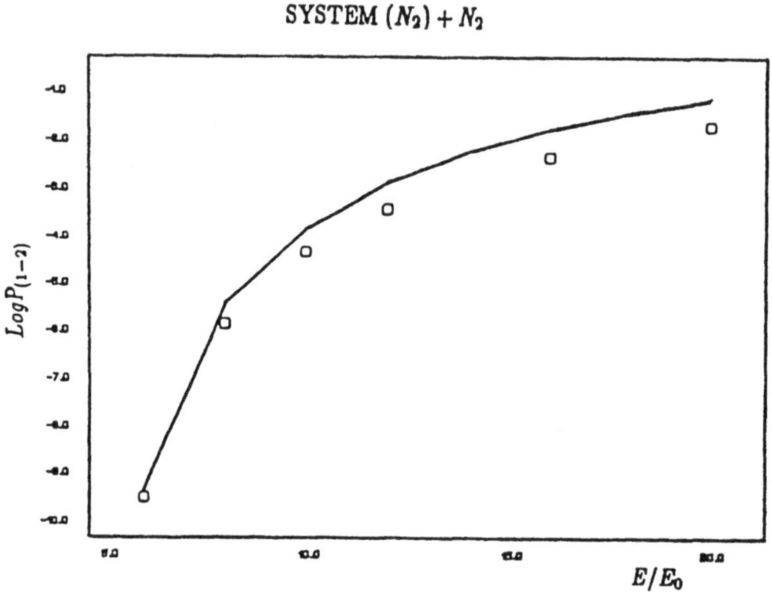

The algebraic method has been applied to several systems[11,13] and the sequence of transformations has been found to be more or less relevant depending upon the system under consideration. In all cases, the farther we go in the sequence of transformations the closer we get to the exact semiclassical results. As a function of the collision energy, the transition probabilities show the correct behaviour for all the cases considered so far. We have also studied the convergence properties of the transformations[9] and the results after a couple of transformations agree within 5% with the exact semiclassical results.

Acknowledgement

I am grateful to Jorge Hernández for his help in the computational part of this work.

REFERENCES

1 J. M. Jackson and N. F. Mott, *Proc. Roy. Soc.* A **137**, (1932) 703.

2 D. Secrest and B. R. Johnson, *J. Chem. Phys.* **45**, (1966) 4556.

3 A. P. Clark and A. S. Dickinson, *J. Phys. B: Atom. Molec. Phys.* **6**, (1973) 164.

4 D. Rapp, *J. Chem. Phys.* **32**, (1960) 735.

5 M. S. Child, *Molecular Collision Theory* (Academic Press, London, 1974).

6 Albert Messiah *Quantum Mechanics* (Wiley, London, 1962) Vol II, Ch. XIX.

7 J. Wei and E. Norman, *Proc. Am. Math. Soc.* **15**, (1964) 327.

8 B. Gazdy and D. A. Micha, *J. Chem. Phys.* **82**, (1985) 4926; J. Récamier, *Chem. Phys. Lett.* **133**, (1987) 259.

9 J. Récamier, D. A. Micha and B. Gazdy, *Chem. Phys. Lett.* **119**, (1985) 383; J. Récamier, D. A. Micha and B. Gazdy, *J. Chem. Phys.* **85**, (1986) 5093.

10 Y. Alhassid and R. D. Levine, *Phys. Rev. Lett.* **54**, (1985) 739.

11 José Récamier and Jesús Ortega, *Molecular Physics* (to appear) (1991)

12 R. Gilmore, *Lie groups, Lie algebras, and some of their applications* (Wiley, New York, 1974) ch. V.

13 José Récamier, *Int. Jour. Quantum. Chem. Symp.* **24**, (1990) 655.

V. ELECTRON IMPACT COLLISIONS

Ab initio studies of e^- - He_2^+ collisions

B M McLaughlin
Institute for Theoretical Atomic and Molecular Physics,
Harvard-Smithsonian Center for Astrophysics,
60 Garden Street, Cambridge, MA 02138, USA

C J Gillan and P G Burke
Department of Applied Mathematics and Theoretical Physics,
The Queen's University of Belfast, Belfast BT7 1NN, UK

J S Dahler
Departments of Chemistry, Chemical Engineering and Materials Science,
University of Minnesota, Minneapolis, MN 55455, USA

The R-matrix method has been used to perform *ab initio* calculations on low energy elastic scattering by electrons from He_2^+ molecular ions . A two state approximation has been used in which the X $^2\Sigma_u^+$ and the A $^2\Sigma_g^+$ states of He_2^+ have been retained in the close coupling expansion. Structure calculations have been performed at SCF, full CI and MCSCF-CASSCF level, using a (4s,2p,2d) Slater basis to gauge the quality of wavefunctions at various bond separations. Cross sections for elastic scattering have been calculated in which the total symmetry of the collision complex is $^3\Sigma_u^+$. SCF and correlated CI wavefunctions are only used in the scattering calculations carried out at the two fixed internuclear separations 1.8 and 2.0625 a_0. The three lowest lying $np\sigma$ $^3\Sigma_u^+$ resonances at each separation have been detected and fitted to a Breit-Wigner form for their positions and widths. In addition, bound state calculations has been performed on the lowest lying $^3\Sigma_u^+$ state of He_2^* at each of the two internuclear separations. The calculated energy for the a $^3\Sigma_u^+$ bound state of He_2^* is seen to compare favorably with previous MCSCF and SOCI calculations performed by quantum chemists.

1. Introduction

In this paper we report results of *ab initio* calculations of low energy elastic scattering by electron from He_2^+ molecular ions. To our knowledge there have been no previous scattering calculations on this system. There have been no experimental measurements on this process probably due to the difficulty of handling He_2^+ in the laboratory. Experimental studies requires a well defined interaction region since the space charge is limited and it is very difficult to produce He_2^+ ions in their ground vibrational state. He_2^+ ions that are produced from neutral dimers are in hot vibrational and rotational states and therefore must be quenched in order to perform experiments on the ground vibrational state[1]. Hence we have a situation where *ab initio* calculations for cross sections and autoionization widths have been made without the availability of comparison with any experimental data or other suitable theoretical work. Estimates for the autoionization widths arising from the interaction of two metastable helium atoms in their triplet states have been reported previously by Garrison *et al.*[2] We note that only approximate estimates of the autoionization widths/resonance widths for the high lying states $^1\Sigma_g^+$, $^3\Sigma_u^+$, and $^5\Sigma_g^+$ of He_2^* have been made. For these states of the helium dimer Garrison *et al.*[2] parameterized the autoionization widths as exponentially decaying functions of internuclear separation as opposed to calculating them from first principles. As noted by Müller *et al.*[3], detailed knowledge of the $^3\Sigma_u^+$ and $^1\Sigma_g^+$ resonances widths as functions of internuclear separation are required to fully understand the energy spectrum of electrons produced by ionizing collisions between pairs of metastable He(2 ^1S, 2 ^3S) atoms. Autoionization is spin forbidden for the $^5\Sigma_g^+$ entrance channel, which may only be accessed through spin-orbital coupling, and so this symmetry may be neglected as indicated by Hill and co-workers.[4]

This system is important as it opens the avenue to studying a host of related theoretical problems[8]. It will then allow us to study in detail Penning and Associative Ionization in He* - He, and He* - He* collisions, with and without the presence of a photon field. Field free associative ionization and excitation transfer cross sections have previously been studied, within a diabatic representation of the He* - He complex, using a multistate curve crossing model[5]. Recent work on Associative and Penning Ionization in He(2 ^3S) - He(2 ^3S) collisions have been investigated by Bieniek *et al.*[6] using a uniform semi-classical approximation. Estimates have been made by Julienne and Mies [7] of the upper and lower bounds for the rate coefficient of Penning Ionization in He(2 ^3S) - He(2 ^3S) collisions, which are respectively 10^{-9} cm^3/sec and 5×10^{-11} cm^3/sec. Weiner *et al.*[9] have recently published an extensive review of Associative Ionization in several systems. Niehaus[10], has recently reviewed field free spontaneous electron emission from slow atomic collisions at incident energies less than 10 keV. In addition a review of field free chemi-ionization in binary collisions at thermal energies has been made by Klucharev and Vujnović[11].

Our work will make it possible to study photoionization of the excited states of the helium dimer where there are resonances in the resulting cross sections. These resonances can be analyzed and identified with Rydberg states of He_2^* from phase shift calculations in e^- - He_2^+ scattering. It may then be possible to resolve the controversy that exists at present in experiments on photon-assisted He^* - He collisions [12,13].

We note that in models of the ejecta of the supernova SN 1987A[14], He^+ ions formed by the radioactive decay of ^{56}Co are removed by radiative and dielectronic recombination, charge transfer, the radiative association process $He^+ + He \rightarrow He_2^+ + h\nu$ and by reactions with CO. It is desirable to know accurately cross sections as well as rate constants for these processes. In particular radiative recombination cross sections and rate coefficients may be obtained from *ab initio* photoionization cross sections calculations on the ground and excited $n = 2$ states of the helium atom[15]. Quantitative calculations of the autoionization widths/resonance widths $\Gamma(R)$, as functions of the internuclear separation variable R, will enable the investigation in detail of the dissociative recombination (DR) process, $e^- + He_2^+ \rightarrow He^* + He$. It is understood that the $^3\Sigma_g^+$ and $^1\Sigma_g^+$ are the two lowest lying electronic states providing the primary routes for DR but other diabatic states of the helium dimer, namely, $^3\Pi_u$, $^1\Pi_u$, $^3\Sigma_u^+$ and $^1\Sigma_u^+$ are possible routes[16]. Finally it is our intention to study vibrational excitation of the ground state of the He_2^+ molecular ion, which will provide a rich source of information on this molecular complex.

2. Details of calculations

We have applied the R-matrix method to the electron-molecular ion complex as outlined by Gillan *et al.*[17] in order to solve the scattering equations. This method has proved, among others, to be very successful in the *ab initio* study of low energy electron molecule collisions. In particular, Tennyson *et al.*[18] and Tennyson and Noble[19] applied the method to the e^- - H_2^+ system. Shimamura *et al.*[20] have extended this work and completed a detailed study of the Feshbach resonances converging to the excited state. In addition, Tennyson[21] has used the R-matrix method to study bound and continuum states of the e^- - CH^+ system. Recently Tennyson and co-workers[22] have completed work on bound states of the e^- - HeH^+ system and have implemented the method of Seaton[23] to calculate the lowest 33 bound electronic states of the HeH molecule.

In the next three sections we outline the most important aspects of our cross section calculations of the scattering process. We begin by considering the representation of the target state wavefunctions and then in the next section we detail the solution of the scattering equations. Finally, we give an explicit discussion of our choice of the $\mathcal{L}^2$ functions which are important constituents of the R-matrix basis states.

2.1 *Target Wavefunction Calculations*

Our target eigenstates are formed from a set of orthogonal molecular orbitals. Each molecular orbital is expanded in a basis of Slater type orbitals (STO), which are centered on the atomic nuclei and known as the linear combination of atomic orbitals (LCAO). We have chosen the (4s, 2p, 2d) double zeta plus polarization (DZ+P), STO basis of Reagen *et al.*[24].

In order to test the sensitivity of the scattering calculations to the target representations used we have employed two different approximations for the target states. Initially we carried out an self consistent field (SCF) calculation on the $1\sigma_g^2 1\sigma_u^1$ X-state of He_2^+ in the STO basis. Due to linear dependence we were forced to delete one of the σ_g and one of the σ_u basis functions in the SCF procedure. This yielded the orbital set

$$1\sigma_g \rightarrow 6\sigma_g,\ \ 1\sigma_u \rightarrow 6\sigma_u,\ \ 1\pi_u \rightarrow 3\pi_u,\ \ 1\pi_g \rightarrow 3\pi_g,\ \ 1\delta_g,\ \ 1\delta_u. \qquad (1)$$

We then Schmidt-orthogonalized the open shell virtual orbitals to the bound orbitals and used the resulting orthogonal orbital set for the expansion of both target states.

Our first approximation was to use single configurations for both the X $^2\Sigma_u^+$ and A $^2\Sigma_g^+$ states. In this approximation the X state was represented by its SCF form. We found that the representation of the A state in the common orbital basis was close to that obtained from an SCF calculation on the A state configuration itself. In the second approximation we began by performing full configuration interaction (CI) calculations[25] for both target states in the common orbital set. The dimension of the configuration state function (CSF) space for each state was 562. This representation was therefore at the STO basis set limit and independent of the form of representation of the molecular orbitals that we had chosen. Using these target state expansions would have rendered the solution of the scattering equations difficult due to the algorithm used in our codes for Hamiltonian matrix construction.

In order to make the inner region Hamiltonian manageable it was necessary to reduce the number of CSFs. Accordingly, we evaluated the first order spin-reduced density matrix for the X state (in the full CI approximation), diagonalized this, and then transformed the initial orbital set to natural orbitals, cf Mulliken and Ermler [26]. The new orbital set was truncated by omitting all natural orbitals, with occupation numbers less than 10^{-3}. The truncated set consisted of the following orbitals:

$$1\sigma_g \rightarrow 3\sigma_g,\ \ 1\sigma_u \rightarrow 2\sigma_u,\ \ 1\pi_u,\ \ 1\pi_g\ . \qquad (2)$$

The performance of a full CI calculation in the truncated natural orbital set (2) yielded 30 CSFs per state with a negligible degradation in absolute energy when compared with full CI in the orbital set (1). We noted that the degradation in energy was larger for internuclear separations different from the equilibrium separation. Clearly, the use of natural orbitals derived from the ground state introduces an extra approximation in the expansion of the excited state. However, we believe that in the present calculation this approximation is satisfactory.

2.2 *The Scattering Calculation*

All of the calculations reported here have been performed with the nuclei held fixed in space. In the fixed nuclei R-matrix theory there are two regions of Euclidean configuration space. The inner region is a hypersphere defined by $r_i < a \ \forall\, i$, where r_i is the radial distance of electron i from the center of mass of the target. The outer region in the electron scattering problem for an N electron target is defined by $r_i < a$ for $i = 1, N$ and $r_{N+1} > a$. The parameter a is the R-matrix radius which we have chosen to be 10 a_0, a value large enough that the target state charge distributions are enveloped by the hypersphere. A multi-center configuration interaction type expansion of the scattering wavefunction is chosen to represent the e^- - He_2^+ complex in the finite inner region. Thus, the inner region wavefunction is expanded in a basis ψ_k,

$$\psi_k(\mathbf{r}_1, \mathbf{r}_2, \mathbf{r}_3, \mathbf{r}_4) = \mathcal{A}\sum_{ij} \bar{\phi}_i(1,2,3)\eta_j(4)\alpha_{ijk} + \sum_j \chi_j(1,2,3,4)\beta_{jk} \tag{3}$$

where $\mathcal{A}$ is the anti-symmetrization operator. The functions $\bar{\phi}_i$ are eigenfunctions of the total spin operator, S^2, and its z-component, S_z, formed by coupling the products of the spin function of the continuum electron with the target electronic states ϕ_i. The η_j functions are continuum molecular orbitals on the center of gravity of the molecule which are non-zero on the boundary of the inner region. The second summation is over square integrable $\mathcal{L}^2$ functions, χ_j, descriptive of states with all of the electrons in bound orbitals. The composition of this second summation is discussed in detail in the following section. In our calculations we have retained in (3) the lowest two states of the He_2^+ molecular ion, these being the X $^2\Sigma_u^+$ ground state and A $^2\Sigma_g^+$ excited state. These two-state calculations reported here will serve as a comparison for future multistate calculations on this system. The coefficients α_{ijk} and β_{jk} are determined by diagonalizing the operator $(H + L)$ in the basis ψ_k in the inner region. Thus,

$$\langle \psi_k \mid H + L \mid \psi_j \rangle = E_k \delta_{kj}, \tag{4}$$

where H is the electronic Hamiltonian for the target plus projectile system and L is a surface projection operator introduced by Bloch [27] which is so defined that $(H + L)$

is Hermitian. The setting up and diagonalization of operator $(H + L)$ in the inner region is carried out using a modified version of the IBM Alchemy quantum chemistry program suite [28,29], now upgraded to run on Cray architectures.

In our scattering calculations the continuum molecular orbitals, η_j, representing the scattered electron, are formed by taking suitable linear combinations of the target molecular orbitals and additional continuum basis functions, the radial parts of which are obtained by numerically solving the model scattering problem

$$\left[\frac{\mathrm{d}^2}{\mathrm{d}r^2} - \frac{l(l+1)}{r^2} + 2V_0(r) + k_j^2\right] u_{ij}(r) = 0 \tag{5}$$

subject to the fixed boundary conditions

$$u_{ij} = 0, \qquad \left[\frac{a}{u_{ij}(r)} \frac{\mathrm{d}u_{ij}(r)}{\mathrm{d}r}\right]_{r=a} = 0 \tag{6}$$

Here V_0 is the spherical part of the static potential of the SCF target ground state of He_2^+. The use of the fixed boundary condition requires the introduction of the Buttle correction[30], when computing the R-matrix. In the present work we have found that it is sufficient to retain three partial waves for each continuum symmetry in our calculation. We chose the continuum basis to be complete up to 11 Rydbergs, a condition that required the use of about nine continuum functions per partial wave.

The outer region collision problem is solved by adopting a single center, no exchange, close coupling expansion of the wavefunction. R-matrix propagator[31] and accelerated Gailitis expansion methods [32] are employed in the solution. To evaluate bound states the external region problem is solved for negative scattering energies, that is, with all channels closed. The R-matrix is then matched onto exponentially decaying wavefunctions. For the scattering process the external region problem is solved for positive energies and the R-matrix matched onto outgoing wave solutions. The T-matrices obtained by this procedure are then employed in standard formulae to produce differential and integrated cross sections.

2.3 *The $\mathcal{L}^2$ Functions for SCF and Full CI Target Representations*

When a single configuration representation was used for each target state we retained the following subset of orbitals from the ground state SCF calculation in our scattering work:

$$1\sigma_g,\ 2\sigma_g,\ 1\sigma_u,\ 2\sigma_u,\ 1\pi_u,\ 2\pi_u,\ 1\pi_g,\ 2\pi_g,\ 1\delta_g,\ 1\delta_u\ . \tag{7}$$

The orbital subset $1\sigma_g$ and $1\sigma_u$ contains those orbitals occupied in both states while its complement, within (7), defines the virtual orbital space. The $\mathcal{L}^2$ functions in equation (3) consisted of two parts with the following makeup:

1. Three electrons in either of the target state orbitals with the fourth electron occupying any of the virtual orbitals from the set specified in (7).
2. Any one of the electrons from the X state configuration promoted into any of the virtual orbitals from (7). The fourth electron is also allowed in the virtual orbital space.

This approximation is known as static exchange plus polarization (SEP) and has been used in previous R-matrix work [17] on other systems.

For scattering calculations with a full CI representation of the target states the truncated natural orbital set (2) was used. In this case to be consistent with the use of full CI, the $\mathcal{L}^2$ terms consisted of all possible arrangements of four electrons in the orbital space. Our scattering calculation is therefore completely correlated within this truncated orbital space. Naturally, only those configurations which are consistent with the symmetry and equivalence restrictions of the scattering state in question are constructed in either approximation.

3. Results and Discussion

In Fig. 1 we display our calculated potential energy curves for the X and A target states of the He_2^+ molecular ion using the SCF approximation. For comparison purposes the accurate multi-configuration self consistent field (MCSCF) results[33] are also shown for the same two states. At the SCF level, we note that single configuration representation of the A state, in particular is inadequate near the united atom limit ($R \rightarrow 0$) and gives no avoided crossing in the A $^2\Sigma_g^+$ state at internuclear geometries for the range 1.0 – 1.5 a_0. Our SCF results for the X state are in excellent agreement with the SCF Gaussian type orbital (GTO) work of Maas*et al.*[37] over the range of internuclear separations considered here. In Fig. 2 we show results for the same states obtained in a complete CI calculation where all possible arrangements of three electrons are allowed from the molecular orbital set (1). Figs. 1 and 2 contain the correction to the Kahn and Jordan[33] data that was pointed out by Yu and Wing[36] for the $^2\Sigma_u^+$ ground state. In our full CI calculations performed out of the SCF molecular orbital set (1), at internuclear separations less than $R \approx 1.5a_0$, we observe that the expansion coefficients of the SCF molecular orbitals become extremely large, an indication of severe linear dependence in the basis set. In the CI technique only configuration coefficients are varied as those of the molecular orbitals are already fixed by the SCF method. Results therefore are untrustworthy at internuclear geometries less than $\approx 1.45a_0$ for the A state using this CI procedure and have not been included here. The X state results are in suitable agreement with the work of Kahn and Jordan[33] and have been included for bond separation $R \geq 1.3a_0$.

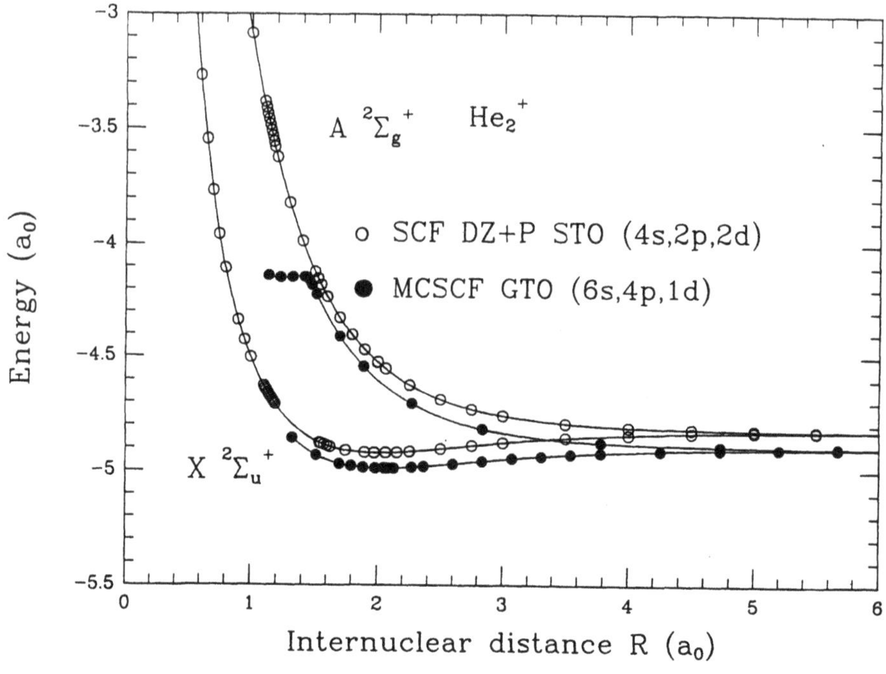

Figure 1: Comparison of the SCF potential energy calculations using the (4s,2p,2d) Slater basis for the X and the A states of He_2^+ with MCSCF (6s,4p,1d) Gaussian basis calculations of Kahn and Jordan.[33]

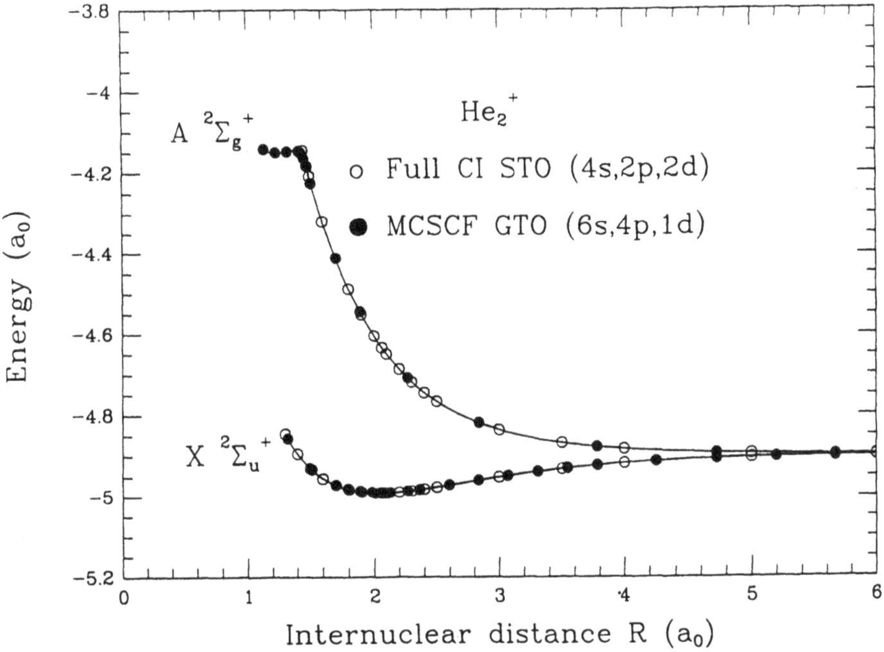

Figure 2: Full CI potential energy calculations for the X and A states of He_2^+ using the (4s,2p,2d) Slater basis compared with the MCSCF (6s,4p,1d) Gaussian basis calculations of Kahn and Jordan.[33]

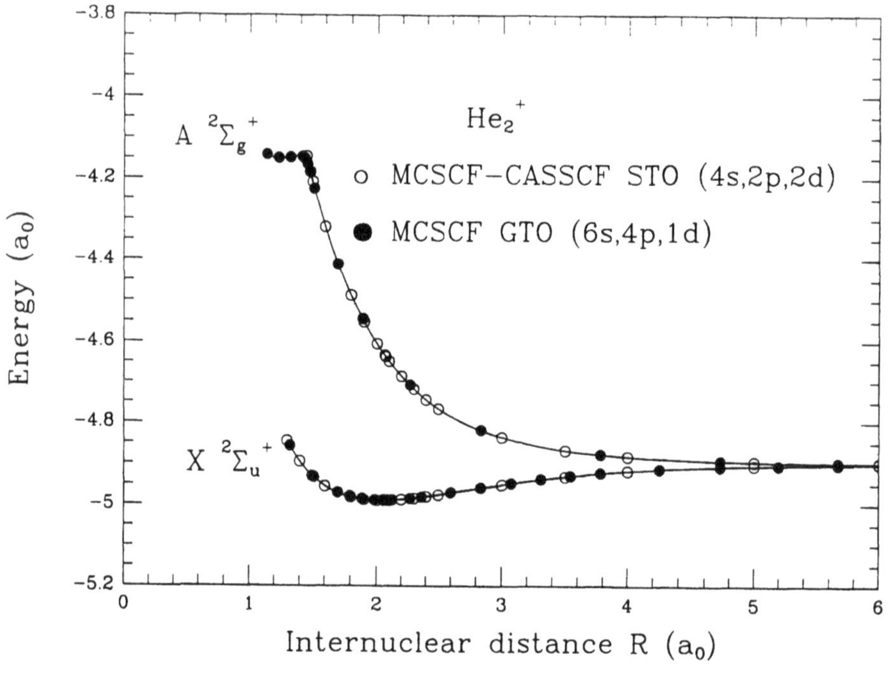

Figure 3: MCSCF-CASSCF potential energy calculations for the X and A states of He_2^+ using the (4s,2p,2d) Slater basis compared with the MCSCF (6s,4p,1d) Gaussian basis calculations of Kahn and Jordan.[33]

One must therefore use the complete active space self consistent field (CASSCF) with the MCSCF procedure[34,35], or variants of this approach, where both configuration and molecular orbital coefficients are optimized. Using the CASSCF procedure one gets better orbitals automatically. However in the case of this (4s,2p,2d) Slater basis there still are problems if ones use the whole set for the $^2\Sigma_u^+$ and $^2\Sigma_u^+$ symmetries at internuclear geometries towards the united atom limit($R \to 0$), where there is severe linear dependence. The main attraction of this (4s,2p,2d) Slater basis is in the advantage of doing CASSCF on the whole orbital set(1), as opposed to a complement of the set, which is the normal way structure calculations are performed when very large Slater basis sets are employed. It is for this reason that our full CI and CASSCF results are equivalent here.

The energies of the X and A states of He_2^+ in both the MCSCF-CASSCF and the full CI approximations are identical to six decimal places for $R \geq 1.45a_0$ using this (4s,2p,2d) Slater basis. For this reason we have chosen to plot them separately in Figs. 2 and 3. For comparison purposes we have plotted the results of Kahn and Jordan[33] in both these figures. We note that in both the MCSCF-CASSCF and full CI calculations using the (4s,2p,2d) Slater basis of Reagen *et al.*[24], the severe linear dependence in the basis set for internuclear geometries $R < 1.45a_0$ gives erroneous results for the A state. We have therefore not included them here. The X state results are in suitable agreement with the work of Kahn and Jordan[33] and have been included for bond separations $R \geq 1.3a_0$. The MCSCF-CASSCF calculations were carried out using the Alchemy II suite of codes[54]. Our scattering work requires the use of a common set of molecular orbitals to represent both the X and A states. One therefore can only use this (4s,2p,2d) Slater basis for scattering work at bond separations greater than $R \approx 1.5a_0$.

There would appear to be a severe lack of information on the transition dipole moment $\mu_{X\to A}(R)$ coupling the X and A states at all internuclear separations for the He_2^+ molecular ion. This is probably due to several anomalies. In the adiabatic regime avoided crossings occur in the $^2\Sigma_g^+$ states at internuclear geometries less than $R \approx 1.45a_0$[38]. In the diabatic regime the A $^2\Sigma_g^+$ state correlates in the united atom limit ($R \to 0$) to the $Be^+(1s2p^2)$ state[45,46]. At internuclear separations greater than approximately $6.0a_0$ the $A^2\Sigma_g^+$ and the $X^2\Sigma_u^+$ states become degenerate. The quantity $\mu_{X\to A}(R)$ is of importance because it allows cross section calculations to be performed on the radiative association process, $He^+ + He \to He_2^+ + h\nu$ and the photo-dissociation process, $h\nu + He_2^+ \to He^+ + He$. Cross section and rate coefficient results will therefore be sensitive to the accuracy of the values obtained for the transition moment, since they depend on the evaluation of the matrix of this quantity between bound and continuum nuclear wavefunctions. This fact has been noted in

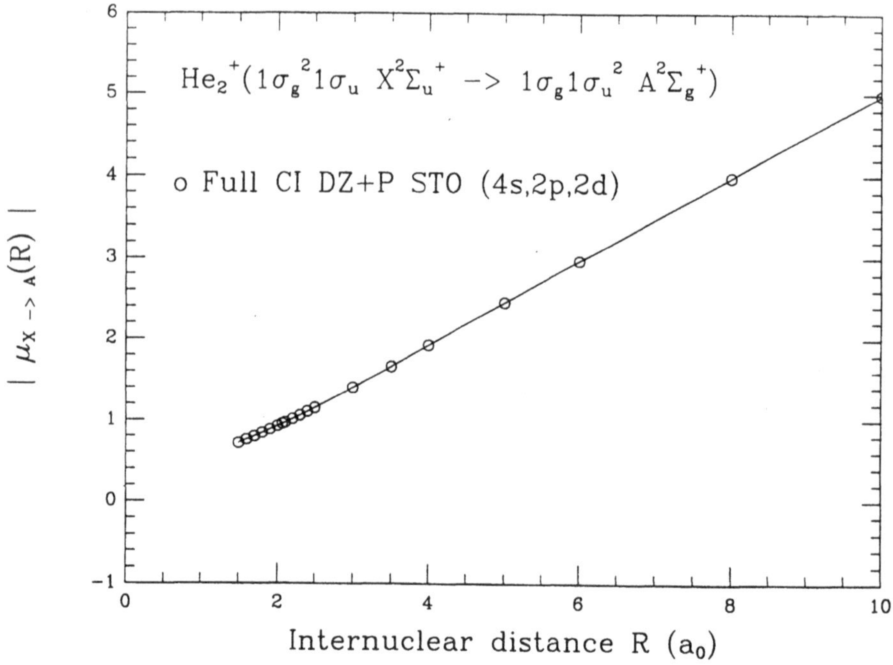

Figure 4: Magnitude of the transition dipole moment coupling the X and A states of the He_2^+ molecular ion. The calculations are carried out at the full CI level with the (4s,2p,2d) Slater basis. Results are displayed here only for $R \geq 1.5a_0$ due to the inadequacies of this basis set representing the excited A state of He_2^+ at lower bond separations.

work on radiative association[39,40] in He^+ - H collisions and radiative quenching[41] in $He(2s^1S)$ - He collisions. It has also been indicated in photodissociation calculations on molecules of astrophysical interest[42].

We note that functionally the MCSCF-CASSCF and the full CI approximations are equivalent here. We have therefore used the full CI method to calculate the transition dipole moment coupling the X and the A state in the He_2^+ molecular ion for internuclear geometries $R \geq 1.5a_0$. The full CI results were obtained using a modified version of the Alchemy I codes[28,29]. Our results are displayed in Fig. 4 for internuclear distances $1.5a_0 \leq R \leq 10a_0$ and illustrate clearly that the asymptotic value of the transition dipole moment is $\sim R/2$ in harmony with theoretical predictions. We note that calculations for the energies of the X and the A states including the transitions moment $\mu_{X \to A}(R)$ are currently being performed using much larger Slater bases[35] with similar techniques. Comparison with this work at a later stage will therefore serve as a guide to the limitations of this (4s,2p,2d) Slater basis used in the present work. The work carried out here indicates that this (4s,2p,2d) Slater basis is only suitable for calculations of the X $\to$ A transition moment at bond distances $R \geq 1.5a_0$.

It is worth noting that for internuclear distances $R \geq 1.5a_0$ the SCF-MO-LCAO wavefunctions qualitatively represent the X and the A states of He_2^+. However the full CI and MCSCF-CASSCF calculations at bond separations $R \geq 1.5a_0$ are in excellent agreement with previous work. In particular, for the X $^2\Sigma_u^+$ ground state there is good agreement with the larger basis set calculations of Sunil *et al.*[43], Kahn and Jordan[33], Yu and co-workers[47,44] and Bauschlicher *et al.*[48] A useful guide on the accuracy of the wavefunction for the X state is provided by comparing the value obtained for the dissociation constant D_e at equilibrium bond length with values available in the literature. Our results presented in table I for the dissociation constant D_e and the equilibrium bond separation r_e are in harmony with the theoretical and experimental values of previous work. Since the A state is repulsive there is no such similar comparison test and one must be guided by the results of previous calculations.

Table II is a comparison of the absolute energy of the ground state at the internuclear separation $R = 2.0625a_0$, with the values obtained by previous work, using various methods. It is clearly seen that our calculated full CI and MCSCF-CASSCF values, at this value of internuclear separation are in excellent agreement with the work carried out by others and are functionally equivalent.

Cross section calculations were carried out at the two internuclear separations, 1.8 a_0 and 2.0625 a_0. The direct potential in our outer region calculations contained, in addition to the dominant Coulomb term, contributions from the ground and excited state quadrupole moments as well as the dipole moment coupling the X and A states. At $R = 2.0625a_0$ which is relatively close to the $X^2\Sigma_u^+$ state equilibrium separation,

TABLE I. The dissociation energy $D_e(eV)$ and the equilibrium internuclear separation $r_e(a_0)$. The calculated values from our target representations are compared with existing ones for the ground state of He_2^+ ($1\sigma_g^2 1\sigma_u^1$ X $^2\Sigma_u^+$).

Method	$D_e(eV)$	$r_e(a_0)$
SCF-GTO (Maas *et al.*[37])	2.622	2.0050
SCF-STO (Present work)	2.648	2.0057
CI-GTO (Maas *et al.*[37])	2.459	2.0457
MCSCF-GTO (Kahn and Jordan[33])	2.451	2.0447
CI-STO (Liu[52])	2.469 ± 0.006	2.0440
CI-GTO/STO (Bauschlicher *et al.*[48])	2.469 ± 0.003	2.0428
CI-STO (Present work)	2.521	2.0444
MCSCF-CASSCF-STO (Present work)	2.521	2.0444
Expt. Estimates (Huber and Herzberg[53])	2.34 *and* 2.55	2.0424

TABLE II. Comparison of the calculated absolute SCF, full CI and MCSCF-CASSCF energies with other quantum chemical results for the X $^2\Sigma_u^+$ state of He_2^+. The internuclear separation is R=2.0625 a_0.

Method	Absolute Energy (a_0)	Basis set
SCF-GTO (Maas *et al.*[37])	−4.92222	6s,4p,1d
SCF-STO (Liu[52])	−4.92270	4s,3p,2d,1f
SCF-STO (Present work)	−4.92260	4s,2p,2d
MCSCF-GTO (Kahn and Jordan[33])	−4.98954	6s,4p,1d
CI-GTO (Maas *et al.*[37])	−4.98904	6s,4p,1d
CI-STO (Liu[52])	−4.99385	4s,3p,2d,1f
CI-CGTO (Baushlicher *et al.*[48])	−4.99384	6s,5p,3d,2f,1g
CI-STO (Present work)	−4.99093	4s,2p,2d,
MCSCF-CASSCF-STO (Present work)	−4.99093	4s,2p,2d,

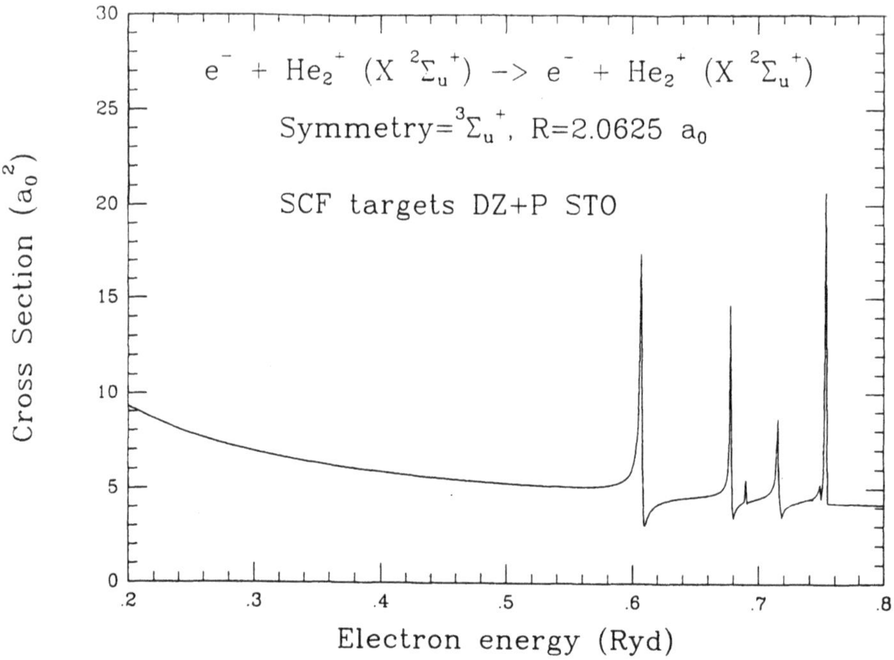

Figure 5: Cross sections for elastic scattering as a function of the incident electron energy. The symmetry is $^3\Sigma_u^+$ and the bond separation is 2.0625 a_0. SCF target wavefunctions are used to represent the X and A states of the He_2^+ molecular ion. The prominent Rydberg series is the $np\sigma$; the weaker one is the $nf\sigma$.

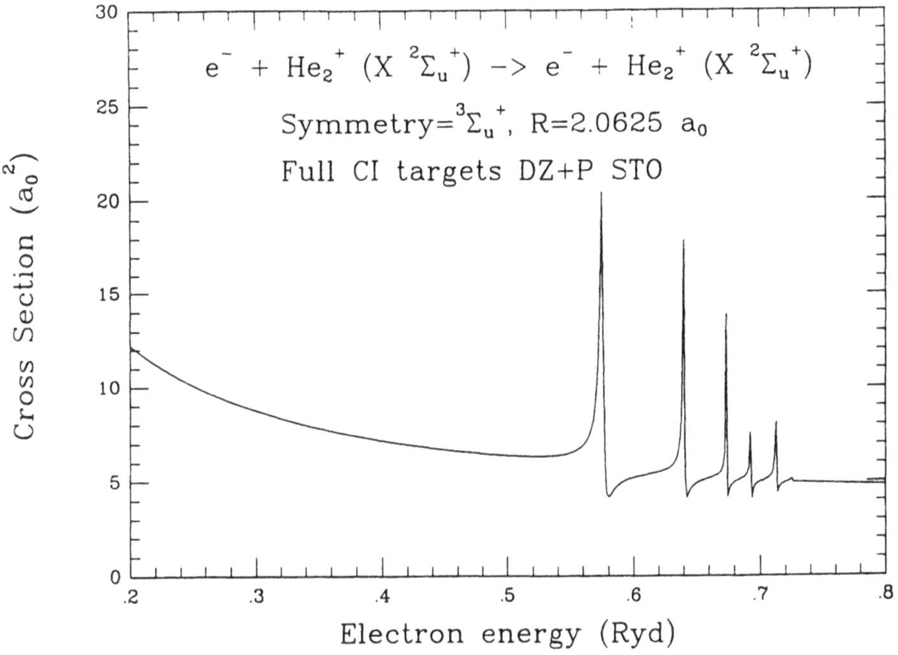

Figure 6: Cross sections for elastic scattering as a function of the incident electron energy. The symmetry is $^3\Sigma_u^+$ and the bond separation is 2.0625 a_0. Correlated CI target wavefunctions are used to represent the X and A states of the He_2^+ molecular ion. The prominent Rydberg series is the $np\sigma$; the weaker one is the $nf\sigma$.

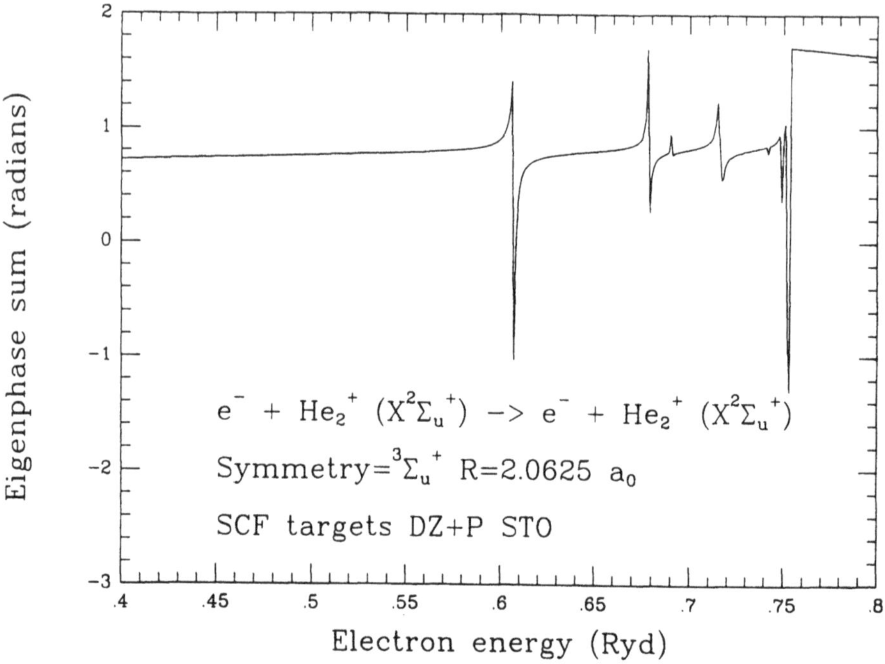

Figure 7: Eigenphase sum for elastic scattering as a function of the incident electron energy. The symmetry is $^3\Sigma_u^+$ and the bond separation is 2.0625 a_0. SCF target wavefunctions are used to represent the X and A states of the He_2^+ molecular ion. The prominent Rydberg series is the $np\sigma$; the weaker one is the $nf\sigma$.

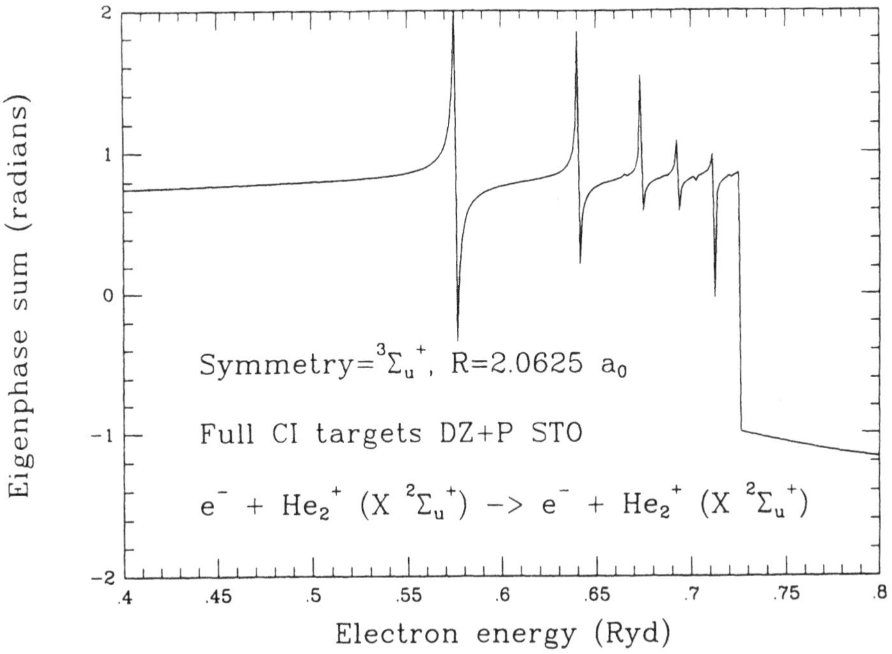

Figure 8: Eigenphase sum for elastic scattering as a function of the incident electron energy. The symmetry is $^3\Sigma_u^+$ and the bond separation is 2.0625 a_0. Correlated CI target wavefunctions are used to represent the X and A states of the He_2^+ molecular ion. The prominent Rydberg series is the $np\sigma$; the weaker one is the $nf\sigma$.

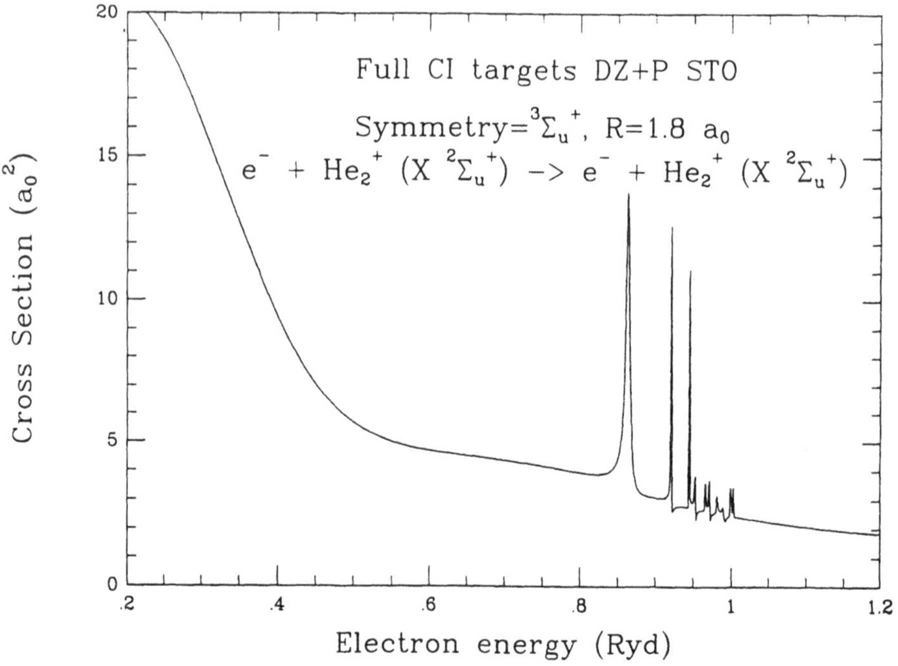

Figure 9: Cross sections for elastic scattering as a function of the incident electron energy. The symmetry is $^3\Sigma_u^+$ and the bond separation is 1.8 a_0. Correlated CI target wavefunctions are used to represent the X and A states of the He_2^+ molecular ion. The prominent Rydberg series is the $np\sigma$; the weaker one is the $nf\sigma$.

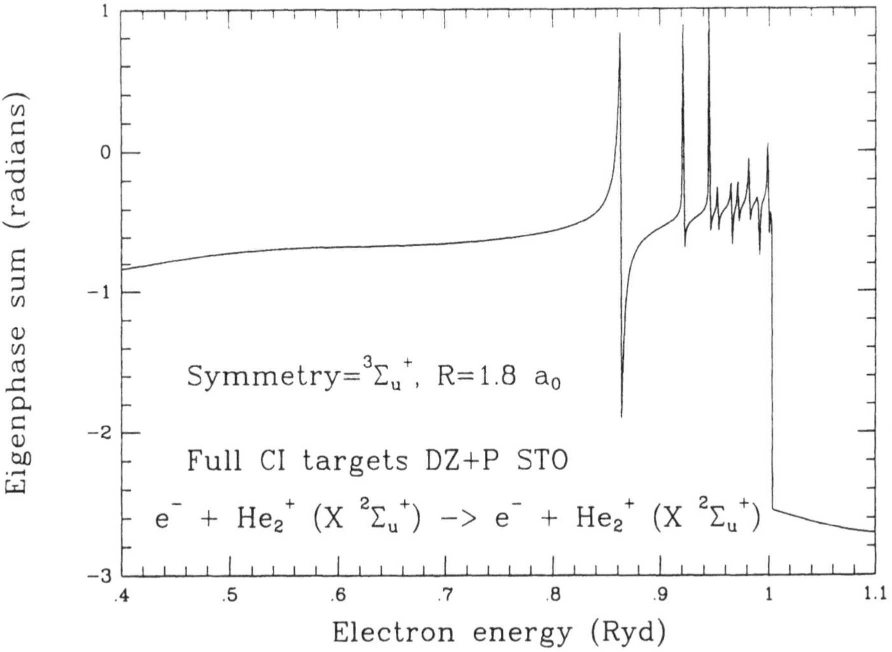

Figure 10: Eigenphase sum for elastic scattering as a function of the incident electron energy. The symmetry is $^3\Sigma_u^+$ and the bond separation is 1.8 a_0. Correlated CI target wavefunctions are used to represent the X and A states of the He_2^+ molecular ion. The prominent Rydberg series is the $np\sigma$; the weaker one is the $nf\sigma$.

TABLE III. Positions, widths, effective quantum number ν and quantum defect μ of the three lowest lying $1\sigma_u 1\sigma_g^2 np\sigma$ resonances in the $^3\Sigma_u^+$ scattering symmetry, for elastic scattering on the X $^2\Sigma_u^+$ state of He_2^+ using CI wavefunctions. The internuclear separation is R=2.0625 a_0.

Position (Ryd)	Width(Ryd)	ν	μ
0.5756	0.296^{-2}	2.5857	0.4243
0.6404	0.199^{-2}	3.4330	0.5670
0.6740	0.460^{-3}	4.2186	0.7814

we performed two scattering calculations. In one we used the single configuration representation of the target states while in the other the full CI representation, from the truncated natural orbital set (2) was used. Figs. 5 and 6 compares the cross sections and Figs. 7 and 8 the eigenphase sums for the two different target state approximations at the separation value of 2.0625 a_0 . At the other separation of 1.8 a_0 we used only the full CI representation of the target states. Fig. 9 illustrate the cross sections for this internuclear distance of 1.8 a_0. Whereas Fig. 10 illustrate the eigenphase sum in the energy region which includes the resonances in this symmetry for the same internuclear distance.

The eigenphase sums $\eta(E)$ shown in the figures were fitted to the Breit-Wigner form for several overlapping resonances, namely,

$$\eta(E) = \eta_0(E) + \sum_{i=1}^{M} \tan^{-1} \frac{\Gamma_i}{2(E_r^i - E)} \qquad (8).$$

Here $\eta_0(E)$ is the background phase shift at scattering energy E, E_r^i is the position of the $i^{\rm th}$ resonance, Γ_i the width and M is the number of overlapping resonances. The resonances obtained were fitted automatically using the computer code of Tennyson and Noble[49], with a linear form for the background phase shift. Analysis of the eigenphase sum yields the effective quantum number ν and the quantum defect μ for each resonance at this fixed internuclear separation. The fitted parameters from our scattering calculations using full CI target state representations are presented in table III. The results are for the internuclear separation 2.0625 a_0 and for the three lowest lying $1\sigma_g 1\sigma_u^2 np\sigma\ ^3\Sigma_u^+$ Rydberg states that were found in the elastic scattering cross section.

We note that the following resonance processes are occurring in the $^3\Sigma_u^+$ symmetry:

$$e^- + \mathrm{He}_2^+(1\sigma_g^2 1\sigma_u X^2\Sigma_u^+) \rightarrow \mathrm{He}_2(1\sigma_g 1\sigma_u^2 np\sigma, nf\sigma, \ldots\ ^3\Sigma_u^+) \qquad (9)$$

$$\downarrow$$

$$\mathrm{He}_2^+(1\sigma_g^2 1\sigma_u X^2\Sigma_u^+) + e^- \qquad (10).$$

Thus during the elastic scattering process there is a rearrangement of the He_2^+ core with the continuum electron attaching itself to the repulsive unstable excited A $^2\Sigma_g^+$ He_2^+ state. These quasi-bound states therefore consist of an unstable excited He_2^+ core surrounded by a diffuse Rydberg orbital. Two resonance series are clearly visible in our scattering cross sections and eigenphase sums. The more prominent is the $np\sigma$ series which we have analyzed in detail; the weaker is the $nf\sigma$ series. The $nf\sigma$ series was impractical to fit for numerical reasons and was not analyzed in detail here.

TABLE IV. Comparison of autoionizing widths Γ obtained using CI and SCF wavefunctions in the scattering approximation at R=2.0625 a_0 for the $^3\Sigma_u^+$ symmetry. The widths of the three lowest lying He_2^* $(1\sigma_g 1\sigma_u^2 np\sigma\ ^3\Sigma_u^+)$ quasi-bound states are compared.

Resonance	Γ_{CI} (Ryd)	Γ_{SCF} (Ryd)
(3pσ)	0.296^{-2}	0.175^{-2}
(4pσ)	0.199^{-2}	0.802^{-3}
(5pσ)	0.460^{-3}	0.544^{-4}

TABLE V. Calculated absolute energies, in Hartrees, of the lowest lying $^3\Sigma_u^+$ state of He_2^*. The energy of the $a\,^3\Sigma_u^+$ state is compared with the MCSCF calculations of Sunil *et al.*[43], the SA-MCSCF/SOCI calculations of Yarkony[50], and the SOCI ones of Konowalow and Lengsfield[51], at the two internuclear separations R=1.8 and 2.0625 a_0. Our results are obtained by performing a bound state R-matrix scattering calculation.

		$a^3\Sigma_u^+$ state
Method		Energy (a_0)
	$R = 1.8(a_0)$	
R-matrix		−5.09290
MCSCF		−5.13420
SA-MCSCF/SOCI		−5.14220
SOCI		−5.14417
	$R = 2.0625(a_0)$	
R-matrix		−5.13126
MCSCF		−5.13847
SA-MCSCF/SOCI		−5.14625
SOCI		−5.14805

Table IV presents a comparison of $np\sigma$ $^3\Sigma_u^+$ autoionizing widths calculated using CI and SCF target state representations at the internuclear separation $R = 2.0625a_0$. The substantial difference between the two sets of values demonstrates the importance of using suitably correlated wavefunctions in any theoretical model employed to obtain these widths. Therefore caution must be exercised in using results for any system obtained within the framework of an SCF representation.

Table V gives our results for the $a^3\Sigma_u^+$ He_2 bound state energy and compares this with detailed quantum chemistry work. Where necessary, for direct comparison, we have interpolated the potential energy curves of other authors. Our absolute energies are in satisfactory agreement with the MCSCF calculations of Sunil *et al.*[43], the state averaged MCSCF/second-order CI (SA-MCSCF-SOCI) calculations performed by Yarkony[50] and second order configuration interaction (SOCI) calculations of Konowalow and Lengsfield[51].

4. Conclusions and Future Directions

In conclusion, we have presented results on low energy elastic scattering by electron from He_2^+ molecular ions. It has also been demonstrated that sophisticated *ab initio* target wavefunctions and a better basis set are required for bond lengths less than $R \sim 1.5a_0$ to account properly for the anomalies in the excited states of the He_2^+ molecular ion. From our scattering work we have also been able to calculate the energies of the $a^3\Sigma_u^+$ bound state of the He_2^* molecule at two internuclear separations. *Ab initio* calculations of the lowest three $np\sigma$ resonances widths of the $^3\Sigma_u^+$ scattering symmetry are presented. In the future as an extension to this work we shall investigate other internuclear separations, so that nuclear vibrational motion may be incorporated into the study of this system in an adiabatic or non-adiabatic approach. Other scattering symmetries will also be studied in order to obtain total elastic and inelastic scattering cross sections. Future calculations will employ a multistate approximation that includes other low lying states of the He_2^+ target. We also plan to carry out *ab initio* studies of photoionization cross sections for several of the low lying excited singlet and triplet states of the helium dimer.

Acknowledgments

We would like to thank the National Science Foundation, the Minnesota Supercomputer Institute, and the Science and Engineering Research Council in the United Kingdom for their financial support during the course of this research. We thank Professor Ronald Gentry, Dr. Hans Peter Helm and Professor Brian Mitchell for useful discussions on the feasibility of performing electron scattering experiments for this molecular complex, and for enlightening discussions on the prospect of measurements for vibrational excitation cross sections in the X state of the He_2^+ molecular ion by electron impact. The computations were performed on the Cray-X-MP/4P/16 MW at the Rutherford Appleton Laboratory, the Cray-X-MP/4P/16 MW, and the Cray-2/4P/512 MW at the Minnesota Supercomputer Center. BMMcL would like to thank the Minnesota Supercomputer Institute for the support of a visiting research fellowship during the course of this work. The MCSCF-CASSCF calculations were performed on the IBM 3090-600J using the Alchemy II codes[54], at the Cornell National Supercomputer Facility, a resource of the Center for Theory and Simulation in Science and Engineering at Cornell University, which receives funding by the National Science Foundation, New York State, the IBM Corporation and members of the Corporate Research Institute. This work was completed during a short term visit by CJG to the Institute for Theoretical Atomic and Molecular Physics at the Harvard-Smithsonian Center for Astrophysics, which is funded by a grant from the National Science Foundation.

References

[1]W. R. Gentry, private communication (1989).

[2]B. J. Garrison, W. H. Miller and H. F. Schaefer III , J. Chem. Phys. **59**, 3193 (1973).

[3]M. W. Müller, W. Bussert, M. W. Ruf and H. Hotop, Phys. Rev. Lett. **59**, 2279 (1987).

[4]J. C. Hill, L. L. Hatfield, N. D. Stockwell, and G. K. Walters, Phys. Rev. A **5** 189 (1972).

[5]J. S. Cohen , Phys. Rev. A **13**, 86 (1976); **13**, 99 (1976).

[6]R. J. Bieniek, M. W Müller and M Movre, J. Phys. B. **23**, 4521 (1990).

[7]P. S. Julienne and F. H. Mies, J. Opt. Soc. Am. B **6**, 2257 (1989)

[8]B. M. McLaughlin, C. J. Gillan, P. G. Burke and J. S. Dahler, Nucl. Instrum. & Methods B **53**, 518 (1991).

[9]J. Weiner, F. Masnou-Seeuws and A. Giusti-Suzor, Adv. At. Mol. Opt. **26**, 209 (1990).

[10]A. Niehaus, Phys. Rep. **186**, 149 (1990).

[11]A. N. Klucharev and V. Vujnovic, Phys. Rep. **185**, 55 (1990).

[12]K. T. Gillen, Phys. Rev. A **39**, 2248 (1989).

[13]P. Monchicourt, P. Padel, D. Dubreuil and J. J. Laucagne, Phys. Rev. A **40**, 1147 (1989);**40**, 6706 (1989).

[14]S. Lepp, A. Dalgarno and R. McCray, Astrophys. J. **358**, 262 (1990).

[15]B. M. McLaughlin, P. Scott and J. S. Dahler, J. Phys. B. , in preparation to be submitted (1991).

[16]S. L. Guberman, *Physics of Ion-Ion and Electron-Ion Collisions*, ed. by F. Brouillard and J. W. McGowan, (Plennum Press, New York and London 1983) 167.

[17]C. J. Gillan, O. Nagy, P. G. Burke and L. A. Morgan, J. Phys. B. **20**, 4585 (1987).

[18]J. Tennyson, C. J. Noble and S. Salvini, J. Phys. B. **17**, 905 (1984).

[19]J. Tennyson and C. J. Noble, J. Phys. B. **18**, 155 (1985).

[20]I. Shimamura, C. J. Noble and P. G. Burke, Phys. Rev. A **41**, 3545 (1990).

[21]J. Tennyson, J. Phys. B. **21**, 805 (1988).

[22]B. K. Sarpal, S. E. Branchett, J. Tennyson and L. Morgan, J. Phys. B. in process of publication (1991).

[23]M. J. Seaton, J. Phys. **18**, 2111 (1985).

[24]P. N. Reagen, J. C. Browne and F. A. Matsen, Phys. Rev. **132**, 304 (1963).

[25]I. Shavitt, *Modern Theoretical Chemistry : Vol. 3, Methods of Electronic Structure Theory*, Edited by H. F. Schaefer III, (Plenum, New York and London, 1977) 189.

[26]R. S. Mulliken and W. C. Ermler, *Polyatomic Molecules: Results of Ab Initio Calculations* (Academic Press, New York, 1981).

[27]C. Bloch, Nucl. Phys. **4**, 87 (1957).

[28]A. D. McLean, *Proceedings of the Conference on Potential Energy Surfaces in Chemistry*, ed. by W. A. Lester Jr. (IBM, San Jose, 1971) 87.

[29]C. J. Noble, Daresbury Laboratory Report DL/SCI/TM33T (1982).

[30]P. J. A. Buttle, Phys. Rev. **162**, 719 (1967).

[31]K. L. Baluja, P. G. Burke and L. A. Morgan, Comp. Phys. Commun. **27**, 299 (1982).

[32]C. J. Noble and R. K. Nesbet, Comp. Phys. Commun. **33**, 399 (1984).

[33]A. Kahn and K. D. Jordan, Chem. Phys. Lett. **128**, 368 (1986).

[34]A. C. Wahl and G. Das, *Modern Theoretical Chemistry : Vol. 3, Methods of Electronic Structure Theory*, Edited by H. F. Schaefer III, (Plenum, New York and London, 1977) 51.

[35]R. E. Farren and K. P. Kirby, private communication (1991).

[36]N. Yu and W. H. Wing, Phys. Rev. Lett. **59**, 2055 (1987).

[37]J. G. Maas, N. P. F. B. van Asselt, P. J. C. H. Nowark, J. Los, S. D. Peyerimhoff and R. J. Buenker, Chem. Phys. **17**, 217 (1976).

[38]A. Metropoulos, C. A. Nicolaides and R. J. Buenker, Chem. Phys. **114**, 1 (1989).

[39]B. Zygelman, A. Dalgarno, M. Kimura and N. F. Lane, Phys. Rev. A **40** 2340 (1989).

[40]B. Zygelman and A. Dalgarno, Astrophys. J. **365** 238 (1990).

[41]B. Zygelman and A. Dalgarno, Phys. Rev. A **38** 1877 (1988).

[42]E. Dishoeck, Ph. D. thesis, *Photodissociation and excitation of Interstellar Molecules : Calculations and Astrophysical Applications*, Leiden, (1984).

[43]K. K. Sunil, J. Lin, H Siddiqui, P. E. Siska, K. D. Jordan and R. Shephard, J. Chem. Phys. **78**, 6190 (1986).

[44]N. Yu and L. Adamowicz, J. Chem. Phys. **90**, 4392 (1989).

[45]M. Barat, D. Dhuico, R. Francois, R. McCarroll, R. D. Piacentini, and A. Salin, J. Phys. B. **5**, 1343 (1972).

[46]A. Metropoulos, Z. Naturforsch. **44a**, 683 (1989).

[47]N. Yu, W. H. Wing and L. Adamowicz, Phys. Rev. Lett. **62**, 253 (1989).

[48]C. W. Bauschlicher Jr., H. Partridge and D. Ceperley, Chem. Phys. Lett. **160**, 183 (1989).

[49]J. Tennyson and C. J. Noble, Comp. Phys. Commun. **33**, 421 (1984).

[50]D. R. Yarkony, J. Chem. Phys. **90**, 7164 (1989).

[51]D. D. Konowalow and B. H. Lengsfield III, J. Chem. Phys. **87**, 4000 (1987).

[52]B. Liu, Phys. Rev. Lett. **27**, 1251 (1971).

[53]K. P. Huber and G. Herzberg, *Molecular Spectra and Molecular Structure IV: Constants of Diatomic Molecules* (Van Nostrand, New York and London, 1979).

[54]The Alchemy I and II suite of programs were originally developed by P. S. Bagus, B. Liu, A. D. McLean, M. Yoshime and B. Lengsfield III.

ALIGNMENT AND ORIENTATION PARAMETERS FOR ELECTRON EXCITATION OF XENON

T. ZUO, R. P. McEACHRAN and A. D. STAUFFER
Department of Physics, York University, Toronto, Canada M3J 1P3

ABSTRACT

The excitation of xenon atoms by electron impact is carried out in a fully relativistic version of the distorted-wave method. Atomic states are represented by Dirac-Fock wavefunctions while the distorted waves are obtained by solving the Dirac equations with scattering state boundary conditions. Alignment and orientation parameters for excitation of the two lowest states with total angular momentum J = 1 are presented and compared with recent experimental results. It is shown that spin flip is more easily measured via angular correlation measurements in the collision frame rather than by coherence analysis of the emitted photon.

1. Introduction

It is well known experimentally[1] that electrons elastically scattered from xenon atoms at low energies are spin polarized. This is due to relativistic effects, primarily spin-orbit coupling, and can be theoretically described most naturally via the Dirac equations.[2] If the xenon atoms are excited in the collision then the possibility of observing in coincidence both the scattered electron and the photon emitted when the excited atom decays leads to detailed information about the excited state of the atom after the collision. This information can be obtained by measuring the angular distribution or the polarization of the emitted photon in coincidence with the the scattered electron and the measured quantities are known collectively as alignment and orientation parameters.[3] Relativistic effects will also play an important role in determining these parameters. For example, the spin of the electron cannot change during the collision in a description that uses the ordinary Schrödinger equation. In addition, these phenomena are only observable if the various fine structure levels are resolved. However, the fine structure splitting results from a coupling of the spin and orbital angular momentum of the atomic electrons which is also a relativistic effect. Thus we have decided to calculate such parameters in a consistently relativistic description. In particular, we have used this description to calculate the electron excitation of xenon atoms to the $6s\,[3/2]_1^o$ and $6s\,[1/2]_1^o$ states in a distorted-wave formulation.

2. Theory

We start from the usual distorted-wave T-matrix[4] but use Dirac-Fock wavefunctions for the bound states. These are readily calculated using the MCDF program of Grant *et al.*[5] The intermediate coupled wavefunctions for the excited states are given by

$$6s[1/2]_1^o = c_1\,(5\bar{p}6s)_{J=1} - c_2\,(5p^3 6s)_{J=1} \tag{1a}$$

$$6s[3/2]_1^o = c_2\,(5\bar{p}6s)_{J=1} + c_1\,(5p^3 6s)_{J=1} \tag{1b}$$

where $(5\bar{p}6s)_{J=1}$ and $(5p^3 6s)_{J=1}$ are the j-j coupled Dirac-Fock wavefunctions of the first two excited configurations with total angular momentum J=1. The coefficients $c_1 = 0.9974$ and $c_2 = 0.0710$ are obtained from the MCDF program. Alternately, we can express these wavefunctions in the L-S coupling scheme as

$$6s[1/2]_1^o = -a_2\,|\,{}^1P_1\rangle + a_1\,|\,{}^3P_1\rangle \tag{2a}$$

$$6s[3/2]_1^o = a_1\,|\,{}^1P_1\rangle + a_2\,|\,{}^3P_1\rangle \tag{2b}$$

where $|\,{}^1P_1\rangle$ and $|\,{}^3P_1\rangle$ are the singlet and triplet L-S coupled wave functions, respectively, for the $5p^5 6s$ configuration with J = 1. The coefficients have the values $a_1 = 0.7734$ and $a_2 = 0.6339$.

The distorted waves are calculated by solving the Dirac equations for a scattered electron in the field of a distortion potential which is chosen as the static potential of the final excited state. This choice has been shown to consistently yield the best results.[6] Exchange is treated exactly via the antisymmetrization of the total wavefunction. These equations are identical to the elastic scattering case[2] except for the specific choice of potential. The formulas for the T-matrices reduced to radial form along with the details of the numerical methods used are given by Zuo.[7]

In cases where the incident electrons are unpolarized and no spin analysis of the scattered electrons is carried out, there are four independent parameters which can be measured in addition to the differential cross section. In coherence analysis experiments they are chosen to be the Stokes parameters P_1, P_2 and P_3 of photons emitted perpendicular to the scattering plane along with P_4, the linear polarization of photons emitted in the scattering plane perpendicular to the direction of the incident electron. In order to describe the appearance of the final state[3], these quantities can be converted into the parameters $(L_\perp^+, \gamma, P_l^+, \rho_{00})$ by

$$L_\perp^+ = -P_3 \tag{3a}$$

$$\gamma = \frac{1}{2}\arg(P_1 + iP_2) \tag{3b}$$

$$P_l^+ = (P_1^2 + P_2^2)^{1/2} \tag{3c}$$

$$\rho_{00} = (1 + P_1)(1 - P_4)/(4 - (1 - P_1)(1 - P_4)) \tag{3d}$$

Alternately, in angular correlation measurements where the angular distribution of the emitted photon is measured in coincidence with the scattered electron, the four parameters are usually chosen to be (λ, χ, $\cos\varepsilon$, $\cos\Delta$) and are given in terms of the T-matrix elements by, for example, Bartschat and Madison.[6] Andersen *et al*[3] give detailed formulas connecting these different sets of basic parameters. In particular, we have the relation

$$\rho_{00} = \tfrac{1}{2}(1 - \lambda)(1 - \cos\varepsilon) \tag{4}$$

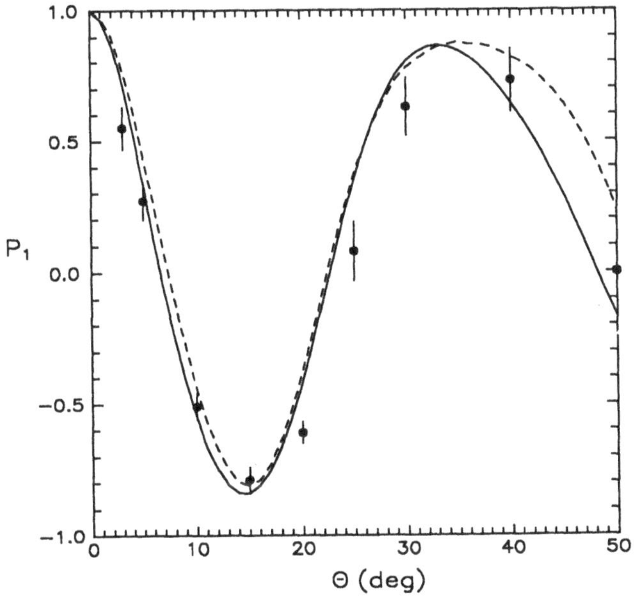

Figure 1. P_1 Stokes parameter for the electron impact excitation of the xenon $6s[3/2]_1^0$ state at 30 eV. ——, present results; - - -, calculations of Bartschat and Madison;[6] •, experimental results of Corr *et al.*[8]

3. Results

To date the only reliable measurements of orientation and alignment parameters for xenon are from the coherence analysis experiments of Corr *et al* [8] who only report results for the 6s $[3/2]_1^o$ state. They have published data for P_1, P_2 and P_3 for incident electrons at 30 eV. In addition there have been semi-relativistic calculations of these parameters by Bartschat and Madison.[6] We compare our results to the experimental measurements and previous calculations in Figures 1 to 3. In general, the agreement between these three sets of data is very good.

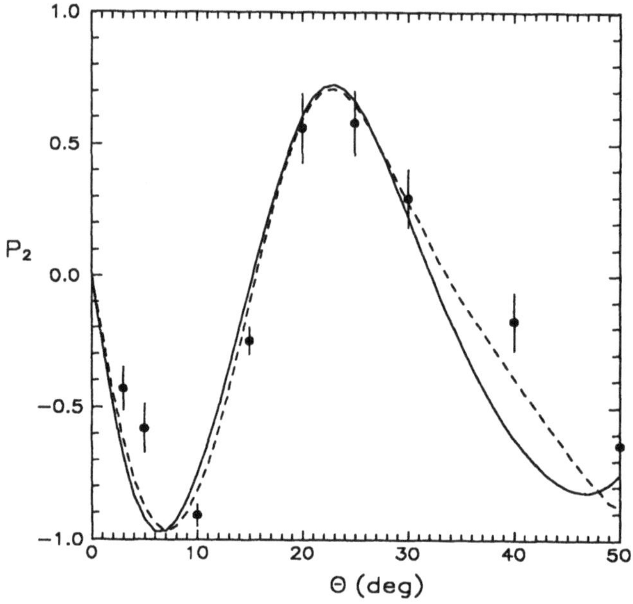

Figure 2. Same as Figure 1 but for the P_2 Stokes parameter.

The fourth parameter, which can be taken to be ρ_{00}, is very small and cannot be distinguished from zero in the present experiments.[8] Since this parameter is a measure of the spin flip of the scattered electron, it is unfortunate that an accurate value for it cannot be determined. As an alternative measure of spin flip, we suggest that $\cos\varepsilon$ be determined. In the absence of spin-dependent forces, this parameter is unity. However, our calculations show that significant deviations from this value can occur at certain scattering angles. To emphasize this point we have done model calculations based on the L-S coupled formulas of Bartschat and Madison.[6] We display the resulting magnitudes of ρ_{00} and $1 - \cos\varepsilon$ as a function of the parameter a_2 from Eq 2b in Figures 4 and 5. From these figures it is obvious that $\cos\varepsilon$ is a much more sensitive quantity than ρ_{00} for studying transitions in xenon. In our present fully relativistic calculations a_2 has the value 0.6336 and 1 - cos ε attains a value of 0.35 at an electron scattering angle of 34 degrees. At this angle, the differential cross section is large enough so that experimental measurements can be made with presently existing equipment.

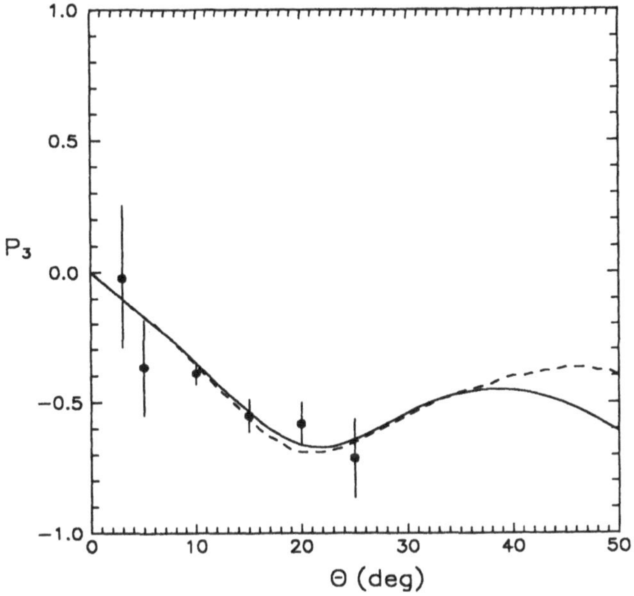

Figure 3. Same as Figure 1 but for the P_3 Stokes parameter.

In principle, cos ε can be determined from angular correlation measurements. However, in the expression for the angular correlation function, this parameter always occurs in the combination (1 - λ) cos ε.[3] From Eq. 4, it is obvious that if ρ_{00} is small, then (1 - λ) will also be small whenever cos ε differs appreciably from unity. Thus a direct measurement of cos ε from angular correlation measurements does not seem possible for this particular transition. Since $(1 - \lambda) = 2\,\sigma(1)/\sigma$, where $\sigma(1)$ is the differential cross section for the excitation of the magnetic sublevel of the $[3/2]_1^o$ state with M = 1 and σ is the sum of the differential cross sections for the excitation of all three possible sublevels, a direct determination of this quantity should be possible from a time-reversed experiment.[3] In this type of experiment, the M = 1 sublevel of the excited state is prepared by optical pumping and is followed by a superelastic collision with an electron. It should be emphasized that these experiments are not coincidence experiments and hence the measured signal is correspondingly larger. On the other hand, population densities of the excited states obtained via optical pumping are much less than ground state densities. Such experiments have been successfully caried out for alkali atoms.[9]

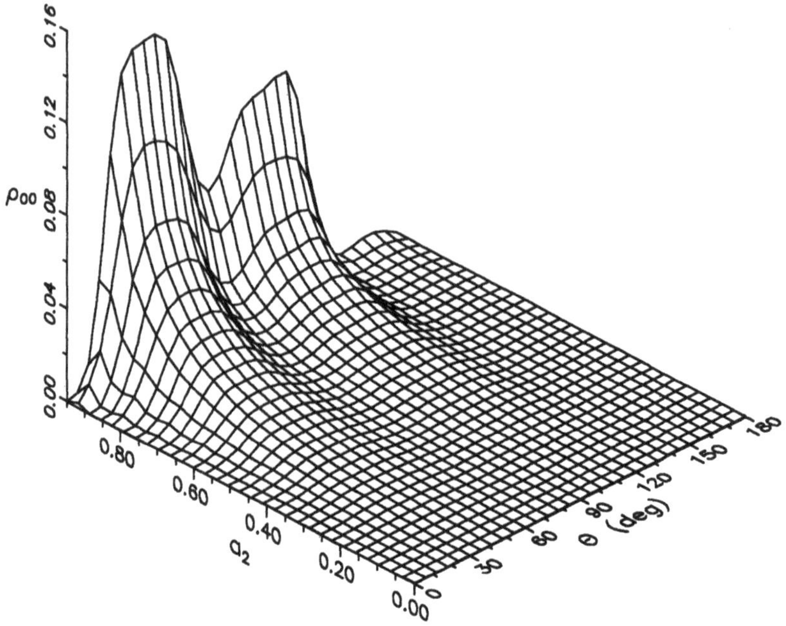

Figure 4. Variation of the parameter ρ_{00} as a function of the coefficient a_2 for the excitation of the $6s[3/2]_1^0$ state of xenon by 30 eV electrons.

Obtaining (1 - λ) directly in this way would provide an accuracy much greater than is possible from angular correlation measurements, particularly when this quantity is small. Having obtained (1 - λ) independently, angular correlation measurements can then be used to determine cos ε. Thus the presence of spin flip could be demonstrated experimentally.

4. Conclusions

We have shown that calculations of alignment and orientation parameters using a fully relativistic approach are straightforward and yield accurate results. Such a method is the natural one for scattering from heavier atoms where spin-orbit coupling plays a major role in determining the excited states of the atom. It is also the preferred method when the spin of the electron is separately determined.

We have also suggested an experimental means of determining the spin flip for the transitions in question. This method circumvents the limitations of the present experimental approaches.

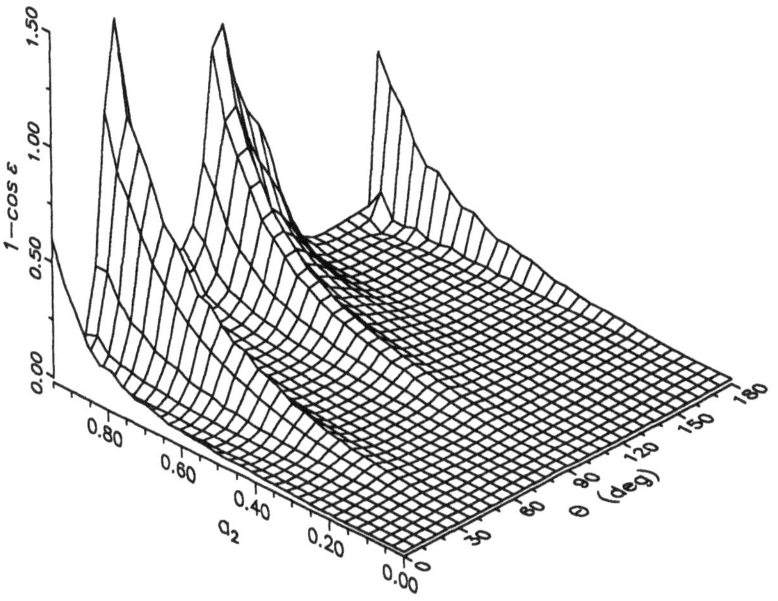

Figure 5. Same as Figure 4 but for the parameter 1-cos ε.

5. References

1. O. Berger, W. Wübker, R. Möllenkamp and J. Kessler, *J. Phys.* *B* **15** (1982) 2473; O. Berger and J. Kessler, *J. Phys.* *B* **19** (1986) 3539.
2. R. P. McEachran and A. D. Stauffer, *J. Phys.* *B* **19** (1986) 3523.
3. N. Andersen, J. W. Gallagher and I. Hertel, *Phys. Rep.* **165** (1988) 1.
4. Y. Itikawa, *Phys. Rep.* **143** (1986) 69.
5. I. P. Grant, B. J. McKenzie, P. H. Norrington, D. F. Mayers and N. C. Pyper, *Comput. Phys. Commun.* **21** (1980) 207.
6. K. Bartschat and D. H. Madison, *J. Phys.* *B* **20** (1987) 5839.
7. T. Zuo, *Ph.D. Thesis* (York University, 1991).
8. J. J. Corr, P. Plessis and J. W. McConkey, *Phys. Rev.* *A* **42** (1990) 5240.
9. J. J. McClelland, M. H. Kelley and R. J. Celotta, *J. Phys.* *B* **20** (1987) L385.

EXPLORING THE IONIZATION THRESHOLD LAW WITH HYDROGEN: PAST, PRESENT, AND FUTURE

M.S. Lubell and X.Q. Guo
Department of Physics, City College of the CUNY, New York, NY, 10031, USA

ABSTRACT

Recently, we completed an investigation of electron-impact ionization of atomic hydrogen in which we determined experimentally the spin-dependent asymmetry Δ defined by $\Delta = [R(\uparrow\downarrow) - R(\uparrow\uparrow)] / [R(\uparrow\downarrow) + R(\uparrow\uparrow)]$, where $R(\uparrow\downarrow)$ and $R(\uparrow\uparrow)$ are respectively the ionization rates for incident and atomic electron spins antiparallel and parallel. The results of our studies reveal statistically significant deviations from the predictions of the conventional Wannier threshold theory for two-electron escape. In this paper we discuss the principles of the experiment, summarize the data we obtained, and suggest some qualitative explanations for the energy dependence of the asymmetry we observed. In addition, we briefly examine the results of several other ionization threshold measurements carried out previously in different laboratories and present new interpretations of the published data from these experiments. Finally, we propose a new approach to the complementary study of double photodetachment of the negative hydrogen ion which has the potential for further elucidating the nature of the threshold behavior of two-electron escape.

1. Introduction

1.1. *Historical Background*

The threshold law for double escape of two electrons from a charged center[1] has been the subject of extensive theoretical[2-18] and experimental[19-33] investigation for almost four decades. Reduced to its most elemental form, the problem presents one of the clearest examples of the influence of electron correlations, and as such it has the potential for providing valuable insight into one of the underlying constructs of modern atomic physics. Until recently, all experimental evidence[19-31] appeared to be consistent with the conventional Wannier approach[2-17] to threshold behavior, which predicts a power-law scaling of the total ionization cross section with the kinetic energy available for double escape. The classical Wannier law with its quantum mechanical extensions takes the form

$$\sigma \propto E^{m} \tag{1}$$

with m given by

$$m = \frac{1}{4}\left(\frac{100Z - 9}{4Z - 1}\right)^{\frac{1}{2}} - \frac{1}{4}, \tag{2}$$

where $+Ze$ is the charge of the residual ionic center from which the electrons escape. For $Z = 1$, m has the value 1.127. As Greene and Rau have shown,[11] the expression for m is valid for all two-electron final states except the $^1P^e$ and $^3S^e$ configurations, for which the threshold behavior is suppressed by a higher value of m. Specifically, for $Z = 1$, the energy dependence of the cross sections for these two states is characterized by an exponent of 3.881 or 3.381, depending upon the detailed assumptions made in the calculation.[8,13]

As Rau[1] has noted, the relative simplicity of all threshold laws derives from an assumption, originally advanced by Wigner,[34] that the energy dependence of the cross section in the threshold region is insensitive to the detailed knowledge of the "reaction zone," a region of space in which the particles are close together and hence highly interactive. This assumption, also known as the ergodic hypothesis, permits an analysis of threshold behavior to be restricted to the mechanics of the escape process alone, instead of requiring a detailed understanding of the dynamics of the collision as a whole. This purely statistical or classically chaotic approach to the early stages of the break-up of an atomic system near threshold has been challenged by Temkin and his co-workers[18] and will probably undergo further scrutiny by workers in the field of quantum chaos.

Experimental tests of the classical Wannier threshold law and its quantum mechanical extensions generally fall into four categories — measurements of the energy dependence of total cross sections, energy-partitioning functions, angular distributions, and spin-dependent asymmetry parameters. The first two of these, historically the earliest as well, while in some sense technologically simpler than the last two, are more prone to systematic error and hence poorer sesitivity, since they require an independent judgment of the way in which detector acceptances and analyzer sensitivities vary with energy. By contrast, the experimental investigations of the energy dependence of angular distributions and spin asymmetries are essentially free from such difficulties. A comprehensive examination of threshold behavior nonetheless should encompass all four approaches, particularly since the range of validity of the underlying assumptions is open to question. During the last twenty-five years, in fact all four have been used in a number of very sophisticated experiments, concentrating on hydrogen, helium, and the alkali atoms.

1.2. *Spin-Dependent Asymmetry Relations*

Recently we completed a study of the spin-dependence in electron-impact ionization of atomic hydrogen,[32,33,35] in which we employed crossed beams of polarized particles. As has been the case in other ionization asymmetry experiments, the quantity we measured was the ionization counting-rate asymmetry, Δ, defined by

$$\Delta = \frac{R(\uparrow\downarrow) - R(\uparrow\uparrow)}{R(\uparrow\downarrow) + R(\uparrow\uparrow)}, \tag{3}$$

where $R\,(\uparrow\downarrow)$ and $R\,(\uparrow\uparrow)$ are respectively the ionization counting rates, normalized to incident beam intensities, for antiparallel and parallel spin configurations of the incident and atomic electrons. For the electron-hydrogen collision, the counting-rate asymmetry, Δ, is related to the cross-section asymmetry, A, by the general expression

$$\Delta = AP_e P_H (1 - F_2)|\cos\alpha|, \tag{4}$$

where $\boldsymbol{P_e}$ and $\boldsymbol{P_H}$ are the polarization vectors of the electron and hydrogen beams respectively, α is the angle between these vectors, F_2 is the fraction of events due to molecules, and A is defined by

$$A = \frac{\sigma(\uparrow\downarrow) - \sigma(\uparrow\uparrow)}{\sigma(\uparrow\downarrow) + \sigma(\uparrow\uparrow)} \tag{5}$$

with $\sigma(\uparrow\downarrow)$ and $\sigma(\uparrow\uparrow)$ the corresponding ionization cross sections for spins antiparallel and parallel. Since the low-energy electron hydrogen problem can be treated non-relativistically to great accuracy, the cross-section asymmetry A can be expressed as

$$A = \frac{|f||g|\cos\theta}{\bar{\sigma}}, \tag{6}$$

where f is the direct-scattering amplitude, g is the exchange amplitude, θ is their relative phase, and $\bar{\sigma}$ is the spin-averaged cross section given by

$$\bar{\sigma} = \tfrac{1}{4}|f + g|^2 + \tfrac{3}{4}|f - g|^2, \tag{7}$$

$\sigma_s = |f + g|^2$ being the singlet cross section and $\sigma_t = |f - g|^2$, the triplet. Alternatively A can be written as

$$A = \frac{\sigma_s - \sigma_t}{4\bar{\sigma}} = \frac{1 - r}{1 + 3r}, \tag{8}$$

where r is the ratio of the triplet to singlet cross section. In the context of the conventional Wannier theory, Eqs. 3-8 demonstrate that the counting-rate asymmetry should be constant throughout the energy region for which the threshold law is valid.

2. The Electron-Hydrogen Spin-Dependent Asymmetry Experiment

2.1 *Overview of Experimental Configuration*

Our experiment employed a crossed-beams geometry, as illustrated in Fig. 1. Polarized electrons were produced by photoemission from a <100> GaAs crystal with the use of 780-nm circularly polarized light generated from a GaAlAs diode laser and prepared by a combination of a half-wave ($\lambda/2$) and a quarter-wave ($\lambda/4$) retardation plate, as shown in the figure. Before reaching the interaction region, the extracted electron beam passed through a set of electrostatic electron-optical elements comprising a 90° spherical bender, a series of lenses and steerers, and a 180° hemispherical monochromator, the overall geometric orientation of which ensured that the electron polarization vector, $\boldsymbol{P_e}$, was pointing either parallel or antiparallel to the direction of the hydrogen beam. The usable electron current, which varied between 30 and 150 nA, was monitored for normalization purposes by a Faraday cup located just beyond the interaction region.

The polarized atomic hydrogen beam originated in an rf dissociation source and underwent high-field state selection in a pair of permanent hexapole magnets which also served to focus the $m_S = +1/2$ atoms at the interaction region. Upon emerging from the second magnet, the atoms entered a small solenoidal field which adiabatically rotated their spins (and consequently their polarization vector, $\boldsymbol{P_H}$) into a longitudinal direction either parallel or antiparallel to the direction of the hydrogen beam. Additional coils maintained the magnetic field in the longitudinal direction and trimmed it to ~100 mG at the crossed-beams interaction site. Prior to entering the interactioon region, the hydrogen beam was collimated to a radial dimension of 3 mm, resulting in a usable average target density of ~10^{10} atoms/cm^3. For normalization purposes, the hydrogen beam was sampled at the end of its line by a quadrupole mass analyzer (QMA). For purposes of "real-time" background subtraction, the hydrogen beam was equipped with a mechanical chopper located between the two hexapole magnets, as shown in Fig. 1

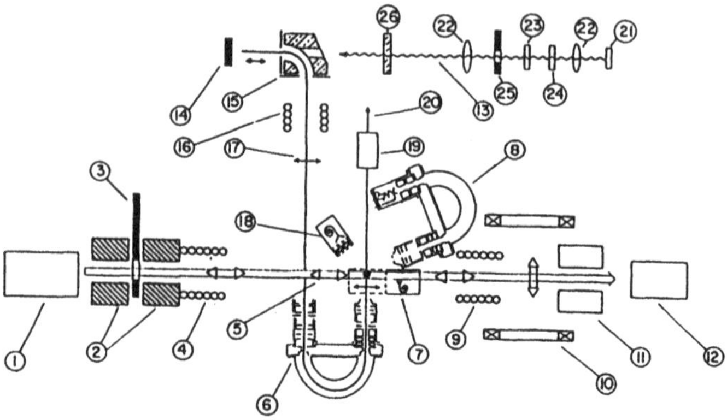

Fig. 1. Schematic layout of the experiment showing (1) hydrogen source, (2) hexapole magnets, (3) hydrogen beam chopper, (4) solenoid, (5) hydrogen polarization vector, (6) electron monochromator, (7) ion detector, (8) electron spectrometer, (9) solenoid, (10) adiabatic spin rotator, (11)Stern-Gerlach polarimeter, (12) QMA, (13) 780-nm laser light, (14) GaAs crystal, (15) 90° spherical bender, (16) solenoid, (17) electron polarization vector, (18) Lyman-alpha detector, (19) Faraday cup, (20) exit to Mott polarimeter, (21) GaAlAs diode laser, (22) lenses (23) $\lambda/4$-plate, (24) $\lambda/2$-plate, (25) laser beam flag, and (26) vacuum window. Item (11), used for the determination of P_H, was removed for the ionization asymmetry measurements as were items (8), (18), and (20).

For ion detection, a channel electron multiplier (CEM) was used, preceded by a grid which prevented electrons or negative ions from reaching the CEM entrance cone. The grid and cone voltages were varied to ensure that the signal was free from photon contamination under the final operating conditions. In order to eliminate the influence of stray electric fields on the ions drifting downstream from the crossed beams site, the interaction region and the CEM were enclosed in a stainless steel box, the dimensions of which ensured that a 4π acceptance was still presented to ions having maximum recoil energy.

2.2. *The Polarized Electron Beam*

We have previously published detailed descriptions of the operating principles and characteristics of the GaAs polarized electron beam[36,37] used in our electron-hydrogen studies. Of particular importance for the threshold investigations we present in this paper was the enhanced lifetime of the crystal which we achieved by embedding a cesium dispenser in the extraction anode. As we noted in one of our previous papers,[37] we were able to extend the typical lifetime of the exponential decay from 12 hours to more than 400 hours by using the anode dispenser to provide constant *in situ* cesiation of the GaAs crystal at a very low level.

In the course of our polarized electron source studies we found that the mean energy of the extracted beam increased by 50-100 meV over a period of one crystal lifetime. This behavior is illustrated in Fig. 2, which shows the typical time dependence of the difference between the apparent energy, $e\,(\Delta V)$, determined from a meter reading of the relative cathode voltage, and the actual energy, E, determined from calibration measurements based upon ionization of H_2. In Fig. 2 we have denoted the difference between these two energies by E_{offset}, which for completeness we write in equation form as

$$\overline{E}_{offset} = e\,(\Delta V) - \overline{E}. \tag{9}$$

The monotonic decrease in the offset energy apparent in Fig. 2 can be understood with the aid of the energy level diagram shown in Fig. 3.

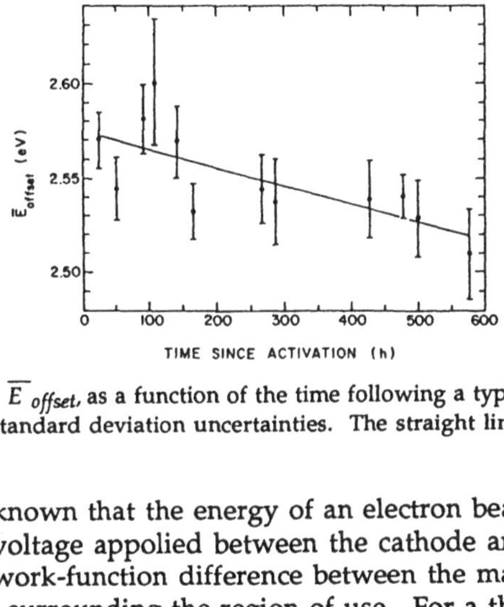

Fig. 2. Energy offset, $\overline{E}_{offset}$, as a function of the time following a typical activation. The error bars represent one standard deviation uncertainties. The straight line is the result of a linear fit to the data.

It is well known that the energy of an electron beam depends not only on the external voltage appolied between the cathode and the region of use, but also on the work-function difference between the material of the cathode and the material surrounding the region of use. For a thermionic cathode, in which we assume for simplicity that the electrons escape only from the Fermi level, the energy of the beam, E, is given by

$$E = e\,(\Delta V) - \Delta\phi, \tag{10}$$

where $\Delta V = V_r - V_c$ is the external accelerating voltage applied between the region of use (r) and the cathode (c), and $\Delta\phi = \phi_r - \phi_c$ is the work-function difference for the respective materials. For a thermionic cathode, the work function, ϕ_c, is relatively stable, and the energy of the beam need only be measured at infrequent intervals. For a GaAs photocathode, by contrast, changes in surface conditions can cause significant changes in the beam energy.

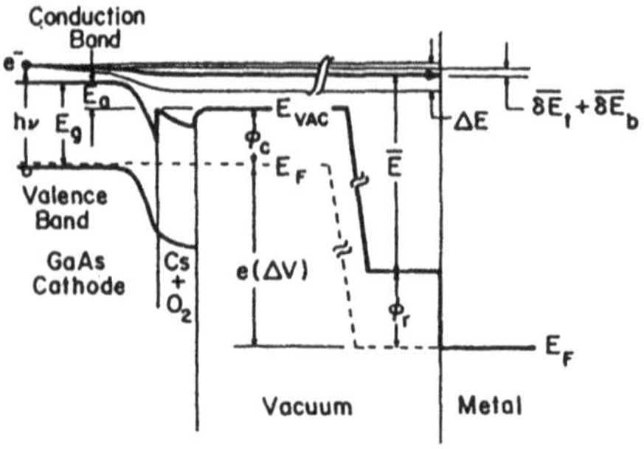

Fig. 3. Schematic drawing of the energy levels of the GaAs-(+Cs+O_2)-vacuum-metal interface pertinent to analysis of beam energy. The symbols used are $h\nu$ for photon excitation energy, E_g for band-gap energy, E_a for electron affinity, ϕ_c for cathode work function, ΔV for accelerating potential between cathode and region of usse, E_{VAC} for vacuum energy level, E_F for Fermi energy level, ϕ_r for region of use work function, δE_t for mean energy loss in bulk thermalization, δE_b for mean energy loss in band bending region, E for mean energy of electrons, and ΔE for energy spread of electrons.

From Fig. 3, we see that the counterpart to Eq. 10 for a GaAs cathode can be written as

$$\bar{E} = e(\Delta V) + h\nu - (\overline{\delta E_t} + \overline{\delta E_b}) - \phi_r\,, \tag{11}$$

where $\bar{E}$ is the mean energy of the electron beam at the region of use, $h\nu$ is the energy of the photons incident on the GaAs crystal (assumed to be greater than the bandgap energy, E_g), δE_t is the mean energy loss due to thermalization of the excited electrons in the bulk crystal, and δE_b is the mean energy loss of the escaping electrons as they pass through the band bending region near the crystal surface. The bulk-thermalization and band-bending energy loss processes not only cause shifts in the mean energy of the escaping electrons, but also produce a broadening in their energy distribution, thereby making the yield of the extracted electrons quite sensitive to the value of the electron affinity of the photocathode, E_a, shown negative in Fig. 3. As the crystal ages, the energy level of the vacuum, E_{VAC}, rises relative to the conduction band energy, and the affinity becomes increasingly less negative. In the process, the low-energy tail of the electron distribution becomes truncated causing the yield to decrease and the mean energy to rise. Therefore, the expression for the energy-offset,

$$\bar{E}_{offset} = e(\Delta V) - \bar{E} = (\phi_r - h\nu) + (\overline{\delta E_t} + \overline{\delta E_b})\,, \tag{12}$$

obtained from Eqs. 9 and 11 must be modified to include a time-dependent term that reflects the the effects of the surface contamination aging process. We alter Eq. 12 to read

$$E_{offset} = (\phi_r - h\nu) + (\overline{\delta E_t} + \overline{\delta E_b}) - \varepsilon(t), \tag{13}$$

where $\varepsilon(t) > 0$ accounts for the effect of the truncation of the low-energy tail.

For experiments in which high resolution in energy is of paramount importance, the temporal dependence included in Eq. 13 makes it imperative that calibrations of the beam energy be performed frequently compared to the time rate of change of ε. Clearly the more rapidly the crystal ages and the shorter its lifetime, the more rapidly ε changes and the more often calibration measurements must be carried out. Thus it follows that a crystal lifetime of >400 hours is far preferable to one of <12 hours — the latter being typical of many other polarized electron experiments — not only for efficiency of operation but also for precision of measurement.

In order to achieve a high degree of confidence in our knowledge of the beam energy, we performed calibration measurements every 24 to 48 hours by turning off the rf power in the hydrogen source and measuring the molecular ionization rate at energies between 0.3 and 1.0 eV above threshold. For our electron energy spread, characterized by a Gaussian halfwidth, σ, of 75 meV, the convoluted ionzation cross section over this energy range is nearly linear, all the vibrational and autoionizing structure remaining unresolved.[37] We therefore were able to use a linear extrapolation of our measurements to an apparent threshold value of 15.50 eV as a means of calibrating the energy of the beam to a typical accuracy of ±30 meV. We note that although the energy drift was substantial enough to require periodic calibration, it was sufficiently small that the polarization of the beam remained effectively unchanged.[37]

2.3 *The Polarized Hydrogen Beam*

In any crossed beams experiment the effects of changes in the overlap geometry must be considered with great care. In our experiment, the collimated hydrogen beam was characterized by a diameter of 6 mm, while the electron beam at the interaction region was characterized by a diameter of approximately 3 mm. Thus, the entire electron beam was always intercepted by the hydrogen beam, but only a portion of the hydrogen beam was sampled by the electron beam. Radial inhomogeneities in the hydrogen beam polarization were therefore a potential source of systematic error that had to be studied. To this end we carried out detailed experimental and theoretical investigations of the properties of the polarized hydrogen beam, which we have described at length in a previous paper.[38]

In brief, our experimental hydrogen beam studies comprised high-field state-selection measurements carried out with aid of a Stern-Gerlach polarimeter located at the end of the beam line, as illustrated in Fig. 1, and density measurements carried out with the aid of vacuum gauges and a residual gas analyzer located at various sites along the beam line. We compared the results of these experimental measurements with the results obtained from several theoretical analyses including one based upon a 10 million trajectory ray-tracing calculation. With the confidence gained from the good agreement among the theoretical and experimental results, we then used the ray-tracing analysis to predict the radial dependence of the density and low-field polarization of the beam at the interaction region, as shown in Fig. 4. Examination of the figure quickly reveals that the polarization of the hydrogen beam weighted by its relative density and averaged over the intersection volume defined by the 3-mm diameter of electron beam varies by less than one part per hundred across the entire 6-mm hydrogen beam diameter. Thus we conclude that drifts in the overlap geometry could only have affected the relative precision of the counting-rate asymmetry, Δ, at a level below $\pm$ 1%.

2.4. *Data Acquisition and Analysis*

In carrying out our asymmetry measurements, we sought to reduce systematic effects by reversing the direction of the electron polarization at frequent intervals. To this end, under computer control, we rotated the $\lambda/2$ and $\lambda/4$ plates through 360°, the former in eight steps of 45° and the latter in four steps of 90°. During data acquisition, for each data run *j*, we time ordered the retardation-plate orientations to produce eight helicity sets, each of which contained four circular polarization settings alternating between left and right. Thus for each run we generated eight sets of data, in each of which $\boldsymbol{P}_e$ and $\boldsymbol{P}_H$ alternated between antiparallel and parallel configurations with each configuration appearing twice. The alternating pattern within each data set is evident in Fig. 4, which shows a typical set of 32 measurements of the normalized ion counting rate for a single data run at an incident electron energy of 14.82 eV. We generated the final asymmetry value for each energy that we investigated by averaging the results from a number of such data runs in accordance with a procedure that allowed us to check for the presence of systematic errors. Several times during the experiment, we also checked for systematic effects by reversing the magnetic field in the interaction region, thereby reversing the direction of $\boldsymbol{P}_H$.

For each of the eight data sets within a given run *j*, we calculated a single "real" asymmetry, $\Delta_R{}^{(i)}$ and two "false" aymmetries $\Delta_{F+}{}^{(i)}$ and $\Delta_{F-}{}^{(i)}$, which should be zero in the absence of systematic effects. We calculated $\Delta_R{}^{(i)}$ from

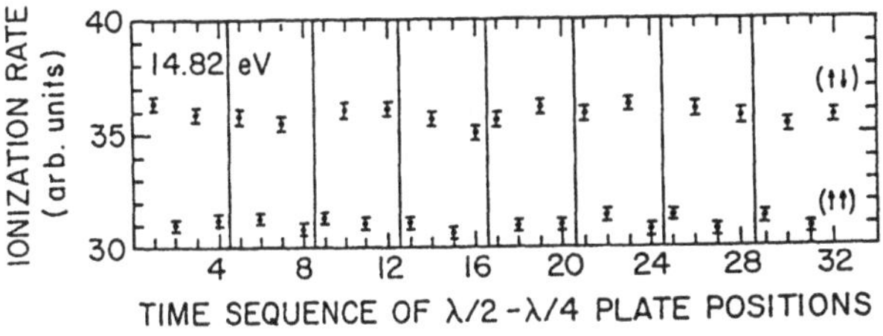

Fig. 4. Ion counting rate normalized to incident electron and atom beam intensities for a set of 32 measurments consituting one data run at an incident electron energy of 14.82 eV. The asymmetry for spins antiparallel and parallel is very evident. The vertical lines drawn separate the eight sets of data for which real and false asymmetries were separately calculated.

the average of the rates $R\,(\uparrow\downarrow)_1{}^{(i)}$ and $R\,(\uparrow\downarrow)_2{}^{(i)}$ for the two antiparallel configurations and the average of the rates $R\,(\uparrow\uparrow)_1{}^{(i)}$ and $R\,(\uparrow\uparrow)_2{}^{(i)}$ for the two parallel configurations. We generated the two "false" asymmetries by taking sums and differences for the antiparallel and parallel configurations in accordance with the prescription

$$\Delta_{F\pm}^{(i)} = \frac{\left[R(\uparrow\downarrow)_1^{(i)} - R(\uparrow\downarrow)_2^{(i)}\right] \pm \left[R(\uparrow\uparrow)_1^{(i)} - R(\uparrow\uparrow)_2^{(i)}\right]}{\left[R(\uparrow\downarrow)_1^{(i)} + R(\uparrow\downarrow)_2^{(i)}\right] + \left[R(\uparrow\uparrow)_1^{(i)} + R(\uparrow\uparrow)_2^{(i)}\right]}. \tag{7}$$

We used the false asymmetries to search for the presence of spurious helicity-correlated effects, experimental noise insufficiently represented by the calculated statistical uncertainties, and tuning drifts of the beams or the electronic instrumentation. By construction, the Δ_{F+}'s are most sensitive to quasilinear drifts, while the Δ_{F-}'s are most sensitive to quasiquadratic drifts. Therefore we combined all the Δ_{F+}'s for all data sets and all data runs irrespective of energy to generate the histogram shown in Fig. 5(a). Similarly, we combined all the Δ_{F-}'s to generate the histogram shown in Fig. 5(b). Consistent with the observation which we had made during data-taking that drift-related noise appeared intermittently during the course of our measurements, we found that to high accuracy we could fit each histogram with the sum of two Gaussians, a dominant one (labeled 1 in each of the panes of Fig. 5 having a narrow width corresponding to "drift-free" operation and a subordinate one (labeled 2) having a broad width corresponding to "drift-hindered" operation. From a comparison of the relative heights of the curves labeled 2 in Figs. 5(a) and 5(b), we can see that quasilinear drifts were more prevalent than quasiquadratic ones. Comparison of the relative heights of curves 1 and 2 in each pane, however, shows that both drifts occurred very infrequently.

In order to determine an effective uncertainty for the true distributions given by the histograms, we used a combination of the variances of the individual Gaussians appropriately weighted for such statistical mixing. We thus concluded that all sixteen false asymmetries were not only very small and but also consistent with zero with essentially ideal chi-square confidence levels. In order to preclude an accidental bias of our results at any given energy due to finite sample size, however, we elected to cut our data based upon a truncation of the false asymmetry histograms at the level of 2.5 standard deviations from zero, a criterion that was satisfied by more than 90% of the data. We thus reduced the contributions of the broad distributions substantially while only removing 3% of the narrow distributions. The sixteen false asymmetries for the truncated data are shown in Fig. 6, and as might be expected from the nature of the histograms of Fig. 5, the results demonstrate that the experiment was free from systematic effects at a very high confidence level. Taken as a group, the Δ_{F+}'s average to $+32(24)\times10^{-5}$ and the Δ_{F-}'s, to $-57(24)\times10^{-5}$, demonstrating that the experiment was free from systematic effects at a level of $\pm5\times10^{-4}$.

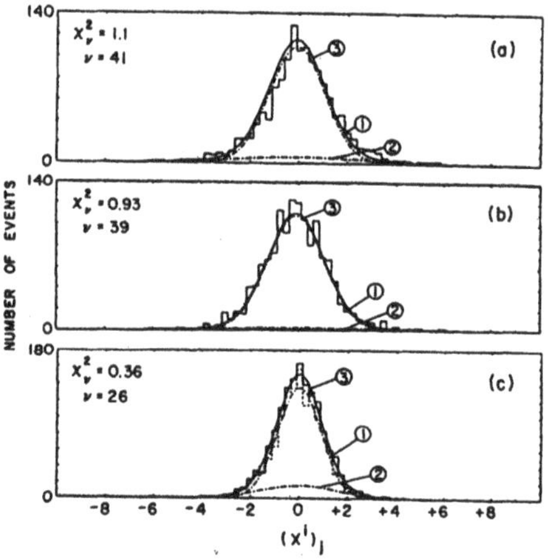

Fig. 5. Analysis of statistical errors: (a) Histogram of normalized false asymmetries, $(x^i)_j = (\Delta_{F+}{}^i)_j/(\sigma_{F+}{}^i)_j$, for all data sets i and all runs j, irrespective of energy, where $(\sigma_{F+}{}^i)_j$ is the statistical uncertainty for $(\Delta_{F+}{}^i)_j$; (b) histogram analogous to that in (a) for Δ_{F-} false asymmetries; (c) histogram (solid) of normalized real-asymmetry residuals, $(x^i)_j = [(\Delta_R{}^i)_j - (\overline{\Delta_R})_j]$, where $(\overline{\Delta_R})_j$ is the weighted mean of the real asymmetry for the j th data run, and histogram (dashed) of real-asymmetry residuals for truncated data set as explained in the text. In each case the curve labeled 3 is obtained from a fit of the data by the sum of two Gaussians labeled 1 and 2.

We carried out our analysis of the real-asymmetry uncertainties in a manner analogous to that which we used for the false asymmetries. The histogram of real-asymmetry residuals, shown solid in Fig. 5(c), again is well represented by the sum of two Gaussians. From its construction, the real asymmetry is sensitive to both quasilinear and quasiquadratic drifts, and as would be expected under such circumstances the height of curve 2 in Fig. 5(c) is greater than that in either Fig. 5(a) or Fig. 5(b). After false-asymmetry truncation, the real asymmetry residuals assume the slightly narrower histogram shown dashed in Fig. 5(c). The lack of bias introduced by the truncation procedure is indicated by the preservation of the symmetry in the truncated histogram, the peak of which remains unshifted horizontally from that of the untruncated histogram at the statistical accuracy of 0.1 standard deviations.

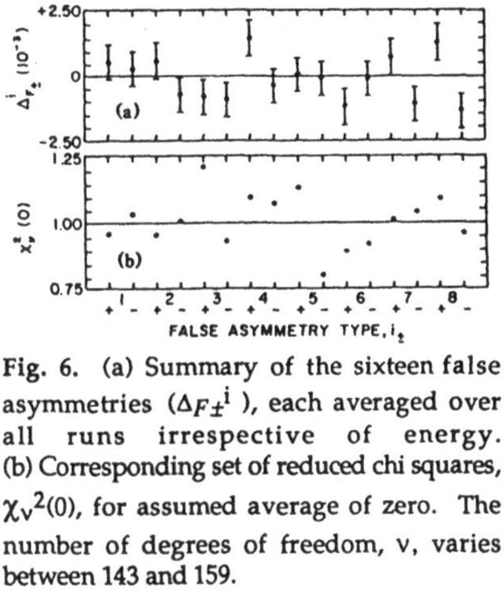

Fig. 6. (a) Summary of the sixteen false asymmetries ($\Delta_{F\pm}{}^{i}$), each averaged over all runs irrespective of energy. (b) Corresponding set of reduced chi squares, $\chi_\nu{}^2(0)$, for assumed average of zero. The number of degrees of freedom, ν, varies between 143 and 159.

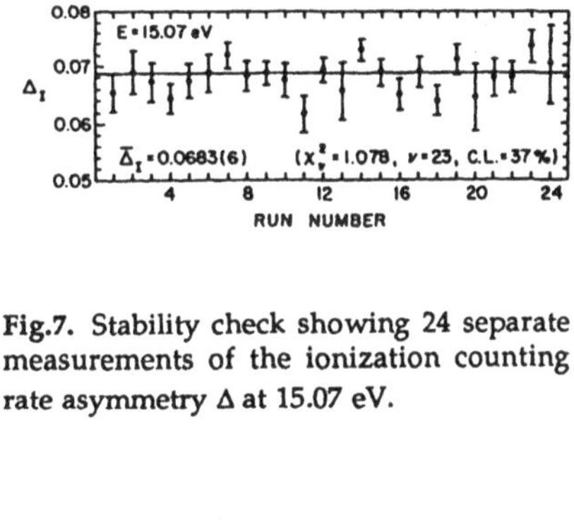

Fig.7. Stability check showing 24 separate measurements of the ionization counting rate asymmetry Δ at 15.07 eV.

In carrying out our asymmetry measurements, we accumulated data at different energies in a quasi-random sequence. In order to check for stability of operation, however, we frequently returned to the measurement of Δ at an energy of 15.07 eV. The results, shown in Fig. 7, indicate that our apparatus was indeed stable at a level appropriate to the accuracy of the measurements. We also examined the data for any energy as well as any temporal dependence of the false asymmetries. At the statistical accuracy of the experiment we found no evidence of any such systematic effects.

We examined our ionization data further for possible contamination due to quenched metastable states or field-ionized Rydberg atoms in the vicinity of the CEM detector. By varying the potentials applied to the CEM and its grid, we were able to rule out the former unequivocally. In order to place a limit on the latter, we measured the ionization rate at values of the mean electron energy just below threshold as well as just above and compared the results with the convoluted rate dependence predicted by the Wannier law, which we took as adequately approximate for this purpose. For our first data point above threshold (at 13.67 eV) — the one that should be most sensitive to Rydberg effects given its small ionization cross section — we found no evidence of Rydberg contamination at a 5% level in event rate, the limit of accuracy of the measursement. This null result is not surprising since the electric field present in the detection region was only 200 V/cm, a value for which only Rydberg atoms with principal quantum number greater than 41 could be field ionized.[39]

2.5. *Results and Conclusions*

We present the results of our ionization counting rate asymmetry measurements in Fig. 8. As we have noted in a previous paper, the results shown for the truncated data set are virtually indistinguishable from those obtained with the untruncated set. Two features of the asymmetry results displayed in Fig. 8 are clearly evident. First, structure is present: a linear fit over the full above-threshold range shown (13.6-15.3 eV), for example, results in a reduced chi square of 4.82 for 17 degrees of freedom, corresponding to a confidence level of $\sim 5\times10^{-10}$. Second, within the first 0.5 eV above threshold, a range for which previous cross section studies have claimed to verify the Wannier power-law behavior, the asymmetry has a generally positive slope: a fit of a constant function to the data shown in the hatched region results in a reduced chi square of 2.32 for 6 degrees of freedom, corresponding to a confidence level of 3%. By contrast, a linear fit (curve *a* in Fig. 8) in this region produces a reduced chi square of 0.78 for 5 degrees of freedom, corresponding to a confidence level of 56%. Both the structure and the non-zero slope are inconsistent with the Wannier prediction of a constant value for Δ.

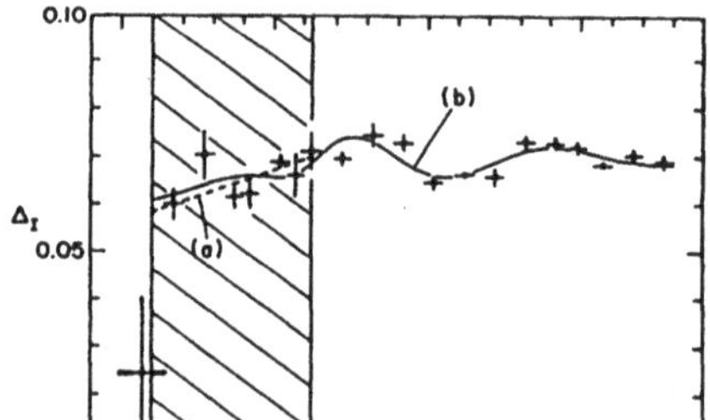

Fig. 8. Measured ionization counting-rate asymmetries, Δ, for incident electron energies between 13.57 and 15.27 eV. The vertical and horizontal error bars represent one-standard-deviation uncertainties. The heavy, outer bar on the 13.57-eV point illustrates the characteristic energy spread of the electron beam. Curve *a* is a linear fit in the restricted (hatched) region for which the Wannier law has been verified in total cross section measurements [Ref. 19]. Curve *b* is an example of a fit of the Coulomb-dipole theory.

For completeness, we also compared our results to the predictions of the Coulomb-dipole theory[18] which predicts rapid oscillations in both the ionization cross section and the associated asymmetry. Since Temkin first introduced the theory, the functional behavior of the Coulomb-diplole threshold law has undergone a number of changes, and at present the precise form is somewhat uncertain. Curve *b* in Fig. 8 illustrates the result of an eight-parameter convoluted fit to one of Temkin's model functions. Although the general shape of the data is reproduced, the reduced chi square associated with the fit has a rather unsatisfactory value of 1.88 for 11 degrees of freedom, corresponding to a confidence level of only 2%. Fits of other model functions fared no better. We therefore conclude that either the specific functional forms are deficient, or the 1.7 eV range over which we applied them is inappropriate. It is also possible, although somewhat unlikely, that the pathology of the functions prevented us from determining an asolute chi-square minimum given the availabilty of only nineteen data points.

The positive slope displayed by Δ within the first 0.5 eV above threshold and the structure evident for the next 1.0 eV suggest that three-body dynamics are affecting the behavior of the asymmetry in the threshold region, calling into question the ergodic *ansatz* of Wigner[33] and the predictions of the Wannier law which rely only on the asymptotic behavior of the wave function. While the origin of the structure may be complex and difficult to trace in detail, the positive slope can be understood at least qualitatively in rather simple terms.

From the definition of Δ, given by Eq. 3, and its connection to the singlet and triplet cross sections expressed through Eqs. 4 and 8, we see that an increase in the value of D requires that the singlet cross section increase faster than the triplet. Moreover, when $\Delta = 0$, the singlet and triplet cross sections are equal, a condition that is well known to occur at sufficiently high values of the incident electron energy. The equality of the singlet and triplet ionization cross sections at energies above 500 eV for hydrogen, for example, is indicative of the lack of any need to place symmetry restrictions on the spin portion of the two-electron wave function. For such kinematic conditions, the two electrons have distinctly different characteristics in configuration space, and spin symmetry or anti-symmetry is not required to satisfy the Pauli exclusion principle. By contrast, in the case of the ground state of the helium atom, where the two bound electrons have identical configuration space properties, the spinor is constrained to be antisymmetric under particle exchange and hence singlet in character.

For threshold ionization, the behavior of Δ can also be examined in the context of the configuration space properties of the two electrons. We can gain insight into these properties if, for the moment, we somewhat imprecisely picture the ionizing collision as comprising two separate processes, one involving the dynamic reaction of the atom to the incident electron and the other involving the momentum transfer from the incident electron to the bound electron resulting in ionization. In the case of the first process, the atom reacts by becoming polarized, the degree of which depends upon the atomic polarizability and the time-averaged electric field the atom experiences during the collision. If the polarization is large, the two electrons will have only a small probability of sharing the same configuration space properties when ionization occurs. If, however, the polarization is reduced, the probability of similar configuration space properties will be enhanced, and the two-electron spinor will have to display an increased antisymmetric or singlet character. It is precisely such a reduced polarization that occurs when the velocity of the incident electron is raised. Under these circumstances the collision time shrinks and the time-averaged electric field becomes smaller. Thus, as the incident electron energy rises in the region just above threshold, the probability of singlet scattering grows relative to that of triplet scattering, and the asymmetry, Δ, assumes a positive slope as shown in Fig. 8.

As the incident energy continues to rise, the effects of the second process — the momentum transfer from the incident to the atomic electron — must be considered. Indeed, in the high-energy limit the energy of the incident electron becomes so large that the magnitude of the momentum transfer usually will not provide the ionized atomic electron with enough momentum to allow the two final-state electrons to have the same configuration space character. In the transition from the threshold region to the high-energy region the asymmetry, Δ, must therefore reach a peak, the height and location of which will depend upon the specific characteristics of the atom. As we showed in a previous paper,[35] for hydrogen the peak occurs at an incident electron energy slightly greater than 15 eV. We note that although the precision of the measurement was marginal, a similar effect was observed in the case of lithium[26] several years ago, although the authors did not comment on it. Very recently, however, a more accurate measurement revealed the peaking property very prominently in the ionization asymmetry for cesium.[31] Curiously, the peaking behavior seems to be absent in the case of sodium or potassium.[25,26,40]

3. Reexamination of Previous Threshold Studies

As we noted at the outset of this paper, the threshold law for double escape has been studied experimentally at great length. Motivated by the violations of the conventional Wannier theory, which we observed in our hydrogen asymmetry investigations, we decided to reexamine the conclusions reached in several previous experiments. For this purpose we selected two total cross section studies, double photodetachment of H^- [24] and

K^-,[28] and two energy partitioning studies, both involving impact ionization of He.[23,27] In all four cases, the experimental groups reported consistency with the predictions of the Wannier law. Our reexamination, however, suggests that other conclusions may be reached.

The original results of the double photodetachment studies are shown in Figs. 9(a) and 10(a) along with the best fits of the Wannier power law presented in the published papers. In each case the experimenters applied a chi-square analysis in an attempt to discriminate between the Wannier and Coulomb-dipole theories. In each case they found that their data could not be used to make such a distinction. In the context of the present discussion we will focus solely on the conclusions they drew about the Wannier law.

The K^- experimenters,[28] Bae and Peterson, obtained a power law exponent of 1.16±0.05, in good agreement with the Wannier prediction of 1.127. The reduced chi square for the Wannier fit was 0.69 for 34 degrees of freedom, corresponding to a 91% confidence level. (We note that with 50% representing an ideal confidence level, the 91% value suggests that the uncertainties for the K^- data points might have been overestimated.) The H^- experimenters,[23] Donahue, Bryant and collaborators, obtained a power law exponent of 1.15±0.04, also in good agreement with the Wannier prediction of 1.127. The reduced chi square for the Wannier fit in their case was 1.15 for 66 degrees of freedom, corresponding to a confidence level of 19%. The H^- experimenters also tried to fit their data with a linear law and concluded that the Wannier exponent of 1.127 was 250 times more likely to be true than an

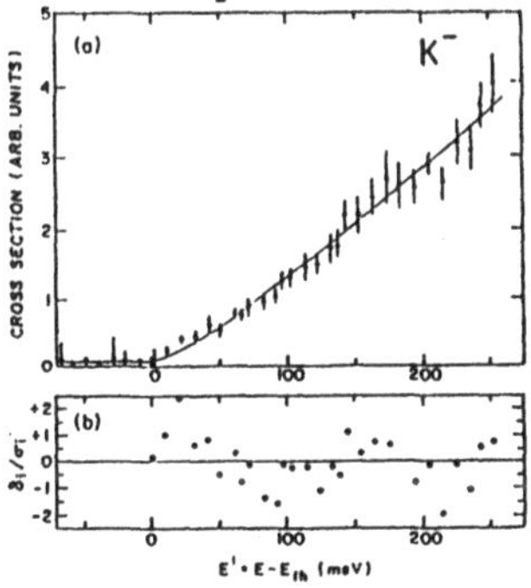

Fig. 9. (a) Cross section for double photodetachment of the K^- ion as a function of excess photon energy above threshold, taken from Ref. 28. The curve is a least squares fit of the Wannier power-law function, also taken from Ref. 28. (b) Normalized residuals corresponding to the data points in (a).

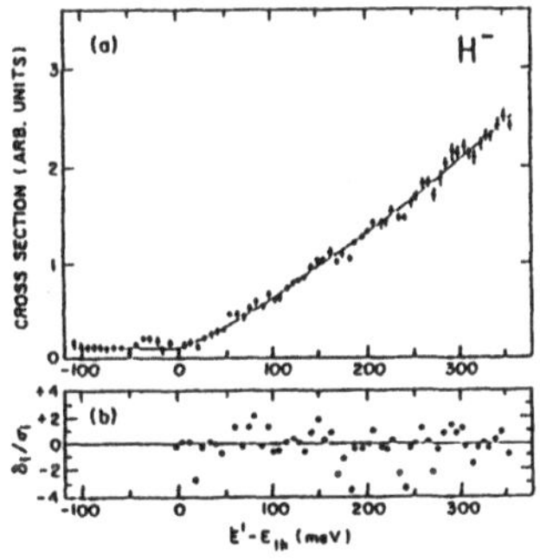

Fig. 10. (a) Cross section for double photodetachment of the H^- ion as a function of excess photon energy above threshold, taken from Ref. 23. The curve is a least squares fit of the Wannier power-law function, also taken from Ref. 23. (b) Normalized residuals corresponding to the data points in (a).

exponent of unity. Both the K^- and H^- experimenters concluded their analysis with the claim that their results were consistent with the Wannier threshold law, the validity of the claim resting solely on the respective chi square analyses.

How valid is the claim from the standpoint of statistics? Unfortunately neither group of experimenters recognized that a chi square analysis is a "one way" test. It is a sufficient test for ruling out a fit, but it is not a sufficient test for validating a fit. A valid fit must not only produce an acceptable chi square, but it must also produce a random ordering of the residuals. In order to provide a visual aid for an analysis of the ordering of the residuals, we have plotted in Figs. 9(b) and 10(b) the respective normalized residuals, δ_i / σ_i, for each of the data points shown in Figs. 9(a) and 10(a). In the case of the K^- experiment, Fig. 9(b) immediately reveals the presence of an ordering. In the case of the H^- experiment, the conclusion is not as obvious from a visual inspection of Fig. 10(b), although a quantitative analysis will reveal the ordering quite clearly, as will shall see.

We carried out a quantitative analysis of the ordering of the residuals for both experiments by applying four different statistical tests — a runs test,[41,42] an analysis of the Durbin-Watson statistic,[43,44] a search for autocorrelation in the normalized residuals,[42] and a search for autocorrelation in the differences of the normalized residuals.[42] In the case of the K^- data, all four tests conclusively showed that the residuals were not randomly ordered. In the case of the H^- data only the runs test, for which the mesh size of the data was insufficiently fine, failed to show a non-random ordering. For both cases, the Durbin-Watson analysis resulted in a 95% probability for non-random ordering with the two autocorrelation analyses giving the same results. Thus with very high probability both experiments reveal the presence of structure, assuming of course that both experiments were not plagued by systematic difficulties of a similar nature. In fact, if we treat the H^- and K^- studies as independent experimental tests of the same physical principle, we must conclude that with a joint probability of only 2.5×10^{-3} do the experiments support the Wannier law.

We turn to a reexamination of the two energy-partitioning experiments[23,27] carried out for electron impact ionization of He. Both of these experiments investigated how the available excess energy, $E' = E - E_{th}$, was partitioned between the two outgoing electrons when the incident electron energy, E, was slightly above the threshold value, E_{th}, required for ionization. Since the two outgoing electrons are assumed to be indistinguishable, the distribution function, $f(x)$, is constrained to be symmetric about the point $x = 0.5$, where $x = E_1 / E'$ is the fraction of the excess energy carried away by either one of the electrons. At the time the first experiment was carried out in 1978 by Pichou and collaborators,[23] the distribution function had already been predicted[3,5] to be uniform on the basis of quantum mechanical extensions of the classical Wannier theory. The data obtained in this first experiment seemed to indicate some significant departure from the prediction of a uniform distribution, but the authors argued that within the uncertainties of their measurement their results showed "that the excess energy is randomly shared between the two electrons up to 3.6 eV." Curiously, they had previously noted in their paper that the error bars displayed in the pertinent figure were "rather pessimistic," based upon the "good reproducibility of the results presented...regardless of experimental conditions." If they really subscribed to this observation, as the clarity of the structure evident in the pertienent figure further suggests, they should not have concluded that the distribution was uniform.

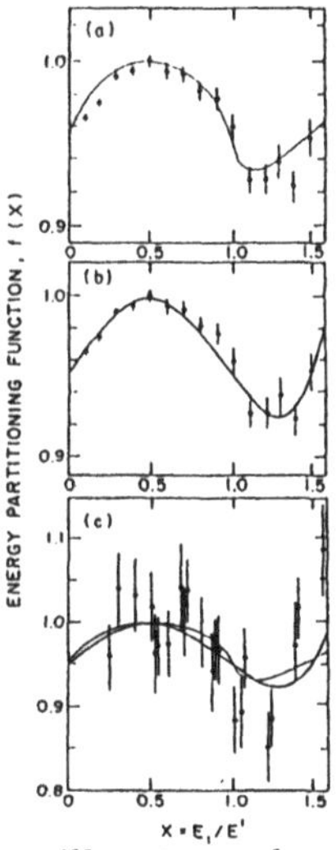

Fig. 11. Energy partitioning function $f(x)$ study showing in (a) the (CRE) data of Hammond *et al.* (Ref. 27) along with the theoretical Wannier prediction developed by Read (Ref. 14); in (b) the data of (a) along with a quartic fit illustrating the symmetry with respect to $x = 0.5$; in (c) the (EL) data of Hammond *et al.* (Ref. 27) along with Read's theory (thin line) and quartic fit to the (CRE) data (heavy line).

We will not employ our hindsight further regarding this first energy partitioning study, because the experiment and theory have already been superseded. Instead, we turn to the second study which was reported by Hammond and collaborators[27] in 1984. By that time, Read[14] had shown classically that the Wannier theory actually predicted a partitioning function that was 5% larger at $E_1 = E'/2$ than at $E_1 = 0$ or E'. The function predicted by Read is shown as the curve in Fig. 11(a). Hammond and collaborators actually carried out two separate types of experiments, one in which the analyzing energy was fixed and the incident energy was varied and one in which the incident energy was fixed and the analyzing energy was varied. The results from the first type, the constant residual energy (CRE) method, are shown as the data points in Fig. 11(b), while the results from the second type, the energy loss (EL) method, are shown as the data points in Fig. 11(c). The authors claimed that Fig. 11(a) illustrates "good agreement between [the function calculated by Read] and the form of f found experimentally." Clearly the figure demonstrates the contrary, since the data points at low x fall substantially below the calculated curve. The authors recognized this deficiency since they stated further, "The main difference is that the experimentally determined results are not symmetric about $x = 0.5$ We believe that this asymmetry is caused by the background yield not varying linearly with the incident energy and by small changes in the incident beam as the energy is varied."

We decided to examine the data closely for evidence of any lack of symmetry about $x = 0.5$. To this end we attempted to fit a quartic function to the data and obtained the curve shown in Fig. 11(b). The reduced chi square for this fit is 1.02 for 12 degrees of freedom, corresponding to a nearly ideal confidence level of 51%. Since the limited data set makes an analysis of the ordering of the residuals virtually impossible, we are forced to conclude that there is no evidence to support the assertion that the data presented are asymmetric with respect to $x = 0.5$. An equally plausible conclusion is that the theoretical curve shown in Fig. 11(a) is not precisely correct. Keeping in mind that the Wannier theory is already known to be a fairly good approximation, we suggest that the energy partitioning results of Hammond and collaborators do not provide the needed agreement to validate the theory as a precise description of the double escape process.

4. An Impetus for the Future

The results we obtained for the hydrogen ionization asymmetry measurements and the results obtained elsewhere for lithium and cesium argue that dynamics must be taken into account in the threshold behavior of two-electron escape, contradicting the Wannier theory[2-17] and Wigner's ergodic hypothesis.[33] This argument is reinforced by the results we obtained from the re-analysis of four other threshold studies.[23,24,27,28,] We believe, however, that additional experiments are essential for determining the precise role of three-body dynamics in threshold problems. To this end, we suggest that a logical first step would be a remeasurement of double photodetachment of H^- with greatly increased precision and greatly reduced backgrounds. Based upon advances in laser and accelerator technologies achieved during the last decade we believe that the original Los Alamos experiment[24] can be improved in statistical accuracy by more than one order of magnitude and in energy resolution by at least a factor of four. In addition, through the use of electron coincidence techniques we believe that the severe background problems which plagued the original experiment can practically be eliminated. Future experiments on other atomic systems would clearly be desirable as well, and we hope that this paper will serve to motivate them.

Acknowledgments

Many people contributed to the work we have described in this paper. We particularly wish to thank our scientific collaborators, D.M. Crowe, M. Eminyan, M. Frankel, J. Friedman, K. Rubin, J. Slevin, F.C. Tang, and A. Vasilakis. We also acknowledge stimulating discussions with C. Bottcher, H.C. Bryant, J.B. Donahue, C.H. Greene, J. Peterson, J. Macek, M. Mittleman, W. Raith, A.R.P. Rau, and A. Temkin. Finally we thank the U.S. National Science Foundation and the City University for the financial support they provided.

References

1. See reviews by A.R.P. Rau, in *Electronic and Atomic Collisions*, ed. J. Eichler, I.V. Hertel, and N. Stolterfoht (Elsevier, Amsterdam, 1984) p. 711; *Comments At. Mol. Phys.* **14** (1984) 285; F.H. Read, in *Electron Impact Ionization*, ed. T.D. Märk and G.H. Dunn (Springer Verlag, Berlin, 1985) p. 42.
2. G.H. Wannier, *Phys. Rev.* **90** (1953) 817.
3. I. Vinkalns and M. Gailitis, in *V Int. Conf. on the Physics of Electronic and Atomic Collisions, Leningrad, USSR, July 17-23, 1967, Abstracts of Papers*, ed. I.P. Flaks (Nauka, Leningrad, 1967) p. 648.
4. R. Peterkop, *J. Phys. B* **4** (1971) 513; **16** (1983) L587; R. Peterkop and A. Liepinsh, *J. Phys. B* **14** (1981) 4125.
5. A.R.P. Rau, *Phys. Rev. A* **4** (1971) 207; *J. Phys. B* **9** (1976) L283.
6. T.A. Roth, *Phys. Rev. A* **5** (1972) 476.
7. U. Fano and C.D. Lin, in *Atomic Physics 4*, ed. G. zu Putlitz, E.W. Weber, and A. Winnacker (Plenum, New York, 1975) p. 47.
8. H. Klar and W. Schlecht, *J. Phys. B* **9** (1976) 1699.
9. U. Fano, *Phys. Rev. A* **22** (1980) 2660.
10. H. Klar, *J. Phys. B* **14** (1981) 3255, 4165; *Z. Physik A* **307** (1982) 75.
11. C.H. Greene and A.R.P. Rau, *Phys. Rev. Lett.* **48** (1982) 533; *J. Phys. B* **16** (1983) 99; *Phys. Rev. A* **32** (1985) 1352.
12. A.D. Stauffer, *Phys. Lett.* **91A** (1982) 114.
13. J.M. Feagin, *J. Phys. B* **17** (1984) 2433.
14. F.H. Read, *J. Phys. B* **17** (1984) 3965.
15. D.S.F. Crothers, *J. Phys. B* **19** (1986) 463.
16. M. Gailitis and R. Peterkop, *J. Phys. B* **22** (1989) 1231.
17. S. Watanabe *J. Phys. B* **24** (1991) L39.
18. A. Temkin, *Phys. Rev. Lett.* **16** (1966) 835; **49** (1982) 365; *J. Phys. B* **7** (1974) L450; *IEEE Trans. Nucl. Sci.* **30** (1983) 1106; *Phys. Rev. A* **30** (1984) 2737; A. Temkin and Y. Hahn, *Phys. Rev. A* **10** (1974) 708; M.K. Srivastava and A. Temkin, *Phys. Rev. A* **43** (1991) 3570.
19. J.W. McGowan and E.M. Clarke, *Phys. Rev.* **167** (1968) 43.
20. C.E. Brion and G.E. Thomas, *Phys. Rev. Lett.* **20** (1968) 241.
21. P. Marchand, C. Paquet, and P. Marmet, *Phys. Rev.* **180** (1969) 123.

22. S. Cvejanovic and F.H. Read, *J. Phys. B* 7 (1974) 1180, 1841; D. Spence *Phys. Rev. A* **11** (1975) 1539.
23. F. Pichou *et al.*, *J.Phys. B* **11** (1978) 3683.
24. J.B. Donahue *et al.*, *Phys. Rev. Lett.* **48** (1982) 1538.
25. M.H. Kelley *et al.*, *Phys. Rev. Lett.* **51** (1983) 2191.
26. G. Baum *et al.*, *J. Phys. B* **18** (1985) 531.
27. P. Hammond *et al.*, *J. Phys. B* **18** (1985) L141.
28. Y.K. Bae and J. Peterson, *Phys. Rev. A* **37** (1988) 3254.
29. H. Kossmann, V. Schmidt, and T. Andersen, *Phys. Rev. Let.* **60** (1988) 1266.
30. P. Schlemmer *et al.*, *Phys. Rev. Lett.* **63** (1989) 252.
31. G. Baum *et al.*, in *XVII Int. Conf. on the Physics of Electronic and Atomic Collisions, Brisbane, Australia, July 10-16, 1991, Abstracts of Contributed Papers* (to be published).
32. X.Q. Guo *et al.*, *Phys. Rev. Lett.* **65** (1990) 1857.
33. M.S. Lubell *et al.*, in *Proc. of the Int. Symp. on Correlation and Polarization in Electronic and Atomic Collisions, August 2-4, 1989, Stevens Institute of Technology, Hoboken, NJ, NIST Spec. Pub.* **789**, ed. P.A. Neill, K. Becker, and M.H. Kelley, p. 95.
34. E.P. Wigner, *Phys. Rev.* **73** (1948) 1002.
35. D.M. Crowe *et al.*, *J. Phys. B* **23** (1990) L325.
36. F.C. Tang *et al.*, *Rev. Sci. Instrum.* **57** (1986) 3004.
37. X.Q. Guo *et al.*, *Rev. Sci. Instrum.* **61** (1990) 1858.
38. N. Chan *et al.*, *Z. Physik D* **10** (1988) 393.
39. T.F. Gallagher, *Rep. Prog. Phys.* **51** (1988) 51.
40. D. Hils and H. Kleinpoppen, *J. Phys. B* **11** (1982) L283.
41. S. Siegel, *Nonparametric Statistics for the Behavioral Sciences* (Mc Graw Hill, New York, 1956).
42. J.V. Bradley, *Distribution-Free Statistical Tests* (Prentice Hall, Englewood Cliffs, NJ, 1968).
43. J. Durbin and G.S. Watson, *Biometrika* **37** (1950) 409; **38** (1951) 159.
44. J. Johnson, *Econometric Methods, 2nd Edition* (Mc Graw Hill, New York, 1972)

VI. EXCITED STATES

Interference and Correlation Involving Dielectronic Processes in Energetic Ion-Atom Collisions

N. Stolterfoht[+], X. Husson, D. Lecler,
Laboratoire de Spectroscopie Atomique, ISMRa, F-14050 Caen Cedex, France

R. Köhrbrück, B. Skogvall,
Hahn-Meitner-Institut Berlin, Glienickerstr. 100, D-1000 Berlin 39, Germany

S. Andriamonje,
Centre d'Études de Physique Nucléaire et de Physique de Particules,
F-33107 Gradignan, France

J.P. Grandin
Centre Interdisciplinaire de Recherche avec les Ions Lourds, F-14040 Caen, France

Abstract

Electron correlation and interference effects are analyzed in energetic ion-atom collisions treated within the semi-classical approximation. Electron-correlation phenomena are associated with dielectronic processes which are due to two-electron transitions produced by the electron-electron interaction. As an example, double excitation of He and Li is discussed in view of interferences between the nucleus-electron and electron-electron interaction. For Li this interference is expected to be influenced by configuration interaction involved in the states $1s2s^2\,{}^2S$ and $1s2p^2\,{}^2S$. Experimentally, double excitation of Li is studied for the high-energy collision system 95-MeV/u Ar^{18+} + Li.

1 Introduction

In the past few years, energetic ion-atom collisions have received a great deal of attention with respect to electron-correlation effects which are due to the mutual interaction of two electrons occurring in the Coulomb field of the nucleus involved.[1-3] Detailed information about the collision mechanisms may be obtained when electron-correlation effects interfere with two-electron processes produced by the nucleus-electron interaction. Particular attention has been devoted to transfer-excitation,[4-6] as well as to double ionization[7-9] and excitation[10-13] to study interferences between events produced by electron-electron and nucleus-electron interactions. A remarkable feature of ion impact is that anti-matter particles can be used to investigate such interference effects for projectiles of opposite charge but equal mass. Studies of He double ionization[9] involving antiproton impact have been performed at the Low Energy Antiproton Ring (LEAR) facility at CERN (Geneva). These measurements have created a lively discussion about electron-correlation effects in ion-atom collisions.[1,2,14-16]

The formal analysis of electron correlation is based on the concept that the nucleus-electron interaction is associated with one-body operators whereas the electron-electron interaction is attributed to two-body operators.[17] The electron-electron interaction is produced by the Coulomb force which is also responsible for the

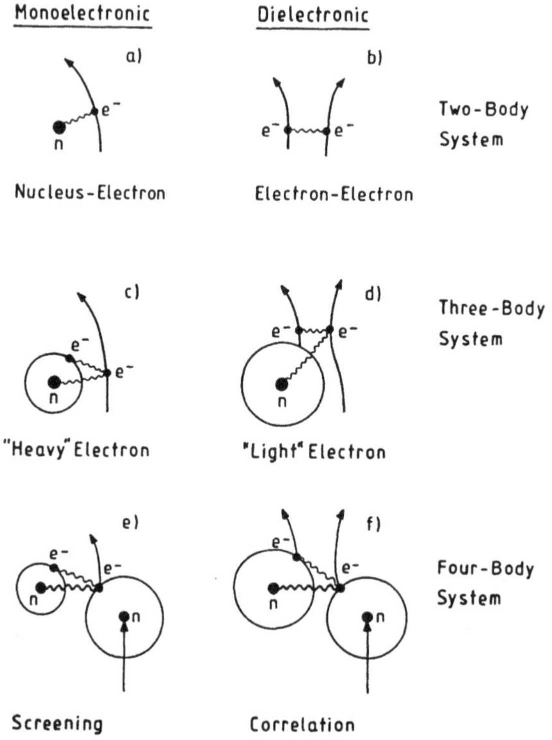

Figure 1 Processes demonstrating monoelectronic and dielectronic processes. The pictures (a) and (b) refer to the nucleus-electron and electron-electron interaction. In (c) and (d) the corresponding processes are shown with an electron bound at the nucleus, referring to a three-body system. In (e) and (f) the analogous mechanisms are attributed to screening and electron correlation in the four-body system.

nucleus-electron interaction. However, as shown in Figure 1 the nucleus is much heavier than the electron so that in the mutual nucleus-electron interaction, it appears as if the electron is the only one which reacts. In this case it is understandable that the nucleus-electron interaction exhibits a single-particle characteristic. In contrast to this feature the electron-electron interaction has a two-particle characteristic. When the nucleus is replaced by an electron, the scattering center is accelerated producing a receding Coulomb field. However, part of the electron-electron interaction may also be represented by a single-particle potential. It is emphasized that this part is associated with the presence of an attractive nucleus. Due to the coupling with the nucleus, the state of the bound electron may remain unaltered, although it participates in the interaction with the incident electron. The electron remains in the ground state when the energy transferred during the collision does not exceed the energy of the first excited state.

If the energy transfer to the bound electron is not sufficient, this energy will be "absorbed" by the nucleus. Such a bound electron seems to be part of the nucleus and, thus, it appears as a "heavy" particle whose interaction exhibits a single-particle characteristic, i.e., whose potential is monoelectronic. The two-particle aspect of the electron-electron interaction becomes important when the bound electron is removed from its ground state. The removed electron acts as a "light" particle which produces the receding Coulomb field, i.e., whose potential is dielectronic. Hence, two-electron events produced by the electron-electron interaction are referred to as *dielectronic processes*[18-20] which are considered as the essential phenomena associated with electron correlation in ion-atom collisions. It is noted that the term *dielectronic* was introduced by Massey and Bates[21] many years ago.

In this work, the dielectronic processes are discussed in conjunction with the interplay between the electron-electron and nucleus-electron interactions. Hence, interference effects between both interactions are considered in collisions. The theoretical method has been discussed before.[20] The analysis is restricted to excitation processes occurring at one atomic center. For more details about electron correlation in ion-atom collisions the reader is referred to previous review articles.[18,19,22,23] It is noted that, in this work, atomic units are used if not otherwise stated.

2 Dielectronic Processes

As mentioned before the concept of electron correlation is based on the partition of the total Hamiltonian into terms of one- and two-body operators. An introduction to the method of partitioning the electron-electron interaction is given in the textbook by Condon and Shortley[17] from which additional information may be obtained. The one-body operators are associated with the interactions postulated within the framework of the independent-particle model (IPM) whereas the two-body operators represent interactions leading beyond that model. Thus, the electron-electron interaction is partitioned into

$$V^e = W^e + V^c \tag{1}$$

$$\text{where} \quad W^e = \sum_j w_j \qquad \text{and} \qquad V^c = \sum_{j<k} q_{jk} \tag{2}$$

The effective potential W^e is given by one-electron operators w_j representing the mean field postulated within the independent-particle model and the residual interaction V^c consists of two-electron operators q_{jk}. In the following, W^e and V^c will also be referred to as *monoelectronic* and *dielectronic* interaction, respectively. The dielectronic potential may be used to partition the full Hamiltonian

$$H = H^o + V^c \tag{3}$$

where the IPM Hamiltonian $H^o = T^e + V^n + W^e$ incorporates the interactions which are represented by one-body operators only.

The dielectronic processes are due to mutual two-electron transitions produced by the electron-electron interaction. Hence, they are produced by the dielectronic interaction V^c. These processes occur in the collision system as well as in the separated atoms after the collision, when the system is prepared in a state which is non-stationary with regard to the interaction V^c, e.g., when it is prepared in a single-configuration state. The occurrence of two-electron processes is associated with the fact that the dielectronic interaction V^c couples states which differ in precisely *two* spin orbitals describing the active electrons. It is anticipated that the dielectronic processes play a central role in the field of dynamic electron correlation. As the dielectronic processes have been discussed before[18,19] only a brief description is given here.

In the following, channel diagrams are presented to discuss the dielectronic processes for an atomic three-body system. Moreover, the three-body phenomena are applied by analogy to four-body systems relevant for ion-atom collisions. It should be mentioned also that channel diagrams have already been used by Fano[24] many years ago. Since then, these diagrams have been applied in numerous studies to clarify configuration interaction in conjunction with autoionization (see, e.g., Ref. 25,26). In accordance with that previous work it is noted that electron-correlation effects are associated with configuration interaction (CI).

Configuration interaction can be visualized in the channel diagram of single-configuration states for He given in Figure 2 indicating energy levels of the

Φ_j^o of the IPM Hamiltonian H^o from eq. (3). Couplings occur between states which differ in two spin orbitals involving bound and continuum electrons. The term "closed channel" here refers to doubly excited states whereas the term "open channel" refers to single- and double-ionization states. Since the single-configuration states Φ_j^o are not eigenstates of the full Hamiltonian, the interaction between the different states is possible. From Figure 2 it is seen that interactions occur within the channels and across the channel borders. *Bound-state CI* occurs within the closed channel. *Free-bound state CI* occurs between the closed channel and single ionization channels. *Interchannel CI* occurs within the single ionization channels[25]. The coupling between the single and double ionization channel is referred to as *transchannel CI*. Finally, coupling within the double ionization (DI) channel is referred to as *DI interchannel CI*.

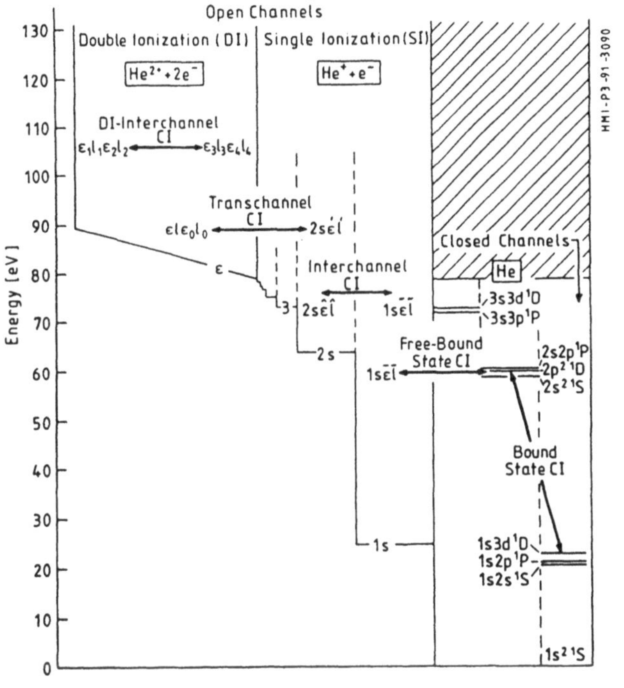

Figure 2: Channel diagram for single configuration states of He indicating various types of configuration interaction (CI).

Due to the energy conservation in the unperturbed system, the interactions along horizontal arrows are favored. However, interactions along nearly vertical arrows are also possible. The selection rule for the CI coupling is that the two single-configuration states have the same total angular momentum and parity. If the spin-orbit coupling scheme is valid, as it is for the case of He, then the orbital angular momentum and the spin must also be equal.

3 Double Excitation Involving Electron Correlation

3.1 *General Considerations*

In the past few years, considerable interest has been devoted to the processes of double ionization and double excitation of He by fast ions.[7-9] McGuire[1] predicted an interference between the two-step process involving two nucleus-electron interactions and the two-step process involving a single nucleus-electron interaction followed by an electron-electron interaction. In lowest order the amplitudes for the first mechanism[9] denoted TS-1 is proportional to the projectile charge Z_p whereas the amplitude for the second mechanism denoted TS-2 is proportional to Z_p^2. Therefore, an interference term is produced depending on Z_p^3 and, hence, on the charge sign. The observation of this interference[9] in double ionization of He created stimulating controversies[1,2,14,16] about electron correlation effects in ion atom collisions.

Recently, particular attention has been paid to the process of double excitation.[27] In fact, double excitation of He has previously been studied in various laboratories since the mid 1960s (see, e.g., Bordenave-Montesquieu et al.[28] and references therein). Measurements devoted to the charge state-dependence of double excitation of He were carried out more recently.[10-13] To interpret the double excitation data various mechanisms are to be considered. Figure 3a shows different paths leading to double excitation via the closed channel and the single-ionization channel. The successive nucleus-electron interactions are represented by two connected open lines. Also, in Figure 3a, shake-up and resonant free-bound state CI are incorporated and a path is added representing bound state CI, i.e., between the states 1s3d ^1D and $2p^2$ ^1D. Moreover, it was found that initial state CI plays a role.[29]

It is expected that the dielectronic processes produced by free-bound state CI should be particularly important since it is resonant. The corresponding path represents the TS-1 mechanism[9] which proceeds via the single ionization channel initiated by nucleus-electron interaction followed by a process similar to dielectronic capture. It should be realized that the interference associated with the TS-1 process produces asymmetric line shapes known as Fano-line profiles.[24] These line profiles make it difficult to extract cross section for double excitation from the experimental data. Hence, there is a conceptual problem in the interpretation of the experimental results in terms of double excitation.

The problems involved in the interpretation of the double excitation are expected to be significantly reduced when the target atom Li is used. The Li spectrum exhibits a relatively small background of continuum electrons due to direct ionization, and the autoionization lines are found to have no asymmetric shapes.[30,31] Hence, the corresponding path involving continuum electrons is probably less important. Instead, shake up and bound-state configuration interaction are expected to gain importance. This can be seen from the corresponding channel diagram is given in Figure 3b.

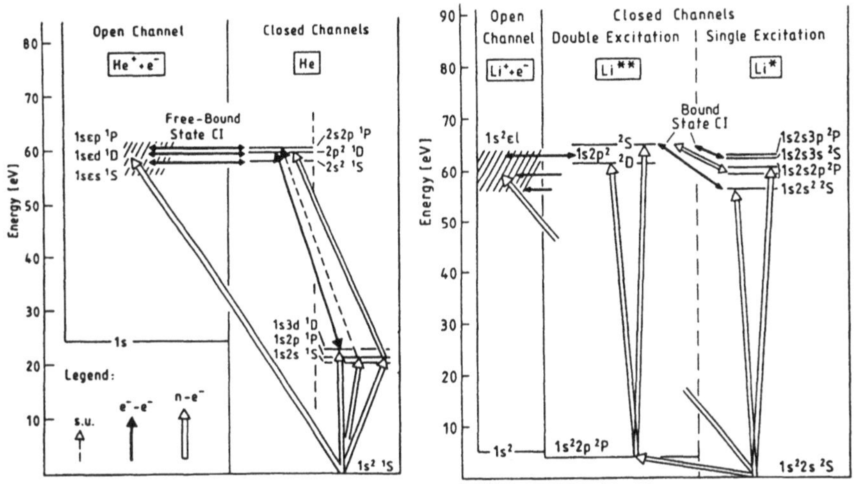

Figure 3: Channel diagrams for single configuration states of He (a) and Li (b). The transitions induced by a nucleus-electron (n- e⁻) interaction are indicated by open lines and the transitions due to dielectronic processes (e⁻- e⁻) and shake-up (s.u.) are indicated by solid and broken lines, respectively.

The comparison of the diagrams in Figure 3 emphasizes also that the double excitation of He affects two equivalent electrons in the 1s orbitals whereas double excitation of Li involves the non-equivalent electrons from the orbitals 1s and 2s. For Li this has the consequence that the configurations produced by single and double excitation are close in energy and, thus, may interact significantly. Figure 3b shows that the energy gap between the states $1s2s^2\,{}^2S$ and $1s2p^2\,{}^2S$ is rather small so that CI effects are expected to be important. It is noted that an interference term depending on Z_p^3 is created as the excitation probabilities for these states depend differently on Z_p. In lowest order, the amplitudes for the production of the states $1s2s^2\,{}^2S$ and $1s2p^2\,{}^2S$ are proportional to Z_p and Z_p^2, respectively. It is noted that the amplitudes for the corresponding states $2s^2\,{}^1S$ and $2p^2\,{}^1S$ of He are both proportional to Z_p^2. Hence, in He the production of a Z_p^3 interference term by bound-state CI is not very likely.

Therefore, the essential difference in the double excitation of He and Li lies in the influence of the bound-state CI on a possible Z_p^3 term.

3.2 *Results for Double Excitation of Li*

The foregoing discussion shows that the double excitation of Li is expected to be influenced by bound-state CI. As noted already, bound-state CI may occur during the collision and/or after the collision partners have separated. In the first case which corresponds to scattering correlation, it is favorable to emphasize the dynamic aspect of the electron correlation effects and, hence, to use the channel diagrams based on the IPM states Φ_j^o (e.g. Figure 3). But these diagrams have the disadvantage that the asymptotic states (in particular, the final state) are not well represented. When electron correlation is important after the collision, we deal with atomic correlation which can be accounted for by projecting the single-configuration state considered here onto a correlated final state.

Alternatively, it may be favorable to use basis states which incorporate electron correlation, i.e. states which diagonalize (at least within a given channel) the full Hamiltonian H. Figure 4 shows diagrams for Li-like systems which involve multi-configurations for the final states. It is important to point out that a Z_p^3 term may not

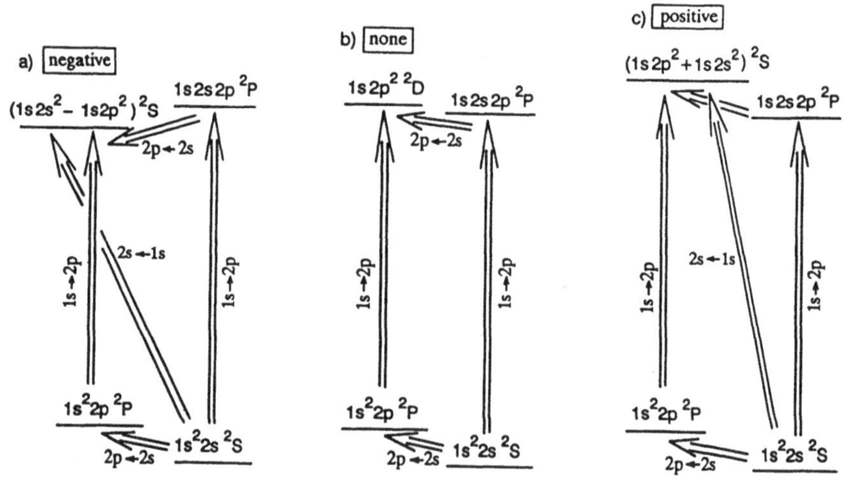

Figure 4: Diagram involving multi-configuration final states in transitions populating doubly excited states in Li-like atoms. The first component in the final multi-configuration states is supposed to be the dominant component. The diagrams a), b), and c) involve destructive, none, and constructive interferences, respectively.

only be produced when different mechanisms interfere (e.g. TS-1 and TS-2) but also when their creation of the corresponding components occurring in the final state depends differently on Z_p. It was noted above that the amplitudes for the production of the components $1s2s^2\ ^2S$ and $1s2p^2\ ^2S$ are in lowest order proportional to Z_p and Z_p^2, respectively. When the transition probability is evaluated the contributions for the different components are added coherently and, hence, the Z_p^3 term is created. The remarkable feature of Li double excitation is that at least three final states exist which exhibit either destructive, none, or constructive interference (Figure 4) involving the same Z_p^3 term. Thus, a single Li autoionization spectrum may yield the same information which is otherwise achieved by using projectiles of opposite charges.

The experiments were performed at the accelerator facility GANIL in Caen using the electron spectroscopy apparatus described previously.[32] Fast projectiles of 95-MeV/u Ar^{18+} were applied to ensure small perturbations. Figure 5 shows autoionization spectra of Li observed at different angles. A small continuous background was subtracted. Most lines exhibit a non-isotropic angular distribution indicating strong alignment effects, e.g., the analysis of the ^2P states shows the nearly exclusive population of the magnetic quantum number $M_L=1$.

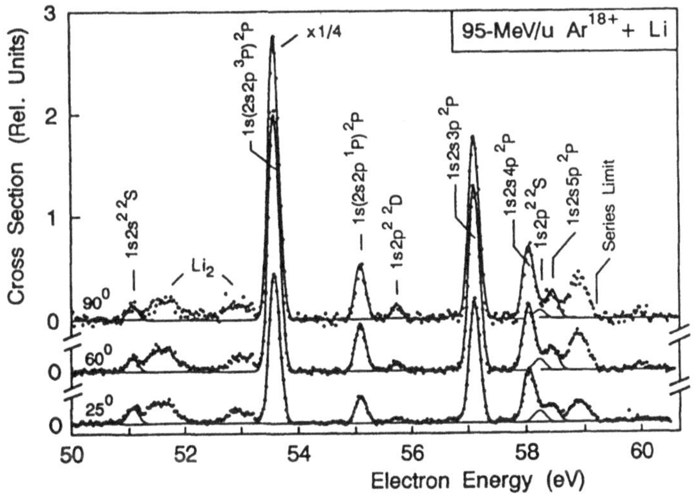

Figure 5 Auger-electron spectrum of core-excited Li produced in 95-MeV/u Ar^{18+} + Li collisions. The electron spectra are obtained for different observation angle as indicated

In Figure 5 the most intense line labeled $1s(2s2p\ ^3P)\,^2P$ is due to the single-electron transition 1s-2p. The single-excitation data are taken as a reference for the angle integrated cross sections given in Table I. Note that the labels $1s2s^2\ ^2S$ and

$1s2p^2\ ^2S$ indicate the main components only and that the corresponding states are represented by a coherent superposition of these components. This superposition give rise to the Z_p^3 interference term. As noted above, if the mixing is limited to two-components only, one expects interference terms of equal absolute value but of opposite sign, i.e. negative for $1s2s^2\ ^2S$ and positive for $1s2p^2\ ^2S$. Within the two-state coupling model the $1s2p^2\ ^2D$ state is free of interference effects.

Table I. Relative cross sections (in %) with respect to the summed cross sections for $1s(2s2p\ ^3P)^2P$ and $1s(2s2p\ ^1P)^2P$.

$1s2s^2\ ^2S$	$1s2s2p\ ^2P$	$1s2p^2\ ^2D$	$1s2p^2\ ^2S$
1.6(1)	100	0.91(7)	1.0(1)

The analysis of the experimental data[31] is still preliminary but it is likely that interference effects are visible. For instance, due to statistical rules the intensity of the $1s2p^2\ ^2S$ is expected to be smaller by a factor of 5 than that of for $1s2p^2\ ^2D$. However, the intensity of the $1s2p^2\ ^2S$ state is seen to be as large as that for $1s2p^2\ ^2D$ (Table I). In this case constructive interference effects may play a role as expected from the diagram in Figure 4.

Further work is needed to interpret the experimental data. Recent work indicates that interference effects involving electron correlation in double excitation may cancel when treated in second-order perturbation theory.[20,33] In this case the double excitation data would provide information about interference effects involving third- or higher-order contributions. For future studies it appears as a challenge to verify double excitation by means of interference effects including a breakdown of the frozen-orbital approach and scattering correlation treated in third or higher order.

References and Footnotes

[+]Work performed also at the Hahn-Meitner-Institut Berlin, D-1000 Berlin 39, FRG

1. J.H. McGuire, Phys. Rev. Lett. **49** (1982) 1153, J.H. McGuire, Nucl. Instr. Meth. B10 (1985) 17, J.H. McGuire, Phys. Rev. **A 36** (1987) 1114 and J.H. McGuire, **High-Energy Ion-Atom Collisions,** Lecture Notes in Physics **294**, ed. by D. Berényi and G. Hock (Springer Verlag, Heidelberg, 1988) p. 415
2. J.F. Reading and A.L. Ford, **Electronic and Atomic Collisions,** Invited Papers, edited by H.B. Gilbody, W.R. Newell, F.H. Read and A.C. Smith (North Holland, Amsterdam,1988) p. 693

3. N. Stolterfoht, **Spectroscopy and Collisions of Few-Electron Ions**, edited by M. Ivascu, V. Florescu, and V. Zoran (World Scientific, Singapore, New Jersey, London, 1989) p. 342
4. J.A. Tanis, E.M. Bernstein, M. Clark, W.G. Graham, R.H. McFarland, T.J. Morgan, B.M. Johnson, K.W. Jones, and M. Meron, Phys. Rev. Lett. **31** (1985) 4040 and J.A. Tanis, E.M. Bernstein, M.W. Clark, W.G. Graham, R.H. McFarland, T.J. Morgan, A. Müller, M.P. Stöckli, K.H. Berkner, P. Gohil, A.S. Schlachter, J.W. Stearns, B.M.Johnson, K.W. Jones, M. Meron, and J. Nason, **Electronic and Atomic Collisions**, Invited Papers, edited by D.C. Lorents, W.E. Meyerhof and J.R. Peterson, (North Holland, 1985) p. 425
5. T.J.M. Zouros, D. Schneider, and N. Stolterfoht, Nucl. Instr. Meth. **B31** (1988) 349
6. W. Fritsch and C. D. Lin, Phys. Rev. Lett **61** (1988) 690
7. H. Knudsen, L.H. Andersen, P. Hvelplund, G. Astner, H. Cederquist, H. Danared, L. Liljeby, and K.G. Rensfeld, J. Phys. B **17** (1984) 3545, H. Knudsen, L.H. Andersen, P. Hvelplund, J. Sørenson, and D. Ciric, J. Phys. B **20** (1987) L253, and references therein
8. A.K.Edwards, R.M. Woods, A.S Beard, and R.L. Ezell, Phys. Rev. **A37** (1988) 3697
9. L.H. Andersen, P. Hvelplund, H. Knudsen, S.P. Møller, K. Elsener, K.G. Rensfeld, and E. Uggeerhøj, Phys. Rev. Lett. **57** (1986) 2147 and L.H. Andersen, P. Hvelplund, H. Knudsen, S.P. Møller, A.H. Sørensen, K. Elsner, K.G. Rensfeld, and E. Uggerhøj, Phys. Rev. A **36** (1987) 3612
10. M. Prost, M. Morgenstern, D. Schneider, and N. Stolterfoht, Proceedings of the Xth Conference of the Physics of Electronic and Atomic Collisions, Abstract of Papers, edited by (Commissariat de Paris, 1977) p 994 and M. Prost, Diplomarbeit, Berlin, 1977, unpublished
11. P.W. Arcuni and D. Schneider, Phys. Rev. A **36** (1987) 3059
12. J.O.P. Pedersen and P. Hvelplund, Phys. Rev. Lett. **62** (1989) 2373
13. J.P. Giese, M. Schultz, J.K. Swenson, H. Schöne, M. Benhenni, S.L. Varghese, C.R. Vane, P.F. Dittner, S.M. Shafroth, and S. Datz, Phys. Rev. A **42** (1990) 1231
14. R.E. Olson, Phys. Rev. A **36**, (1987) 1519
15. L. Végh, Phys. Rev. A 37 (1988) 992
16. J.H. McGuire and J. Burgdörfer, Phys. Rev. **A36** (1987) 4089
17. E.U. Condon and C.H. Shortley, **The Theory of Atomic Spectra** (University Press, Cambridge, 1964)
18. N. Stolterfoht, Physica Scripta **42** (1990) 192
19. N. Stolterfoht, Nucl. Instrum. Methods, to be published
20. N.Stolterfoht, Lecture Notes in Physics (Springer Verlag, Heidelberg, 1991) to be published
21. H.S.W. Massey and D.R. Bates, Rept. Prog. Phys. **9** (1942) 62

22. J.H. McGuire and J.C. Straton, **The Physics of Electronic and Atomic Collisions**, Invited Papers, edited by A. Dalgarno, R.S. Freund, P.M. Koch, M.S. Lubell, and T.B. Lucatorto (American Institute of Physics, New York,1990) p. 280
23. J.F. Reading and A.L. Ford, **Comments on Atomic and Molecular Physics**, Vol **23** (1990) 301
24. U. Fano, Atomic Physics (Plenum Press, New York, 1969) p. 209
25. W. Mehlhorn, **Atomic Inner Shell Processes**, ed. B. Crasemann (Plenum Press, New York, 1985) p. 119
26. N. Stolterfoht, Phys. Reports **146** (1987) 215
27. J.H. McGuire and N.C. Deb, in **Atomic Physics with Positrons**, edited by E.A.G. Armour and J.W. Humbentson, NATO Advanced Study Institutes, Ser. B, Vol. **169** (Plenum, New York,1988)
28. A. Bordenave-Montesquieu, A. Gleizes, and P. Benoit-Cattin, Phys. Rev. A **25** (1982) 245
29. T. Ishihara, private communication (1990)
30. R. Ziem, R. Bruch, and N. Stolterfoht, J. Phys. B **8** (1975) L480
31. N. Stolterfoht, X. Husson, D. Lecler, R. Köhrbrück, B. Skogvall, S. Andriamonje, and J.P. Grandin, to be published
32. N. Stolterfoht, Z. Physik **248** (1971) 81
33. J.C. Straton and J.H. McGuire, Phys. Rev. A, in press

CORRELATION EFFECTS IN DOUBLY EXCITED RYDBERG STATES

Yukap Hahn
Department of Physics, University of Connecticut
Storrs, CT. 06269

ABSTRACT

Accurate description of doubly excited Rydberg states (DERS) of neutral atoms and near-neutral ions is difficult because of the strong electron-electron correlation effects, specially when two or more electrons are in high Rydberg states. Even when only one electron is in high Rydberg states, the correlation is found to be important if the spectral density in that region is high. The coupling of DERS to the continuum is relatively weak, but the Auger decay widths of these states are sensitive to the electron-electron correlation. No reliable theoretical approach is currently available which can treat the DERS with uniform accuracy, although MCHF has been applied with success in some cases. We report on the convergence study of MCHF for DERS, in He and Ba. The correlated HF and time-independent mean field theories are being developed for DERS. Relevance of this study to the Wannier problem is also discussed.

1. Introduction

Production of doubly excited Rydberg states (DERS) and study of their properties (energies, widths etc.) have been of much current interest.[1–5] Such states can be excited by a variety of methods; a stepwise excitation by multiple laser irradiation, synchrotron irradiation producing systems with inner-shell holes, and electron-ion and ion-atom collisions. Thus, dielectronic recombination, resonant collisional excitation and excitation-Autoionization in electron-ion collisions involve DERS and their cascade decays. Resonant-transfer excitations produce DERS, while hollow ions with multiple inner-shell holes are created by synchrotron irradiation and by high energy ion-atom collisions. Generation of DERS with two or more electrons in high Rydberg states has proven difficult when the final states are reached by stepwise laser irradiation. On the other hand, planetary atoms with a single electron in high Rydberg state have been excited and their properties studied in detail in recent years[1,2] .

Theoretical treatment of DERS is in an early stage of development, and currently no acceptable procedure is available which is uniformly applicable to all situations. In view of increased experimental effort on this problem in recent years, a review of the theoretical status is warranted. The DERS with two electrons in the excited states ($n_1l_1n_2l_2$) may be divided roughly into two types; (a) $n_2 >> n_1$, which we denote in the following as planetary atoms, and (b) $n_2 = n_1$, the binary atoms (sometimes called the Wannier ridge states). The conventional procedures such as MCHF and variational principles (VP) worked well for low-lying DERS with $n \lesssim 5$, but are not suitable for $n \gtrsim 5$ for two basic reasons: (i) The minimum character of the spectrum of DERS is no longer present since there are many quasi-discrete states which lie below the particular state d of interest and also one or more continua of spectrum into which d is embedded. (ii) Density of the discrete spectrum is very high, due to the long range Coulomic character of the core potential. (Obviously, spectra for negative ions will be simpler, but the correlation among the electrons is relatively stronger.)

The strong configuration interaction (CI) in MCHF suggests that perhaps the zeroth order central field approximation for DERS may be inadequate. There have been much work done on the new classification of DERS using the O_4 group[3] and hyperspherical coordinates.[4]. However, the radial correlations in such schemes are difficult to implement.

In the following we discuss the DERS using the coordinates r_1, r_2 and r_{12}, and limiting to the simple case of $L = S = 0$. The results can easily be extended to more general situations. Three theoretical approaches of interest are a modified form of the multiconfiguration Hartree-Fock (MCHF), the homogeneous and correlated HF (HCHF), and inhomogeneous and correlated HF (ICHF) methods. They are designed to handle the various difficulties of DERS. The basic ideas involved are simple but their applications are more difficult. We present the theories and some preliminary results we have obtained thus far.

2. Theoretical Approaches for DERS.

We noted in Sec. 1 two principal difficulties of treating DERS, namely (i) presence of many states below d in energy and resulting loss of the extremum property, and (ii) high density of states in the neighborhood of d, with strong CI. Three theoretical approaches to be discussed below are (1) the conventional MCHF, (2) the HCHF and (3) the ICHF methods, all of which are essentially based on the HF procedure with node-counting to resolve the difficulty (i). Difficulty (ii) is more complicated to treat, and depends very much on the n value.

2.1. MCHF.[6,7]

This is a well-established method for bound state problems, and has been used extensively also for DERS, so long as the final results (Auger and radiative decay widths) are not too senstive to CI. The mixing of configurations for the particular state d may be represented as

$$\begin{aligned} \Psi &= \Psi_{HFd} + \sum_{d'd''} \frac{1}{E - H_{d'd''}} (V_{d''d}) \Psi_{HFd'} \\ &\equiv \Psi_d^0 + \frac{1}{\Delta} V \Psi_{d'}^0 \quad (2.1) \end{aligned}$$

in obvious notation and where the denominator is a matrix which has to be inverted. However, for higher energy levels, the method becomes less effective.

The widths of DERS are apparently more sensitive to CI than the DERS energies, and are closely related to the problem in variational principles (VP) of Kohn, Schwinger, etc. in scattering theory.[8]. Because of the presence of the continuum spectrum around E, the calculated K matrix is not bounded and can have irregularities. To circumvent this difficulty, the stabilization procedure[9] was developed, for example; we write the variational wave function for scattering as

$$\begin{aligned} \Psi_t &= (S_t + K_t C_t) + \sum_\alpha c_\alpha X_\alpha \\ &\equiv \Phi_t + F_t, \quad (2.2) \end{aligned}$$

where X_α are square-integrable trial functions. A variational functional is defined as

$$[K] = <\Psi_t | H - E | \Psi_t> + K_t. \quad (2.3)$$

Variation of $[K]$ with respect to K_t and c_α gives the amplitude which has in general the behavior shown in Fig. 1, where a is a nonlinear parameter in the trial function Ψ_t. The probability of predicting the correct amplitude increases as the number of terms N in the trial function increases. But, the fluctuation in the amplitude becomes also more violent, and even in the limit of large N the VP has a finite probability of yielding any value of K, between $+ \infty$. Evidently, this fluctuating feature is intrinsic to scattering problems, where the spectrum of the system around E is dense, and a continuum of states lie below and above E. Different versions of VP for scattering all share this deficiency.

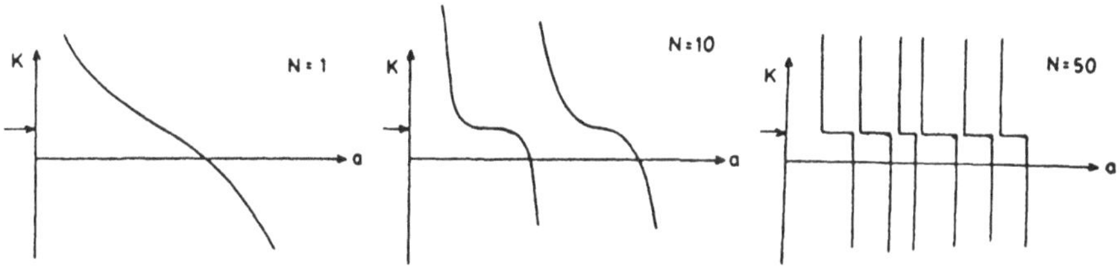

Figure 1. The behavior of the K matrix element determined by a variational principle in scattering theory is illustrated as a function of the number of linear parameters N in the calculation and for a nonlinear parameter a in the trial function.

In order to remedy this undesirable feature, a drastic reformulation of the variational principle was required. Thus a variational bound principle for scattering was derived[10], which has the desired extremum property and no fluctuations in K. The original spectrum had to be split into two parts, each of which is then treated separately. On the other hand, the stabilization procedure[9] is a slight modification of the original VP; instead of directly varying the functional $[K]$, the NxN energy matrix

$$< X_\alpha | H | X_\beta >$$

is first diagonalized to generate a variational spectrum E_α of H. Then, for a specific energy $E = E_\alpha$, we may set

$$< X_\alpha | H - E | \Psi_t> = 0 \qquad (2.4)$$

and obtain the amplitude

$$K_t = - < X_\alpha | H - E | S_t > / < X_\alpha | H - E | C_t >. \quad (2.5)$$

In the calculation of the widths for DERS, the first-order distorted wave theory is applied, as

$$\Gamma = \Sigma A_a + \Sigma A_r , \quad (2.6)$$

where the Auger transition probabilities A_a are generally much larger than the radiative transition probabilities A_r, and where

$$A_a = | < \Psi_d | V | \Phi_t > |^2 \quad (2.7)$$

The CI of the intermediate states d in A_a for example may be expressed as

$$A_a = | < \Phi_t | V | \Psi_d^0 + \frac{1}{\Delta} \mathbb{V} \Psi_d^0 > |^2 . \quad (2.8)$$

Comparing this with (2.5), we expect that fluctuations similar to that experienced in the variational principle calculation for scattering would also be present in (2.8). Indeed, preliminary study of the DERS in He** and Ba** shows such fluctuations; as more states are added in the mixing, the magnitude of the fluctuation in the calculated widths increases[11]. Therefore, to extract meaningful result from the calculation with a finite N, a suitable "averaging" procedure is needed that sets the criteria for choosing the "proper" configurations for mixing.

Overall, the MCHF is expected to be effective for $n_{1,2} = n \lesssim 10$, because of rapid increase in the density of levels with n. Even then, mixing of a large number of configurations may be necessary. As discussed above, with large N, we expect the resulting widths to converge in most cases, but occasional fluctuations may occur as N is increased by a small steps. Since in MCHF, the number of configurations N which are being mixed is a discrete integer and not continuous, such fluctuations would show up abruptly. Nevertheless, the widths may be taken 'on the average'.

2.2. HCHF.[12]

An alternative approach to DERS was proposed[12] recently, in which the conventional HF procedure was augmented with a set of correlation functions. Here the difficulty (i) of Sec. 2 is again handled by the HF and node-counting, while the density of states problem (ii) is treated in terms of a correlation function ϕ. Thus for example, let the trial function in this correlated HF as

$$\Psi_t = \Psi_{HF} \phi \; ; \quad (2.9)$$

where

$$\phi = \sum_\alpha c_\alpha f_\alpha \quad (2.10)$$

with the predetermined functions $f_\alpha(r_{12};r_1 r_2)$ and $\Psi_{HF} = \psi_a(r_1)\, \psi_b(r_2)$. Then

$$< \psi_\alpha | H - E | \Psi_t > = 0 \quad (2.11)$$

gives the usual HF equation for ψ_b, with nonlinear (self-consistent) direct and nonlocal exchange potentials. A similar equation is derived for ψ_a. These equations depend implicitly on the parameters c_α. On the other hand,

$$< \Psi_{HF} f \,|\, H - E \,|\, \Psi_t > \; = 0 \qquad (2.12)$$

for all o gives a matrix equation that determines the coefficients c_α, as

$$c_\alpha = - \sum_\beta < \Psi_{HF} f_\beta \,|\, H - E \,|\, \Psi_{HF} > / < \Psi_{HF} f_\alpha \,|\, H - E \,|\, \Psi_{HF} f_\beta > . \qquad (2.13)$$

Note the similarity between (2.13) and (2.5).

As stressed earlier, the important feature of the above procedure is that the node-counting of the HF functions is retained so that we are not necessarily solving for the lowest energy state. To further expedite the theory and extend its applicability to heavier atoms, we may introduce pseudopotentials for the ionic core, such as Ba^{++} and Cs^{++}, where the effect of the core is expected to be less critical, provided the long range core polarization potentials are included. The HCHF may be useful for $n \lesssim 30$, where the radial wave functions with many nodes are still numerically tractable.

Two other options may be mentioned here on the choice of the trial function Ψ_t. Instead of (2.10), O may be determined from a nonlinear equation of the type (2.11). Or more than one Ψ_{HF} wave functions may be introduced, in which case we obtain a set of coupled equations of the type (2.11) for each HF component.

2.3. ICHF.[13]

For DERS with very large n values, $n \gtrsim 30$, the above approaches are not practical, even if they converge. The high Rydberg states must be treated as a continuum, a situation analogous to scattering problems and the quantum defect approach. We may adopt the time-independent mean field theory of collisions. This method was originally developed for nuclear reactions involving many fermions, but should be applicable to DERS, specially when n is very large, where $E = -2/n^2 \longrightarrow 0$, and for E above the ionization threshold ($E = 0$). The theory is summarized here in the distorted wave (DW) formalism. The distorted quantities are denoted by tildes. The scattering amplitude is written in the form

$$T_{fi} = < \tilde{\Phi}_f \,|\, \tilde{V}_{i,f} \,|\, \tilde{\Phi}_i > \; + \; < \tilde{\Phi}_f \,|\, \tilde{V}_f \, G^+ \, \tilde{V}_i \,|\, \tilde{\Phi}_i >$$

$$\equiv T^{DWB} + T^{DWC} , \qquad (2.14)$$

where we require that $(H - E)\, \tilde{\Phi}_{i,f} = \tilde{V}_{i,f} \, \tilde{\Phi}_{i,f}$. That is, $\tilde{\Phi}$ depend on $\tilde{V}$. The main interest is in the second correction term DWC. We rewrite it as

$$T^{DWC} = < \tilde{\Phi}_f \,|\, \tilde{V}_f \,|\, \tilde{F}_i^+ >$$

$$= < \tilde{F}_f^- \,|\, \tilde{V}_i \,|\, \tilde{\Phi}_i > , \qquad (2.15)$$

where

$$(E^+ - H)\, \tilde{F}_i^+ = \tilde{V}_i \tilde{\Phi}_i \qquad (2.16)$$

$$(E^- - H)\, \tilde{F}_f^- = \tilde{V}_f \tilde{\Phi}_f , \qquad (2.17)$$

with the outgoing/incoming wave boundary conditions. E^+ has an imaginary part which makes all the scattering functions square-integrable. The physical amplitude is obtained by analytic continuation in E^+, Im E $\longrightarrow$ 0. The special feature of ICHF is that F can be obtained by inhomogeneous HF equations derived from (2.16) and (2.17), with the correlation function included. Thus, we let

$$\tilde{F} = F_{HF}{}^* \phi \ , \quad (2.18)$$

where

$$F_{HF} = F_1 F_2 \quad (2.19)$$

$$\phi = 1 + \sum_{\alpha} c_{\alpha} f_{\alpha} \ , \quad (2.20)$$

and where the f's are predetermined correlation functions. Then, we obtain for F_1 and F_2

$$< F_2 | \, E^+ - H \, | \, \tilde{F} \, > \; = \; < F_2 \, | \, \tilde{V} \, \tilde{\Phi} \, > \quad (2.21)$$

which is the inhomogeneous HF equation, and for the c's

$$< F_{HF} f_{\alpha} | \, E^+ - H \, | \, \tilde{F} > \; = \; < F_{HF} \, f_{\alpha} | \, \tilde{V} \, \tilde{\Phi} > \quad (2.22)$$

give a matrix equation with the inhomogeneous term which can be solved for the coefficients c . Again, c are nonlinearly mixed with F_{HF}. Obviously, this approach may be applied to cases in which E can be above or below the ionization threshold. The resulting amplitude and the widths should be interpreted as an average over the many DERS levels near E.

As with HCHF, we can also generalize (2.18) by either deriving a nonlinear equation for O of the type (2.21), or replacing the c 's with additional HF-type functions and obtain a set of coupled equations for F_{HF}'s.

3. Numerical Results

In the following preliminary result obtained using MCHF, HCHF and ICHF will be presented. For neutral atoms and near-neutral ions, the Auger widths are generally much larger than the radiative widths, by as much as two orders of magnitudes. Therefore, we concentrate only on the Auger widths.

3.1. He** by MCHF[11]

Table I contains the mixing coefficients for the states ns^2 , which are mixed with np^2 and nd^2. d1lst of all, we note that the configuration interaction (CI) becomes stronger as n is increased; for n > 4, the designation of the states as ns^2 for example is no longer meaningful. It is also apparent that as n increases, higher l configurations are desired. Finally, the widths are quite sensitive to mixing, but the ns^2 configurations are found to be less sensitive to CI than other states.

After many trials, a simple recipe for mixing configurations for DERS is derived, which seems to work in most cases with minimal number N of configurations to be mixed:

(a) Start with a complete set of two-state (N = 2) mixing, and check the signs of their configuration mixing coefficients (CMC) with respect to that of the state d of interest.

(b) From each series of the types (l_1 l_2, choose two states which showed the opposite CMC signitures in step (a) and lie close to the state d.

(c) Mix together all the pairs from all the series chosen in step (b), but again requiring that each pair retains their opposite signitures of CMC. The magnitude of all the CMC should be larger than 0.005, to have the mixing meaningful. This recipe is being further tested and refined by comparing with the available widths of DERS in He. Although the final accuracy of the widths calculated this way may be limited, if successful, this is one possible way to treat the binary and planetary states of large n.

n	ns	np	nd	Width(SC)	Width(MC)
2	0.86	0.55	-0.01	14.1	22.9
3	0.76	0.63	0.16	21.2	23.6
4	0.69	0.68	0.25	25.5	19.9
5	0.65	0.70	0.29	2.3	2.6
3s3d				33	36 (31/20)
2s3s				4.3	7.3 (3-25/6)

Table I. The widths, in $10^{13}sec^{-1}$, and the mixing coefficients for the ns^2 of He** are calculated in SCHF and MCHF. The increase in the strength of correlation with increasing n is apparent from the magnitudes of the mixing coefficients. Sample DERS used for testing the mixing recipe are also given. The numbers in the brackets in the last two rows are the currently available theoretical/experimental widths, respectively.

3.2. Ba** by MCHF[11].

We specifically consider the planetary states of Ba** for the series 7sns, 8sns, 7snd and 8snd, for which experimental data have been available[1] for several years, but no adequate explanations have been given. There are many configurations which lie near these states, and the widths are found to be extremely sensitive to CI. We examined the sensitivity by gradually increasing the number of mixed states N. Figure 2 shows the general trend and the fluctuations for 7s13s of Ba**. As expected, the fluctuation in the widths becomes more violent as we mix more states. This is similar in behavior to the VP in scattering theory.

Table II contains the result of calculation of the widths for Ba** using the recipe. The sets given in this table are the only combinations that satisfied the recipe. Note the drastic change in 8s13d. The uncertainty in the experimental data for the ns series is roughly at $\pm$30% for the ns series and $\pm$ 50% for the nd series.

We also extend the calculation to the ns^2 states of Ba**. Since the mixing is not as strong in this case, we estimate that the total widths calculated here should be reliable perhaps to $\pm$50%.

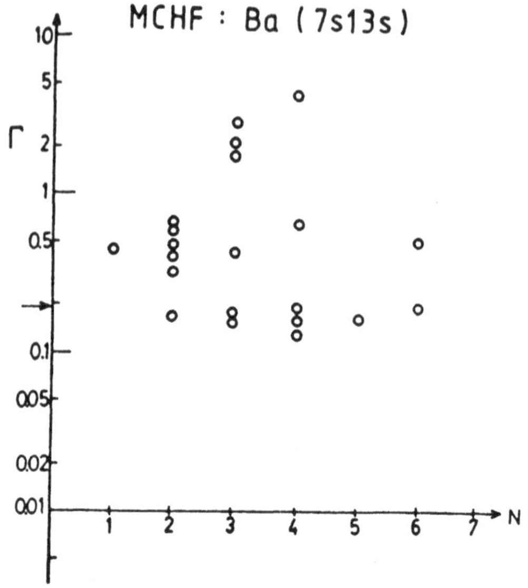

Figure 2. The total Auger widths (in $10^{13}sec^{-1}$) vs the number of configurations mixed in the MCHF calculation for the Ba** 7s13s are shown, to illustrate the convergence property of Γ in N. Note specially the increasingly large fluctuation in Γ as N increases.

n	n′	Width (SC)	Width (MC)	Exp.
7s	13s	0.51	0.46	0.91
7s	13d	0.47	0.52	0.1
8s	13s	1.7	1.2	1.2
8s	13d	1.2	0.16	0.2
7s	7s	26	25	---
8s	8s	16	17	---

Table II. The widths of the DERS of Ba**, nsnℓ with ℓ=0 and 2, are given in uits of $10^{13}sec^{-1}$. They are calculated by MCHF using the simple recipe for mixing derived in sec. 3.1.

3.3. Ba** by HCHF[12].

We summarize in Table III the preliminary result of calculation using the correlation function ϕ = 1 and introducing the pseudopotentials for Ba + *. Some cross polarization potentials are also included. The energies are compared with other calculations[5] and recent fits using quantum defect theory. The agreement is satisfactory, and the correlation effect may be included to further improve the wave function. The ns^2 series is usually weakly mixed, but the good agreement seen in Table III for very high n states is unexpected. Of course, some core distortions are already included in the pseudopotentials used.

Calculations with the correlation function ϕ and the testing of the resulting wave functions in terms of the A's are in progress.

states	RCN	SC/MFA	Aymar
5d2	0.8903	0.8693	
5d5f	0.7219		
6s6p	0.9766		
6s4f	0.7903		
6p4f	0.6096		
6s2	1.0798	1.0771	1.1196
6p2	0.7985	0.7802	
6d2	0.4571	0.4310	
7s2	0.5239	0.5058	0.526
7p2	0.4281	0.4078	
7d2	0.2864	0.2654	
8s2	0.3157	0.2980	0.312
8p2	0.2709	0.2535	
8d2	0.1967	0.1800	
9s2	0.2120	0.1970	0.212
9p2	0.1875	0.1723	
9d2	0.1436		
10s2	0.1524		0.151
11s2	0.1150		0.113
12s2	0.0899		0.0878
13s2	0.0722		
14s2	0.0593		
15s2	0.0496		

28s2	0.0114		
29s2	0.0105		
30s2	0.0098		

Table III. The energies (in Ry) of the DERS of Ba** are calculated by HCHF, with the correlation function ϕ = 1. The correlation effect on the energies is at the 10% level for the ns^2 series, but the widths will be more sensitive to in single configuration HF with an empirically adjusted parameter.

3.4. He** by ICHF.[13]

Instread of the correlation functions f in F_t, we may add more of the F_{HF}'s. This will result in a set of coupled HF equations for F_{HF}. A crude two-state coupled equations have been set up for He**, and the solutions are being studied. The 'average' excitation energies and widths are evaluated as functions of Im E and Re E. The self-consistency requirement on the solutions consumes a large amount of computer time.

Obviously, the DERS in the limit $E \longrightarrow 0$ is the Wannier ionization threshold problem[14] We are studying this problem by going[15] from $E < 0$, while the scattering picture describes the process from the region $E > 0$. Since the DERS are somewhat more readily tractable, the current experimental and theoretical efforts on DERS are justified.

4. Discussion

The problem of DERS is far from being completely understood, because of the strong electron-electron correlations. Some of the recent attempts to study this problem are summarized here. The MCHF procedure with a simplifying recipe seems to be effective for the DERS with $n < 10$. without such a prescription, the method becomes too complicated and unstable to apply. The simplifying recipe discussed in sec. 3.1 is not perfect and still in a developing stage, and is being tested and refined. The HCHF approach may be effective for DERS with $n \lesssim 30$, where the node-counting and the correlation function makes up the strong CI. On the other hand, for $E = 0$, the ICHF with complex energies may be versatile enough to handle the DERS in the threshold region with very large n.

Finally, the questions of labelling the DERS are yet to be addressed in the context of MCHF, HCHF and ICHF. The approach of using an alternate set of quantum numbers (LSKT) has been successful in some cases, but further refinements are needed to handle binary atoms and to be useful in practical computation of the widths.

Acknowledgments

The material reviewed here is based on the collaborative efforts of many people, in particular, B. Giraud and J. Pascale of Saclay, and I. Nasser of King Fahd University. I would also like to thank the organizers of this workshop, Dr. Cisneros and Dr. Morgan for providing a relaxed atmosphere to have many fruitful discussions on a wide range of subjects of current interest.

4. References

1. L. A. Bloomfield et al, Phys. Rev. Lett. 53, 2234 (1984)
2. J. Boulmer et al, J. Opt. Soc. Amer. B4, 805 (1987)
   L. Pruvost et al, J. Phys. B23, L95 (1990)
   B. Carre et al, Saclay preprint, 1990.
3. D. R. Herrick, Adv. Chem. Phys. 52, 1 (1983)
   D. R. Herrick and O. Sinanoglu, Phys. Rev. A11, 97 (1975)
4. A. R. P. Rau, Rep. Prog. Phys. 53, 181 (1990)
   C. D. Lin, Adv. Atom. Molec. Phys. 22, 77 (1986)
5. M. Aymar, J. Phys. B22, 2359 (1989)
   P. Hamacher and J. Hinze, J. Phys. B22, 3397 (1989)

M. Zubek et al, J. Phys. B22, 3411 (1989)
Y. K. Ho, J. Phys. B15, L691 (1982)
6. R. G. Cowan, Theory of Atomic Structure and Spectra (Univ. Calif. 1981)
7. C. Fischer, Hartree-Fock methods for Atoms (Wiley, 1977)
8. C. Schwartz, Ann. Phys. (N.Y.) 6, 156 (1959) and 6, 170 (1959).
9. F. E. Harris, Phys. Rev. Lett. 19, 173 (1967)
10. Y. Hahn and L. Spruch, Phys. Rev. 153,
Y. Hahn, Phys. Rev. C1, 12 (1970)
11. I. Nasser and Y. Hahn, Phys. Rev. A
12. Y. Hahn and J. Pascale (to be published)
J. Pascale, Phys. Rev. A28, 632 (1983)
13. B. Giraud, Physica 19D, 112 (1986)
Y. Hahn and B. Giraud (unpublished)
14. G. H. Wannier, Phys. Rev. 90, 817 (1953)
15. A. Temkin and Y. Hahn, Phys. Rev. A9, 708 (1974)

SYMMETRIES AND DECAY MECHANISMS OF DOUBLY-EXCITED RESONANT STATES

J. M. ROST and J. S. BRIGGS
Fakultät für Physik, Albert-Ludwigs-Universität,
Hermann-Herder-Str.3, D-7800 Freiburg, W.Germany

Abstract

Five approximate quantum numbers introduced by Herrick to classify atomic doubly-excited resonant states are shown to be particular combinations of three quantum numbers of molecular orbitals of the separable two-centre Coulomb problem. Consideration of the motion of the electronic centre-of-mass around the saddle point of the two-centre potential leads to propensity rules for radiative and non-radiative decay of doubly-excited states.

1. Introduction

In a series of papers (for a review see Herrick[1]) a scheme of classification for doubly-excited atomic resonances was proposed. In this scheme states can be assigned some (or all) of five quantum numbers. Three of these, K, T and A arose from a reduction of the $O_4 \times O_4$ product group of two non-interacting electrons and subsequently Lin[2] was able to correlate hyperspherical potential curves with these three quantum numbers. In alternative group theory reductions, Kellman and Herrick[3] identified sequences of levels grouped according to rovibrational structure. This led to many papers suggesting approximate rotation-vibration Hamiltonians as models for a two-electron atom.

All of this work suggested, but did not identify, some approximate separability of the two-electron problem in terms of set of collective variables. This separability and the corresponding collective co-ordinates have since been identified by Feagin and Briggs[4]. The collective co-ordinates are those of the relative vector $\mathbf{R}$ of the two electrons and the vector $\mathbf{r}$ of the electronic centre-of-mass (ECM) with respect of the nucleus.

A rotational transformation from the laboratory-fixed to the body-fixed frame provides automatically the rotational structure of resonances. The treatment of the body-fixed vector $\mathbf{R}$ as an adiabatic co-ordinate provides similarly vibrational structure in the adiabatic potentials obtained by averaging over the ECM motion in $\mathbf{r}$. Finally, and most importantly, the $\mathbf{r}$ motion itself is separable in MO prolate spheroidal λ, μ, φ co-ordinates with associated good quantum numbers n_λ, n_μ and $n_\varphi \equiv m$. It has already been shown that the Herrick K, T and A quantum numbers can be related to the parabolic quantum numbers n_1, n_2, m, A obtained for the bound electron when the other is moved to infinity. These are also related asymptotically to the n_λ, n_μ and m quantum numbers. Recent work has shown that a vibrational quantum number v_2 introduced by empirical reasoning[3] can be identified with $2n_\lambda + m$, when the ECM motion is expanded about the saddle point of the three-body Coulomb potential for fixed R. Indeed as will emerge below the two features of the separable ECM motion that result in Herrick's quantum numbers are

(i) the asymptotic $R \to \infty$ behaviour in terms of parabolic quantum numbers and

(ii) the ECM motion in the neighbourhood of the saddle point given by $\lambda = (r_1 + r_2)/R \approx 1$ and $\mu + (r_1 - r_2)/R \approx 0$ in prolate spheroidal coordinates.

2. The correlation diagram (CD)

As is well-documented[5] the two-centre problem is separable in the λ, μ, φ co-ordinates and correspondingly molecular orbital (MO) states have quantum numbers n_λ, n_μ, m. These MO assignments are in one-to-one correspondence with parabolic SA $(n_1 n_2 m)$ and spherical UA (nlm) quantum numbers as $R \to \infty$ and $R \to 0$ respectively. In the SA limit one-electron parabolic states are produced and the additional designation $A = (-1)^{n_\mu}$ is necessary to specify the linear combination (LCAO), $\Phi_1 + A\Phi_2$, which forms an MO with correct nodal structure. One notes in the correlation diagram (fig. 1) that some MO are depressed smoothly in energy as R decreases from infinity, whereas others are first depressed and then rise in energy so that the MO energy possesses a minimum. This is the phenomenon of promotion (increase of principal quantum number between SA and UA) known to be of fundamental importance for the formation of molecules and in ion-atom collisions. This feature also plays a major role in the classification of MO for the two-electron atomic problem as described below.

The evenness or oddness of n_μ expressed by $A = (-1)^{n_\mu}$ is fundamental to the behaviour of MO energies, although this does not appear to have been pointed out hitherto. For this reason we present, in Fig. 1a and b the separate correlation diagrams for $A = +1$ and $A = -1$ respectively. As discussed extensively in Rost and Briggs[6], $A = -1$ states have a node on the saddle ($r_1 \approx r_2$) of the two-centre potential, whereas $A = +1$ MO have an anti-node there. Accordingly $A = +1$ MO have more "molecular" character and this is seen qualitatively in Fig. 1 by the fact that the bundle of $A = +1$ MO emanating from a given SA hydrogenic N level are more strongly split in energy at finite R than their $A = -1$ counterparts. The latter are always promoted i.e. $n > N$ while only $A = +1$ MO can correlate SA principal shells N with UA principal shells n = N.

The qualitative appearance of the CD, that of lower levels of a given bundle crossing the remaining members (for $A = +1$ several higher bundles may be crossed) as they are promoted is readily understood[5] and is crucial for the formation both of molecular states and resonant electronic states of atoms. For large R, before molecule formation takes place, the shift of atomic levels is due to the Stark effect. Levels with

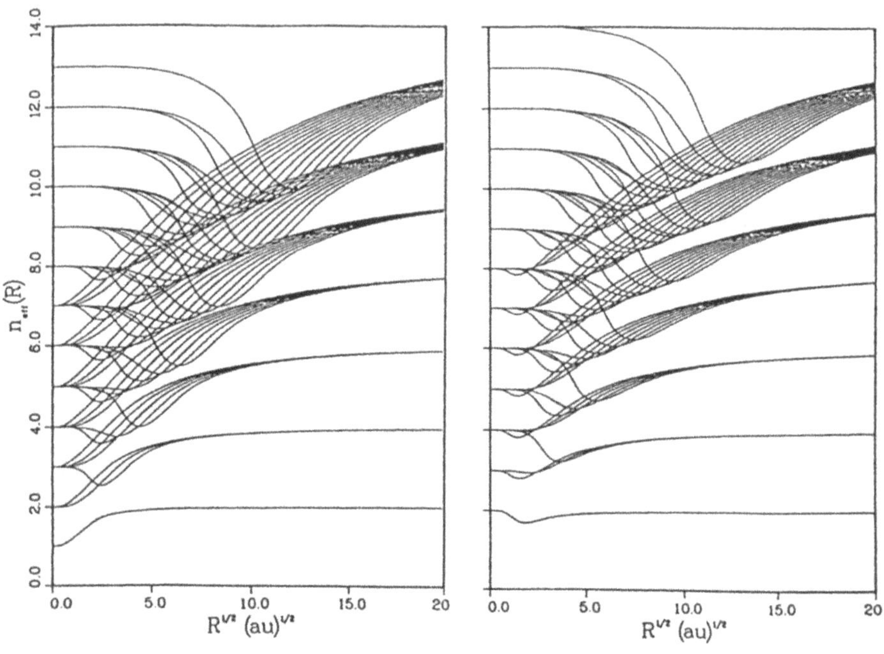

(a) MO with A = 1 **(b)** MO with A = -1

Figure 1: The correlation diagram of the two-centre Coulomb problem for MO with $m \leq 2$ up to the manifold $N = 7$ of the separated atom. The energy is given in terms of the effective quantum number $n_{\text{eff}} = [2/\mathcal{E}_{im}(R^{1/2})]^{1/2}$.

the minimum value of the "electric" quantum number $k = n_1 - n_2$ for a given m are depressed most. These MO are those in which the two particles of like charge are roughly at 180^0 with the particle of unlike charge between them. Hence the two electric dipoles are maximally oppositely directed to give minimum energy. By contrast states with maximum k have particles of like charge maximally on the same side of the third particle, leading to maximum energy configuration of interacting dipoles. The promotion condition is

$$n - N = [\tfrac{1}{2}(l - m + 1)] \equiv [\tfrac{1}{2}n_\mu] \equiv \begin{cases} n_2, & A = +1 \quad (1a) \\ n_2 + 1, & A = -1 \quad (1b) \end{cases}$$

where $[x]$ is the integer value of x. From (1) it is readily seen that MO maximally depressed by the Stark effect at large R have n_2 maximum and $n_1 = 0$ to give maximum promotion at small R. It is also clear from (1) that all $A = -1$ MO are promoted and additionally that promotion is directly proportional to the number of "angular" nodal surfaces n_μ. One also notes that promotion involves a minimum change in binding energy between UA and SA limits but a maximal change in the character of the MO. Such MO will be called "elastic" and are uniquely described by $n_2 \neq 0$. By contrast all MO with a *minimum* of angular excitation $n_2 = 0$ will be called "stiff". To this class belong all non-promoted $A = +1$ MO with the minimm value of $n_\mu = 0$ and the $A = -1$ states with the corresponding minimum value $n_\mu = 1$ in accordance with Eq. 1b. For a given total number of nodal surfaces the greater is n_μ the more elastic is the MO.

We now discuss the $A = +1$ CD in more detail. Three features will be stressed which describe the saddle behaviour of the MO. The first has already been discussed in connection with the definition of diabatic MO curves[6a,b] asymptotic to the $R \to \infty$ 3-body breakup threshold. Each stiff MO, with positive energy gradient at $R = 0$, has fixed quantum numbers $(n_\lambda, 0, m)$ and is the precursor of a complete in-saddle sequence of MO whose other members are all promoted MO with quantum numbers $(n_\lambda, n_\mu = \text{even}, m)$. The promoted MO show avoided crossings with other members of the sequence.[6c] The diabatic connection of all avoided crossings connects the stiff precursor MO near $R = 0$ with the $R \to \infty$ 3-body break-up threshold. Note that the precursor MO are those with positive gradient near $R = 0$ in Fig. 1a and are therefore easily recognised.

The second characteristic of MO distinguishing their behaviour in the saddle region $\mathbf{r} \approx 0$ concerns the (λ, φ) part of the $\mathbf{r}$ motion in the plane perpendicular to $\hat{R}$. To describe saddle motion it is convenient to adopt cylindrical co-ordinates $\mathbf{r} = (\rho, z, \varphi)$. Then if we approximate in the region of the saddle point $\lambda \approx 1, \mu \approx 0$ we have

$$\rho = \tfrac{R}{2}[(\lambda^2 - 1)(1 - \mu^2)]^{\frac{1}{2}} \to R((\lambda - 1)/2)^{\frac{1}{2}} \quad (2a)$$

$$z = \tfrac{R}{2}\lambda\mu \to \tfrac{R}{2}\mu \quad (2b)$$

$$\varphi = \varphi. \quad (2c)$$

In contrast to the motion parallel to $\hat{R}$ (described by μ or z), the motion in the plane perpendicular to the $\hat{R}$ axis (described by (λ, φ) or (ρ, φ)) is bound near $\mathbf{r} = 0$. As may be readily shown the motion is that of an isotropic 2-dimensional harmonic oscillator with hamiltonian

$$H_{\perp} = -\frac{1}{2}\left(\frac{1}{\rho}\frac{d}{d\rho}\rho\frac{d}{d\rho} + \frac{1}{\rho^2}\frac{d^2}{d\varphi^2}\right) + \frac{1}{2}\omega^2\rho^2 \tag{3}$$

where $\omega = 4R^{-3/2}$. The energies are given by $\hbar\omega(v_2 + 1)$ with

$$v_2 = 2n_\rho + m. \tag{4}$$

According to (2) the saddle nodes (n_ρ, m) go over to MO nodes (n_λ, m) and hence the degeneracy of the two-dimensional isotropic oscillator is expressed as the MO saddle degeneracy condition

$$v_2 = 2n_\lambda + m \tag{5}$$

for various combinations of n_λ and m.

3. MO interpretation of Herricks' quantum numbers

As discussed[4], the total wavefunction limits the states of total LSt symmetry that can be built upon a given MO internal state with fixed $(n_\lambda n_\mu m)$ quantum numbers and $t = n_\mu + m = l$. For example all σ_g MO $(m = 0, t = 0)$ give total states with $(-1)^{S+L} = +1$. Therefore the series $^1S^e, ^3P^o, ^1D^e, \ldots$ is possible for increasing total angular momentum L. Similarly for σ_u MO $(m = 0, t = 1)$ we have the complementary spin series $^3S^e, ^1P^o, ^3D^e, \ldots$. For MO with $m \neq 0$ the situation is more complicated. Each member of a series consists of a pair of states with the same angular momentum $L > m$ but with opposite parity and spin. For example, every π_g MO $(m = 1, t = 0)$ gives series $(^1P^e, ^3P^o), (^1D^e, ^1D^o) \ldots$. Similarly each π_u MO gives rise to series of pairs of states $(^3P^e, ^1P^o), (^3D^e, ^3D^o) \ldots$. Hence one sees already how the MO picture automatically leads to rotor series whose existence was postulated empirically by Herrick.

One aspect of the MO ordering of two-electron states should be mentioned explicitly. On the basis of ordering states according to group-theoretical considerations Herrick noticed that in zeroth-order, pairs of states e.g. $(^1P^e, ^3P^o)$ or $(^1D^e, ^1D^o)$ are degenerate and he explained this in analogy to the familiar Λ-doubling found in diatomic molecules (Λ corresponds to our m quantum number). Since our MO treatment of 2-electron atoms is identical to that of diatomic molecules, the occurrence of degenerate pairs $(^1P^e, ^3P^o)$ or $(^1D^e, ^1D^o)$ is *exactly* the Λ-doubling phenomenon. As in molecules the degeneracy is lifted when rotational coupling to other MO of different body-fixed m is taken into account.

In a single-channel adiabatic approximation each state $(LS\pi)$ is associated with a potential curve in which vibrational states corresponding to increasing degrees of excitation in R may be defined. However, the vibrational structure of two-electron states introduced by Herrick on the basis of a triatomic molecular model involves three normal modes and only the "breathing" v_1, longitudinal mode (symmetric stretch) corresponds to R. The other two modes appear when the exact λ, μ internal motion is approximated in the region of the saddle point where $\lambda \sim 1, \mu \sim 0$. The corresponding asymmetric stretch (v_3, MO coordinate μ) and bending (v_2, MO coordinate λ) modes are illustrated in fig. 6 of ref. 6a.

Herrick introduced five quantum numbers K, T, A (called by him ν) on the basis of SO_4 group theory and v_2 and d on the basis of the triatomic molecular model of two-electron atoms. All of these five quantum numbers are contained in the three quantum numbers $(n_\lambda n_\mu m)$ unique to each MO.

The MO explanation of the origin of Herricks' quantum numbers is simple. It also explains a feature that Herrick repeatedly found puzzling, namely, why are quantum numbers based on SA wavefunctions and symmetries ($Z \to 0$ limit) valid for classifying resonant states involving intermediate values of R The answer lies in the fact that the MO quantum numbers $n_\lambda n_\mu m$ are valid for all R. In particular for $R \to \infty$ they can be expressed uniquely in terms of the parabolic quantum numbers $(n_1\ n_2\ m)$ or $(N\ k\ m)$ of the SA state to which each MO is asymptotic. The "extra" quantum number $A = (-1)^{n_\mu}$ is necessary to specify the correlation (this is the ν quantum number of Herrick). It also specifies the LCAO form of the MO in terms of SA states in which electron 1 or electron 2 is bound i.e.

$$|n_\lambda n_\mu m> \approx |n_1 n_2 m>_1 + A|n_1 n_2 m>_2 \qquad \text{for } R \to \infty \tag{6}$$

Herricks' other two labels, d and I $(= L - m)$, were derived on the basis of an ad hoc model of a two-electron atom as a triatomic molecule. In particular he was able to organise states into I supermultiplets of states where those corresponding to a given value of v_2 are almost degenerate. We have established that the v_2 quantum number, corresponding to a bending vibration[3] is nothing more than the quantum number $2n_\lambda + m$ describing degenerate saddle-point bending vibrations. Then the degeneracies for fixed v_2 in Herricks' supermultiplet classification have their origin in the underlying degeneracy of MO for fixed $2n_\lambda + m$. In addition Herricks' classification according to $I = L - T$ with $L - T \geq 0$ arises naturally since $T = m$ and since $\mathbf{L} \cdot \hat{R} = m$ then $L - m \geq 0$. In the MO picture $I = L - m$ describes the difference of total angular momentum L and MO angular momentum m. Hence, in an $I = 0$ multiplett, for $m = 2$ the saddle degeneracy first involves D states, for example the ${}^1D^e, {}^1S^e$ pair built on the saddle-degenerate MO (5gδ,4gσ) with $v_2 = 2$.

There remains the quantum number d, particularly favoured by Herrick. Indeed he was also able to give an interpretation, if not a justification of this number group-theoretically. It will emerge that classification according to d, for fixed N is *exactly* the same as classification according to n_λ denoting the number of elliptical MO nodes. Already in section 2 we have noted that n_λ denotes the "stiffness" of MO and that there is little mixing of MO with different n_λ. The quantum number d arises out of the reduction $SO_4 \times SO_4 = SU_2 \times SU_2 \times SU_2 \times SU_2$ according to irreducible representations in which the operator $\mathbf{J}^2_>$ is invariant. Here $\mathbf{J}^2_>$ is the angular momentum operator with the eigenvalue $d = \max(J_+, J_-)$ where $\frac{1}{2}(\mathbf{L}+\mathbf{B})^2 = J_+(J_+ + 1)$ and $\frac{1}{2}(\mathbf{L}-\mathbf{B})^2 =$

$J_-(J_- - 1)$. In this representation $d = \frac{1}{2}(K + N + T - 1)$. The MO interpretation follows from $K = n_2 - n_1$, $T = m$, to give

$$d = n_2 + m. \tag{7}$$

The real meaning for intra-shell resonances however relies on the fact that for fixed N, d gives the number of λ nodes i.e. $n_\lambda = n_1 = N - n_2 - m - 1 = N - d - 1$.

To summarise we have established that all of Herricks' *five* quantum numbers, arrived at by various physical arguments supported by the SO_4 group classification express different aspects of the MO character through the *three* unique MO numbers $n_\lambda n_\mu m$. The connection is

$$\begin{aligned} T &= & m & \\ A &= & (-1)^{n_\mu} & \\ K &= & n_2 - n_1 &= [n_\mu/2] - n_\lambda \\ v_2 &= & 2n_1 + m &= 2n_\lambda + m \\ d &= & n_2 + m &= [n_\mu/2] + m. \end{aligned} \tag{8}$$

4. Propensity rules

Based on a consideration of the underlying MO nodal structure of two-electron resonant states which has been confirmed recently[7], propensity rules for radiative and non-radiative decay of resonant states (i.e. for their widths) have been given[6b]. Here they will be briefly re-iterated to provide more details of their derivation.

The propensity rules are based on the observation that, to the extent that states of given total symmetry are built upon a single MO, transitions between these states are governed by *selection rules* for transitions between the corresponding MO. In this way the propensity rules arise naturally out of the theory and are not dependent upon empirical deductions. Their simplicity is partly due to the structure of the operators responsible for non-radiative (auto-ionisation) and radiative decay.

The process of autoionisation is the direct analogue of dissociation $H^+ + H(n_1 n_2 m)$ in H_2^+. It proceeds in the MO picture via the non-adiabatic couplings involving the operator ∇_R. These terms separate into the radial and rotational coupling types well-known in molecular and ion-atom collision physics. Non-radiative transitions between MO are induced by off-diagonal matrix elements and are governed by the following selection rules

$$\Delta t = 0 \tag{9a}$$

$$\text{Radial} \quad < \Phi_f | \partial / \partial R | \Phi_i >; \quad \Delta m = 0 \tag{9b}$$

$$\text{Rotational} \quad < \Phi_f | l_\pm | \Phi_i > / R^2; \quad \Delta m = \pm 1. \tag{9c}$$

We note that the combination of (9a) and (9b) implies $\Delta A = 0$ for radial coupling. Similarly (9a) and (9c) imply $\Delta A \neq 0$ for rotational coupling.

These selection rules alone are not sufficiently restrictive to classify states according to the magnitude of their widths. For that a consideration of the facility with which internal saddle $n_\lambda n_\mu$ excitation can couple to vibrational R motion is necessary. Decay occurs preferentially[6b] by radial coupling within saddle sequences corresponding to

$$\Delta n_\mu = 2, \quad \Delta n_\lambda = 0, \quad \Delta m = 0. \tag{10}$$

Within the saddle sequence Eq. 9 implies that a state of the N-manifold decays preferentially to a state of the $(N-1)$-manifold which then dissociates to $e^- + H(N-1)$ or $e^- + He^+(N-1)$ if this is energetically possible. This is also consistent with the Hellman-Feynman theorem for radial coupling

$$< \Phi_f|\partial/\partial R|\Phi_i >= (\mathcal{E}_f - \mathcal{E}_i)^{-1} < \Phi_f|\partial h/\partial R|\Phi_i > \tag{11}$$

showing that decay occurs preferentially to states closest in energy and is particularly effective beteen states which exhibit an avoided crossing. This also explains why decay rates of in-saddle ($A = +1$) sequences of states are much faster than those of side-saddle ($A = -1$) states. The radial coupling is strongest at the avoided crossings between N and $(N-1)$ manifolds and since the $A = +1$ series have much sharper avoided crossings (see fig.1) the radial couplings are correspondingly greater according to Eq. (11).

All saddle sequences end in a precursor MO having $n_\mu = 0$ or $n_\mu = 1$. For states built on these MO decay is forbidden according to the propensity rule $\Delta n_\mu = 2$. These states must decay by the weaker rotational transition

$$\Delta n_\mu = 1, \quad \Delta n_\lambda = 0, \quad \Delta m = \pm - 1 \tag{12}$$

within a UA n-manifold. These transitions occur with $\Delta A \neq 0$ and therefore imply a change from an anti-node at the saddle, or vice versa. Finally, the least probable transitions are those with

$$\Delta n_\lambda \neq 0 \tag{13}$$

occurring ***between*** different saddle sequences. That these transitions do not lead to large dissociation widths is plausible when one remembers that excitation in the λ mode corresponds to bending vibrations where the ECM executes stable vibrations perpendicular to $\hat{R}$. These vibrations couple weakly to symmetric stretch R vibrations leading to decay. By contrast $\Delta n_\mu = 2$ transitions correspond to reduction in the asymmetric stretch ECM vibration along $\hat{R}$ which couples directly to the symmetric stretch R vibration. Application of the non-radiative propensity rules are given in ref. 6b.

The key observation leading to propensity rules for radiative transitions is that the dipole operator $\mathbf{D} = \mathbf{r}_1 + \mathbf{r}_2 = 2\mathbf{r}$ i.e. the dipole transition involves only the ECM motion. This gives the selection rule $\Delta t = \pm 1$ directly. The linearised ECM dynamics around the saddle point as the motion of an isotropic harmonic oscillator (3) has already explained the quasidegeneracies in the two-electron spectrum. It can also be shown how a linearisation of the ECM motion around the saddle leads to the additional selection rule

$$\Delta v_2 = 0, \pm 1 \tag{14}$$

which can be expressed in terms of the change in MO quantum numbers by the identification $v_2 = 2n_\lambda + m$. This rule incorporates different versions of the radiative propensity rule given by ourselves[6b], by Sadeghpour and Greene[8] and by Gou et al.[9].

Acknowledgement

This work was supported by the DFG within the Sonderforschungsbereich 276 at the University of Freiburg.

References

1. D. E. Herrick *Adv.Chem.Phys.* **52** (1983) 1
2. C. D. Lin *Phys.Rev.Lett.* **51** (1983) 1348
3. M. E. Kellman, D. R. Herrick *Phys.Rev.* **A22** (1980) 1536
4. J. M. Feagin, J. S. Briggs (a) *Phys.Rev.Lett.* **57** (1986) 984
   (b) *Phys.Rev.* **A37** (1988) 4599
5. J. D. Power *Proc.Roy.Soc.* **A 274** (1973) 663
6. J. M. Rost, J. S. Briggs (a) *J.Phys.* **B22** (1989) 3587
   (b) *J.Phys.* **B23** (1990) L339
   (c) *Chem.Phys.Lett.* **177** (1991) 321
7. J. M. Rost, J. S. Briggs, J. M. Feagin *Phys.Rev.Lett.* **66** (1991) 1642
8. H. R. Sadeghpour, C. H. Greene *Phys.Rev.Lett.* **65** (1990) 313
9. B. C. Gou, Z. Chen, C. D. Lin *Phys.Rev.A (1991 to appear)*

ATOMIC PHYSICS AT THE SSC

H.C. BRYANT
Department of Physics and Astronomy
The University of New Mexico
Albuquerque, NM 87131 USA

ABSTRACT

The first results from the SSC could well be major advances in fundamental atomic physics, rather than in fundamental particles, simply because the linac accelerating H^- up to relativistic velocities for injection into the low energy booster ring will come on line about 4 years before two 20 TeV protons will have close encounters of any kind. In its own way, a high quality, intense, beam of H^- ions with energies of the order of GeV is just as unique in the physical world as is the beam of ultra high energy protons being produced by the big ring. Both beams allow us to make observations at the frontiers; in the linac case, the unique probes available would allow us to open a new window on the one- and two-electron atom. Moreover, the use of laser detachment techniques could have useful applications to accelerator physics.

1. Introduction

"A donde fueres, haz lo que vieres,"
-or- "Go with the flow."
-Anon.

The advent of the SSC machine is widely expected to give a strong impetus to high energy particle physics in the U.S. What may not, however, have been recognized is that the H^- linac injector at the new facility could also mean a major advance for an emerging branch of atomic physics involving the study of atoms and ions moving at near luminal velocities: relativistic atomic physics.

Table I presents the SSC linac requirements and parameters.[1, 2] The present design calls for a beam energy of 600 MeV, but provision has been made for an energy upgrade to 1000 MeV. Table II gives the main parameters[3, 4] of the Collider.

---

[1]G.E. Jamieson and G. Leifeste, private communication.

[2]Jerry M. Watson, "The SSC Linac", invited paper. 1990 Linear Accelerator Conference, Albuquerque, NM Sept. 10-14, 1990. Available as a report from the SSC Lab, 2550 Beckleymeade Ave., Dallas, TX 75237.

[3]From Particle Properties Data Booklet. April 1990. North-Holland. Available from Berkeley and CERN. (See M. Aguilar-Benitez et al. Physics Letters B239 April 1990).

[4]"Site-Specific Conceptual Design of the Superconducting Super Collider" SSCL-SR-1055, M.G.D. Gilchriese and K. Metropolis, eds. SSC Lab, 2550 Beckleymeade Avenue, Dallas, TX 75237, July 1990. See also Physics Today, March 1991, page 51.

Table I
SSC Linac Requirements
(LEB is low energy booster ring)

Filling collider ring
- 25 mA H^- current during macropulse
- 6.6 μs macropulse (3-turn LEB injection)
- 1×10^{10}/LEB bunch
- <0.5 π mm-mrad (t, rms, norm) emittance
- 10 Hz repetition rate

Test beam operation
- 25 mA during macropulse
- 35 μs macropulse (16-turn LEB injection)
- 5×10^{10}/LEB bunch
- <4 π mm-mrad (t, rms, norm) emittance
- 10 Hz repetition rate

Linac Parameters
- 600 MeV kinetic energy (provision for 1000 MeV)
- 1.22 GeV/c momentum (1.697 GeV/c)
- 0.7925, β, (0.875)
- 1.6397, γ (2.064)
- 2.34 ns, micropulse spacing
- 0.105 ns, micropulse width
- 0.75 mm, radius of beam, rms
- 0.2 π mm mr (rms) emittance
- 99% of beam within 1 MeV window
- 150 m total length (249 m)

Table II
Collider Parameters

Physics Start Date	1999
Particles collided	pp
Maximum beam energy	20 TeV
Luminosity	$10^{33} cm^{-2} s^{-1}$
Circumference	87.12 km
Interaction regions simultaneous	4 max
	8 max. total
Bunch length	6 cm

Since the late seventies[5], following proof of principle at a 4 MeV Van de Graaff[6], work has been going on at the 800 MeV linear accelerator (LAMPF) at Los Alamos using laser beams and other probes, such as strong fields and thin foils, to study the structure and electromagnetic interactions of the simplest of atomic systems H° and H^-. The basic idea is to take advantage of the relativistic kinematics of an atom moving with a large β (at 800 MeV, $\beta/c = 0.84$) to Doppler-shift beams from ordinary pulsed lasers into the vacuum ultraviolet, corresponding to excitation energies of the neutral hydrogen atom and its negative ion. Enormous electric fields can be induced in the atom's frame using modest laboratory magnets. Recently, continuously tunable, intense beams from the CO_2 laser have been used to study for the first time multiphoton processes in the H^- ion.

To give a clearer idea of the power of relativistic kinematics in laser-ion beam studies, consider the relativistic Doppler formula,

$$E=\gamma E_L(1+\beta \cos \alpha) . \tag{1}$$

Here the center-of-mass photon energy, E, is given in terms of its laboratory energy, E_L, with α being the angle between the laser beam and the particle beam, such that for head-on collisions $\alpha=0$. At 800 MeV, since $\gamma=1.853$ and $\beta=.842$, one can continuously tune a fixed-frequency laser through the range,

$$0.29\ E_L \leq E \leq 3.4\ E_L .$$

Furthermore the laser intensity, I, in the center of mass (Watts/cm^2) is related to the lab intensity, I_L, by the square of the Doppler factor. That is,

$$I = \gamma^2(1 + \beta \cos \alpha)^2 I_L , \tag{2}$$

so that one can get an intensity gain at $\alpha=0$ of 11.65!

Finally, the barycentric electric field E produced by a transverse laboratory magnetic field B is given, in S.I. units, by

$$E = \gamma\beta c B , \tag{3}$$

so that a 1 Tesla lab field can result in a center of mass electric field of 4.7 MV/cm.

Before the development of relativistic beam techniques, the resonance structure of the H^- ion in the vacuum ultraviolet was essentially unknown except for a few electron-hydrogen scattering measurements. With our new methods we were able to demonstrate a rich structure of doubly-excited resonances in H^- and to study their behavior

---

[5] H.C. Bryant et al., Phys. Rev. Letters 38, 228 (1977).

[6] H.C. Bryant, P.A. Lovoi and G.G. Ohlsen, Phys. Rev. Letters, 27, 1628 (1971).

in electric and magnetic fields.[7] We were also able to do similar studies on the low-lying states of H°. Precision checks on the isotropy of space-time are also possible using our technique.

The work already done and being planned for LAMPF may be regarded as prelude to the eventual flowering of these endeavors at the SSC.

2. Diagnostics, Monitoring and Tailoring for the H^- Linac at the SSC

The use of lasers to probe relativistic atomic beams, developed at LAMPF over the past twelve years, can be applied to the diagnosis, monitoring, and tailoring of the H^- linac beam at the SSC. The basic idea is to direct a laser beam at the H^- beam so that its Doppler-shifted frequency is centered on a well-defined feature in the absorption spectrum. This feature could be the well-known Feshbach resonance in H^- , for example, at an excitation energy of 10.9264(6) eV, whose intrinsic width is some 30 microvolts, or it could be a hydrogen resonance line excited by a 2-step process in which the H^- is first photodetached using a precursor infrared beam and the ground-state H° is then excited. Even a three-step process might be contemplated, in which the complication of 3 separate laser beams might be exchanged for higher resolution and more convenient laboratory wave lengths as well as a higher signal-to-noise ratio. By selective photodetachment or excitation, small regions of the overall phase space could be studied.

The energy resolution in the center of mass of a system moving at βc in a beam whose momentum dispersion is $\delta p/p$, and where the rms angular uncertainty is $\delta\alpha$, is given by

$$\frac{\delta E}{E} = \left\{ \left(\frac{\delta E_L}{E_L}\right)^2 + \left(\frac{\beta \sin\alpha}{1+\beta\cos\alpha}\right)^2 (\delta\alpha)^2 + \left(\frac{\beta^2+\beta\cos\alpha}{1+\beta\cos\alpha}\right)^2 \left(\frac{\delta p}{p}\right) \right\}^{\frac{1}{2}} . \tag{4}$$

Let us assume now for example that the laboratory energy resolution of the laser line is negligibly narrow, so that $\delta E_L/E_L \sim 0$, and that we are exciting the ^1P Feshbach resonance at 10.926 eV just below the threshold for $\gamma + H^- \rightarrow H°(2) + e$. By using a fixed-frequency laser, the observed angular width of the resonance would reflect both the angular uncertainty $\delta\alpha$ and the momentum spread $\delta p/p$ of the beam. At LAMPF $\delta\alpha$ can be as low as 10 microradians and $\delta p/p$ about 10^{-4}.

If it were possible to operate near $\cos\alpha = -\beta$ (the "Doppler free" angle)[8], sensitivity to $\delta p/p$ would disappear.

---

[7] P.G. Harris, H.C. Bryant, A.H. Mohagheghi, R.A. Reeder, H. Sharifian, C.Y. Tang, J.B. Donahue, C.R. Quick, D.C. Rislove, W.W. Smith, J.E. Stewart. Phys. Rev. Letters 65, 309-312 (1990).

[8] H.C. Bryant, Electronic and Atomic Collisions, N. Oda and K. Takayanagi, eds., North Holland Pub. Co. 1980, pp. 145-160.

If it were possible to operate near $\alpha = 0$, sensitivity to $\delta\alpha$ would disappear.[9]

If we wished to monitor both variables at once we would have to use one of the hydrogen lines, by first photodetaching (to either the ground state or an excited state) and then exciting the line. Such a technique has been used to observe the narrowing of the momentum spread at LAMPF.[10]

In Figure 1 we show as an example the laboratory intersection angle α required to excite the Feshbach resonance below n=2 in H^- using garden-variety pulsed lasers. The use of pulsed lasers gives a large signal to noise ratio. The availability of a tunable laser in the lab with sufficient intensity would be a great boon for this work because it would allow for continuous monitoring. A uv laser that could operate essentially CW while the H^- beam were present, for example, is a hollow cathode Cu^+ laser which would offer unique possibilities: 1) Several uv lines together very near to the 4th YAG line (259.06, 259.90 and 260.03 nm) would allow step-wise tuning. 2) Single mode operation with 10-15 MHz FWHM. 3) A long laser pulse that overlaps the beam pulse. 4) The excellent beam quality of the CW gas laser. On the other hand, no commercial model of such a laser has been made. Research is going on to develop such a laser in the framework of the NSF-Hungarian Academy of Sciences cooperation. The feasibility of such a laser for the SSC should be investigated.

In order to permit flexibility in monitoring the linac beam as it is accelerated, space should be provided between accelerating sections to insert laser beams with precision optics so that the angle α may be altered at will with high resolution. The resulting electrons could be bent out of the beam at any point into a detector with a weak magnetic field. By operating near 90° and focusing the laser beam with a cylindrical lens in the transverse direction, one could examine separately small subsets of the larger diameter beam. Also one could monochromatize by photodetaching only those ions within a prescribed momentum bite.

We believe these techniques are quite promising and powerful. There are also many variations that could be studied to find the optimum arrangement for a given application. Therefore, at this point, it is important to design the space around the linear accelerator in such a way as to allow laser access to it. Equipment we are developing for use at LAMPF could be transferred over to experiments at the SSC linac.

Further development of appropriate laser systems would also be called for.

3. Atomic Physics Research

In addition to the diagnostics and other applications described above, it appears that unique basic atomic physics could also be performed at the SSC linac. Among the reasons why such work is so exciting, we can list

- A more intense (milli- rather than micro-amps!), better quality beam with

---

[9]D.B. Holtkamp and C.R. Quick, Nuclear Instruments and Methods in Physics Research, A287, 348 (1990).

[10]P.G. Harris et al., Nuclear Instruments and Methods...A292, 254 (1990).

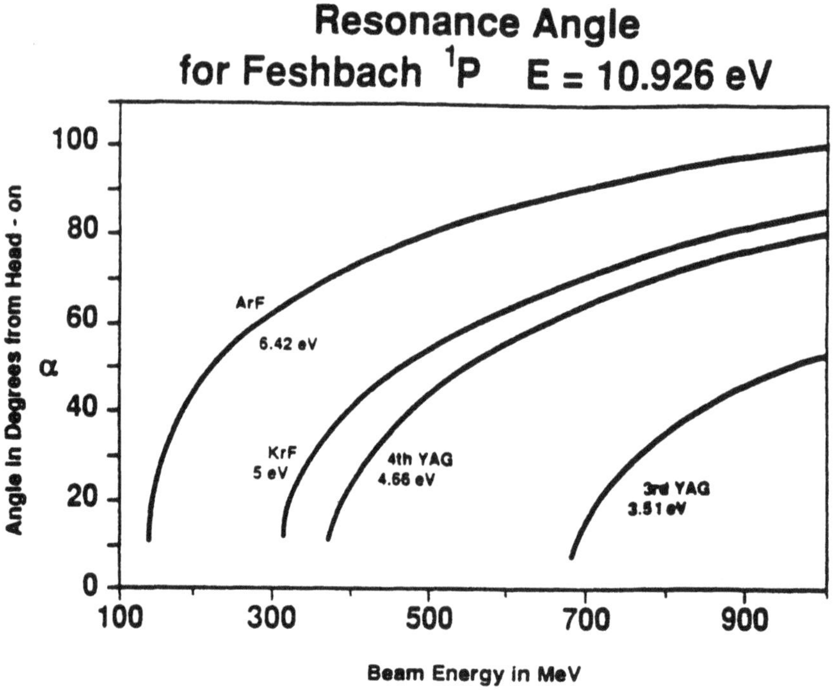

Fig. 1 Angle of excitation of the Feshbach resonance at 10.926 eV as a function of H^- kinetic energy for several "garden-variety" laser lines of high intensity.

possibly a higher kinetic energy than the LAMPF beam (800 MeV). Even at 600 MeV, the beam would be very attractive.

- A chance to develop an atomic physics facility from scratch rather than retrofitting it to an existing accelerator where compromises must be made, as was the case at LAMPF. With very little additional cost in the initial stages atomic physics experimental space and facilities can be accommodated, which later on would be extremely expensive to build into the system.
- More visibility and support both technically and financially, since it would be part of a "flagship of American science".

Of course, in the study of something as rich in phenomena as atomic physics, many new and unexpected ideas will undoubtedly arise, but we can only plan based on what we already know. Therefore we sketch below some of the kinds of physics one could do using the H^- beam at the SSC.

a) High Resolution Spectroscopy of H^-

Recently[7] [11]we have been able to study the highly correlated, doubly-excited states in H^- by first exciting them through the process

$$\gamma + H^- \rightarrow H^{-**} \rightarrow H^\circ(n) + e \ ,$$

with subsequent motional field stripping of $H^\circ(n)$ in an appropriately-chosen magnetic field. A dissertation written on this work won the Louis Rosen prize for the best done at LAMPF in the past year[12]. The cases for which n = 4,5,6 and 7 were studied. See Fig. 2. With better signal-to-noise, the levels studied could be pushed up to much higher n's so that the systematics could be established. We could achieve a much clearer signal by introducing a second laser beam, rather than the stripping field, to label the final H° state, through, for example, its promotion to a higher n state, which could be unambiguously identified in an electron spectrometer e.g.

$$\gamma + H^\circ(n) \rightarrow H^\circ(12).$$

We demonstrated a "proof-of-principle" for this technique[13] in a short experimental run at LAMPF in August 1990 (unpublished). And we obtained interesting data[13] on the n=2 channel in a run October 3-11, 1990.

By the application of external fields to the interaction region or by multiphoton

---

[11]P.G. Harris, H.C. Bryant, A.H. Mohagheghi, R.A. Reeder, C.Y. Tang, J.B. Donahue, C.R. Quick. Physics Review A 42, 6443-6465, (1990).

[12]P.G. Harris, Ph.D. Dissertation, May, 1990, UNM "Observation of High-Lying Resonances in the H^- Ion" (LA-11843-T).

[13]Collaborators in this effort included E. MacKerrow, M. Halka, A.H. Mohagheghi, C.Y. Tang, C.R. Quick, J.B. Donahue, J. Tiee, S. Cohen, H.C. Bryant.

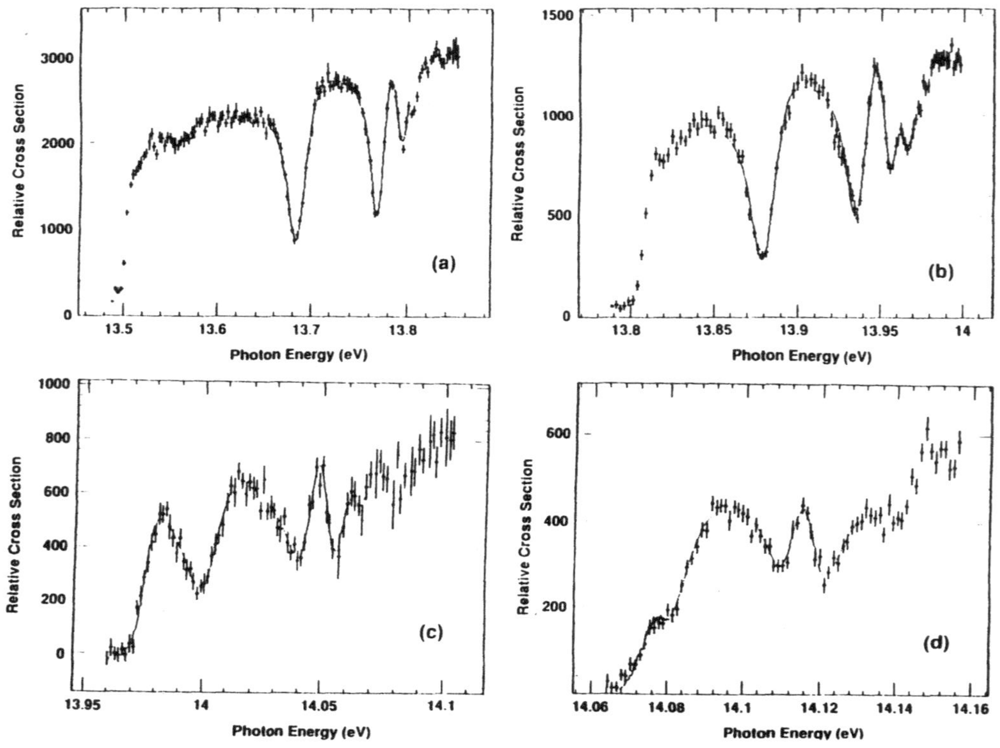

Fig. 2. High-lying resonances in the ^1P continuum of H^-. The energies of these resonances can be fitted to a remarkably simple formula reminiscent of that of Balmer for neutral hydrogen. The motional electric field is adjusted to strip $H^*(n)$ as follows: (a)$n \geq 4$, (b)$n \geq 5$, (c)$n \geq 6$, (d)$n \geq 7$.

excitation, a thorough picture of the resonance structure of H^- could be mapped out.

b) Multiphoton Studies

Currently our experimental program includes the study of multiphoton detachment of H^- using a CO_2 laser beam. Preliminary results are already available,[14 15 16] and work is continuing. With the CO_2 beam, with a lab photon energy of 0.117 eV, at 1000 MeV we should be able to study detachment with photon numbers ranging from 2 to 24.

Because of the fundamental simplicity of the atomic system involved, along with the interesting complication of 2 electrons, these measurements merit precision work, for which conditions such as laser intensity and focal spot are carefully controlled.

This work is likely to continue to excite high interest for many years.

c) Strong-field effects

Our studies of H^- in strong electric fields have yielded surprises[17] and such work should continue. A near-luminal H^- ion moving through a modest laboratory magnetic field experiences enormous electric fields in its barycentric frame yet is essentially undeflected because of its high magnetic rigidity.

d) Passage through thin foils and channeling

A recent Ph.D. dissertation[18] in our group was written on the study of the excitation of H°(n) by the passage of H^- through carbon foils ranging in thickness from 20 $\mu g/cm^2$ to 300 $\mu g/cm^2$. In this case the foil delivers an intense perturbation to the H^- ion for times of the order of a femtosecond.

Further studies are contemplated[19] using very thin oriented crystals of Si or sapphire (Al_2O_3) in which channeling may be expected to occur for intact atomic systems.

---

[14] C.Y. Tang et al., Phys. Rev. A 39, 6068 (1989).

[15] C.Y. Tang et al., International Conference on Multiphoton Processes (Invited Paper) Paris, September 24-28 (1990). Published in Multiphoton Processes edited by G. Mainfray and P. Agostini. Service de Physique des Atomes et des Surfaces, Centre d'Etudes de Saclay, 91191 Gif-sur-Yvette Cedex. 1990. pp. 69-78.

[16] W.W. Smith et al., J. Opt. Soc. Am. B 8, 17-21 (1991).

[17] H.C. Bryant et al., Phys. Rev. Letters 58, 2412 (1987). J.D. Stewart et al., Phys. Rev. A 38, 5628 (1988). P.G. Harris, et al., Phys. Rev. A. 41, 5968-5973 (1990).

[18] A.H. Mohagheghi, Ph.D. Dissertation, September 1990, UNM, "Interaction of Relativistic H^-Ions with Thin Foils" LA. See A.H. Mohagheghi et al., Physical Review A 43, 1345-1365 (1991).

[19] Károly Rózsa, a collaborator from the Hungarian Academy of Sciences, has been investigating the manufacture of oriented crystal foils of submicron thickness.

Such work may have practical applications.

e) Searches for a preferred frame

Tests of special relativity based on the exquisitely well-known energy levels of atomic hydrogen can be contemplated for a high quality 1000 MeV H^- beam. At LAMPF we have checked the Doppler[20] formula to the 36th power of β. Conditions for such tests would be unique for the SSC.

4. A More Detailed Example

Let us apply some of the techniques alluded to above to the search for a preferred frame of reference at the SSC linac. In this case one might expect some sort of natural "speedometer" that would give an increasingly larger signal as one increased one's speed with respect to the preferred frame, presumably a frame more or less at rest with respect to the local (galactic) center of mass. Therefore, one might expect the relativistic Doppler formula (which is essentially time dilation) would display some pathology most noticeable for systems with near luminal velocities in the preferred frame. How well could we check the Doppler formula at the SSC?

a) Method

First a beam of monochromatized H°(1s) atoms would be generated by photodetachment via the Feshbach resonance using the 4th harmonic of the YAG (See fig. 1). Using, for example, a supercollimated beam line as designed for the HIRAB facility at LAMPF,[21] and a well collimated laser, we can obtain $\delta\alpha \sim 10^{-6}$ radian, so that angular spread may be neglected in eq. 4. Neglecting also the laser line width in the laboratory and taking $\delta E/E$ for the Feshbach resonance as 2.7×10^{-6} (see Fig. 3) we find the momentum spread of H°(1s) atoms resulting from photodetachment through the resonance is

$$\frac{\delta p}{p} = \begin{Bmatrix} 3.6\times10^{-6} & at & 600 MeV \\ 3.4\times10^{-6} & at & 1000 MeV \end{Bmatrix} \tag{5}$$

[20] D.W. MacArthur et al., Phys. Rev. Letters 56, 282 (1986).

[21] James N. Knudsen, HIRAB Beamline Optics, LA-UR-90-3321, Los Alamos National Laboratory. Sept 20, 1990.

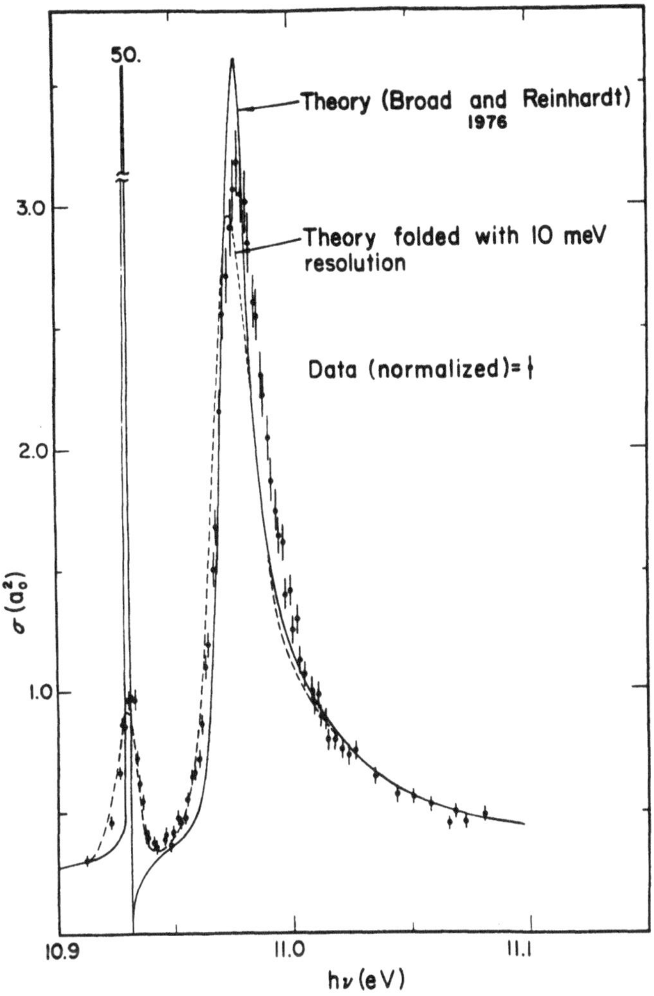

Fig. 3. The ^1P doubly excited resonances in H^- near n=2. The resonance is known as the "Feshbach" resonance and the one on the right as the "slope". The data are from reference 22 and the theory from 23.

Next, the beam of H°(1s) atoms would be excited to the H°(3p) state by a second, well collimated YAG 4th harmonic laser beam. Taking $\delta E/E$, conservatively, as 10^{-6}, the H°(3p) is confined now to $\delta P/P$ of about 1.3×10^{-6} for both 600 and 1000 MeV beams.

Finally, using a beam collimation of $\delta\alpha \sim 10^{-6}$ radian, we excite the H°(3p) to the following states with energy resolution near the "Doppler free" angle of $\approx \gamma\beta\delta\alpha$ using the YAG second harmonic:

Beam Energy	Final State	$\delta E/E$
600 MeV	H° (13d)	1.3×10^{-6}
1000 MeV	H° (6d)	1.8×10^{-6}

These final states can be detected with very little background by motional field stripping.

b. Results

The proper analysis of data taken under the above conditions would require fitting all measurements to the Doppler expression using maximum likelihood analysis. One could then determine the Robertson parameter[20] g_o to a few ppm from

$$E=(\gamma E_2/g_o)(1+\beta\cos\alpha)\ , \tag{6}$$

for example. Angles would have to be measured with microradian accuracy and wavelengths to at least 1 ppm in the lab.

We can present the implication for our proposed measurements in another, more dramatic, way: From the above we find that for a particular β, γ can be determined to a precision of roughly 2×10^{-6}, assuming $g_o \equiv 1$. Now expand $\gamma \equiv (1-\beta^2)^{\frac{1}{2}}$ in powers of β.

$$\gamma=1+\frac{1}{2}\beta^2+\frac{3}{8}\beta^4+\frac{15}{48}\beta^6+\frac{105}{384}\beta^8+\ldots, \tag{7}$$

or

$$\gamma=\sum_{n=0}^{\infty} C(2n)\beta^{2n},\ where\ C(2n)=\frac{2n-1}{2n}C(2n-2)\ ,\ and\ C(0)=1. \tag{8}$$

From this point of view a computation shows that if we truncate the power series at β^{2n}, the value of n required for the exact value of γ to differ from the truncation by 2×10^{-6} is $2n = 50$ for 600 MeV and $2n = 88$ for 1000 MeV! As mentioned above[20], our best measurement so far can be interpreted as certifying the expression for γ to the β^{32} term.

5. Further Remarks

It is perhaps worthwhile pointing out the possibility of the production of very high energy hydrogen atoms (and perhaps even H^-) by the method of laser stimulated radiative recombination of protons[22]. A feasibility study of the production rates in one of the 200 GeV test beams should be done before further speculation, however.

6. Conclusion

If the H^- linac at the SSC were made available for atomic studies, there are many fascinating measurements that could be performed. These measurements could map out accurately doubly excited structures in H^-, and the response of both H^- and H° to strong electromagnetic fields that are presently inaccessible by conventional techniques. In addition, a relativistic H° beam could be used as a unique probe of the isotropy of space-time.

Acknowledgements

This paper is based on a proposal submitted November 1, 1990, to the Texas National Research Laboratory Commission.

I wish to thank the many people who have contributed to the work reviewed, as can be seen from the bylines of the papers cited. Finally, I wish to thank the organizers of this wonderful conference: Profs. Carmen Cisneros and Tom Morgan.

This work was supported by the Division of Chemical Sciences, Office of Basic Energy Sciences, Office of Energy Research, U.S. Department of Energy.

---

[22] T. Morgan, "Laser Stimulated Radiative Recombination of Protons," in this volume.

NUMERICAL EVALUATION OF THE INTEGRAL OF A PRODUCT OF THREE 6-DIMENSIONAL SPHERICAL HARMONICS

A. AMAYA-TAPIA

Instituto de Física, UNAM, Laboratorio de Cuernavaca.
Apdo. postal 139-B, Cuernavaca, Mor. CP 62191, MEXICO

and

E. CHACON

Instituto de Física, UNAM, Apdo. Postal 20364
México, D. F., CP 01000, MEXICO

ABSTRACT

We discuss the integral of a product of three 6-dimensional spherical harmonics over the surface of the unit 6-sphere, and write a FORTRAN program that evaluates numerically this integrals. The hyperspherical harmonics we use are those introduced by Chacón and collaborators and they have the scheme clasification $O(6) \supset SO(2) \times SU(3) \supset SO(3) \supset SO(2)$. We discuss the relevance of this integrals for the study of three body systems with arbitrary masses. And also show that if two body interactions V(1,2) are expanded in terms of products of a hyperradial function times a harmonic, i. e., $R(\rho)\mathbf{Y}(\Omega)$, the matrix elements of V(2,3) and V(3,1) are proportional to those of V(1,2) when they are taken with respect to the same states.

1.Introduction

Recently Chacón et al[1]. had derived an analytical expresion for hyperspherical harmonics clasified by the chain O(6) ⊃ SO(2) × SU(3) ⊃ SO(3) ⊃ $SO(2)$,introduced by Dragt[2]. In this work we discuss the integral of a product of three 6-dimensional spherical harmonics over the surface of the unit 6-sphere. Also we present a FORTRAN program for the numerical evaluation of this integral. One of the fields in which this integrals may be useful is in the analysis of three body systems wich continues to be a frequent subject of study in non relativistic quantum mechanics[3]. Lately several analysis had been made use of hyperspherical harmonic basis in studying three body systems with various types of interactions[4].

The six dimensional Laplacian may be expresed in terms of a set of 6-dimensonal spherical coordinates consisting of a hyperradius ρ plus five angles denoted globally with$\mathbf{\Omega}$

$$\nabla^2 = -\frac{1}{\rho^5}\frac{\partial}{\partial\rho}\rho^5\frac{\partial}{\partial\rho} + \frac{1}{\rho^2}\Lambda^2(\mathbf{\Omega}), \tag{1}$$

where $\Lambda^2(\mathbf{\Omega})$, the Casimir operator of the six dimensional orthogonal group O(6), depends only on the five angles $\mathbf{\Omega}$, and derivatives with respect to them [5].

The hyperspherical harmonics, denoted as $\mathbf{Y}(\mathbf{\Omega})$, are eigenfunctions of Λ^2, namely,

$$\Lambda^2\mathbf{Y}_s(\mathbf{\Omega}) = \lambda(\lambda+4)\mathbf{Y}_s(\mathbf{\Omega}), \tag{2}$$

where the label s stands for a set of five quantum numbers. One of them is λ, whose role is to identify the irrep of O(6) to which $\mathbf{Y}_s$ belongs. Two more quantum numbers are L and M, related to irrep of the physical rotation groups SO(3) ⊃ $SO(2)$.

Ideally, the two remaining labels should be related to irreps of subgroups of O(6) containing SO(3). To introduce one more label, Dragt [2] introduced a classification scheme, in wich the chain of groups and corresponding labels for their irrep are

$$\begin{array}{ccccccc} O(6) & \supset & SO(2)\times SU(3) & \supset & SO(3) & \supset & SO(2) \\ \lambda & & \nu & & L & & M \end{array} \tag{3}$$

Due to the fact that the groups SO(2) and SU(3) are complementary [6] whithin the irreducible representations $[\lambda]$ of O(6), the irrep labels of SU(3) are given in terms of λ and ν; in fact, the Young partition for SU(3) is $[\lambda, \frac{1}{2}(\lambda-\nu)]$. Thus the scheme of classification (Eq. 3) lacks one quantum number, and the Dragt harmonics carry an aditional multiplicity index q, having as a consequence that the set of functions $\mathbf{Y}_{\lambda\nu qLM}(\mathbf{\Omega})$, which is to be discussed in this article is not orthogonal with respect to the index q.

2. Relation of the states with three body systems.

In the treatment of three body systems interacting via two body forces, attention is focused on the intrinsic Hamiltonian H_I which depends only on two relative vectors **x** and **y** and their canonical conjugate momenta. We define them and CM vector through the following formulas [7] ,

$$\mathbf{x} = \sqrt{\mu_{12}}\,(\mathbf{r}_1 - \mathbf{r}_2),$$
$$\mathbf{y} = [\frac{(m_1+m_2)m_3}{M}]^{\frac{1}{2}}\,[\frac{m_1\mathbf{r}_1 + m_2\mathbf{r}_2}{m_1+m_2} - \mathbf{r}_3], \tag{4}$$
$$\mathbf{z} = \frac{1}{M}(m_1\mathbf{r}_1 + m_2\mathbf{r}_2 + m_3\mathbf{r}_3)$$

where $\mathbf{r}_i$ and m_i are the position vector and the mass of the particle number i, μ_{ij} is the reduced mass of particles i and j,

$$\mu_{ij} = \frac{m_i + m_j}{m_i m_j}, \tag{5}$$

and M is the total mass of the system. In terms of this coordinates the kinetic energy in the body frame

$$T = -\frac{\hbar^2}{2}(\nabla_x^2 + \nabla_y^2), \tag{6}$$

involves a sum of two 3-dimensional Laplacians, which are equivalent to a 6-dimensional Laplacian. By replacing the vectors **x** and **y** by a set of 6-dimensional spherical coordinates (ρ, Ω), the two 3-dimensional Laplacians transforms into the 6-dimensional Laplacian defined in Eq. 1

A method of analysis based on Raleigh's principle, frequently used for three body systems [8] , consists in setting up the matrix of the intrinsic Hamiltonian H_I with respect to a complete set of functions depending on the relative vectors, the matrix been truncated to some finite order. The diagonalization of this matrix supplies then approximate eigenvalues and eigenfunctions for the Schrodinger equation $H_I\Psi = E\Psi$. Taking as basis a system of functions which are products of a hyperradial function $R_n(\rho)$ times a Dragt harmonic $\mathbf{Y}_{\lambda\nu qLM}(\Omega)$, it is clearly seen that the matrix elements of the Hamiltonian written in terms of spherical coordinates, with respect to this basis involve, insofar as the angular variables concerns, the overlap integral of two $\mathbf{Y}'s$ and the integral of a product of three $\mathbf{Y}'s$. The last mentioned integral comes from the fact that the two-body interactions in the Hamiltonian depend on ρ and Ω, so they can be expressed as linear combinations of terms $R_n(\rho)\mathbf{Y}_{\lambda\nu qLM}(\Omega)$.

The preceding discussion thus indicates the importance, for the three-body theory, of having available the integrals of products of 2 and 3 hyperspherical harmonics over the surface of a 6-dimensional sphere.

3. Behaviour of states under transposition of particles.

By a cyclic permutation of particle number indexes on the vectors defined in Eq. 4 we can have two new set of relative vectors. One of them is the set

$$\mathbf{x}' = \sqrt{\mu_{23}}\,(\mathbf{r}_2 - \mathbf{r}_3),$$
$$\mathbf{y}' = [\frac{m_1(m_2+m_3)}{M}]^{\frac{1}{2}}\,[\frac{m_2\mathbf{r}_2 + m_3\mathbf{r}_3}{m_2+m_3}] - \mathbf{r}_1. \tag{7}$$

Note that the interaction V_{12} depends only on $\mathbf{x}$, while V_{23} (and V_{31}) depends on complicated combinations of $\mathbf{x}$ and $\mathbf{y}$. But in the new relative vectors, V_{23}depends only on $\mathbf{x}'$, while V_{12} (andV_{31}) depends on complicated combinatons of $\mathbf{x}'$ and $\mathbf{y}'$. After some simple algebra, we can deduce

$$\mathbf{x} = \mathbf{x}'cos'\varphi' - \mathbf{y}'sin\varphi'$$
$$\mathbf{y} = \mathbf{x}'sin\varphi' + \mathbf{y}'cos\varphi', \tag{8a}$$

with

$$tan\varphi' = -[\frac{Mm_2}{m_1 m_3}]^{\frac{1}{2}}. \tag{8b}$$

Therefore, $x^2+y^2 = x'^2+y'^2$, i.e. calling ρ' and $\mathbf{\Omega}'$ the 6-dimensional spherical coordinates based on the vectors $\mathbf{x}'$ and $\mathbf{y}'$, we have $\rho = \rho'$. Another consequence of Eq. 8 is

$$(\pm i\mathbf{x} + \mathbf{y}) = e^{\pm i\varphi'}(\pm i\mathbf{x}' + \mathbf{y}'). \tag{9}$$

Now, in Eq. 4.2 of ref. 1 it was shown that the 6-dimensional solid harmonics $Y_{\lambda\nu qLM}(\mathbf{x},\mathbf{y}) = \rho^{\lambda}\, Y_{\lambda\nu q\mathbf{LM}}(\mathbf{\Omega})$ are homogeneous polynomials of degree $\frac{1}{2}(\lambda+\nu)$ on the components of the vector $(-i\mathbf{x}+\mathbf{y})$, and of degree $\frac{1}{2}(\lambda-\nu)$ on the components of $(i\mathbf{x}+\mathbf{y})$. From Eq. 9 we thus come to the conclusion that $Y_{\lambda\nu qLM}(\mathbf{x},\mathbf{y}) = \mathrm{e}^{i\nu\varphi'}\, Y_{\lambda\nu q\mathbf{LM}}(\mathbf{x}',\mathbf{y}')$ and as $\rho = \rho'$ we have in terms of the hyperspherical harmonics

$$\mathbf{Y}_{\lambda\nu qLM}(\mathbf{\Omega}) = e^{i\nu\varphi'}\mathbf{Y}_{\lambda\nu qLM}(\mathbf{\Omega}'). \tag{10}$$

Using this result we see that the expansion of the interaction $V_{23}(\mathbf{x}') = \mathbf{V}_{23}(\rho,\mathbf{\Omega}')$ with respect to the basis $R_n(\rho)\mathbf{Y}_{\lambda\nu qLM}(\mathbf{\Omega})$ is a task no more difficult than the expansion of the interaction $V_{12}(\mathbf{x}) = \mathbf{V}_{12}(\rho,\mathbf{\Omega})$ itself.

Naturally, a similar conclusion applies also for the third set of relative vectors. For this case, a result analogous to Eq. 10 exists, with φ' and $\mathbf{\Omega}'$ on the rigth hand side replaced by φ" and $\mathbf{\Omega}$",where $tan\varphi" = -\sqrt{Mm_1/(m_2m_3)}^{\frac{1}{2}}$.

4. Analytic expressions for Dragt harmonics.

These were given in Sect. 4 of Ref. 1, and are here reproduced in Table 1. The $\mathbf{Y}_{\lambda\nu qLM}(\mathbf{\Omega})$ depend on five angles denoted as $\theta_1, \theta_2, \theta_3, \alpha$ *and* γ. The first three are Euler angles Ref. 9 specifying the orientation of an intrinsic frame with respect to the Lab frame, and the dependence on them is through the representation functions $D^L_{MK}(\theta_1, \theta_2, \theta_3)$ of $SO(3)$, defined as in Ref. 9. The angle α has the domain $0 \leq \alpha \leq 2\pi$, and it enters in the harmonic functions through the simple factor $e^{i\nu\alpha}$. The fifth angle γ , has the domain $0 \leq \gamma \leq \frac{\pi}{4}$, and it enters through the function $\Gamma(\gamma)$ defined in Table 1. The measure on the surface of the unit 6-dimensional sphere is taken to as

$$d\Omega = sin\theta_2 \; sin4\gamma \; d\theta_1 \; d\theta_2 \; d\theta_3 \; d\alpha \; d\gamma. \tag{11}$$

There are, in all, 5 summations in the expression for $\mathbf{Y}(\mathbf{\Omega})$, whose limits are indicated in Table 1; a prime over a summation symbol means that in the corresponding sum the dummy index jump by two from one value to the next. There are non vanishing contributions to $\Gamma(\gamma)$ only when $\sigma_0 \leq \sigma_1$.

λ is a non negative integer, ν must be in the set$\{\lambda, \lambda-2, \lambda-4, ..., -\lambda\}$, and L should be in the interval $0 \leq L \leq \lambda$. The symbol ϵ takes the values 0 or 1, when $\lambda - L - \epsilon$ is an even or odd non-negative integer, respectively, and the possible values of the integer q are deduced from Eq. 3.12 of Ref. 1, namely,

$$max\{0, \frac{1}{2}(\lambda-\nu) - L\} \leq 2q \leq min\{\lambda - L - \epsilon, \frac{1}{2}(\lambda-\nu) - \epsilon\}. \tag{12}$$

If no integer q satisfies Eq. 12, there is no harmonic function having the assumed values λ ,ν and L

The factor $B(l_1)$ in Table 1 replaces the b in Eq. 4.12a of Ref. 1; the present expresion for $B(l_1)$ corrects an error that crept in the b given on Ref. 1. The symbol $\langle l_1 m_1; l_2 m_2 | LK\rangle$ in Table 1 stands for a Clebsch-Gordan coefficient of SO(3).

Once we have presented in complete detail the analytic structure of the $\mathbf{Y}_{\lambda\nu qLM}(\mathbf{\Omega})$, we proceed to discuss the integral of a product of three of them.

TABLE 1

$$\mathbf{Y}_{\lambda\nu qLM}(\theta_1,\theta_2,\theta_3,\alpha,\gamma)=e^{i\nu\alpha}\sum_{K=-L+\epsilon}^{K=L-\epsilon}{}' D^{L*}_{MK}(\theta_1,\theta_2,\theta_3)\ \Gamma_K^{\lambda\nu qL}(\gamma);$$

$$\Gamma_K^{\lambda\nu qL}(\gamma)=\sum_{l_1=\epsilon}^{L}\ \sum_{m_1=\mu_0}^{\mu_1}{}'\ \sum_{\sigma=\sigma_0}^{\sigma_1} A^{\lambda\nu qL}_{l_1 m_1\sigma}(K)$$

$$\cdot(\cos\gamma-\sin\gamma)^{m_1+\frac{\lambda-K}{2}-r}(\cos\gamma+\sin\gamma)^{-m_1+\frac{\lambda+K}{2}-r}\quad;$$

$$A^{\lambda\nu qL}_{l_1 m_1\sigma}(K)=B(l_1)\quad C(l_1,m_1)\quad C(L+\epsilon-l_1,K-m_1)$$

$$\cdot\langle l_1m_1;L+\epsilon-l_1,K-m_1|LK\rangle\quad D(\sigma,l_1)\quad S(\sigma,l_1);$$

$$B(l_1)=(-)^{l_1}\left[\frac{(2l_1+1)(2L+2\epsilon-2l_1+1)(L-l_1)!(l_1-\epsilon)!}{(2l_1+1)!!(2L+2\epsilon-2l_1+1)!!}\right]^{\frac{1}{2}};$$

$$C(l,m)=(-)^{\frac{l+m}{2}}\frac{\sqrt{(l+m)!(l-m)!}}{2^l(\frac{l+m}{2})!(\frac{l-m}{2})!};$$

$$D(\sigma,l_1)=\frac{2^r(\lambda+1-r)!}{(q-r+\sigma)![\frac{(\lambda-L-\epsilon)}{2}-q-\sigma]!}\cdot\frac{(\frac{\lambda-L-\epsilon}{2})!}{(\lambda-L-\epsilon+1)!};$$

$$S(\sigma,l_1)=\sum_{t=t_0}^{t_1}\frac{(\frac{-\lambda+L-\epsilon-1}{2})_t}{t!(r-\sigma-t)!(l_1-\epsilon-\sigma+t)!(\sigma-t)!(L-l_1-r+\sigma+t)!};$$

$$r=-l_1+\epsilon+L-\frac{\lambda-\nu}{2}+2q+2\sigma;$$

$$\mu_0=max(-l_1,-L-\epsilon+l_1+K),\quad \mu_1=min(l_1,L+\epsilon-l_1+K);$$

$$\sigma_0=max(0,l_1-\epsilon-L+\frac{\lambda-\nu}{2}-2q),$$

$$\sigma_1=min(\frac{\lambda-L-\epsilon}{2}-q,l_1-\epsilon-L+\frac{\lambda-\nu}{2}-q);$$

$$t_0=max(0,\sigma+\epsilon-l_1,l_1+r-L-\sigma),\quad t_1=min(\sigma,r-\sigma);$$

5. Integral of a product of three Dragt harmonics.

Let us introduce the notation

$$(\lambda'\nu'q'L'M'|\lambda"\nu"q"L"M"|\lambda\nu qLM) \equiv$$

$$\oint \mathbf{Y}^{*}_{\lambda'\nu'q'L'M'}(\Omega)\ \mathbf{Y}_{\lambda"\nu"q"L"M"}(\Omega)\ \mathbf{Y}_{\lambda\nu qLM}(\Omega)\ d\Omega, \tag{14}$$

with $d\Omega$ given in Eq. 11. Inserting in 14 the expression for $\mathbf{Y}(\Omega)$ given at the first line of Table 1, the integration over α can be done inmediately, and that over Euler angles is performed using Eq. 4.62 of Ref. 9. We thus obtain,

$$(\lambda'\nu'q'L'M'|\lambda"\nu"q"L"M"|\lambda\nu qLM) = \delta_{\nu',\nu+\nu"}\ \frac{16\pi^3}{2L'+1} < LM; L"M"|L'M' >$$

$$\cdot \sum_{KK'K"}{}' < LK; L"K"|L'K' > \mathbf{I}_{K'K"K}(\lambda^i\nu^i q^i L^i). \tag{15}$$

where $\mathbf{I}_{K'K"K}$ stands for the integral of a product of three functions $\Gamma(\gamma)$ of Table 1, namely

$$\mathbf{I}_{K'K"K}(\lambda^i\nu^i q^i L^i) \equiv \int_0^{\frac{\pi}{4}} \Gamma_{K'}^{\lambda'\nu'q'L'}(\gamma)\ \Gamma_{K"}^{\lambda"\nu"q"L"}(\gamma)\ \Gamma_{K}^{\lambda\nu qL}(\gamma)\ sin4\gamma\ d\gamma. \tag{16}$$

A remarkably property of Eq. 15 is that all the dependence in the rotational quantum numbers M are contained in a Clebsh-Gordan coefficient and that the sums over K's are independent of them. According to the seleccion rule for Kronecker products Ref. 10 of irrep of O(6), insofar as its λ-dependence goes, the "$3-\lambda$" symbol in Eq. 15 is non-vanishing only when λ' takes one of the values in the set

$$[\lambda+\lambda", \lambda+\lambda"-4, ..., |\lambda-\lambda"|]. \tag{17}$$

Taking now from Table 1 the expression for $\Gamma(\gamma)$ and introducing it into Eq. 16, we conclude that

$$\mathbf{I}_{K'K"K}(\lambda^i\nu^i q^i L^i) = \sum_{u} A^{\lambda'\nu'q'L'}_{l'm'\sigma'}(K')\ A^{\lambda"\nu"q"L"}_{l"m"\sigma"}(K")\ A^{\lambda\nu qL}_{lm\sigma}(K)\ I(a,b), \tag{18}$$

where u stands for all 9 subindexes in the A's; a and b are defined as

$$a = m_1' + m_1" + m_1 + \frac{1}{2}(\lambda' + \lambda" + \lambda - K' - K" - K) - r' - r" - r,$$

$$b = -m_1' - m_1" - m_1 + \frac{1}{2}(\lambda' + \lambda" + \lambda + K' + K" + K) - r' - r" - r \qquad (19),$$

and

$$I(a,b) \equiv \int_0^{\frac{\pi}{4}} (cos\gamma - sin\gamma)^a \; (cos\gamma + sin\gamma)^b \; sin4\gamma \; d\gamma$$

$$= \frac{2^{\frac{b}{2}+2}}{(a+2)(a+4)} \, {}_2F_1(-\frac{b}{2}, \frac{a}{2}+1; \frac{a}{2}+3; \frac{1}{2}). \qquad (20)$$

Let us remark that $\mathbf{Y}_{00000}(\mathbf{\Omega}) = 1$, so when $\lambda" = \nu" = q" = L" = M" = 0$ the "$3-\lambda$" symbol in Eq. 14 reduces to the overlap symbol

$$(\lambda'\nu'q'L'M'|\lambda\nu qLM)$$

$$= \delta_{\lambda'\lambda}\delta_{\nu'\nu}\delta_{L'L}\delta_{M'M}(\frac{16\pi^3}{2L+1})\sum_K{}' \int_0^{\frac{\pi}{4}} \Gamma_k^{\lambda\nu q' L}(\gamma)\Gamma_K^{\lambda\nu qL}(\gamma) sin4\gamma \; d\gamma$$

$$= \delta_{\lambda'\lambda}\delta_{\nu'\nu}\delta_{L'L}\delta_{M'M}(\frac{16\pi^3}{2L+1})\sum_u\sum_K{}' \; A^{\lambda\nu q' L}_{l_1' m_1' \sigma'}(K) A^{\lambda\nu qL}_{l_1 m_1 \sigma}(K) I(\bar{a},\bar{b}), \qquad (21)$$

with $\mu = (l_1 m_1, \sigma, l_1', m_1', \sigma')$, $\bar{a} = m_1' + m_1 - K + \lambda - r' - r$, $\bar{b} = -m_1' - m_1 + K + \lambda - r' - r$. The values of the overlaps are needed if one wishes to orthogonalize the basis. The value of the overlap symbol should be zero whenever one of the quantum numbers λ, ν, L, M is distinct from its primed counterpart.

6. Examples

We proceed to examine some particular cases for the "$3-\lambda$" symbol, whose results appear on tables 2 and 3. The former table is an illustration of the computation of the overlap symbol Eq. 21 with our program, when all double prime labels are zero, $\lambda = \lambda'$=6, $L = L' = 4$ and ν, ν',q,q' can take all possible values. As the overlap matrix is symmetric, we present only the result for $q' \geq q$. The overlaps when $q = q'$ are needed if one wishes to normalize the states to unity. It is possible to calculate some particular "$3-\lambda$" symbol values from the known closed formula [11]

$$(\lambda'\nu'q'00|\lambda"\nu"q"00|\lambda\nu q00) = \frac{8\pi^3}{\frac{\lambda'}{2}+1}\langle\frac{\lambda}{4}\frac{\nu}{4}; \frac{\lambda"}{4}\frac{\nu"}{4}|\frac{\lambda'}{4}\frac{\nu'}{4}\rangle^2, \qquad (22)$$

where for these states ν is in the set $\{\lambda, \lambda-4, \lambda-8, ..., -\lambda\}$ and $q = \frac{1}{4}(\lambda-\nu)$ is here superfluos. In table 3 we present calculations for $\lambda = 60, \lambda" = 20, \lambda' = 40, \nu' = 0$. Up to the number of significative figures showed in table 3, both results, using our program and formula 22, coincides exactly.

TABLE 2

INTEGRALS OF THREE HYPERSPHERICAL DRAGT HARMONICS:

<LA' NU' Q' L'|LA" NU" Q" L"|LA NU Q L>

For LA"=NU"=Q"=L"=0; LA'=LA=6; L'=L"=4

<NU'	Q'|	NU	Q>	
< -6	1	-6	1>=	.79582776D+04
< -4	1	-4	1>=	.22737936D+05
< -2	0	-2	0>=	.13436053D+04
< -2	1	-2	0>=	-.24805021D+04
< -2	1	-2	1>=	.22944645D+05
< 0	0	0	0>=	.95085915D+04
< 0	1	0	0>=	-.49610042D+04
< 0	1	0	1>=	.95085915D+04
< 2	0	2	0>=	.22944645D+05
< 2	1	2	0>=	-.24805021D+04
< 2	1	2	1>=	.13436053D+04
< 4	0	4	0>=	.22737936D+05
< 6	0	6	0>=	.79582776D+04

TABLE 3

INTEGRALS OF THREE HYPERSPHERICAL DRAGT HARMONICS:

<LA' NU' Q' L'|LA" NU" Q" L"|LA NU Q L>

For LA'=40; LA"=20; LA=60; NU'=0; L'=L"=L=0

<LA'	NU'|	LA"	NU"|	LA	NU>	
< 40	0	20	20	60	-20>=	0.49204411D-01
< 40	0	20	16	60	-16>=	0.27062426D+00
< 40	0	20	12	60	-12>=	0.76914263D+00
< 40	0	20	8	60	-8>=	0.14813117D+01
< 40	0	20	4	60	-4>=	0.21348316D+01
< 40	0	20	0	60	0>=	0.24016856D+01
< 40	0	20	-4	60	4>=	0.21348316D+01
< 40	0	20	-8	60	8>=	0.14813117D+01
< 40	0	20	-12	60	12>=	0.76914263D+00
< 40	0	20	-16	60	16>=	0.27062426D+00
< 40	0	20	20	60	-20>=	0.49204411D-01

References

[1] E. Chacón, O.Castanos and A.Frank, *J. Math. Phys.*, **25**, (1984) 1442.

[2] A. J. Dragt, *J. Math. Phys.* **6**, (1965) 533.

[3] *Few-Body Problems in Particle, Nuclear, Atomic and Molecular Physics* (Few-Body Systems, Suppl.2), J. L. Ballot and M. Fabre de la Ripelle, eds. (Springer, Wien-New York, 1987).

[4] J. Botero and C. H. Green, *Phys. Rev. Lett.* **56**, (1986) 1366; S.V.Petrov, S.S.Jarovoy and Yu.A.Babaev, *J. Phys. B: Atom. Molec. Phys.* **20**, (1987) 4679; C.D.Lin and X.H.Liu, *Phys. Rev. A* **37**, (1988) 2749.

[5] R. C. Whitten, *J. Math. Phys.*, **10**, (1969) 1631.

[6] M. Moshinsky and C. Quesne, *J. Math. Phys.* **11**, (1970) 1631.

[7] J. Raynal and J. Revai, *Nuovo Cim.*, **68A**, (1970) 612.

[8] L. Pauling and E. B. Wilson, *Introduction to Quantum Mechanics* (McGraw-Hill, New York, 1935), Sec. 26.

[9] M. E. Rose, *Elementary Theory of Angular Momentum* (J. Wiley, New York,1957)

[10] F. D. Murnaghan, *The Theory of Group Representations* (Dover, New York, 1963) p.276.

[11] H. Klar and M. Klar, *J. Phys. B: Atom. Molec. Phys.* **13**, (1980) 1057.

If the reader is interested on more information about our program or a copy of it he should contact any of the authors.

RELATIVISTIC CI FOR ATOMS

R. Jáuregui, C. F. Bunge and E. Ley-Koo

Instituto de Física, UNAM
Apdo. Post. 20-364, 01000 México D.F., MEXICO

ABSTRACT

A general purpose program to carry out open-ended atomic configuration interaction calculations using a Dirac bispinor basis has been developed. The radial functions are linear combinations of Slater-type orbitals, however, any other suitable radial function should be applicable. In a series of calculations for high-Z one electron atoms,we examine the accuracy of approximating the "small" component radial function Q_a by $Q_a = \frac{1}{2mc}(\sigma \cdot \mathrm{p})P_a(r)$, where P_a is the corresponding "large" component radial function. Similar basis are used in a series of many-electron calculations including a correlated He calculation.

I. Introduction

Variational methods play a crucial role in the construction of approximate solutions of the nonrelativistic Schrödinger's equation. However, since the pioneering work of Kim[1], it has been observed that the direct extension of these techniques to the relativistic Dirac atomic problem leads to catastrophic results. The presence of spurious roots and the variational collapse of the bound-electron solutions into the negative energy continuum are two of the problems most currently found[1−3].

In the last decade several procedures within the independent-particle model Dirac-Hartree-Fock (DHF) scheme have been proposed to avoid these difficulties[3−8]. The recovery of the correct nonrelativistic limit as the speed of light is increased as well as the appropiate behaviour of the wavefunction at the nucleus have been found to be good guides towards a satisfactory solution of the problem[5–7]. These conditions are usually implemented either by the use of projection operators[8] or by direct restrictions on basis functions[3–6]. Among the latter type, one of the most successful types of basis sets is one in which large and small components of each spinor are functionally related so that there is an appropriate balance of the corresponding kinetic and potential energy expectation values. This condition is currently called 'kinetic balance' [6] and in the nonrelativistic limit it corresponds to bispinors of the for

$$\Psi_{njm_j} = \begin{pmatrix} \frac{P(r)}{r}\Omega_{\kappa jm_j} \\ \frac{iQ(r)}{r}\Omega_{-\kappa jm_j} \end{pmatrix} \tag{1a}$$

with $\kappa = j \pm \frac{1}{2}$ for $j = l \mp \frac{1}{2}$ and

$$Q(r) = \frac{1}{2mc}\left(\frac{d}{dr} + \frac{\kappa}{r}\right)P(r) \tag{1b}$$

Once relativistic effects are taken into account within DHF theory correlation effects can be incorporated via perturbation or variational methods. While perturbative techniques are still very clumsy due to a lack of general strategies to truncate complete nonperturbed basis sets, multiconfigurational Dirac-Fock (MCDF) codes have become quite useful and are widely employed. These codes inherit many of the advantages of nonrelativistic multiconfigurational Hartree-Fock (MCHF), such as the generation of compact wavefunctions, but also carry forward all their intrinsic limitations, such as convergence problems. An attractive alternative would consist in extending standard configuration interaction (CI) calculations into the realm of Dirac's equation.

In this work we report preliminary results of using kinetically balanced sets, in the sense of equation (1), in CI calculations. In the next section, we review the salient features of the CI method and discuss the advantages of open ended CI. In Section III, the use of kinetically balanced basis for hydrogenic atoms is compared with a straight application of the variational principle. Then, in Section IV, we study the consequences of a direct translation of nonrelativistic Slater-type orbital (STO) basis for Hartree-Fock calculations into kinetically balanced Dirac STO's (KBDSTO) for Ne, Ar and Xe. A similar translation for a correlated He wavefunction is carried out in Section V. Finally, in Section VI we summarize our findings.

II. Configuration interaction and related orbital methods

In CI, the solutions to Schrödinger's equation

$$H\Psi = E\Psi \tag{2}$$

are approximated by wavefunctions Φ expanded in terms of symmetry N-electron eigenfunctions:

$$\Phi = \sum_{k,p} \Phi_k^{(p)} a_{kp}, \tag{3}$$

where a_{kp} are variational coefficients, and $\Phi_k^{(p)}$ denotes a configuration state function given as a linear combination of Slater determinants constructed according to the selected symmetry of the approximate wavefunction:

$$\Phi_k^{(p)} = \sum_{\alpha} D_{k\alpha} c_{k\alpha}^{(p)}. \tag{4}$$

Here the index p distinguishes the possible degenerate elements of configuration k. For the nonrelativistic Schrödinger problem the Slater determinants are constructed from an antisymmetrized product of symmetry spinorbitals ψ_{nlmm_s} with equivalence restrictions, viz., the radial functions R_{nl} are independent of the quantum numbers m and m_s:

$$R_{nl}(r) Y_{lm}(\theta, \phi) \sigma_{m_s} . \tag{5}$$

While for the Dirac equation the orbitals ψ_{njm_j} are bispinors with eigenvalues j and m_j. In this work,the radial functions are linear combinations of normalized STO's S_{il},

$$R_{nl} = \sum_i S_{il} c_{iln} \tag{6a}$$

$$S_{il} = N_{il} r^{n_{il}-1} exp(-\alpha_{il} r), \tag{6b}$$

where α_{il} is a so-called orbital exponent, a positive quantity which can be freely varied to optimize the energy, if a variational principle exists. (For molecular calculations the orbitals are characterized by an irreducible representation of the corresponding symmetry group.)

If the configuration space approaches completeness and if a variational principle exists, one would be able, aside from numerical instabilities that might arise, to approach the exact wave function by the variational determination of coefficients a_{kp}. This is the central idea behind CI. However, for the Dirac equation, the conditions for the existence of such a variational principle are not clear and one of the goals of this work is to initiate the study of the consequences of using kinetically balanced basis sets in CI calculations.

In MCHF and MCDF calculations, expansion (3) is restricted to configuration excitations within a physical electron-shell and the basis orbitals are solutions to the MCHF and MCDF integro-differential equations, respectively. These solutions provide optimum orbitals for the given expansion. One may consider different shells separately thus generating as many basis sets, and thereafter carry out CI combining configuration excitations of any number of electron shells. However, for each symmetry, the radial functions used to represent one shell, in general, will be different and nonorthogonal to those radial functions involved in the representation of another shell. Code and computer time limitations imposed by the use of nonorthogonal basis severely restricts the study of core-valence correlation effects in MCHF and MCDF calculations, and this is one major reason for developing relativistic CI. Another important reason is to avoid possible convergence problems in the solution of the self-consistent field MCHF and MCDF equations.

In CI, on the other hand, the only convergence problems are those intrinsic to all orbital expansions of the type (2), namely, the poor convergence in radial and angular expansions due to an equally poor convergence of the harmonic expansion for the reciprocal of the inter-electronic distance [9]. Nevertheless, MCDF calculations have the advantage over CI calculations that, apparently, sinking into negative energy states is unheard of, whereas this has precisely been the main obstacle to relativistic CI.

Open ended CI is a computer implementation of CI which is essentially free from unnecessary restrictions. An efficient and fairly portable implementation of open ended CI has been developed by our research group during the last decade. Practical limitations to achieve the ideal open ended CI are of three types:

(i) From the computer implementation point of view. A sensible implementation requires all array sizes to be given in terms of meaningful parameters. Ideally, all these parameters are generated from the input data for each specific problem. Examples of meaningful parameters are the maximum total number of electrons, the maximum number of electrons in open shells, and the maximum size of the CI expansion.

General algorithms are eventually brought back to reality by numerical instabilities, as pointed out below. In productive programs all input is a compact representation of a few and simple ideas. No practical limitations have been found so far in our present codes.

(ii) From the speed and size of the computer facilities. These include CPU time requirements, cache memory size, central memory size, and disk memory size. In CI, the evaluation of integrals is usually a limiting step in molecular calculations, not in atomic ones. The four index transformation to obtain two-electron integrals in an orthogonal basis is now well understood[10] and presents no problems in atomic calculations. The efficient construction of the Hamiltonian matrix rests upon a complete identification of all equalities among matrix elements. Similar considerations apply to the evaluation of reduced density and transition density matrices. Powerful eigensolvers yield eigenvalues and eigenvectors for matrices of essentially any size once the Hamiltonian matrix is available. The present code, being general for any atomic or molecular state, is believed to be correct over a wide variety of circumstances. CPU limitations even for the largest supercomputers available are anticipated in calculations beyond singles and doubles CI for highly occupied d-shells in transition metals, and highly occupied f-shells in lanthanides and actinides.

(iii) From the point of view of numerical analysis. The necessary accuracy will depend on the depth of the application. In atoms, when using Slater-type functions, the number of radial functions per symmetry and per shell cannot exceed beyond 12-15 without the appearance of numerical instabilities. Also, the concurrent use of both projection operators to evaluate symmetry eigenfunctions and angular momentum orbitals with $l > 10$ requires going from 16-figure arithmetic to a higher-figure arithmetic in fixed-point highly intensive sections of the program. For most computers this means having to resort to quadruple arithmetic; however, in most applications there is no need for $l > 10$.

III. Hydrogenic atoms

Let us consider Dirac's equation for a hydrogen-like atom:

$$H_D\Psi = (c\boldsymbol{\alpha}\cdot\mathbf{p} + c^2(\beta - 1) - \frac{Ze^2}{r})\Psi = E\Psi \tag{7}$$

where Dirac's operators $\boldsymbol{\alpha}$ and β are expressed by the matrices

$$\boldsymbol{\alpha} = \begin{pmatrix} 0 & \boldsymbol{\sigma} \\ \boldsymbol{\sigma} & 0 \end{pmatrix} \text{ and } \beta = \begin{pmatrix} I & 0 \\ 0 & I \end{pmatrix}; \tag{8}$$

here $\boldsymbol{\sigma}$ stands for the three Pauli matrices and I is the 2×2 unit matrix. The central character of the Coulomb potential allows to write the solution in the form

$$\Psi = \begin{pmatrix} \frac{P(r)}{r}\Omega_{\kappa j m_j} \\ \frac{iQ(r)}{r}\Omega_{-\kappa j m_j} \end{pmatrix},$$

so that the large and small radial wavefunctions $P(r)$ and $Q(r)$ satisfy the equations

$$\begin{aligned} \frac{1}{hc}(E + mc^2 + \frac{Ze^2}{r})Q &= (\frac{d}{dr} + \frac{\kappa}{r})P \\ \frac{1}{hc}(E - mc^2 + \frac{Ze^2}{r})P &= (\frac{d}{dr} - \frac{\kappa}{r})Q \end{aligned} \tag{9}$$

There exists numerical evidence[4] showing that, when one considers basis functions of the form

$$\begin{aligned} \phi_i &= r^{\gamma+i}e^{-\lambda r}\begin{pmatrix} 1 \\ 0 \end{pmatrix}, \\ \phi_{M+i} &= r^{\gamma+i}e^{-\lambda r}\begin{pmatrix} 0 \\ 1 \end{pmatrix}, i = 0, ..., M-1, \end{aligned} \tag{10}$$

there is a variational principle so that the energy eigenvalues resulting from a variational calculation satisfy a generalized Hylleraas-Undheim theorem. That is, the 2M eigenvalues split into two sets so that M roots lie in the Dirac's sea below $-mc^2$ and the remaining M roots behave exactly as if the Dirac Hamiltonian were bounded from below. The latter move progressively lower as M is increased, but never cross the exact energies. A spurious root for κ greater than zero is always present.

At the same time, as a consequence of Dirac's equation,

$$Q_a(r) = \frac{1}{(mc + \frac{E}{c} + \frac{Ze^2}{cr})}\left(\frac{d}{dr} + \frac{\kappa}{r}\right)P_a, \tag{11}$$

so that there is a natural balance between the large $P_a(r)$ and the small $Q_a(r)$ radial functions. This relationship may be approximated by demanding

$$Q_a(r) = \frac{1}{2mc}(\frac{d}{dr} + \frac{\kappa}{r})P_a(r)$$

to hold true.

Now, how does the use of basis sets (10) compare with the use of kinetically balanced basis sets (1b)? In order to answer this question we have considered the result that is obtained from an optimization of DSTO's of both kinds. In the case of the ground 1s state of Hg^{+79} when M=4 we obtain for a basis of the form (10)

$$E = -3532.1922\,0147\,8059. \tag{12}$$

with the optimized STO's given by the following set of principal quantum numbers and orbital exponents:

$$1s = 69.3379, 2s = 97.92179, 3s = 115.357, 4s = 135.357.$$

The exact eigenvalue is

$$E = -3532.1922\,0147\,8057. \tag{13}$$

We attribute the difference in the last digit between (12) and (13) to numerical round off.

Direct examination of the variational eigenfunction shows that, as expected, the small component radial function is proportional to the large component with a proportional factor C_{mix}:

$$C_{mix} = -.32219\,56904\,79099 = -\frac{\alpha E}{Z}, \tag{14}$$

where α is the fine structure constant and Z is the nuclear charge. If this condition is directly imposed on the basis functions the resulting variational problem is M-fold degenerate. That is, no negative energy sea solution is present and there is just one four-degenerate eigenvalue.

Now, we shall consider the case where the basis functions Q_a and P_a are proportional to each other but the C_{mix} coefficient is approximated by

$$C_{mix} = -.32220 \tag{15}$$

leads to an energy

$$E = -3532.2537\,, \tag{16}$$

which is well below the true eigenvalue. That is, the precision at which C_{mix} is given is in direct correspondence with the precision at which the energy is determined.

Taking into account this high sensitivity on the C_{mix} coefficients it is not a surprise that the kinetically balanced basis sets give less accurate results as the nuclear charge Z is increased. In Table I we give the results of considering KBSTO's optimizing freely the orbital exponent of $P_a(r)$. No roots appear below $-mc^2$ but the lowest eigenvalue always lies below the true eigenvalue.

According to these results one would expect that, for neutral atoms, the orbital exponents in valence orbitals can be freely optimized with kinetically balanced basis sets. More generally, valence orbitals subjected to not too large effective charges should be well described by kinetically balanced basis sets.

TABLE I. Results obtained for the ground state energy by free optimization of P_a; $Q_a = \frac{1}{2mc}(\sigma.\mathrm{p})\mathrm{P_a}$. (Energies in hartree, errors in μhartree)

Z	EXACT ENERGY	ENERGY OBTAINED	ERROR
1	-0.500 006 656 598	-.500 006 656 897	3×10^{-4}
3	-4.500 539 299	-4.500 539 581	3×10^{-1}
10	-50.006 674	-50.006 712	30
18	-162.7048	-162.7180	13 200
80	-3532.1922	-3801.1922	300×10^{6}

IV. Nonrelativistic to relativistic translation of Hartree-Fock results

Although it is true that relativistic effects induce changes in the nonrelativistic electronic density, it is also true that one would expect these changes to be small for relatively low Z. Thus, one could approximate the DHF orbitals with the nonrelativistic HF orbitals translated into kinetically balanced bispinors, viz., in Eqs.(6) we take the principal quantum numbers n_{jl} ,the orbital exponents α_{jl} and the same coefficients c_{ijn} from the nonrelativistic orbitals and define the small component $Q(r)$ by means of Eq.(1b). Reorthogonalization of these relativistic orbitals changes sligthly the coefficients c_{ijn}. How far in Z one can go before these approximate DHF orbitals become useless?

The results of this traslation for Ne, Ar and Xe ground states are given in Table II. We note that the results are quite good so that nonrelativistic Hartree-Fock orbitals are a good starting point for relativistic Dirac-Fock calculations. The main defect of this translation appears to be the assumption of equal radial parts for orbitals with equal nl quantum numbers and different j. One may attempt to remedy this situation by carrying a full CI within a given shell involving only the corresponding symmetry one-electron functions and using the resulting natural orbitals as the improved DHF orbitals.

TABLE II.Results of the nonrelativistic to relativistic translation of HF (energies in a.u.)

ATOM	*NONRELATIVISTIC*[11]	BALANCED	NUMERICAL *DHF*[11]
Ne	-128.5471	-128.6914	128.6919
Ar	-526.8169	-528.6580	-528.6838
Xe	-7232.1384	-7424.3050	-7446.9002

V. Correlated wavefunction for He ground state

For an N-electron atom, we take as starting point the Hamiltonian

$$H = \sum_{i=1}^{N} H_D(r) + \sum_{i>j} \frac{e^2}{r_{ij}} \tag{17}$$

As we already mentioned, the configurations for a CI calculation are taken to be constructed from Dirac orbitals (1a).

In the particular case of He atom ground state, we considered a nonrelativistic $4s3p2d1f$ optimized basis that yields a correlated wavefunction with energy E_{nr},

$$E^{(nr)} = -2.9028\,1824\,. \tag{18}$$

We translated this basis to a kinetically balanced relativistic basis and considered a full CI calculation which in this case extends to 30 two-electron functions.The accuracy of our results was tested by two different integration methods an analytical one and a numerical one[9]. We obtained the result,

$$\begin{aligned} E^{(r)}_{ana} &= -\,2.9029\,5197\,794047 \\ E^{(r)}_{num} &= -\,2.9029\,5197\,794057\,. \end{aligned} \tag{19}$$

Thus, we found a difference between nonrelativistic and relativistic energy equal to

$$E^{(r)} - E^{(nr)} = -133.7\mu h \tag{20}$$

Meanwhile the difference between DHF and HF energy is[11]

$$\begin{aligned} E^{(DHF)} - E^{(HF)} &= -\,2.8618133 + 2.8616799 a.u. \\ &= -\,133.4\mu h. \end{aligned} \tag{21}$$

indicating that relativistic correlation effects are only $-0.3\mu h$ in this example.

VI. Conclusions

In this work, we have presented preliminary results of CI relativistic calculations using basis sets in which the radial functions P(r) and Q(r) are related in the form

$$Q(r) = \frac{1}{2mc}\left(\frac{d}{dr} + \frac{\kappa}{r}\right)P(r). \tag{1b}$$

We have found that, due to the corresponding screening of nuclear charge, these basis are generally reasonable representations of the valence orbitals.

We also found that, for neutral atoms, the direct translation of a nonrelativistic Hartree-Fock basis to a relativistic basis gives rise to quite good results (Table II). Thus, we conclude that, for this kind of systems, nonrelativistic Hartree-Fock is a good starting point for relativistic Dirac-Hartree-Fock. As we have shown for the He ground state, something similar occurs for correlated wavefunctions of systems where the relativistic correction is expected to be small.

The importance of a correct relation between small and large components prevents the direct application of basis (1b) to systems for which relativistic effects are essential. Thus, we shall study more general basis sets in future work. The excellent results obtained for DHF of similar basis where the $\frac{1}{2mc}$ factor in (1b) is substituted by a coefficient that is self-consistently obtained[11], suggest to study the variational behaviour of this coefficient. However, the use of alternative functional relationships between $P(r)$ and $Q(r)$ is not discarded.

REFERENCES

1 Y. K. Kim, Phys. Rev. **154**(1967)17; **159** (1967)190E.

2 T. Kagawa, Phys. Rev. **A12**(1975)2245; **22**(1980)2340; F. Mark and Rosisky, Chem. Phys. Lett. **74**, (1980)562; F. Mark, H. Lischka and F. Rosisky, Chem. Phys. Lett. **71**(1980)507; O. Matsuoka, N. Suzuki, T. Aoyama and G. J. Malli,

J. Chem. Phys. **73**(1980)1320; Y. Ishikawa, R. Barrety and K. M. Sando, Chem. Phys. Lett. **117**(1985)444.

3 Y. Ishikawa, R. C. Binning, Jr. and K. M. Sando, Chem. Phys. Lett. **101**(1983)111.

4 G. W. F. Drake and S. P. Goldman, Phys. Rev.**A23**(1981)2093.

5 S. P. Goldman and G. W. F. Drake, Phys. Rev.**A24**, (1981)183; Phys. Rev. **A25**(1982)2877; S. P. Goldman, Phys. Rev. **A37**(1988)16.

6 Y. S. Lee and A. D. MacLean, J. Chem Phys. **76**, (1982)735; K. G. Dyall, I. P. Grant and S. Wilson, J. Phys. **B17**(1984)493; **B17**(1984)1201; R. E. Stanton and S. Havriliak, J. Chem. Phys. **81**(1984)1910; I. P. Grant, J. Phys. **B19**(1986)3187; I. P. Grant and H. M. Quiney, Adv. in Atom. and Mol. Phys. **23**(1988)37; H. M: Quiney, I. P. Grant and S. Wilson, J. Phys **B20**, (1987)1413.

7 W. Kutzelnigg, Int. J. Quant. Chem.**25**(1984)107.

8 G. Hardekopf and J. Sucher, Phys. Rev. **A30** (1984)703; J. L. Heully, I. Lindgren, E. Lindroth and A. M. Martenson-Pendrill, Phys. Rev. **A33**(1986)4426; G. E. Brown, Phys. Scripta **36**(1987)209.

9 C.F.Bunge, Phys. Rev. **A40**(1989)477, E. Ley-Koo and C. F. Bunge, Phys. Rev. **A40**(1989)1215.

10 C. F. Bunge, A. V. Bunge, G. Cisneros and J. P. Daudey, Comp. Chem. **12**(1988)91.

11 E.Clementi,"Modern Techniques in Computational Chemistry", ESCOM,Science Publishers,Leiden (1989).

VII. MOLECULES AND MORE

CORRELATION EFFECTS ON ATOMIC MAGNETISM

by

Virgilio Beltrán L.

Instituto de Ciencias Nucleares
Universidad Nacional Autónoma de México, (UNAM)
Circuito Exterior, C.U.
México, D.F. 04510
México

CORRELATION EFFECTS IN ATOMIC MAGNETISM

Spectroscopy is the science of the interaction of radiation and matter. It is based in Planck's fundamental relation $\Delta E=h\nu$. By observing the radiation absorbed or emitted by a sample of matter, usually in the form of a collection of microscopic entities, we may obtain information on its elementary mechanical states in the form of an empirical value ΔE_e of their energy differences. On the other hand, by utilizing de Broglie's relation $p=h/\lambda$ the equations of motion for the particles composing the system may be obtained, the wave equation for stationary states $\mathcal{H}\Psi=E\Psi$ may be derived and a theoretical value ΔE_t for the energy differences can be deduced. Of course we seek to have $\Delta E_e=\Delta E_t$, which ultimately means testing the fundamental relations on which physics is based.

If the sample is immersed in a magnetic field H the energy E depends on its magnitude and, in general, we can express it as a power series in H; i.e., we can write :

$$\Delta E_e = \Delta E_o + a\, H + b\, H^2 + \ldots \qquad \ldots (1)$$

From the term linear in H we obtain the empirical value of the g_J factor

$$g_J^e = a/\mu_B M_J \ , \qquad \ldots (2)$$

where μ_B is the Bohr magneton and M_J is the projection of the elementary angular momentum, in units of $\hbar$, along the direction of the magnetic field.

The theoretical value of this atomic parameter is obtained from the wave equation for the stationary states of the system, $\mathcal{H}\Psi=E\Psi$, and from the definition:

$$g_J^t= (\partial\langle\Psi|\mathcal{H}|\Psi\rangle/\partial H)_{H=0}/\mu_B M_J \qquad \ldots (3)$$

, where $\mathcal{H}$ is the Hamiltonian of the system and μ_B is the Bohr magneton. We seek now, of course, $g_J^e=g_J^t$ and we inquire what exactly is put to test on doing this.

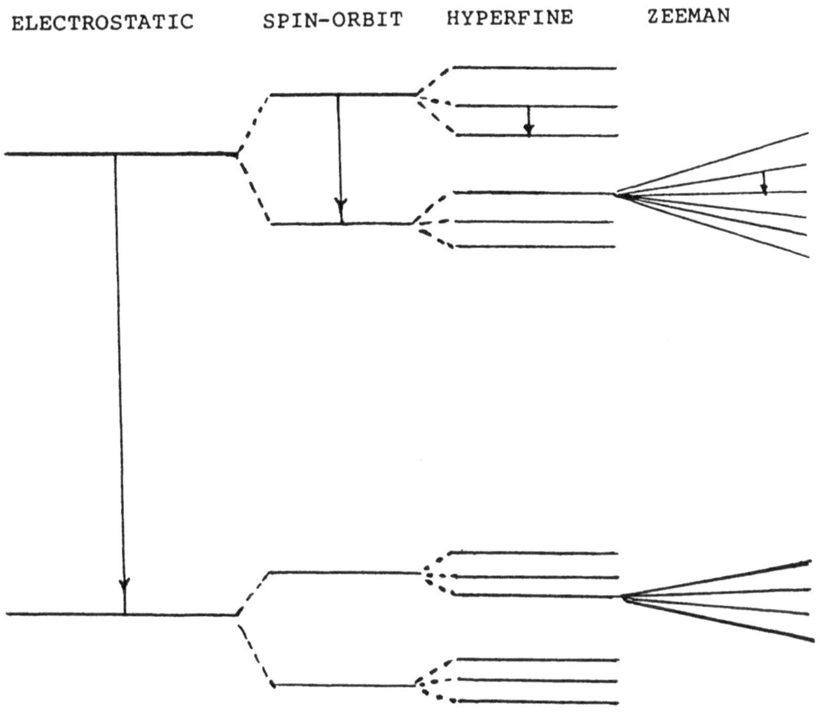

E	30000	10	5	1	cm^{-1}
ΔE/E	10^6	10^2	10	1	ppm

Figure 1

g_J arises mainly from the term in the Zeeman operator $\hat{\mathcal{Z}}=-\mu\cdot H.$, which in an overall energy level diagram accounts for only a few ppm of the total electrostatic binding energy of the atom (Fig. 1). For purposes of calculating g_J it can be completely neglected for building the wave function Ψ of the atom, along with all other terms dependent on magnetic properties. Ψ is then determined by the equation $H_0\Psi=E_0\Psi$, where

$$H_0=\sum(\nabla_i^2/2m + Ze^2/r_i + e^2/r_{ij}) \qquad \ldots (4)$$

contains only the electrostatic interaction.

We know that even with this simplified Hamiltonian the energy eigen-value equation cannot be solved exactly but for one-electron and a few two-electron systems. The corresponding solutions, Ψ_{ex}, are called "exact" solutions. The approximate but well defined Hartree-Fock solutions, Ψ_{HF}, are obtained by substituting the electron repulsion term $\Sigma e^2/r_{ij}$ by a central potential $V(r_i)$. These solutions describe the motion of indistinguishable electrons in a central field composed from that due to the nuclear charge and an average shielding for each electron resulting from all the other electrons of the atom. As far as the electric force is concerned, the electrons move independently of each other since no terms dependent on their mutual distances are left in the approximate Hamiltonian. We may then say that their motions are not Coulomb-correlated. Motion correlation of a different nature is present, however, on account of the exclusion principle which is included on antisymmetrizing the wave function, since this precludes electrons with the same set of quantum numbers from occupying the same place in physical space. We can then also say that the Hartree-Fock solutions for the electrons are Pauli-correlated, but Coulomb-uncorrelated.

Whichever wave function Ψ we use, however, the theoretical g_J will be given, in this approximation, by the expression:

$$g_J^t = (\partial\langle\Psi|\hat{\mathcal{Z}}|\Psi\rangle/\partial H)_{H=0}/\mu_B M_J \quad \ldots (5)$$

Since $\langle\hat{\mathcal{Z}}\rangle$ is very small compared to the interactions which determine the wave function of the atom, Coulomb and Pauli, Ψ will be quite independent of $\hat{\mathcal{Z}}$. The Zeeman operator $\hat{\mathcal{Z}}$ however, since it includes terms which are dependent on structure and on mutual distances between electrons, will yield expectation values which are dependent on the wave function Ψ. Comparison of g_J^t and g_J^e will, then, test $\hat{\mathcal{Z}}$ if Ψ is known, or it will test Ψ if $\hat{\mathcal{Z}}$ is known. If the last is true we can regard the magnetic moment as a probe of atomic structure which senses its details without perturbing it. In particular, if Hartree-Fock wave functions are employed, expectation values g_J^{HF} will be obtained which do not reflect the effect of Coulomb correlation on the wave functions.

The Zeeman Hamiltonian $\hat{\mathcal{Z}}$ is, indeed, known at least to within terms of order α^2. Not all of the terms composing it, however, are sensitive to contributions due to correlation in the wave function. The main one, $\mathcal{Z}_0 = \mu_B H(g_S\hat{S}+g_L\hat{L}_z)$ yields the familiar Landé rational values

$$g_J^o = \{g_S(J^2- L^2+S^2) + g_L(J^2- S^2+L^2)\}/2J^2 \quad \ldots (6)$$

where J^2=J(J+1), L^2=L(L+1) and S^2=S(S+1). These values depend only on angular momenta; i.e., on geometry, and do not reflect any effect from correlation. There will be deviations from these rational values which arise from departures of g_S from 2- the QED correction- and departures of g_L from 1- the isotopic correction- which results from the motion of the nuclear mass. They are taken into account by simply substituting into Eq. (5) the values for g_S and for g_L which are obtained from the expressions:

$$g_S = 2(1+\alpha/2\pi-0.328\alpha^2/\pi^2-1.1763\alpha^3/\pi^3+\ldots)$$
$$= 2(1.001159652193(10)\ldots)_{t,e}$$

and

$$g_L = 1-(m/M)(1+(2/3)\Sigma(\varepsilon_{2p}-\varepsilon_{ns})[\int_0^\infty r\ P_{ns}P_{2p}dr]^2)$$

$$g_L = 1, \quad \text{for S states.}$$

The last one depends on the wave function but only through one-electron integrals; it will then be quite insensitive to the small differences in Ψ resulting from the Coulomb correlation.

One more deviation from the Landé rational values may result from the coupling of states by the spin-orbit interaction (Fig. 2). This depends on energy levels separations, spin-orbit parameter values and the g_J values of the admixed states. It will also be quite insensitive to the Coulomb correlation except, maybe, for the value of the spin-orbit parameter ζ. This effect, however, will show in deviations of g_J amounting to less than 1 ppm.

Most of the effect of Coulomb correlation to g_J results from the contribution to $\hat{\mathcal{Z}}$ from relativistic and velocity dependent interactions; i.e., from the part of the Breit Hamiltonian,

$$\begin{aligned}\delta Z = &-\mu_B H\cdot\Sigma(l_i+2\ s_i)T_i/mc^2 + \mu_B Ze^2\Sigma(\nabla_i(1/r_i)xA_i)\cdot\sigma_i/2mc^2\\ &+\mu_B e^2\Sigma'(\nabla_i(1/r_{ik})xA_i).(\sigma_i+2\sigma_k)/2mc^2\\ &-e^2\Sigma'(A_i\cdot p_k/r_{ik}) + (r_{ik}\cdot A_i)(r_{ik}\cdot p_k/r_{ik}^3)/m^2c^3,\end{aligned}$$

where $A=(1/2)Hxr$, which depends linearly on the magnetic field. This is of the order of α^2. It contributes some tens of ppm to the g_J value, of which only a few ppm are due to correlation.

DEPARTURES FROM RATIONAL VALUES

— From spin-orbit interaction

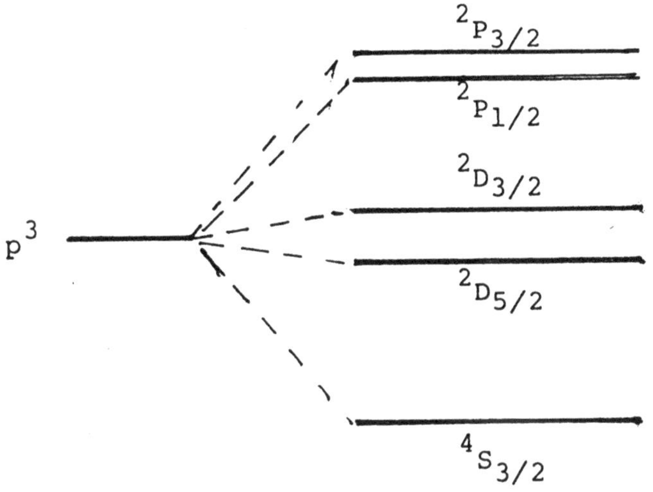

$$\Psi(S,L,J,M) = \Sigma C(S'L')\Psi(S',L',J,M)$$

$$\delta g_J \propto (\zeta/\Delta E_{so})^2$$

$$\delta g_J(^4S_{3/2}) = -(2/3)(\zeta_{2p}/PS)^2$$

$$\delta g_J(^2D_{5/2}) = \delta g_J(^2P_{1/2}) = 0$$

Figure 2

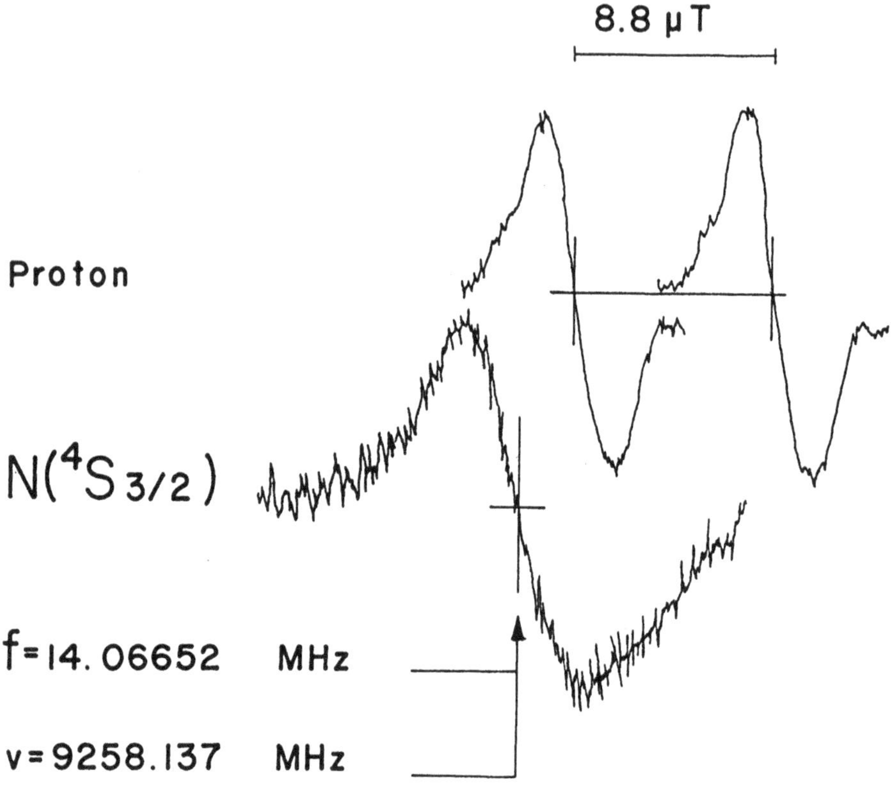

Figure 3

Evaluation of the effect of correlation in atomic magnetism implies then the measurement of Zeeman transitions on several atomic states with ppm accuracy, or better, and reliable calculations of atomic magnetic moments within the Hartree-Fock scheme; p^n atomic configurations being necessary for obtaining a full development of atomic magnetism, since these will include non-zero orbital angular momentum. Meaningful testing of the wave function, on the other hand, requires calculations with methods which include correlation.

This program has been accomplished through measurements and calculations performed over several decades on many states of the light atoms from H to F.[1] He(1s 2s $^3S_1$) is of course the simplest magnetic state of an atom with more than one electron. It was, indeed, the object of very precise measurements and calculations in the pioneering work of V. Hughes *et al.*[2] Being, however, a two s-electrons state it will not show magnetic effects resulting from electron orbital angular momentum. It will not show structural effects resulting from core polarization, either; since this is due to the uneven interaction of an outside electron with those of a closed s subshell. Atomic nitrogen with its $1s^2 2s^2 2p^3$ ground state configuration, on the other hand, is ideally suited for this work . It has 3 unbalanced p electrons which contribute to the magnetic moment, its ground state configuration splits into 5 levels of different symmetries which weigh differently the various contributions to the Zeeman energy term- like the spin-orbit contribution in (Fig. 2)-, and the microwave Zeeman transitions in its ground state are very narrow; comparable in fact with the proton nuclear magnetic resonance lines which are used for measuring the magnetic field (Fig. 3, from Ref. 3). We have recently measured the g_J factors in three states of atomic nitrogen, $^4S_{3/2}$, $^2D_{3/2}$ and $^2D_{5/2}$ by Zeeman microwave spectroscopy.[3] The very important value for the ground $^4S_{3/2}$ states, accurate to 1 ppm, is given in Table 1 together with the results of other measurements made by use of different techniques.

TABLE 1

$g_J(N, {}^4S_{3/2})$	Method of measurement	Year
2.002 117 (3)	Microwave absorption	1954
2.002 134 (5)	Atomic beam	1971
2.002 134 (25)	Optical pumping	1973
2.002 134 652(64)	Laser tecnique	1977
2.002 136 7 (22)	Microwave absorption	1989

The measurement made by Hirsch, Zimmerman, Larson and Ramsey[4] using a laser technique is accurate to 0.02 ppm; i.e. to the order of α^3. This is better than the order of α^2 to which the Breit Hamiltonian is supposed to hold. It may then be considered as exact for purposes of comparison with theoretical results. Accurate Hartree-Fock values of these g_J factors are also known. The difference $g_J^e-g_J^{HF}$ can be taken as due to the effect of correlation. In the $^4S_{3/2}$ state of atomic nitrogen it amounts to 3.5 ppm.

The origin of this difference may be inferred by comparison with He($^3S_1$). $g_J^e-g_J^t$ in this atomic state is essentially zero for this atomic state for any reasonable wave function employed in the calculation; whether it be HF, Pekeris' completely correlated wave function for the $^3S_1$ state, or even with a crude model of the helium atom based on hydrogenic Slater-type wave functions with exponents which minimize the energy. We always obtain the same value. This means that correlation does not affect the magnetic moment of excited helium, at least to the order of about 1 ppm. Since no core polarization can occur in the helium atom and a strong one is expected, and proved to exist by the non-zero hyperfine structure, in N(p^3,$^4S_{3/2}$) we are lead to conclude that the correlation effect observed in g_J(N,$^4S_{3/2}$) is mainly due to core polarization resulting from the 3 unpaired p electrons; i.e., to Pauli correlation. This conclusion is subtantiated by the graph of **Fig. 4**, from Ref. 5, in which $g_J^e-g_J^t$ is plotted for several states of the light atoms. Disregarding the unrepresentative value associated with F($^2P_{1/2}$), these results support the conclusion that there is a contribution to g_J amounting to about 1 ppm per polarizing $2l$ electron. g_J(Li, $^2S_{1/2}$) and g_J(O,$^3P_{1/2}$) show this deviation clearly and the other atomic states in Fig. 4 show it within their error bars. More precise measurements in these states should confirm this number.

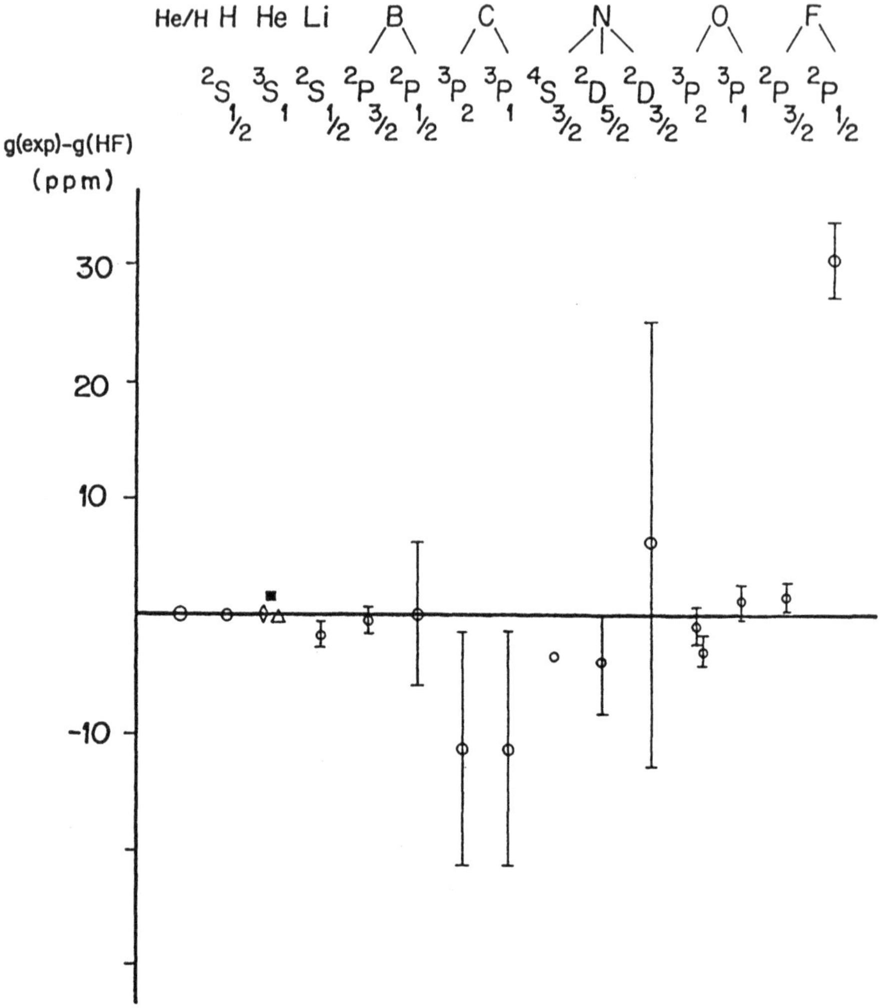
He/H
H
He
Li
B
C
N
O
F
$^2S_{1/2}$
$^3S_1$
$^2S_{1/2}$
$^2P_{3/2}$
$^2P_{1/2}$
$^3P_2$
$^3P_1$
$^4S_{3/2}$
$^2D_{5/2}$
$^2D_{3/2}$
$^3P_2$
$^3P_1$
$^2P_{3/2}$
$^2P_{1/2}$
g(exp)-g(HF)
(ppm)
30
20
10
-10

Figure 4

A theoretical value consistent with this result has been derived by calculating $g_J(N, {}^4S_{3/2})$ with a wave function obtained by the method of interaction of configurations, specifically tailored to yield a non-zero hyperfine structure. The wave function, with 41 ^4S terms, was built from a HF wave function and Slater type orbitals. The basis set of the excited configurations was that derived by Schaefer, Klem and Harris on constructing a polarization wave function for calculating a_J. The values derived for a_J and $g_J(N, {}^4S_{3/2})$ are in very good agreement with the experimental ones; i.e., 96% of a_J and 92% of $g_J^e - g_J^{HF}$ are obtained with this polarization wave function, although only 27% of the correlation energy is obtained. Since polarization is due to exclusion also, this calculation shows that the effect of correlation on the magnetic moments comes from exclusion; i.e., it is due to some "remnant" Pauli correlation, not included in the HF wave function, which is picked up by the same excited configurations responsible for the non-zero value of the hyperfine structure. The internal consistency of these experiments and calculations subtantiate then the quantum mechanics of atomic structure theory and, of course, they also fail to prove wrong the basic principles on which physics is based. The results are shown in Table II.

Atomic energy or magnetic parameter	From Ref.X	Many body theory	Experiment
E - E (HF) (hartree)	-0.0508	-0.1618	-0.1886
a_J (MHz)	10.07	10.49	10.45
$g_J - g_J$(HF) (ppm)	- 2.81	-2.70	-3.06

TABLE II

REFERENCES

1. V. Beltrán-López and T. González, Phys. Rev. **A 2**, 1651 (1970); **4**, 429 (1971) and references therein.
2. W. Perl and V. W. Hughes, Phys. Rev. 91, 842 (1953),W. Perl, Phys. Rev. 91, 852 (1953), and C.W. Drake, V. W. Hughes, A. Lurio, and J. A. White, Phys. Rev. 112, 1627 (1958).
3. V. Beltrán-López, J. Rangel G., A. González-Nucamendi, J. Jiménez-Mier, and A. Fuentes-Maya, Phys. Rev. **A** 39, 58 (1989).
4. J. M. Hirsch, G.H. Zimmerman III, D. J. Larson, and N. F. Ramsey, Phys. Rev. **16**, 484 (1977).
5. V. Beltrán-López and J. Jiménez-Mier, Phys. Rev **A 43**, 4026 (1991)

COLLISIONS OF NEGATIVE IONS AT LOW COLLISION ENERGIES

R. L. CHAMPION
Physics Department, College of William and Mary
Williamsburg, VA 23185, USA

ABSTRACT

Two studies involving negative ion-atom collisions are chosen to illustrate the varied character of such systems. The examples are: O^- + H and SF_6^- + SF_6. For the former, the cross section for electron production is found to rise as the relative collision energy is decreased below the electron affinity of oxygen, indicating that the rate constant for associative detachment may be large at near-thermal energies. In the latter system, which is important to the field of gaseous dielectrics, electron detachment is found to be a relatively unimportant inelastic product channel. For both cases, the results are far from definitive; the studies thus far have raised as many questions as have been answered.

The purpose of this report is to discuss the collisional dynamics observed in low energy collisions of two rather different negative ion-atom systems. These two systems are O^- + H and SF_6^- + SF_6; the range of laboratory collision energies for which various total cross section measurements have been made is from about 10 eV up to 500 eV. These two systems are of interest for quite different reasons and the experiments have been performed using different experimental techniques. Consequently, the two will be presented independently in the brief accounts which follow.

1. O^- + H

1.1 Introduction

The system O^- + H can produce free electrons via associative detachment,

$$O^- + H \rightarrow OH + e \tag{1}$$

or collisional detachment,

$$O^- + H \rightarrow O + H + e. \tag{2}$$

Only the former occurs for low collision energies, viz., for those relative collision energies below 1.46 eV, the electron affinity of O^-. This anion had been considered[1] as a participant in the negative ion chemistry of interstellar molecules but was dismissed as reaction (1) was assumed to slow at temperatures relevant to the interstellar environment. This assumption prompted a calculation[2] of the intermolecular potentials for the molecular ion OH^- which, in turn, led to the suggestion that associative detachment might not be slow. The reason for this suggestion is apparent from the results of those calculations which are illustrated in Figure 1.

The ground electronic state of OH^- is of $^1\Sigma$ symmetry, has a vertical electron affinity of about 1.8 eV and has been well-characterized[3]. It does not cross the $^2\Pi$ state of OH and does not couple strongly to that state[4]; hence collisions at low energies which are attributed to the $^1\Sigma$ state of OH^- are essentially non-reactive. However, in a collision the excited electronic states of the molecular anion which result from $O^-(^2P)$ interacting with $H(^2S)$ must also be considered in the dynamics for (1) and (2). In addition to the $^1\Sigma$ state there are the $^1\Pi$, $^3\Pi$ and $^3\Sigma$ states. As may be seen in Figure 1, the $^1\Pi$ and $^3\Pi$ states were determined to be attractive, intersecting the neutral $OH(^2\Pi)$ at an internuclear separation of about 2.5Å. The combined statistical weights of these two states is 2/3 and the attractive curves are somewhat similar to the induced dipole potential given by $-\alpha e^2/2R^4$ (where α is the polarizability of H). From Figure 1 it appears that no barrier to associative detachment exists and hence one might expect (1) to be fast, with a rate constant in the neighborhood of 2/3 of that predicted by a simple Langevin orbiting model[5]. To date, it seems as if no measurements of rate constants for (1) have been reported. The purpose of the experiment described herein is to measure, for as low a relative collision energy as possible, the cross section for (1). It should be stressed, however, that these experiments can not access the regime of very low energies which is relevant to the question of

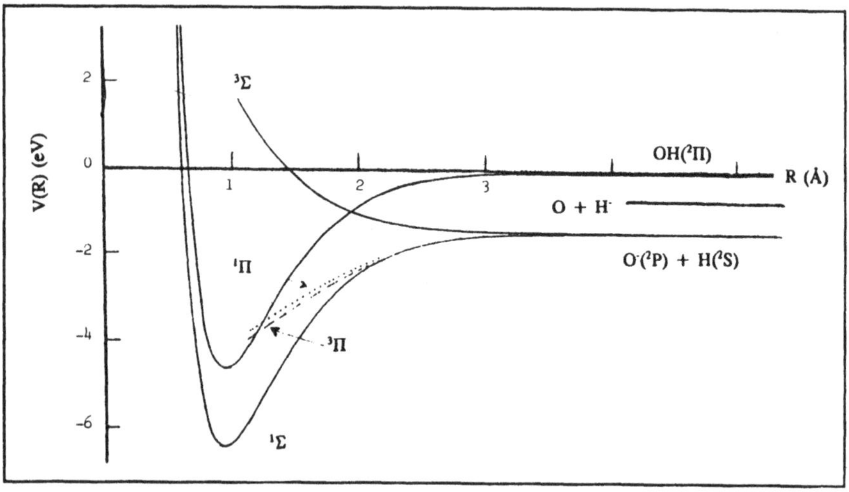

Figure 1 Intermolecular potentials for OH and OH^- (from ref. 2)

interstellar chemistry. Nonetheless, the experimental results do strongly suggest that the rate for (1) is not trivial at the temperature of space.

1.2 Experimental Method

The experiments are carried out on a crossed-beam apparatus which has been used to study H^- + H and which has been described in some detail[6]. The O^- ion beam is extracted from a discharge containing N_2O and Argon. The ion beam passes through a magnetic mass spectrometer and is focussed into the collision region (or interaction zone) which is within a truncated section of a cylindrical electrostatic energy analyzer. The O^- beam enters the analyzer resonantly and intersects orthogonally with the atomic hydrogen beam within the analyzer. Electrons formed in the interaction region are forced through an aperture in the inner analyzer electrode by the transverse electric field of the analyzer section. The products of charge transfer (i.e., H^-) are similarly extracted and subsequently distinguished from product electrons by means of weak magnetic fields. The atomic hydrogen beam is produced by a radio-frequency discharge source. The fraction of molecules which is dissociated can be ascertained by examining, for example, the decrease in the H^- signal from $O^- + H_2 \rightarrow H^- + ...$ as the discharge is turned on. The cross sections for $O^- + H_2 \rightarrow H^- + ...$ and $O^- + H_2 \rightarrow e + ...$ are both known[7] and the cross section for $O^- + H \rightarrow H^- + O$ is zero at lab energies where the dissociation fraction can be determined. In determining absolute cross sections, the target thickness and beam overlap, etc. may also be determined with the above reactants, viz., $O^- + H_2$. There is one potentially serious source of error which can't be quantified, however: If vibrationally excited H_2 molecules which are produced within the r-f discharge survive, and reach the interaction region (even though such molecules must make tens of collisions with the walls of the exit aperture), their role in producing free electrons in $O^- + H_2(v \neq 0) \rightarrow e +$ ... could be important. Unfortunately, neither the relative density nor the cross sections for vibrationally excited H_2 are known at the present. In the discussions below, it is assumed that $H_2(v \neq 0)$ is not present in the "atomic hydrogen" beam when the r-f discharge is on.

1.3 Results

The experimental results for the associative and collisional electron detachment cross sections are given in Figure 2; the two channels, (1) and (2), cannot be distinguished in this experiment. For $E < 1.46$ eV, however, only associative detachment [i.e., (1)] is energetically possible. The increase in the cross section as E is lowered below the electron affinity of oxygen implies, unambiguously, that one or more of the intermolecular potentials which describe O^- + H must be attractive and must couple strongly to the $^2\Pi$ state of OH. Although the $^1\Sigma$ state of OH^- is attractive it probably does not lead to associative

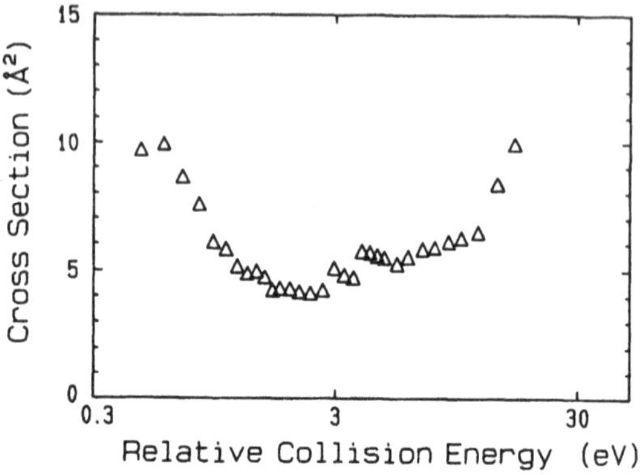

Figure 2 Detachment cross section for $O^- + H$

detachment and, even if it did it could not account for the observed cross section as its relative statistical weight is only 1/12. Hence we are led to conclude that other states, such as those depicted in Figure 1, are attractive and lead to the low energy behavior observed in Figure 2 for the detachment cross section. In fact the Langevin cross section for $\alpha = 0.7 Å^3$ (the polarizability of hydrogen) is about $15 Å^2$ for E = 1 eV; the measured cross section at 1 eV in $6Å^2$. The combined statistical weights of those two states which Huron and Tran Minh[2] calculated to be attractive is 2/3. Their contribution to the associative detachment cross section at E = 1 eV could then be expected to be a maximum of about $10Å^2$. The observation that the cross section is ~60% of that given by the simple Langevin model is also observed for the reactants $Cl^- + H$.[8]

At this point the experimental measurements have not been extended to collision energies below 0.5 eV. The observation that this lowest energy measurement gives a cross section which is below that at E = 0.6 eV may imply that a slight barrier to associative detachment occurs in the incoming channels. If this is true (one data point is completely inadequate information upon which to draw such a conclusion, however), the thermal rate constant for associative detachment could be small.

Before the dynamics for this system are completely understood, several things are needed: First, this type of crossed-beam experiment needs to be extended to lower energies and to be refined (the errors associated with the measurements of Figure 2 are about 50%). Secondly, experiments of the drift-tube variety should be performed at room temperature to directly measure the rate constant for (1). Finally, it is obvious that the $^3\Sigma$, $^3\Pi$ and $^1\Pi$ intermolecular potentials for $O^- + H$ should be carefully reexamined and those for $H^- + O$ calculated in order to assess their role in detachment (and charge transfer) at the higher collision energies.

2. $SF_6^- + SF_6$

2.1 Introduction

An important role for a molecule used as a gaseous insulator or dielectric is to attach and "hold-on" to any free electrons which might otherwise be instrumental in initiating a discharge process. Sulphur hexafluoride is an attractive dielectric gas specifically because of its large cross sections for direct and associative attachment of electrons; these properties have been well-documented[9]. Less well understood are the mechanisms whereby a molecular anion such as SF_6^- decomposes in subsequent collisions with SF_6. There have been several studies[10] which provide indirect evidence suggesting that collisional detachment of SF_6^- is, in fact, a relatively unimportant process when compared to its collisional decomposition into $SF_5 + F^-$.

The purpose of this discussion is to summarize the results of recent direct measurements in which absolute cross sections for collisional decomposition (and charge transfer) of SF_6^- have been determined.[11] An energy level diagram indicating the energetic thresholds for collisionally decomposing ground state SF_6^- into:

$$\begin{aligned} SF_6^- + X \rightarrow\ & SF_6 + X + e && \text{(a)} \\ & SF_5 + F^- + X && \text{(b)} \\ & SF_5^- + F + X && \text{(c)} \end{aligned} \qquad (3)$$

is given in Figure 3.

The experimental quantity to be measured is the cross section, $\sigma_i(E)$, where E is the relative collision energy and i stands for some inelastic channel. For those applications where the anion speed can be described by a drift velocity, v_d, "reaction coefficients," κ_i, can be determined from the cross sections by

$$\frac{\kappa_i}{N} = \frac{1}{m v_d} \int_0^{\infty} f_i(E_{lab})\, \sigma_i(E_{lab})\, dE_{lab} \qquad (4)$$

where E_{lab} is the projectile energy in the lab frame, N the number density, m the projectile mass, and $f(E_{lab})$ is the kinetic energy distribution function descriptive of the system. It is the reaction coefficients given by (4) that can be compared to model calculations based upon drift-tube data.

2.2 Experimental Method

The cross sections for (3) have been determined with an apparatus which employs an ion beam, gas target configuration and by separating or trapping the products from different channels with the use of electric and/-or magnetic fields.[11] The separation of the charged products is possible because of either the large mass difference (e.g., e^- vs. F^-) or the substantial difference in the laboratory kinetic energies of the dissociation products. Such experiments are fairly straightforward <u>except</u> when (i) low collision energies --such as those relevant to gaseous dielectrics-- are involved or (ii) the projectile is a complex molecular anion, such as SF_6^-, which can be formed in any number of excited vibro-rotational states. It is useful to comment on each problem briefly.

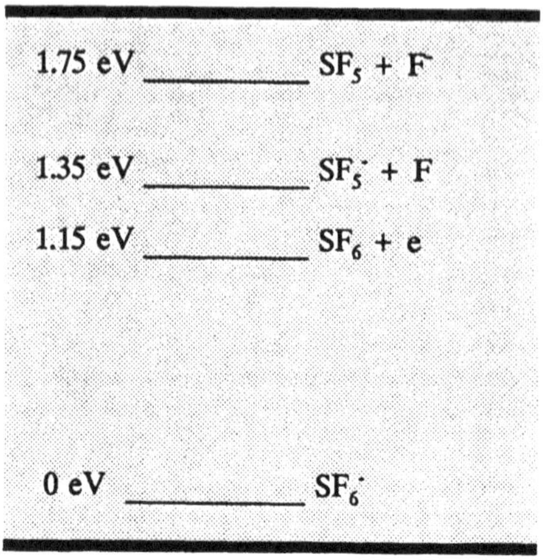

Figure 3. Energy levels for SF_6^- and some of its products.

In ion beam experiments, laboratory collision energies below about 25 eV are quite difficult to achieve owing to space charge limitations and to stray electric fields within the beam optics system. These effects result in a rapidly decreasing primary ion beam intensity as E_{lab} decreases, as well as to increased uncertainties in the trapping efficiencies for the different products and finally to a concomitant reduction in the signal-to-noise ratio in the cross section measurement. One way to attempt to circumvent this problem in the case of SF_6^- is to examine its collisional dynamics for the case where the target is not SF_6, but rather is a species with a small mass, say helium. In this case the relative collision energy is only 4/150 of the laboratory energy, allowing one to sample extremely low relative collision energies. Although helium is clearly not too similar to SF_6, it is possible to gain insight into the collisional dynamics of SF_6^- for low relative collision energies by examining the reactants SF_6^- + He.

The second problem specifically related to SF_6^- is the lack of information concerning the distribution of internal energy in the primary ion beam. Any SF_6^- ion formed by electron attachment must be relaxed collisionally or by a radiative transition if it is to survive. In high pressure environments, collisional relaxation insures that most SF_6^- ions stabilize quickly; this is not the case for pressures typical to ion sources however. Consequently the SF_6^- ion beam has a wide, almost continuous, distribution of internal energies, including those leading to autodetachment. It is with this ill-defined primary ion beam that the current ion-beam, gas target experiments have been performed. The next generation of experiments hopefully will be designed such that the internal energy of the ion beam is better defined than is presently the case.

2.3 Results

The cross sections for electron detachment for $SF_6^- + SF_6$ are given in Figure 4 along with the results for several rare-gas targets. The detachment of SF_6^- by SF_6 is seen to be very small for $E < 60$ eV. In fact the measurement is consistent with the detachment cross section being zero to within the experimental uncertainty in the measurement, namely $\pm 0.1Å^2$.

When we look at the results for the electron detachment of SF_6^- by He, as also presented in Figure 4, we immediately see that the energetic threshold for detachment is below the adiabatic EA of SF_6 ($\sim$1.1 eV). Electron detachment is observed to occur for relative collision energies below the EA, down to as low as 0.2 eV. It is obvious that excited states of SF_6^- are in the primary ion beam.

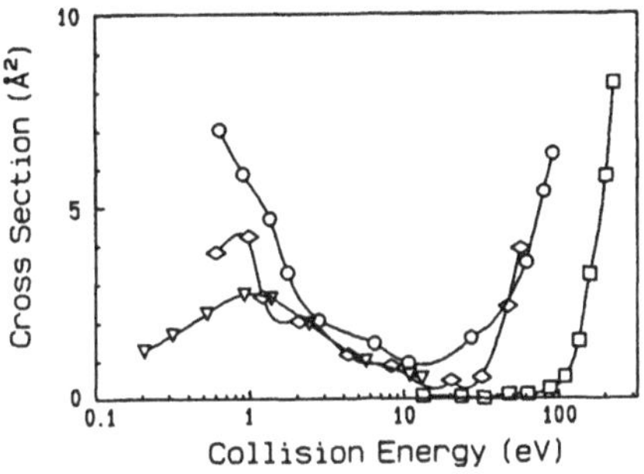

Figure 4. Detachment cross sections for collisions of $SF_6^- + SF_6$(squares), He(triangles), Ne(diamonds), and Argon-(circles).

In contrast to electron detachment, the cross sections for collisional decomposition of SF_6^- into F^- or SF_5^- are both observed to be quite large. For example, the ratios $\sigma_a:\sigma_b:\sigma_c$ for E = 10 eV are $\leq 0.1Å^2:22Å^2:15Å^2$, respectively.

The question arises: why, even with excited SF_6^- ions in the beam, is electron detachment by SF_6 so insignificant for $E \leq 60$ eV? The answer may lie in a statistical argument related to a model of unimolecular decomposition described by Klots[12]. In such a model a collisionally excited molecule subsequently decays, independently of how it was excited, with a decay rate governed by the phase space available to the products. For the product channels given by (3), the two-step process may be represented by:

$$SF_6^- + X \rightarrow SF_6^{-*}(U,J_0) + X\,, \qquad (5)$$

$$\begin{array}{lr} \rightarrow e^- + SF_6 & (a) \\ \rightarrow F^- + SF_5 & (b) \\ \rightarrow SF_5^- + F & (c) \end{array}$$

where U is the total internal energy of the excited SF_6^- product which is partitioned vibro-rotationally and J_0 is the rotational quantum number. The rates for decomposition are (for i = a, b, or c) given by

$$k_i\,(U,J_0) \sim \frac{\gamma_i}{(2J_0+1)\,\rho^0(\epsilon)} \int_0^{U-U_{0i}} \rho^i(\epsilon) \sum_{J=J_{min}}^{J_{max}} \sum_{L=L_{min}}^{L_{max}} (2J+1)\, d\epsilon \qquad (6)$$

where $\gamma_b = \gamma_c = 6\gamma_a$ are symmetry numbers, ϵ is the vibrational energy, $\rho(\epsilon)$ is the vibrational density of states; $\rho^0(\epsilon)$ refers to $SF_6^{-*}(U,J_0)$, and the product density-of-states functions, $\rho^i(\epsilon)$, are for SF_6, SF_5 and SF_5^- for i = a, b, and c respectively. U_{0i} are the decomposition energies given in Fig. 3. The double sum in (6) basically counts the ways the original angular momentum can be partitioned among the products and may be approximated to be $\sim J_0^3$ for computing k_b and k_c. The double sum is simply $2J_0 + 1$ for k_a, however, since s-electrons dominate the detachment mechanism. This factor alone is sufficient to account for the relatively small decomposition rate observed for detachment. At this point the harmonic frequencies (and possibly even the symmetries) of SF_6^-, SF_5 and SF_5^- are not known; consequently the various functions, $\rho(\epsilon)$, which describe their vibrational state densities cannot be determined, and further detailed use of (6) is unwarranted at this time. Moreover, there is no clear way to relate the collision energy, E, to U and J_0.

In summary, as is the case for the reactants O^- + H discussed previously, a reasonably complete understanding of the collisional dynamics for this complicated many-body system is not yet available. The experimental results suggest that a two-step model in which the collisionally-excited molecular anion decays according to statistical model may be an appropriate model to describe the dynamics. Unfortunately, not enough is yet known about the structure of the molecules to proceed further with such a model. From an experimental point of view, one needs to focus on low collision energies and on producing a well-characterized ion beam with which to do the experiment. The former need (accurate cross sections at low collision energies) is especially crucial if one is to calculate reaction coefficients such as those given by (4). This is due to the fact that $f(E_{lab})$ generally decreases extremely rapidly for $E_{lab} \gtrsim$ few eV for applications related to discharge phenomena. While the latter goal of making a well-characterized beam of SF_6^- is probably unattainable, it may be possible to at least make a "cool" primary ion beam in which the mean internal energy is low.

3. Acknowledgements

The work discussed herein was supported in part by the U.S. Department of Energy, Basic Energies Sciences, Atomic Physics Program. Those participating in the experiments are Y. Wang, M. Huels and L. Doverspike of W&M and R. J. Van Brunt and J. Olthoff at NIST.

4. References

1. A. Dalgarno and R. M. McCray, Astro. Jour. **181** (1973) 95.
2. B. Huron and F. Tran Minh, Astron. and Astrophys. **38** (1975) 165.
3. H. Hotop, T. A. Patterson and W. C. Lineberger, J. Chem. Phys. (1974) 1806.
4. P. K. Acharya, R. A. Kendall and J. Simons, J. Chem. Phys. **83** (1985) 3888.
5. A. Herzenberg, Phys. Rev. **160** (1967) 80.
6. M. A. Huels, R. L. Champion, L. D. Doverspike and Y. Wang, Phys. Rev. **A41** (1990) 4809.
7. M. S. Huq, D. Scott, R. L. Champion and L. D. Doverspike, J. Chem. Phys. **82** (1985) 3118.
8. J. P. Gauyacq, J. Phys. **B15** (1982) 2721.
9. See, e.g., O. J. Orient and A. Chutjian, Phys. Rev. **A34** (1986) 1841 and references cited therein.
10. A. V. Phelps and R. J. Van Brunt, J. Appl. Phys. **64** (1988) 4269.
11. Y. Wang, R. L. Champion, L. D. Doverspike, J. K. Olthoff and R. J. Van Brunt, J. Chem. Phys. **91** (1989) 2254.
12. C. E. Klots, J. Phys. Chem. **75** (1971) 1526; Z. Naturforsch. **27A** (1972) 553; J. Chem. Phys. **64** (1976) 4269; Chem. Phys. Lett. **38** (1976) 61.

TRANSPORT COEFICIENTS OF SF_6 IONS IN THEIR PARENT GAS

J. de Urquijo, I., Alvarez, H. Martínez and C. Cisneros
Instituto de Física, Universidad Nacional Autónoma de México, Apdo. Postal 139-B, 62191 Cuernavaca, Mor. México

ABSTRACT

A drift tube-mass spectrometer with variable drift distance has been used to measure the mobility and the longitudinal diffusion coefficients of SF_3^+, SF_5^+, SF_5^-, and SF_6^- in SF_6 for which the extrapolated zero-field reduced mobilities were found to be 0.651, 0.591, 0.592 and 0.542 $cm^2\ V^{-1}\ s^{-1}$ respectively. A wide E/N range from 15-510 Td was covered. A remarkable similarity between the mobility curves of SF_5^+ and SF_5^- has been observed. Appart from some mass-analysed mobility data for SF_5^- and SF_6^- up to E/N ~ 140 Td from other author, the present transport data are believed to be the first measured with simultaneous mass analysis.

1. Introduction

The mobility and diffusion of ions and electrons through a gas under the influence of an external electrostatic field are basic transport processes in low temperature plasmas. For a quantitative understanding of the behaviour of this discharges, knowledge of these transport properties is essential. Among the gases of present or potential use as insulators, sulphur hexafluoride has been widely studied because of its high dielectric strength, which stems for its high electron capture cross section for low energy electrons. Considerable progress has been gained in the determination of electron swarm data, both in theory and experiment (see Refs. 1-3 and references therein). In contrast, mass-analysed transport data for the various ionic species formed in SF_6 under discharge conditions have been scarce. In fact, prior to the present research, the only mass-analysed transport data available were those of Patterson[4] for four negative ion species over a relatively narrow range of the paramenter E/N, from 5-140 Td, where E is the electric field intensity, N is gas number density, and 1 Townsend (Td) = 10^{-17} V cm^2. All other research work was carried out in apparatus lacking mass analysis, thus leading the authors to compare their measured mobilities with those of Patterson. Reviews on this subject have been given by Brand and Jungblut[5], Morrow[6], and de Urquijo et al[7]. The research presented in this paper is part of a long term project dealing with an investigation into ion transport phenomena in electronegative gases.

2. Apparatus and measurement technique

The drift tube-mass spectromenter used in these experiments has been described in detail previously[8]. Hence only a brief description will be presented here. The ions were produced by an axial molecular beam ioniser consisting of a hot filament, an ionising region and a set of electrostatic lenses to focus the beam into a 0.4 mm diameter orifice at the centre of a thin disk of 0.1 mm thickness into a thermalisation region of 3.1 cm length wherefrom they were pulsed by a set of closely spaced mesh grids into the drift space. The ion source position could be varied up to a distance of 38.72 cm from the exit orifice connecting the drift space with the mass analysis region. The mass spectrometer was tuned to resolve the peaks of the selected ion species with a resolution better than 1:200. The accuracy in the setting of E/N was estimated to be less than about ± 1%

SF_6 gas of 99.9% minimum purity was admitted into the drift chamber without further purification. The drift velocity v_d, and the longitudinal diffusion coefficient D_L, were derived from the analysis of arrival time spectra taken at an average of four source positions under the same E/N and N conditions. This latter technique permits the elimination to a large extent of the so called end effects, and of the time involved in the passage of the selected ion species through the mass spectrometer. Gas pressures were varied between 3.3 and 15.8 Pa, at room temperatures in the range 296-302 K.

3. Transport coefficients

Values of the drift velocity (v_d) and of the longitudinal diffusion coefficient (D_L) were obtained for SF_3^+, SF_5^+, SF_5^- and SF_6^- in SF_6 over the combined E/N range 15-510 Td. It is common practice to report reduced mobilities K_0, instead of drift velocities, the former being defined as $K_0=(v_d/N_0)/(E/N)$, where $N_0 = 2.69 \times 10^{19}$ cm^{-3} is the gas number density at STP. No particular pressure dependence of either K_0 or of the product ND_L was found; the reported data are then the mean values. Table 1 displays a summary of the measurement ranges and typical overall uncertainties of the data.

TABLE 1

Uncertainties in the measurement of K_0 and ND_L

Ion	E/N (Td)	Uncertainties (%) K_0	ND_L
SF_3^+, SF_5^+	<210	1.7	
SF_3^+, SF_5^+	>210	2.6	
SF_3^+, SF_5^+	<180		10 - 15
SF_3^+, SF_5^+	>180		20 - 25
SF_5^-	<230	2.2	9 - 16
SF_6^-	<230	1.5	9 - 16
SF_5^-, SF_6^-	>230	2.5 - 3	20 - 25

3.1 *Mobilities*

Present mobilities of SF_3^+, SF_5^+, SF_5^- and SF_6^- are shown plotted together in figure 1 as a function of E/N. All sets of mobility data show a clear trend towards attaining a constant value for E/N less than about 100 Td. An extrapolation to zero field would then give values of the zero-field reduced mobility K_{00}, which are given in Table 2. In connection with this, it must be noted that K_{00} values for SF_5^- and SF_6^- from the present research are in excelent agreement with those found by Patterson.[4] On the other hand, we know of no previous mass-analysed data for either SF_3^+ of SF_5^+ with which to compare the present ones. Comparison with the so called Langevin polarisation limit formula for the reduced mobility[11]

$$K_{00} = 13.853/(\alpha_d\ \mu)^{1/2} \qquad (cm^2V^{-1}s^{-1}) \qquad (1)$$

shows that the predicted values are higher by 10-20% than the measured ones. In this equation, α_d is the dipole polarisability of the nuetral in units of $Å^3$, and μ is the reduced mass of the colliding partners in units of g mol^{-1}. This discrepancy is expected since the above expression is strictly applicable to monoatomic ion-neutral systems under the influence of a purely point-charge induced dipole interaction potential, and at temperatures well below 300 K.

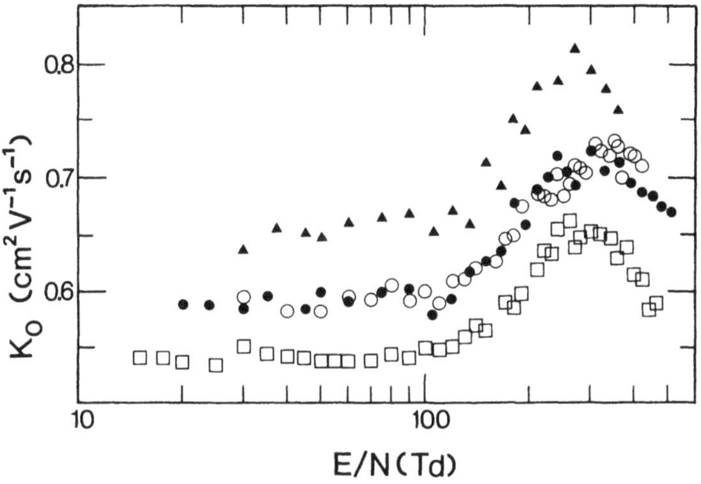

Fig. 1 Reduced mobilities of SF_3^+ (triangles), SF_5^+ (closed circles), SF_5^- (open circles) and SF_6^- (squares) in SF_6.

Notwithstanding the above discrepancy, it is interesting to note that equation (1) predicts that the product $K_{00}\mu^{1/2}$ is a constant. Table 2 also shows this product for the four ion species studied, using the present K_{00} values. It is seen that this product is remarkably constant for SF_3^+, SF_5^+, SF_5^-, where the possibility of resonant charge transfer can be ruled out. However, for the case of SF_6^-, the calculated product is slightly lower by about 4%. Closer agreement to the experimental K_{00} values has been found by Patterson[4,9], who assumed an (n,6,4) interaction potential with n = 12 and by

Brand and Jungblut[5], with n = 9 . Also, a (12,4) core model[10,11], could also explain these lower mobilities.

Referring to figure 1 again, the present data show clear maxima in the neighborhood of E/N ~ 280-300 Td, followed by a decrease of the mobility with E/N. This behaviour is indicative of the transition from the collisional regime dominated by the attractive part of the interaction potential into that in which the repulsive part of the potential becomes predominant at higher energies of relative motion. It is to be noticed that the reduced mobility curves of SF_5^+ and SF_5^- are remarkably similar up to E/N ~ 280 Td.

To the extent of accuracy of the present values, this fact would suggest that the mobility for these ion species is rather insensitive to ion structure, and that the collision process would be dominated by the same interaction potential. However, above E/N ~ 280 Td, the mobility values for SF_5^- are systematically higher than those of SF_5^+ by about 2-3.5%. A comparison and discussion of the present data with those previously reported by other authors can be found elsewhere[8,12].

TABLE 2

Summary of some experimental data

Ion	K_{00} (exp) (cm^2/V s)	$\mu^{1/2} K_{00}$ (exp) $(g/mol)^{1/2}$(cm^2/V s)	ND_L (0) (10^{17}/cm s)
SF_3^+	0.651±0.009	48.4±0.1	4.53
SF_5^+	0.591±0.006	48.7±0.1	4.11
SF_5^-	0.592±0.008	48.8±0.1	4.12
SF_6^-	0.542±0.007	46.3±0.1	3.77

3.2 *Longitudinal diffusion coefficients*

Present values of the mass-analysed, density normalised longitudinal diffusion coefficients, ND_L, for SF_3^+, SF_5^+, SF_5^- and SF_6^- in SF_6 are shown plotted in figure 2 as a function of E/N. Overall uncertainties for these data are also displayed in Table 1. In spite of the relatively large scatter in these data, the trend of the curves at low E/N suggests that they

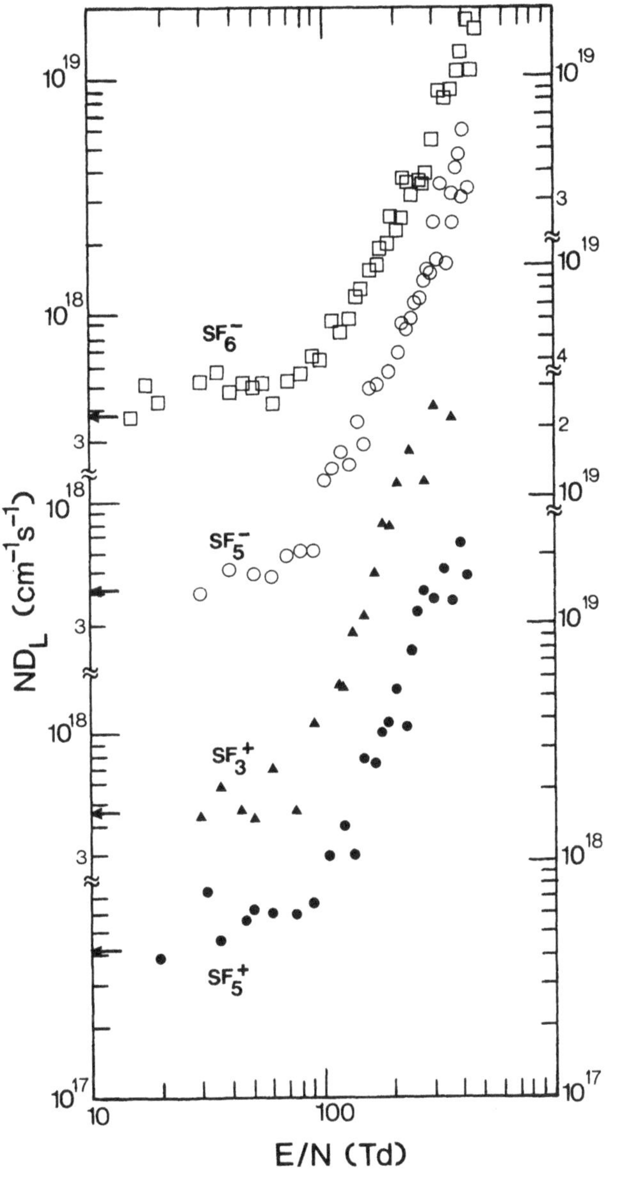

Fig. 2 Density normalized longitudinal diffusion coefficients of SF_5^+, SF_3^+, SF_5^- and SF_6^- in SF_6.

appear to tend to a constant value. Indeed, at vanishingly small fields, Einstein's relation predicts a value of

$$ND_L\ (0) = N_0K_{00}(kT/e) \qquad (2)$$

where k is the Boltzmann's constant, e is the electron charge and T , the gas temperature. Using the present K_{00} values, ND_L (0) is shown by small arrows on the left hand axis of figure 2, and is also given in Table 2.

4. Conclusions

A drift tube-mass spectrometer with a variable drift distance has been used to measure the mobility and longitudinal diffusion coefficients of SF_3^+, SF_5^+, SF_5^- and SF_6^- in SF_6 over the combined E/N range from 15-510 Td. Appart from the K_{00} data from Patterson[4] for SF_5^- and SF_6^- up to E/N ~ 140 Td, the present transport coefficients are believed to be the first measured with simultaneous mass analysis. The mobility curves for SF_5^- and SF_5^+ have been found to be remarkably similar. The present transport data could be valuable for the modelling of gas discharge process, and be useful to probe the interaction potential between these ions and the parent gas molecules.

Acknowledgements

The authors wish to thank Mr. S. Pérez for his technical assistance. This work was supported by DGAPA grant number IN-104889

References

1. T. H. Teich, in "*Electron and Ion Swarms*" (L.G. Christophorou, Ed) (Pergamon, New York, 1981) p.241
2. A. V. Phelps and R. J. Van Brunt, J. Appl. Phys. **64**, 4269 (1988)
3. H. Itoh, Y. Miura, N. Ikuta, Y. Nakao, and H. Tagashira, J. Phys. D: Appl. Phys. **21**, 922 (1988)
4. P. L. Patterson, J. Chem. Phys. **53**, 696 (1970)
5. K. P. Brand and H. Jungblut, J. Chem. Phys. **78**, 1999 (1983)
6. R. Morrow, IEEE Trans. Plasma Sci. **PS-14**, 234 (1986)
7. J. de Urquijo, I. Alvarez, C. Cisneros and H. Martínez, in "*Nonequilibrium Effects in Ion and Electron Transport*"

(J. W. Gallagher et al, Eds.) (Plenum, New York, 1990) p.211-27

8. J. de Urquijo, I. Alvarez, C. Cisneros and H. Martínez, J. Phys. D: Appl. Phys. **23**, 778 (1990)
9. P. L. Patterson, J. Chem. Phys. **56**, 3943 (1972)
10. E. A. Mason, H. O'Hara and F. J. Smith, J. Phys. D: Appl. Phys. **5**, 169 (1972)
11. E. A. Mason and E. W. McDaniel, "*Transport Properties of Ions in Gases*", (Wiley, New York, 1988)
12. J. de Urquijo, I. Alvarez, H. Martínez and C. Cisneros, J. Phys. D: Appl. Phys. In press.

On the Role of "Scarred" Wavefunctions in Photoionization and in Atomic and Molecular Collisions

R.V. Jensen
Department of Physics
Wesleyan University
Middletown, CT 06457

Abstract

The classical dynamics of strongly perturbed or strongly coupled quantum systems, like atoms in intense electromagnetic fields or colliding atoms and molecules, can exhibit chaotic behavior. This microscopic mechanism for ergodic behavior can be used to motivate simplified, statistical descriptions of atomic and molecular processes. However, careful studies of the quantum wavefunctions of a wide variety of model and realistic systems that exhibit classical chaos, have shown that the wavefunctions do not always reflect the ergodic properties of the classical trajectories. In fact, the wavefunctions often exhibit an extraordinary amount of structure that appears to "scar" short, unstable classical periodic orbits. The experimental consequences of these "scarred" wavefunctions in the spectrum of atoms in strong magnetic fields, in the excitation and ionization of atoms in intense electromagnetic fields, and in atom-molecule collisions is discussed.

1. Introduction

In the last 10 years great progress has been made in our understanding of the quantum behavior of classically chaotic systems – the so-called problem of "quantum chaos"[1]. Experimental and theoretical studies of highly excited hydrogen atoms in strong magnetic fields[2,3] and intense microwave fields[4,5] have revealed a number of new physical phenomena in the quantum description of strongly coupled and strongly perturbed nonlinear oscillators that exhibit chaotic behavior in the classical limit. For example, detailed investigations of the electron dynamics in a hydrogen atom exposed to stong electromagnetic fields have shown how the linear quantum dynamics can *mimic* the nonlinear classical chaos for limited times[4,5] and how quantum interference effects can "localize" the wavefunction and suppress the irregular, chaotic motion for longer times[6,7]. In addition, measurements and calculations of the spectra of hydrogen atoms in strong static magnetic fields[2,3] have revealed interesting relationships between the statistics of the irregular spectrum and the predictions of the Random Matrix Theory of nuclear physics.[8]

However, one of the most exciting developments has been the discovery that even though quantum interference effects may suppress the chaotic components of the classical motion, the quantum wavefunctions continue to be highly correlated with unstable classical periodic orbits. Even though the classical description of the electron motion in a highly excited hydrogen atom in an intense microwave field or in a strong magnetic field may be fully chaotic (meaning ergodic and mixing), the quantum wavefunctions nevertheless appear to cling to the remnants of stability represented by the unstable periodic orbits.

These highly structured wavefunctions were called "scars" of the unstable periodic orbits by Heller[9] who first observed them in detailed numerical calculations of the eigenfunctions of a simple model system that is classically chaotic – the frictionless motion of a particle, "billiard ball", moving on a 2-dimensional billiard table shaped like a racetrack or stadium.

Figure 1a shows a typical classical trajectory of a free particle undergoing specular reflection on the walls of *stadium billiard.* The classical motion in this bounded domain (consisting of two semi-circles separated by straight sections) exhibits the irregular, ergodic motion with extreme sensitivity to initial conditions that defines classical chaos. In fact, the classical stadium billiard is one of the few dynamical systems that has been rigorously proven to meet the mathematical requirements of the definition of chaos for almost all initial conditions. Of course the perceptive reader will recognize that for many initial conditions the motion will still be periodic - for example trajectories that bounce vertically between the straight sides or the orbit that bounces horizontally between the centers of the two semi-circles. There are an infinite number of these periodic orbits; however, they represent a small "set of measure zero" of all possible initial conditions and they are all highly unstable. It is often difficult to find these periodic orbits using numerical simulations because they are so unstable and sensitive to the initial conditions.

In stark contrast, Figure 1b shows a contour map of the probability $|\psi|^2(x,y)$ of finding an electron for a highly excited ($\sim$ the 10,000th) eigenstate of the stadium billiard. Although the classical trajectories appear to ergodically cover the stadium domain with uniform probability density, the quantum probability of finding the electron, prepared in this highly excited state (and in many other highly excited states), is localized along ridge-like structures which Fig.1c shows are highly correlated with the path of the "double-diamond", unstable classical periodic orbit.

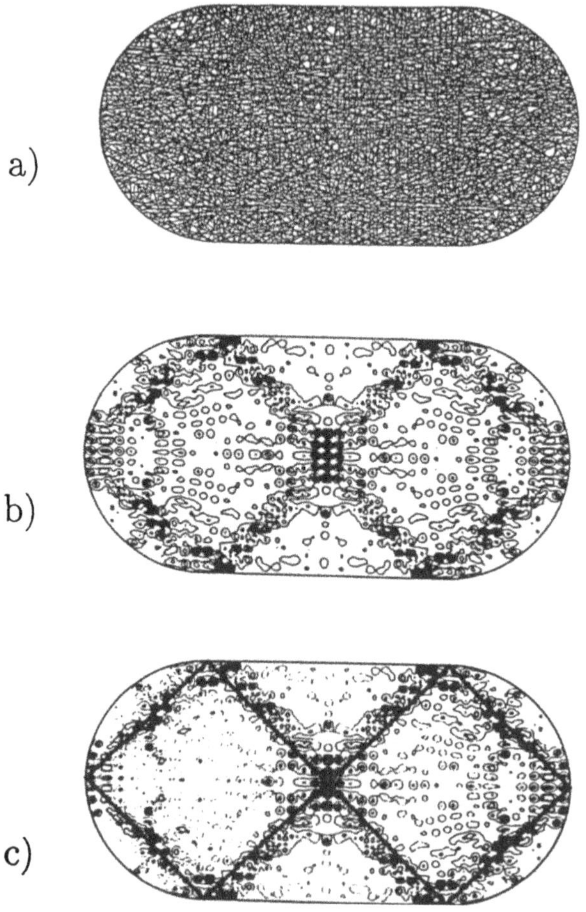

Fig. 1: a) Approximately 10,000 bounces of a typical classical trajectory in the stadium billiard. b) A topographic map of the electron probability for a highly excited eigenstate of the stadium billiard. c) A comparison of the structure of this eigenstate with the path of the unstable, periodic, "double-diamond" orbit. (Figure 1a) was provided by E. Heller and Figs 1b) and 1c) are reprinted from Ref. 9.)

The localized ridges of this quantum eigenstate are said to exhibit a "scar" of the corresponding unstable periodic orbit. Heller's original observations of the highly excited eigenfunctions of the stadium billiard indicated that this scarring of individual eigenfunctions was almost typical behavior in the sense that as many as 50 % of the numerically computed eigenfunctions apear to exhibit distinct scars of one or more short, unstable periodic orbits like the one shown in Fig. 1. Moreover, each short unstable periodic orbits appears to be responsible for a family of scarred eigenstates. (Consequently, there is no one-to-one relationship between periodic orbits and eigenfunctions.)

To clearly demonstrate that these scarred wavefunctions are not unique to the stadium billiard, I have also used Heller's computer codes to calculate the highly excited eigenstates of the *Sinai billiard* which is composed of a square table with a circular scatterer in the center. The classical dynamics of a free particle scattering between the the square walls and the circular barrier has also been rigorously proven to be ergodic and mixing. However, many of the highly excited wavefunctions still exhibit distinct scars of unstable periodic orbits. Figure 2 shows two examples of these scarred wavefunctions displayed in one quarter of the Sinai Billiard.

The existence of these "scarred" wavefunctions has profound experimental implications because the individual wavefunctions of classically chaotic systems can be much more regular than the typical classical trajectories. As a consequence, statistical predictions of the evolution of observables based on the ergodic hypothesis may fail for the quantum system. More simply put, the quantum wavefunctions may not go where the ergodic classical orbits go. Moreover, since these quantum billiards can also be viewed as a simple models for a drum head, or an electromagnetic cavity, or a concert hall, or a compound nucleus, or even a quark bag-model for a nucleon, this peculiar property of the wave mechanical description of a system with chaotic ray dynamics should have important applications to many diverse fields of physics and engineering.

The purpose of this paper is to provoke and stimulate the imaginations of atomic and molecular physicists by providing a "picture book" of some of the scarred wavefunctions that have recently been discovered in strongly coupled and strongly perturbed quantum systems, such as Rydberg atoms in strong magnetic fields, or highly excited hydrogen atoms in intense microwave fields or ground state atoms in super-intense, high-frequency laser fields. In addition, I will briefly examine the possible role of scarred wavefunctions in atom-molecule scattering systems which correspond to open billiard or "pinball" systems created by opening holes in the sides of chaotic billiards like the stadium or Sinai billiard.

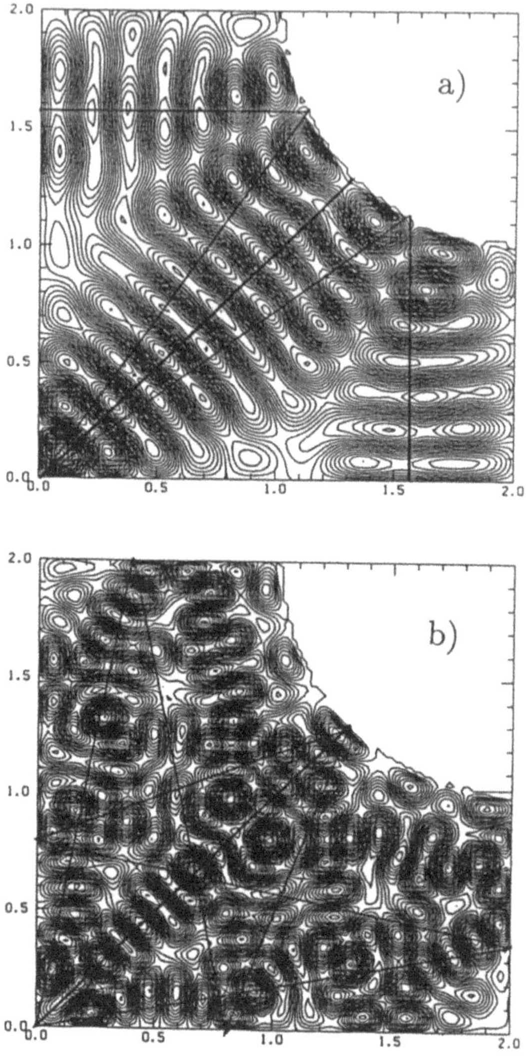

Fig. 2: Two examples of "scarred" eigenfunctions in one quarter of the Sinai Billiard calculated using a modification of Heller's computer code. Some of the short, unstable classical periodic orbits are indicated by the heavy solid lines.

2. Applications of Scars in "Bound" Systems

2a. Rydberg Atoms in Strong Magnetic Fields

The classical description of the electron motion in a highly excited atom placed in a strong magnetic field exhibits a transition from mostly regular motion to chaos when the Lorentz force on the moving electron becomes comparable the the Coulomb binding.[2,3] Since the classical theory is only an approximation to the true quantum mechanical description of this simple physical system, the Rydberg atom in strong magnetic field provides an excellent system for theoretical and experimental studies of "quantum chaos". Unfortunately, the quantum electron dynamics in this bound system cannot be observed directly. The only accessible observables are the energy spectrum and the associated transition matrix elements. Nevertheless, the experimental and theoretical measurements of these observable quantities in the classically chaotic regime have revealed at rich structure that has been related to the properties of short, often unstable, periodic orbits in this strongly perturbed Coulomb problem.[2,3,11]

Specifically, the photoabsorption spectrum of hydrogen atoms[11] and lithium atoms[12], measured by first exciting the atom to the $n = 2$ state and then sweeping the frequency of a laser and measuring the probability of exciting Rydberg states, was found to exhibit a distinct oscillatory structure as a function of energy. When the photoabsorption spectrum was Fourier transformed with respect to energy, these oscillations were found to be directly associated with the periods of classical periodic orbits.[11] Although the original Garton-Tompkins oscillations[13], corresponding to a period of 2/3 times the free electron cyclotron period, were identified with a periodic orbit that is usually stable, many other quasi-Landau resonances have been subsequently correlated with unstable periodic orbits.[2,3,11]

By analogy with the scarred wavefunctions of chaotic billiards, these regular oscillations in the photoabsorption spectrum may be associated with the family of "scarred" wavefunctions associated with short periodic orbits. However, because of the enormous difficulty in numerically solving the Schrödinger equation for the highly excited, 2-dimensional states of the hydrogen atom in a strong magnetic field, this appealing hypothesis was only recently confirmed.[14] For example, Fig. 3 shows a topographic map of the electron probability for a highly excited eigenstates of a hydrogen atom in a strong magnetic field taken from Ref. 14, that shows distinct ridges along the paths of unstable, classical periodic orbits.

Even more recently the the high-precision experimental measurements of the photoabsorption spectrum by Kleppner's group[12] have revealed regular families of extremely longed lived states well up in the continuum (with $E > 0$) for this system. If these resonance states can also be identified with scarred wavefunctions, then these results would provide clear evidence that the phenomenon of quantum scarring of classically chaotic systems extends to open systems as well (which is the subject of some more speculative remarks in Section 3). The formidible theoretical analysis of these latest experiments is currently an active area of research.[12]

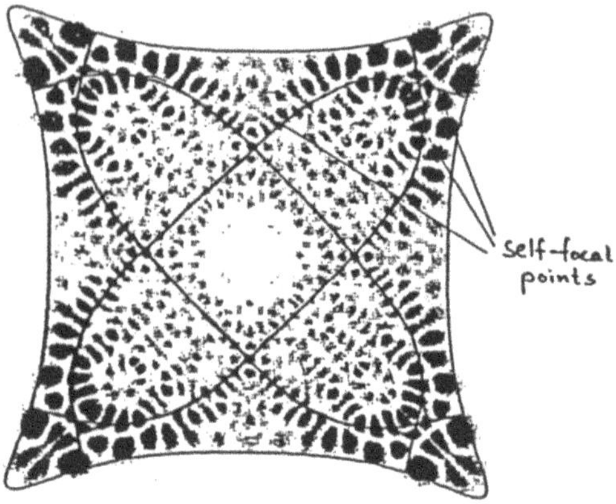

Fig. 3: A topographic map of the electron probability for a highly excited eigenstate of a hydrogen atom in a strong magnetic field in semi-parabolic coordinates is compared with the paths of some short, unstable classical periodic orbits. (Reprinted from Ref. 14.)

2b. Suppression of Microwave Ionization due to Scars of Unstable Periodic Orbits

The classical description of the electron motion in a highly excited hydrogen atom exposed to a strong microwave field also exhibits a transition form regular behavior to chaos when the oscillating electric field becomes comparable to the Coulomb field.[15] In this case the manifestation of the classical chaos in the quantum system is unambiguous – the transition to widespread chaotic motion leads to ionization of the atom. Experimental studies[4] of the microwave ionization of Rydberg atoms over the past 17 years have also revealed many new physical phenomena in this strongly perturbed quantum system. Specifically, the range of experimental parameters appear to be divided into a number of distinct regimes depending on the ratio of the microwave frequency, Ω, to the Kepler orbital frequency, $1/n^3$, which is also approximately equal to the $\Delta n = 1$ quantum transition frequency for large quantum numbers.

The wide range of different behaviors is exhibited in Fig. 4, which compares the experimental measurements[4] of the scaled threshold fields, $F_0 = n^4 F$, for the onset of microwave ionization with the classical and quantum mechanical predictions[7] for a wide rande of scaled frequencies, $\Omega_0 = n^3\Omega$. For low scaled frequencies, $\Omega_0 \ll 1$, both the experimental and theoretical thresholds are close to the static field threshold, $n^4 F \simeq .13$, to rip the electron away from the nucleus. However, as the scaled frequency is increased the threshold fields for the onset of ionization decrease significantly below the static field threshold. For $\Omega_0 \leq 1$ the experimental thresholds are well described by the classical theory of the onset of chaotic ionization including the distinct fluctuations in the ionization thresholds near low-order rational values of Ω_0.

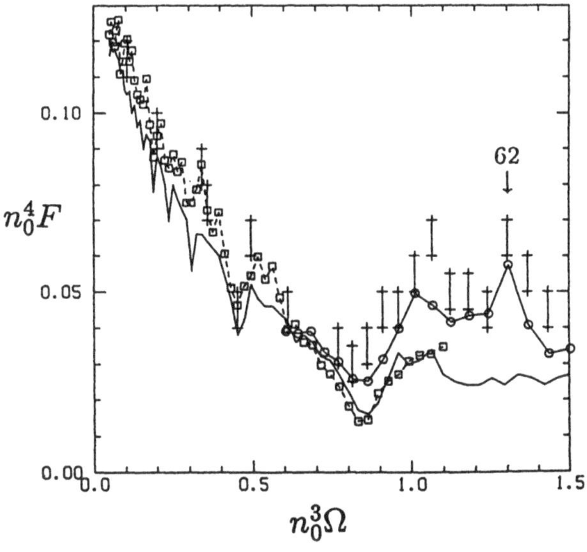

Fig. 4: The experimental measurments[4] of the scaled threshold fields, $n_0^4 F$, for 10% ionization at 9.9 GHz with $n_0 = 32$ to 90 (squares) and 36 GHz with $n_0 = 45$ to 80 (squares) are compared with classical[5] (solid curve) and quantum[7] (crosses) predictions for a one-dimensional model of the experiment. (Reprinted from Ref. 7.)

As the the scaled frequency is increased above $\Omega_0 = 1$ the experimental measurements and the quantum calculations begin to rise above the classical predictions. This tendency for quantum dynamics to be more stable than the classical was originally predicted by Casati et al.[6] by appealing to quantum interference effects related to Anderson localization in solid-state physics. However, this theoretical analysis did not anticipate the large fluctuations in the experimental thresholds like that exhibited near $\Omega_0 = 1.3$. Even if quantum interference tends to suppress the chaotic classical dynamics, why, for example, should one initial quantum state, $n_0 = 62$ in the experiment, be so much more stable than the adjacent quantum states, $n_0 = 61$ and 63?

In a recent Physical Review Letter[10] Sanders, Saraceno, Sundaram and I showed that the physical mechanism responsible for this striking fluctuation in the quantum excitation and ionization is the excitation of a "scarred" wavefunction for $n_0 = 62$ that is highly localized to the vicinity of an unstable classical periodic orbit while the wavefunctions for $n_0 = 61$ and $n_0 = 63$ are delocalized by the perturbing field. Specifically, a numerical integration of the time-dependent Schrödinger equation for a 1-dimensional model of the experiment confirmed the relative stability of the $n_0 = 62$ state. Then by projecting the perturbed wavefunctions onto the classical action-angle phase-space, the structure of the quantum wavefunctions could be compared with the classical dynamics.

A topographic map of the phase-space projection (Husimi distribution) of the quantum wavefunction that is predominantly (98%) excited from the $n_0 = 62$ state is superimposed on a Poincaré section of the classical phase space in Fig. 5a. The electron probability is highly localized in a narrow band of excited states near $n_0 = 62$, with peaks near the unstable period orbit at $\theta = 0$ and 1, and forms a distinct ridge along the stable and unstable manifolds indicated by the line of arrows. The classically regular region outlined by the "heart" shaped trajectories surrounding the stable periodic orbit at $\theta = 0.5$ can also support localized wavefuntions[10]; however, the remarkable feature of the scarred wavefunction in Fig. 5a is that it is localized along the "broken separatrix", defined by the unstable fixed point and the stable and unstable manifolds, where the classical dynamics is fully chaotic.

In contrast the phase space projections of the wavefunctions excited for $n_0 = 61$ and 63 are spread out all over this region of classical phase space with significant overlap with highly excited states with $n > 80$. in Fig. 5b. For example, Fig. 5b shows one of the "delocalized" quasi-energy (qe) states that is excited for $n_0 = 61$.[10]

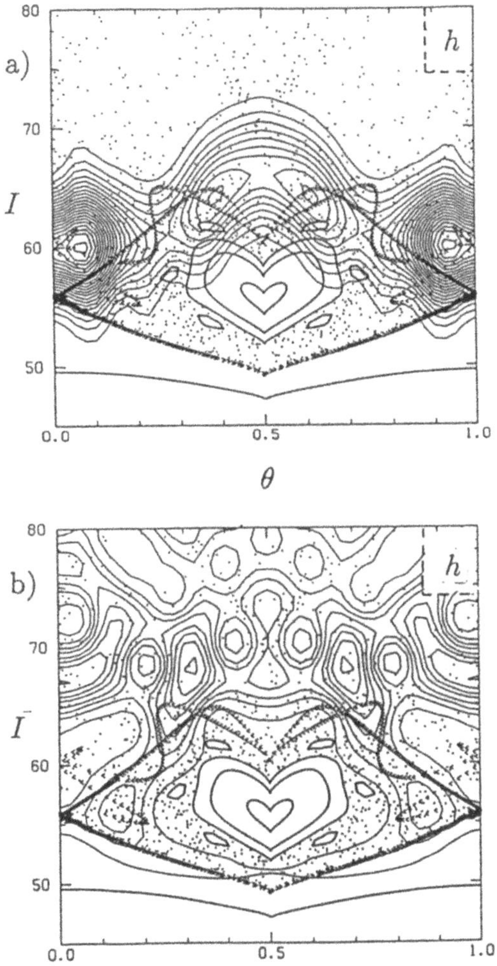

Fig. 5: The level contours (solid curves) of the Husimi distribution in action-angle (I, θ) space is superimposed on a Poincaré section (dots) of the classical dynamics for a) the single quasi-energy state that is dominantly excited for $n_0 = 62$ with $n_0^3\Omega = 1.3$ and $n_0^4 F = 0.05$ and b) one of the qe-states excited for $n_0 = 61$. The lines of arrows show the primary stable and unstable manifolds and their homoclinic oscillations. The relative size of Planck's constant, h, is indicated by the boxed area. (Reprinted from Ref. 7.)

2c. The Role of Scars in the Suppression of Ionization in Intense, High-Frequency, Laser Fields

Recently, Su, Eberly, and Javanainen[16] have performed a numerical study of the quantum behavior of a similar 1-d model with a smoothed Coulomb potential, $-1/\sqrt{1+x^2}$. In very high fields, $n^4F = 5.0$ (a.u.), with frequency $n^3\Omega = .52$ (a.u.), they found that the ionization of the ground state ($n_0 = 1$) is strongly suppressed and that the electron probability exhibits a double peak structure which they associated with the "dichotomous" wavefunction in the double-welled, time-averaged atomic potential studied by Gavrilla et al.[17] However, the peaks of the dichotomous wavefunctions were predicted by this *static*, time-averaged theory[17] to be separated by a distance $2F/\Omega^2$ while Su's peaks are only F/Ω^2 apart.

Our analysis[18] indicates that these multiply peaked wavefunctions result from the *diabatic* excitation of wavefunctions that are localized to the vicinity of unstable and weakly stable periodic orbits embedded in the chaotic classical phase space. These "scarred" wavefunctions provide a new *dynamical* mechanism for the high-field stabilization that accounts for the reduced spacing of the peaks.

In the high-field, high-frequency limit studied by Su et al., the classical motion of the free electron in the external field is only perturbed during the brief times when the electron is reversing its motion in the vicinity of the atomic potential. Consequently, the classical dynamics in the oscillating, Kramers-Henneberger frame of reference is well approximated by a discrete-time, area-preserving map, and the quantum dynamics by the corresoponding quantum map.[18,19] In Fig. 6b we show the phase-space projection (Husimi distribution) of one of the eigenfunctions of the time-evolution operator for one-period of the perturbing field that is peaked on the unstable periodic ($\times$) point at the saddle ($x = -F/\Omega^2 = -18.5$ (a.u)) between the double wells in the time-averaged potential. If this quasi-energy (qe-) state is diabatically excited along with qe-states that are localized near the weakly stable periodic orbits in the each of the double wells near $x = 0$ and $x = -37.0$ (like the one shown in Fig. 6a), then the resulting wavefunction will exhibit three peaks separated by $F/\Omega^2 = 18.5$ as found in Su's calculations.[16]

We are currently extending this analysis[20] to the 3-d hydrogen atom to see if this remarkable suppression of ionization due to the dynamic excitation of highly localized, scarred wavefunctions will persist in experiments with real atoms in intense laser fields.

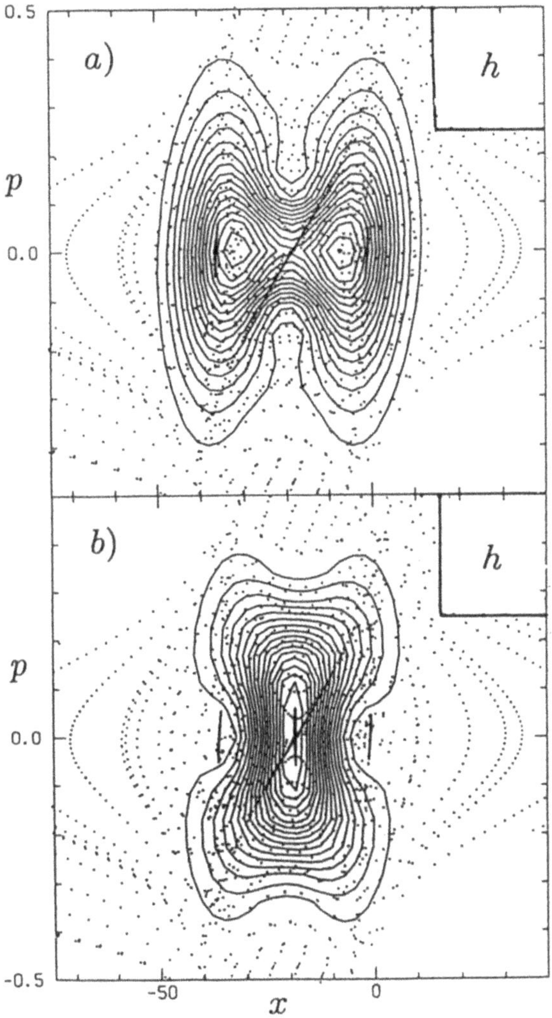

Fig. 6: The level contours of the Husimi distributions[10,18,19] for qe-states that scar a) the weakly stable fixed points at $x \approx 0$ and $\approx 2F/\omega^2 = -37$ and b) the unstable fixed point at $x = F/\omega^2 =$ -18.5 are superimposed on the classical $x - p$ Poincaré section for the map.

3. Applications of Scars in Scattering Problems

The novel results of these investigations of the structure of eigenfunctions in quantum billiards, the spectroscopy of highly excited atoms in strong magnetic fields, the microwave ionization of highly excited hydrogen atoms, and the stabilization of ground state atoms in super-intense, high-frequency laser fields, promise to have many significant applications in two important areas of atomic physics: the spectroscopy of strongly coupled systems and the description of the interaction of intense radiation with matter. However, these initial investigations of "quantum chaos" might appear to have little to say about the third major branch of atomic and molecular physics involving the strong (but transient) coupling in collisional processes.

In fact, the microwave ionization problem can be recast as an example of reactive and "dissociative" scattering.[15] In the experiments[4] of Koch and Bayfield, hydrogen atoms in an atomic beam are carefully prepared in well specified initial states; they enter the "scattering" region defined by the microwave cavity or waveguide; and they emerge in a superposition of excited states and ionized states. If we define an S-matrix for the transition amplitudes coupling the incoming and outgoing channels, then the parameter dependence of this S-matrix will reflect all of the new physical processes revealed by the experimental and theoretical studies of this classically chaotic system, namely channel mixing due to quantum dynamics that mimic classical chaos[4,5], the inhibition of excitation and ionization[6,7] due to "Anderson localization", and the large fluctuations in the quantum transport between channels due to the excitation of "scarred" states[10].

Moreover, the recent experimental measurements by Kleppner's group of the photoabsorbtion of Rydberg atoms in strong magnetic fields[12] have revealed that long lived states persist well up in the continuum. The description of these resonance states, which should also be related to the quantum scarring of unstable periodic orbits in this classically chaotic system, can also be formulated in terms of a scattering problem.

Consequently, the natural extensions of these new results for bound systems promise to open up a whole new frontier for the field of quantum chaos describing quantum dynamics of chaotic classical scattering systems with applications to a wide variety of open systems in atomic and molecular physics and other areas of physics, like nuclear, solid-state, and elementary particle physics where strong collision processes occur. In particular, it will be very interesting to explore the analog of the scarring phenomenon in scattering systems which may be expected to cause striking fluctuations in the quantum transport as measured by the components of the scattering matrix and the differential and total crossections for elastic, inelastic, and reactive scattering.

3a. Quantum Pinball

Consider the fundamental domain displayed in Fig. 2 which corresponds to one quarter of the Sinai billiard. This figure shows the level contours for two eigenstates of the closed Sinai Billiard that appear to exhibit "scars " along unstable periodic orbits. In order to transform this quantum billiard to the new game of "quantum pinball" we simply open the sides of the domain to provide incoming and outgoing channels. Several different possibilities may then be explored using this single model.

For example, if we open two opposite, vertical sides, then an electron entering the scattering region from a " wave-guide" attached to the left hand side will be either reflected by the "funnel" into the propagating modes in the large waveguide or transmitted into one of the propagating modes of the smaller waveguide connected to the narrow side on the right. In this case there are no classical periodic orbits in the scattering region, there is no chaotic scattering, and the exact quantum calculations for the transmission give regular series of steps as a functions of energy that increase by one additional mode when the energy increases enough to allow a new mode to fall below the cut-off of the smaller waveguide. Another possibility is to extend the fundmental domain to the entire lower half of the Sinai billiard shown in Fig. 7. Now if we connect the opposing vertical sides to waveguides this problem corresponds to a simple "constriction". In this case there is a single unstable periodic orbit in the constriction that influences a whole family of classical scattering trajectories that spend more and more time near the periodic orbit. Two such trajectories are shown in Fig. 7.

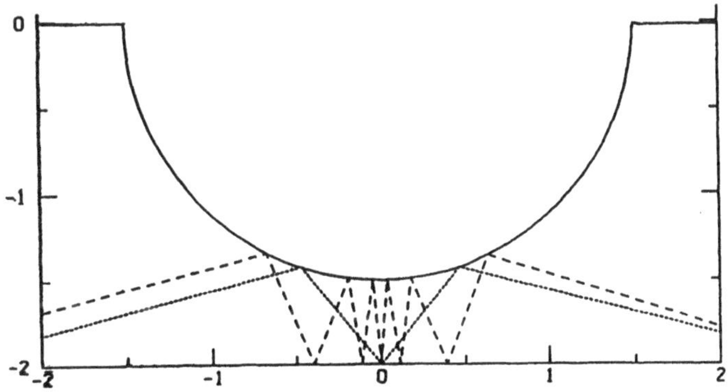

Fig. 7: Two classical trajectories that bounce one or more times along the single, vertical periodic orbit in the "constriction" formed by opening the ends of the lower half of the Sinai Billiard.

Again the scattering is not chaotic and the transmission or conductance of the system exhibits regular steps as a function of energy when new modes can propagate through the narrow region with small, regular oscillations that appear to be associated with the resonance in the constriction. (Although the reflection and transmission at such a junction between "waveguides" is an old problem in microwave engineering, similar problems have also been the focus of much current experimental[21] and theoretical[22] interest in the description of the ballistic transport of electrons in small, mesoscopic ~ 100 Ångström wires.)

However, if we open only the two sides adjacent to the curved section on the top and the right side in Fig. 2 then the classical and quantum dynamics become much more interesting. In this case, which corresponds to a right angle turn in the waveguide or a bend in a mesoscopic wire, there are an infinite number of unstable periodic orbits, the scattering is chaotic, and the exact quantum calculations exhibit irregular fluctuations in the scattering matrix elements and crossections. In this case the unstable classical periodic orbits are expected to have a profound influence on the scattering[23–28]that hopefully can be understood in terms of these orbits, their stability, and the associated resonance wavefunctions that "scar" these orbits like the wavefunction of the bound system displayed in Fig. 2

These studies are currently an active area of theoretical research. In addition, this game of quantum billiards is rapidly gaining popularity in the solid-state community[28,29] where the properties of these scattering systems can be experimentally studied by measuring the conductance fluctuations of different geometric configurations of thin wires as the Fermi energy of the conducting electrons or an applied magnetic field is varied.

3b. Atom-Molecule Collisions

This simple quantum pinball problem is also closely related to a variety of inelastic and reactive processes in atom-molecule collisions, such as $H + H_2$, $F + H_2$, and $H + Cl_2$, that have been extensively studied by physical chemists. In these cases the scattering potentials are smooth with no hard walls but the geometry and the complexity of the chaotic classical trajectories are the same as indicated by the typical scattering potentials displayed in Fig. 8. Moreover, the theoretical chemists have already made the connection between the periodic orbits[30] and scarred resonant states[31] in these systems, however, there appears to be very little clean experimental data on these problems. In fact, the first detailed comparisons of experimental measurements of the $D+H_2$ reactive crossections with theoretical classical, semi-classical, predictions were only recently proclaimed.[32]

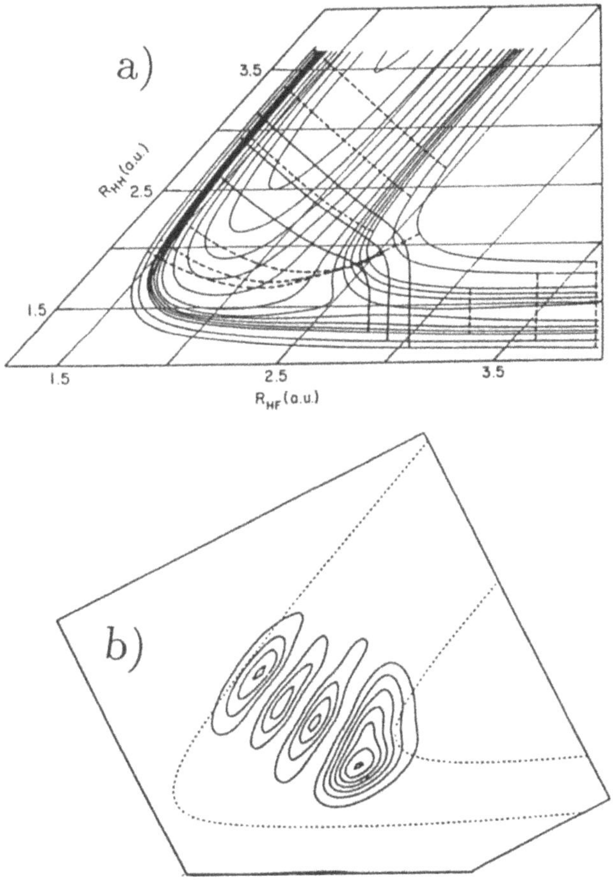

Fig. 8: a) Some of the unstable, classical periodic orbits (indicated by the heavy solid curves are superimposed on the level contours of the 2-dimensional scattering potential for $F + H_2$. (Reprinted from Ref. 30.) b) An example of a quantum mechanical resonance state in the $F + H_2$ potential that correlates well with some of the unstable classical periodic orbits in a). (Reprinted from Ref. 31.)

4. Conclusions

It is intriguing that work of the theoretical chemists may provide the keys to the understanding the remarkable fluctuations in the electron transport and conductance of small mesoscopic wires, while the chemists may have to rely on the solid-state experiments for clean tests of their theories for reactive scattering. This last example provides one more important illustration of the intricate ties between a variety of different fields that arise in the interdisciplinary studies of nonlinear dynamics, chaos, and quantum chaos which are built upon the foundations of new ideas, intuition, and theoretical methods developed in the attempts to understand the behavior of the "simple" hydrogen atom in strong fields.

However, it is important to remember that despite the compelling numerical evidence for the importance of scarred wavefuctuions in strongly coupled or strongly perturbed quantum systems that are classically chaotic, there is not yet a comprehensive theory to explain this remarkable phenomenon. A number of different theoretical approaches have been proposed[33] for the correlation of quantum wavefunctions with classical trajectories using the elegant semi-classical theories of Gutzwiller[34,35] for bound systems and Miller[36,37] for scattering systems. But a detailed theoretical understanding of the general issue of the relationship between the classical, semi-classical, and quantum theory, and the specific issues of of when these localized wavefunction arise and why the scarring of individual wavefunctions is so strong remain open problems.

This work was supported by a National Science Foundation Research Grant (PHY-90139541).

References

1. *Chaos and Quantum Physics*, eds. M.-J. Giannoni, A. Voros, and J. Zinn-Justin, (Elsevier Science Publishers, London, 1990).

2. A. Holle, G. Wiebusch, J. Main, K.H. Welge, G. Zeller, G. Wunner, T. Ertl, and H. Ruder, Z. Phys. D *5*, 279 (1987) and references therein.

3. H. Friedrich and D. Wintgen, Phys. Rep. *183*, 37 (1989); H. Hasegawa, M. Robnik, and G. Wunner, Prog. Theor. Phys. *98*, 198 (1989); D. Delande, in *Chaos and Quantum Physics*, eds. M.-J. Giannoni, A. Voros, and J. Zinn-Justin, (Elsevier Science Publishers, London, 1990).

4. J.E. Bayfield and P.M. Koch, Phys. Rev. Lett. *33*, 258 (1974); K.A.H. van Leeuwen, G. v.Oppen, S. Renwick, J.B. Bowlin, P.M. Koch, R.V. Jensen, O. Rath, D. Richards, and J.G. Leopold, Phys. Rev. Lett. *55*, 2231 (1985); E.J. Galvez, B.E. Sauer, L. Moorman, P.M. Koch, and D. Richards, Phys. Rev. Lett. *61*, 201 (1988).

5. R.V. Jensen, Phys. Rev. A *30*, 386 (1984); R.V. Jensen, S.M. Susskind, M.M. Sanders, Phys. Rep. *201*, 1 (1991) and references therein.

6. G. Casati, B.V. Chirikov, and D.L. Shepelyansky, Phys. Rev. Lett. *53*, 2525 (1984); G. Casati, I. Guarneri, and D.L. Shepelyansky, IEEE Journal of Quantum Electronics, *24*, 1420 (1988) and references therein.

7. R.V. Jensen, S.M. Susskind, and M.M. Sanders, Phys. Rev. Lett. *62*, 1476 (1989),

8. *Quantum Chaos and Statistical Nuclear Physics*, eds. T.H. Seligman and H. Nishioka (Springer-Verlag, New York, 1986).

9. E.J. Heller, Phys. Rev. Lett. *53*, 1515 (1986) and references therein.

10. R.V. Jensen, M.M. Sanders, M. Saraceno, B. Sundaram, Phys. Rev. Lett. *63*, 2771 (1989).

11. J. Main, G. Wiebush, A. Holle, and K.H. Welge, Phys. Rev. Lett. *57*, 2789 (1986); A. Holle, J. Main, G. Wiebush, H. Rottke, and K.H. Welge, Phys. Rev. Lett. *61*, 161 (1988).

12. G.R. Welch, M.M. Kash, C. Iu, L. Hsu, and D. Kleppner, Phys. Rev. Lett. *62*, 893 (1989); C. Iu, G.R. Welch, M.M. Kash, L. Hsu, and D. Kleppner, Phys. Rev. Lett. *63*, 1133 (1989); and C. Iu, G.R. Welch, M.M. Kash, D. Kleppner, D. Delande, and J.-C. Gay, Phys. Rev. Lett. *66*, 145 (1991).

13. W.R.S. Garton and F.S. Tompkins, Astrophys. J. *158*, 839 (1969).

14. D. Wintgen and A. Hönig, Phys. Rev. Lett. *63*, 1467 (1989).

15. R.V. Jensen, Comments At. Mol. Phys. *25*, 119 (1990).

16. Q. Su, J.H. Eberly, and J. Javanainen, Phys. Rev. Lett. *64*, 862 (1990).

17. M. Pont, N.R. Walet, M. Gavrila, and C.W. McCurdy, Phys. Rev. Lett. *63*, 2771 (1989).

18. R.V. Jensen and B. Sundaram, Phys. Rev. Lett. *65*, 1964 (1990).

19. S.-J. Chang and K.J. Shi, Phys. Rev. A *34*, 7 (1986).

20. B. Sundaram and R.V. Jensen, in preparation.

21. B.J. van Weees, H. Van Houten, C.W.J. Beenakker, et al. , Phys. Rev. Lett. *60*, 848 (1988).

22. A. Szafer and A.D. Stone, Phys. Rev. Lett. *62*, 300 (1989).

23. B. Eckhardt, Europhy. Lett. *5*, 107 (1988).

24. P. Gaspard and Rice, J. Chem. Phys. *90*, 2225; 2242; 2255 (1989)

25. R. Blümel and U. Smilansky, Phys. Rev. Lett. *60*, 477 (1988); Physica *D36*, 111 (1989); Phys. Rev. Lett. *64*, 241 (1990).

26. E. Doron, U. Smilansky, and A. Frenkel, to appear in Physica D (1991).

27. E. Doron, U. Smilansky, and A. Frenkel, Phys. Rev. Lett., *65*, 3072 (1990).

28. R.A. Jalabert, H.U. Baranger, and A.D. Stone, Phys. Rev. Lett. *65*, 2442 (1990).

29. M.L. Roukes and O.L. Alerhand, Phys. Rev. Lett. *65*, 1651 (1990).

30. E. Pollak and M.S. Child, Chem. Phys. *60*, 23 (1981).

31. T.C. Thompson and D.G. Truhlar, J. Phys. Chem. *88*, 210 (1984).

32. See news article on the experimental and theoretical results by the research groups of Zare, Miller, and Truhlar in the Jan. 26, 1990 issue of Science (p.413).

33. D. Wintgen , Phys. Rev. Lett. *61*, 1803 (1988);. E.B. Bogomolny, Physica D *31*, 169 (1988); M.L. Du and J.B. Delos, Phys. Rev. A *38*, 1896 (1988).

34. M.C. Gutzwiller, J. Math. Phys. *12*, 343 (1971).

35. M.C. Gutzwiller, in *Chaos and Quantum Physics*, eds . M.-J. Giannoni, A. Voros, and J. Zinn-Justin, (Elsvier Science Publishers,London, 1990).

36. W.H. Miller, Advan. Chem. Phys. *25*, 69 (1974).

37. U. Smilansky, in *Chaos and Quantum Physics*, eds . M.-J. Giannoni, A. Voros, and J. Zinn-Justin, (Elsvier Science Publishers,London, 1990).

VIII. PHOTOIONIZATION AND DETACHMENT

PHOTOIONIZATION OF EXCITED STATES OF ATOMS*

STEVEN T. MANSON
Department of Physics and Astronomy,
Georgia State University, Atlanta, GA 30303 USA

ABSTRACT

The interaction of electromagnetic radiation with excited states of atoms has been the subject of a number of theoretical investigations. From these studies it has been learned that, although the interactions are the same in excited and ground states of atoms, their manifestation can be very different for photoionization owing to the physical size difference between ground and excited states. This difference leads, for example, to excited state oscillator strengths which are dominated by multiple Cooper minima in both $\ell+1$ and ℓ-1 channels. Experimental data is fairly scarce, but where it does exist, good agreement between theory and experiment is generally found. An overview of both the theoretical and the experimental situation is presented.

Photoionization of excited states of atoms has been the subject of numerous theoretical investigations but of relatively few experimental studies.[1] These studies are of interest for a number of reasons. The interaction of electromagnetic radiation with excited atoms provides a laboratory to study atomic continuum wave functions in a region much larger than the ~a_0 probed by ground state photoionization. In addition, although the interactions are the same in ground and excited states of atoms, their manifestation can be very different in the photoionization process owing to the physical size difference between ground and excited states.

Furthermore, excited state photoionization has important applications in atmospheric and astrophysical phenomena, along with any environment hot enough to produce excited states in suffcient quantity. Finally, the process of radiative recombination, the inverse of photoionization, has begun to be explored experimentally.[2] For low energy electrons, the recombination is primarily to excited states whose inverse is excited state photoionization.

In this paper, the flavor of this phenomonology of excited state photoionization, as revealed by theoretical studies, is presented. In addition, the accuracy of the calculated results is assessed by comparison with available experiments. As an example of the phenomenology which is predicted by theory, the calculated photoionization cross section[3] for the excited 9d state of Cs is shown in Fig. 1. The outstanding feature of this cross section is the minima which dominate the oscillator

*Work supported by the United States Army Research Office and the National Science Foundation.

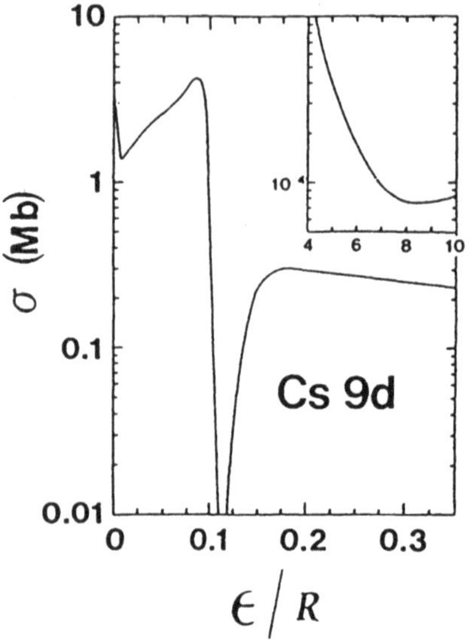

Figure 1. Photoionization cross section for Cs 9d, calculated using Hartree-Slater (HS) wave functions, as a function of photoelectron energy in Rydbergs. The inset shows the higher energy portion of the cross section.

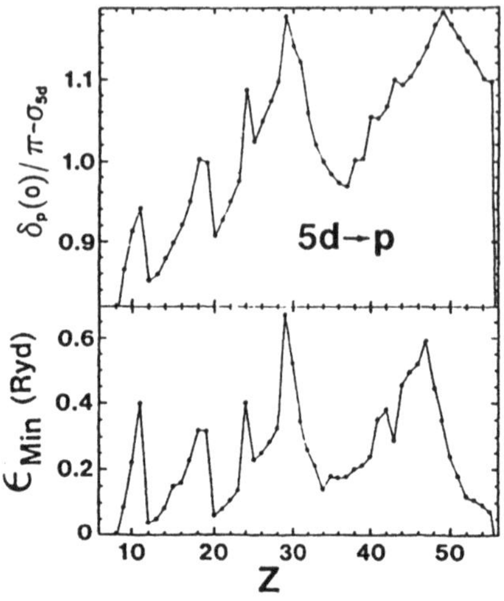

Figure 2. Zeros in 5d→εp dipole matrix elements. The lower curve is the energy at which the minimum occurs in Rydbergs above the threshold plotted against Z. The upper curve is the difference between the p-wave phase shift at the threshold, δp(0) in units of π radians, and the quantum defect of the discrete 5d state, σ_{5d}.

strength distribution in the low energy region. These minima, known as Cooper Minima,[4] are due to zeros in the dipole matrix elements. Cooper Minima are in evidence for ground state photoionization,[5] but they behave rather differently there; there is at most one zero in an $\ell\to\ell+1$ channel and none in $\ell\to\ell-1$ channels.[6] For excited states as many as three zeros in an $\ell\to\ell+1$ channel have been predicted, along with as many as two in an $\ell\to\ell-1$ channel.[7] In Fig. 1, the 9d→εf photoionization transition has three zeros which are seen as minima in the cross section. Note that the middle minimum is extremely deep compared to the others; this is due to the fact that the 9d→εp transition also has a zero right around this energy. Thus, it is clear, that these zeros are very significant in excited state photoionization in the near threshold region.

At this point, it is worthwhile to emphasize that these zeros are not an isolated curiosity, but are found for almost every excited state oscillator strength distribution over a broad range of atomic numbers. As an example, the locus of photoionization energies at which the 5d→εp zero appears[8] is shown in Fig. 2, where it is seen that this zero occurs from Z=9 to Z=55, a not inconsequential portion of the periodic table. This locus forms a fairly complicated curve. Nevertheless, we have been able to understand this behavior in terms of the difference in quantum defects (phase shifts) between the initial discrete and final continuum states involved in the photoionizing transition.[7] The locus of this difference, also shown in Fig. 2, is seen to follow the locus of the minima remarkably well. The details of the argument and the derivation are presented elsewhere.[7]

A next question to be addressed is the accuracy of the calculated results. To get some idea, the photoionization cross section for Cs 6d, calculated in several approximations[9], is shown in Fig. 3. The details of the cross sections differ substantially between central-field Hartree-Slater (HS) and Hartree-Fock (HF) calculations, but quantitatively, the existence of the deep minimum remains. Thus the caveat: the location of the minima are extremely sensitive to the details of the calculation; the existence of the minimum is not.

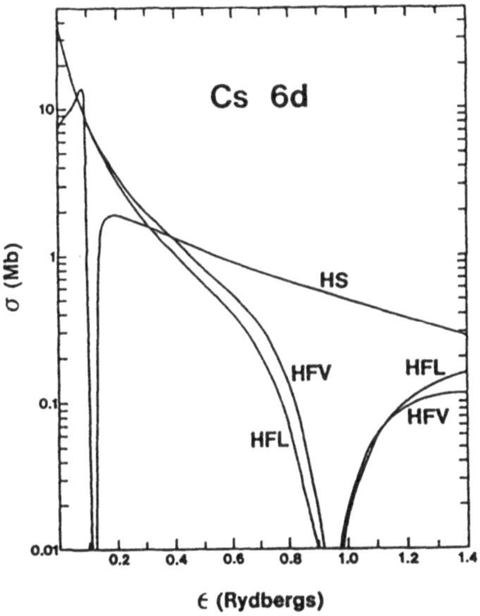

Figure 3. Photoionization cross section for Cs 6d as a function of photoelectron energy in Hartree-Fock length (HFL), Hartree-Fock Velocity (HFV) and Hartree-Slater approximations.

Unfortunately, there are no experimental results as yet in the region where the minima are predicted. Nevertheless, it is of interest to inquire as to the agreement of theory with available experiments. As an example, the situation for Na 3p is shown in Fig. 4, where excellent agreement between theory and experiment is seen.[10] In addition, the situation for Cs 7d is shown[11] in Fig. 5, where excellent agreement is also seen.[12] Thus, it appears that theory does a pretty good job away from the zeros.

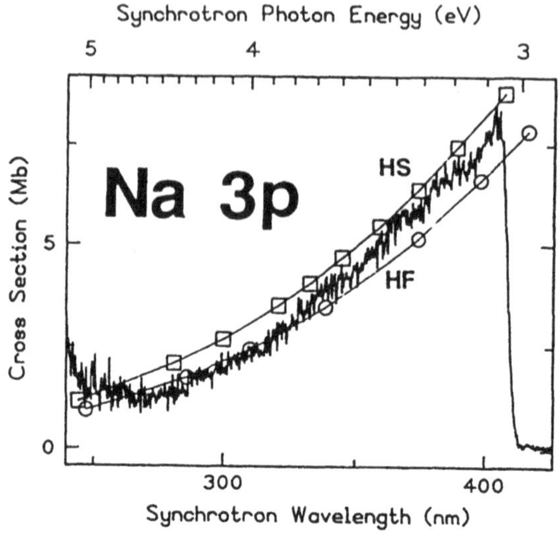

Figure 4. Photoionization cross section for Na 3p. Theoretical results are Hartree-Fock (HF) and Hartree-Slater (HS). The experimental cross section is absolute and not normalised to theory.

Although there is no direct experimental confirmation of the multiple $\ell \to \ell+1$ zeros or the $\ell \to \ell-1$ zeros in the dipole matrix elements for excited state photoionization, there is some suggestion of an $\ell \to \ell-1$ zero. In an experiment on Na 4d photoionization at a single energy[13] the use of circularly polarized light allowed the determination of the $4d \to \epsilon f$ and $4d \to \epsilon p$ cross sections individually; the latter was found to be to small to be seen. The photon energy employed is, however, just in the region where theory predicts a $4d \to \epsilon p$ zero[14], thus a possible $\ell \to \ell-1$ zero. In general, experiment is urgently required to test the theoretical predictions.

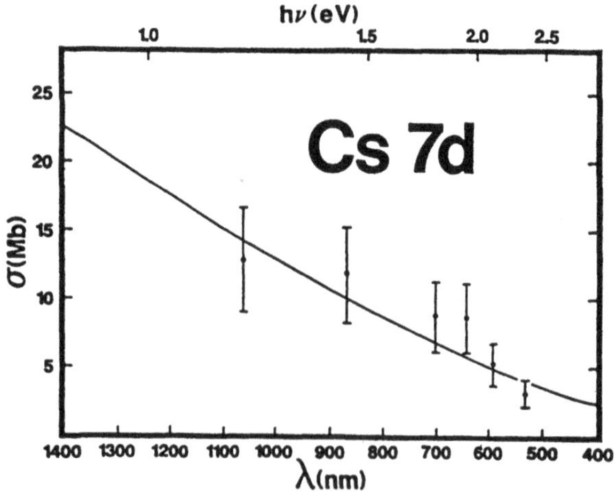

Figure 5. Cs 7d photoionization cross section. The solid curve is the Hartree-Fock result and the points are experimental. Note that only a single theoretical curve is given since the length and velocity results are virtually identical in this range. The experimental cross section is absolute and not normalised to theory.

REFERENCES

1. F. Wuilleumier, D. L. Ederer and J. L. Picque, Advances in Atomic and Molecular Physics. eds. D. R. Bates and B. Bederson(Academic Press, NY, 1988), **V. 23**, pp. 197-286.
2. L. H. Anderson and J. Bolko, J. Phys. **B23**, 3167 (1990).
3. J. Lahiri and S. T. Manson, Phys. Rev. Lett. **48**, 614 (1982).
4. J. W. Cooper, Phys. Rev. **128**, 681 (1962).
5. V. Fano and J. Cooper, Rev. Mod. Phys. **41**, 441 (1968).
6. S. T. Manson, Phys. Rev. **A31**, 3698 (1985).
7. J. Lahiri and S. T. Manson, Phys. Rev. **A33**, 3151 (1986).
8. A. Z. Msezane and S. T. Manson, Phys. Rev. Lett. **48**, 473 (1982).
9. A. Z. Msezane and S. T. Manson, Phys Rev. **A29**, 1594 (1984).
10. J. M. Preses, C. E. Burkhardt, R. L. Corey, D. L. Earson, T. L. Daulton, W. P. Garver, J. J. Leventhal, A. Z. Msezane and S. T. Manson, Phys. Rev. **A32**, 1264 (1985).
11. K. D. Bonin, M. Gatzke, C. L. Collins and M. A. Kadar-Kallen, Phys. Rev. **A39**, 5624 (1989).
12. S. N. Nahar and S. T. Manson, Phys. Rev. **A40**, 6300 (1990).
13. A. V. Smith, J. E. M. Goldsmith, D. E. Nitz and S. J. Smith, Phys. Rev. **A22**, 577 (1980).
14. A. Z. Msezane, J. Lahiri and S. T. Manson, Phys. Rev. **A33**, 4406 (1986).

CORRELATION IN PHOTODETACHMENT

DAVID J. PEGG

Department of Physics, University of Tennessee,
Knoxville, TN 37996, U.S.A.
and
Physics Division, Oak Ridge National Laboratory,
Oak Ridge, TN 37831, U.S.A.

ABSTRACT

Electron correlation plays a major role in all aspects of the photodetachment of an electron from a negative ion. Photodetachment measurements are well suited to investigate the relatively short range forces associated with correlation due to the absence of the long range Coulomb interaction. Measurements of electron affinities, asymmetry parameters and cross sections are described to illustrate the influence of correlation on photodetachment.

1. Introduction

Electron correlation effects, which signature breakdowns in the independent electron model, play important roles in the stabilization of the majority of negative ions (anions) and in determining the manner in which such ions interact with radiation via, for example, the bound-free process of photodetachment. Photodetachment measurements provide a means of investigating the short range and weak forces associated with electron correlation in the absence of the longer range and stronger Coulomb forces that dominate, for example, photoionization. In single photon-single electron detachment, energy and angular momentum are transferred to an electron bound in a negative ion by the absorption of a photon. As a result of this interaction, the electron is detached leaving behind a residual atom which may or may not be in an excited state. Generically, the process can be written as: $h\upsilon + X^- \rightarrow X^{-*} \rightarrow X^* + e^-$. Here allowance has been made for the possible photoexcitation of the negative ion in an intermediate step. A more detailed discussion follows of the role of correlation in the bound, quasibound and unbound states that are encountered in the initial, intermediate and final stages, respectively, of photodetachment.

2. Bound States

The least tightly bound electron in a negative ion is attached to the parent atom by a relatively weak force that arises primarily from polarization and exchange effects - manifestations of electron correlation. Correlation effects result from non-central interactions between pairs of electrons not accounted for in the independent electron model. Most definitions of correlation use the Hartree-Fock approximation as the independent electron model. In this case some correlation, that due to exchange symmetry, is built into the single configuration wave functions. Structure calculations based on the Hartree-Fock model fail in most cases, however, to even predict the existence of negative ions. Since correlation results in configuration mixing, one method of accounting for it is to form a superposition of different configurations to represent the wave function. Such methods are called multiconfigurational (MCHF) or configuration interaction (CI) methods.

The binding energy of the least tightly bound electron in a negative ion is numerically equal to the electron affinity of the parent atom which is defined as the difference in total energies of the atom and the ion. Over 80% of the naturally occurring elements form stable atomic negative ions when an extra electron is attached to the parent atom in its ground state.[1] The prototypical stable atomic negative ion is the H^- ion which is formed in the $(1s^2)^1S$ state. The electron affinity of $H(1^2S)$ is 0.7542 eV. Without correlation, which amounts to 1.1 eV in this case, one would predict this two-electron system to be unstable. In addition to stable ions, there exist an interesting class of negative ions that are metastable. They are formed when an electron is attached to the parent atom in a metastable excited state. The prototypical metastable atomic negative ion is the three-electron ion He^- which is formed in the spin-aligned $(1s2s2p)^4P$ state when an electron is attached to the He atom in the $(1s2s)^3S$ state. The electron affinity of $He(2^3S)$ has been calculated by Bunge and Bunge[2] to be 77.51 ± 0.04 meV. In this case correlation amounts to 0.20 eV. In the Hartree-Fock approximation one would not predict the metastable He^- ion to be bound.

Group I anions (negative ions of the alkalis) are often considered to be effective two-electron systems in the frozen-core approximation. In this model, two highly correlated valence electrons orbit a nucleus screened by two, otherwise inert, core electrons. Under such restrictions, the alkali anions become tractable to theory. In these systems most of the correlation is contained in the valence electrons. In the terminology of configuration interaction (CI) calculations, the ground states of alkali anions consist of a dominant s^2 configuration, predicted by the independent electron model, and a strong admixture of p^2 configurations. Electron affinities have been determined using both ab initio methods that include correlation and semi-empirical model potential calculations. The electron affinity of $Li(2^2S)$, for example, has been calculated by Sims et al.[3] to be 0.607 eV. Again, in the Hartree-Fock approximation, the Li^- ion fails to be bound. The difference in correlation energy between Li^- and Li is 0.73 eV.

Group IIa anions (negative ions of the alkaline earths) can be formed in spin-aligned metastable $(nsnp^2)^4P$ states when electrons are attached to the $(nsnp)^3P$ metastable excited states of the parent atoms. This process is analogous to the formation of He^- the ion in the spin-aligned $(1s2s2p)^4P$ state. These states have been studied theoretically, for example, by Bunge and his collaborators[4] and experimentally by Bae and Peterson,[5] Kvale et al[6] and Hanstorp et al.[7] Alkaline-earth anions can be treated theoretically as effective three-electron systems in the frozen core approximation. The lighter alkaline earths (Be,Mg) do not form stable negative ions but it is now known that the heavier ones (Ca,Sr,Ba) do. The experimental demonstration of the existence of a stable Ca^- ion, for example, was made by Pegg et al.[8] More will be said about this ion in a later section.

3. Quasibound States

A typical stable negative ion has only one bound state. This is a consequence of the relatively short range of the force binding the extra electron and the shallow potential well associated with it. Excited states of negative ions, with few exceptions, lie above the first detachment threshold and are embedded in continuum associated with lower lying bound states of the parent atom. They are unstable, being subject to autodetachment and, in a few cases, radiative decay. The degree of stability of these excited autodetaching states depends on the strength of the interaction coupling them to the continuum. Short-lived negative ion states can be excited in an intermediate step of the photodetachment process if they are optically coupled to the ground state of the ion. The autodetachment of such states can produce resonant structure in the energy dependence of the cross section and asymmetry parameter. Major corrections to the independent electron model are required in the vicinity of these resonances. These autodetaching states are core-excited or doubly-excited states. Double excitation by single photon absorption is, of course, not possible within the independent electron model. The existence and properties of these states depend on the correlation between the two electrons. For relatively low excitation one can model the process by considering one electron moving in the field of an excited atom. For high excitation, where the threshold for double detachment is approached, it appears it is more appropriate to consider the two correlated electrons to move in the field of the residual ion. In the double detachment limit, two highly correlated electrons move slowly in the Coulomb field of the residual ion. This classical unbound three-body system is considered the prototype correlation problem. The Wannier law, which describes the threshold behavior of a three-particle final state, has been tested in single photon-two electron detachment experiments.[9]

Doubly-excited resonances, of both the Feshbach and shape variety, have been observed in the photodetachment studies of H^- by the University of New Mexico-Los Alamos National Laboratory group.[10] A strong shape resonance, $(1s2p^2)^4P$, has been studied in the photodetachment cross section of He^- by Peterson et al.[11] This resonance lies just above the threshold for the opening of the $^3P\epsilon p$ channel. The heavier alkali anions such as Cs^- and Rb^- exhibit Feshbach resonances just below the ^2P detachment thresholds.[12]

4. Unbound States

The final state of the photodetachment process is a continuum (unbound) state consisting of the detached electron moving relative to the residual atom. The dominant long range interaction in this case results from polarization and exchange effects which are manifestations of electron correlation. This situation is fundamentally different from the corresponding case of photoionization where the long range interaction in the final state is dominated by the Coulomb force between the outgoing electron and the residual positive ion. The induced multipole moments resulting from the polarization of the atomic electron charge cloud act back on the outgoing electron transferring to it additional angular momentum not considered in the independent electron model. Such changes can alter the angular distribution of the photoelectrons. The induced polarization, by coupling final state channels, can also result in anomolous behavior in the cross section at certain detachment thresholds. Consider, for example, the photodetachment of an alkali anion such as Li^-. In this case channel mixing produces a Wigner cusp in the cross section for the $^2S\epsilon p$ channel at the threshold for the opening of the $^2P\epsilon s$ channel.[13]

In order to calculate photodetachment cross sections one needs to choose appropriate wave functions to represent the initial (bound) state and the final (continuum) state. It is necessary to take account of correlation in both the bound and continuum states. This has been achieved, for example, by Moores and Norcross[13] in their calculations of the cross sections for photodetachment of the alkali anions. These investigators used CI wave functions for the bound state and close-coupled wave functions to represent the final state. Close-coupling methods had been previously used successfully in studying electron-atom scattering, a process with the same final state as photodetachment.

5. Measurable Quantitites

The rest of this paper will describe experiments, performed by the University of Tennessee-Oak Ridge National Laboratory group, that illustrate the role of correlation in photodetachment. Such experiments are able to test different theoretical approaches to the correlation problem. The experiments employ photoelectron detachment spectroscopy to measure the kinetic energies and angle-dependent yields of electrons ejected in the photodetachment process. These quantities, in turn, can be used to determine electron affinities, asymmetry parameters and cross sections. Electron affinity measurements test bound-state wave functions while asymmetry parameter and cross sections measurements can attest the validity of both bound-state and continuum (unbound) state wave functions. In addition, the latter quantities are sensitive to the presence of quasibound excited states of negative ions.

6. Experimental Procedure

The crossed-beam apparatus and the experimental techniques used to investigate the photodetachment of atomic negative ions have been described in detail in a recent publication by Pegg.[14] A schematic of the apparatus is shown in Figure 1. A fast moving, tenuous beam of negative ions is prepared by sequential charge transfer when a beam of positive ions from an accelerator is passed through an alkali vapor charge exchange cell. The negative ion beam is intersected perpendicularly by a plane-polarized beam of photons from a flashlamp-pumped pulsed dye laser. Following photodetachment events, electrons ejected from the moving ions in the direction of their motion, are collected, energy analyzed and detected using a spherical-sector electron spectrometer placed in the path of the ion beam. The angular distribution of the detached photoelectrons can be determined by measuring their yield as a function of the angle between the electric field vector of the plane-polarized laser beam and the collection direction. This is achieved in the present apparatus by keeping the collection direction fixed in the forward direction and rotating the laser polarization vector using a $\lambda/2$ phase retarder (double Fresnel rhomb). The output power of pulsed dye laser is carefully chosen to avoid saturation on each of the photodetachment processes studied. An injection lens is used to focus the laser beam onto the ion beam in the interaction region. A fraction of the laser light is detected with a photodiode and used for normalization purposes. Similarly the ion beam intensity, as measured in a Faraday cup downstream of the spectrometer, is monitored for normalization. Sets of apertures are used to keep the geometrical interaction volume between the laser and ion beams fixed during the course of the measurements.

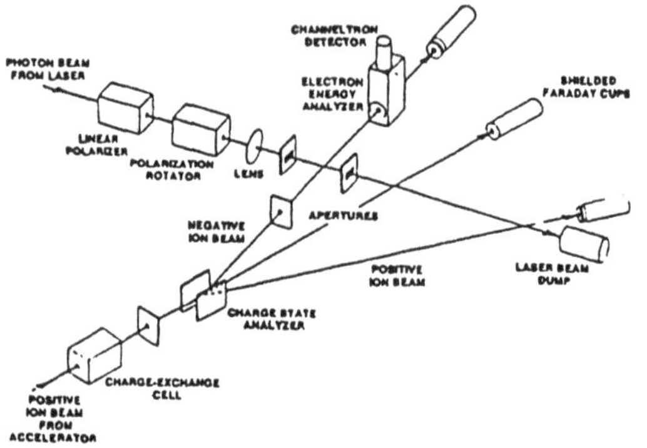

Fig. 1. Schematic of crossed-beam apparatus used for photoelectron detachment spectroscopy

A synchronous detection scheme is employed to extract the photoelectron signal from a rather large electron background associated primarily with collisional detachment of the ions of the beam. A pair of gate-and-delay generators are used to alternately switch on and off two multichannel scalars that record electrons generated in coincidence with the laser being on and off, respectively. As a consequence of the fast moving, unidirectional nature of the ion beam, the kinetic energies, yields and energy and angular distributions of the detached photoelectrons are modified by kinematics. The combination of fast beams and forward-directed electron collection used in the present apparatus permits one to exploit kinematic effects such as shifting and doubling of the peaks in the photoelectron spectra. In the present experiments the absolute cross sections are not measured directly. Instead one measures, under identical geometrical conditions, the ratio of cross sections for photodetaching electrons from a beam of the ions of interest to that of a beam of reference ions. The ratio is then normalized using the known photodetachment cross section of the reference ion. To enhance the precision of the measurements a technique, based on the kinematic shift of spectral lines, has been developed to eliminate the need to know the efficiencies for collection and detection of electrons from the ion beams of the two different species. The energies of the two ion beams are carefully chosen so that the detached electrons from each beam have the same energy in the laboratory frame. Under this condition the efficiencies must be the same for the electrons from each beam and thus cancel in the ratio. In order to make corrections for the differing ion beam densities and solid angle transformation factors, one must know the ion beam velocities. These quantities can be accurately measured by an in-situ analysis of peak separations in the photoelectron spectra.

7. Electron Affinities

Electron affinities can be determined, via the Einstein energy balance equation, by measuring the kinetic energies of electrons detached from negative ions by photons of known energy ie. photoelectron detachment spectroscopy. If the electron is emitted from a moving ion, as in the present experiments, the energy of the ions must be accurately known in order to transform the measured electron energy from the laboratory frame to the ion rest frame. Unfortunately, the energy of a single peak in a photoelectron spectrum is difficult to measure accurately due to the presence of unknown, and essentially unmeasurable, energy offsets associated with, for example, contact and surface potentials. To minimize this problem, in cases where more than one peak exists, one can measure the separations of pairs of peaks and make an assumption that the offset is comparable for each peak and thus cancels in the energy difference. If the photoelectron spectrum contains at least three peaks of known ion frame separation, it is possible to determine simultaneously the energies of both the detached electrons and the ion beam from an analysis of the separations of two different pairs of peaks. The method has been successfully tested on the metastable He^- ion. The electron affinity of $He(2^3S)$ was measured to 76.35 ± 1.53 meV. This value is in good agreement with the best theoretical result of 77.51 ± 0.04 meV.[2]

The method just described was applied by Pegg et al.[8] to the determination of the electron affinity of $Ca(4^1S)$. Prior to this experiment it was generally accepted that the alkaline earth anions were unstable since their formation involved adding the extra electron to a new subshell, where enhanced screening would prohibit stabilization. Figure 2 shows the photoelectron spectrum obtained by detaching electrons from a 60-keV beam of Ca^- ions. The relative intensities and spacing of the peaks unambiguously show that this ion is stably bound in the $(4s^2 4p)^2P$ state. The existence of three peaks in the spectrum permits one to determine the electron affinity of Ca by measuring the separations of two pairs of peaks. The highest energy peak (~5 eV) corresponds to the photodetachment process that leaves the Ca atom in the $(4s^2)^1S$ ground state. The two low energy peaks (~1eV) have a common origin in the process in which the Ca atom is left in the $(4s4p)^3P$ excited state. Peak doubling, such as this, is a kinematic effect. If the velocity of the detached electron in the ion frame is less than the ion beam velocity, electrons ejected in both forward and backward directions in the ion frame will be collected in the forward direction in the laboratory frame. The separation, in the laboratory frame, between two such kinematically-doubled peaks is $\Delta E_L = 4\sqrt{\epsilon E_c}$, where ϵ is the energy of electrons moving with the same velocity as the ions and E_c is the ion-frame energy of the detached electron. The separation of the second and third peaks, both forward-directed peaks, is given by $\Delta E_L = E_e + 2\sqrt{\epsilon}\,[\sqrt{E_c + E_e} - \sqrt{E_c}]$. Here $E_c = E_c(^3P)$ and E_e is the excitation energy of the ^3P state, a quantity well known from optical spectroscopy. The two equations for the separations of the two pairs of peaks contain two unknowns ϵ and E_c. One can therefore eliminate the dependence on the ion beam energy completely and solve for $E_c(^3P)$. The electron affinity is then $E_a = E_\gamma - E_e - E_c(^3P)$, where E_γ is the measured photon energy. The final result of the experiment was $E_a(Ca) = 43 \pm 7$ meV. This value represents the smallest binding energy measured to date for the ground state of a stable atomic negative ion. A MCHF calculation by Fischer et al[15] confirmed that the Ca^- ion is stably bound. In this effective three-electron calculation extensive correlation had to be built into the wave functions to stabilize the ion. Previous calculations failed to predict a stable system. Since these results were published, the Ca^- ion has become a testing ground for different approaches to the correlation problem. These new calculations, using various computational techniques, all predict the Ca^- ion to be stable.[16]

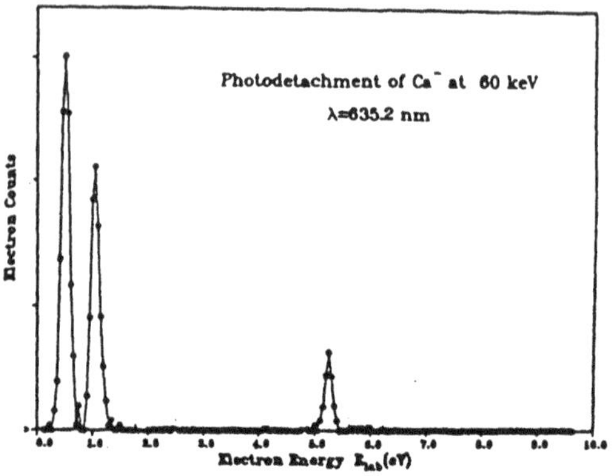

Fig. 2. Spectrum of photodetached electrons from a Ca^- beam. The first two peaks are associated with the $^1S\epsilon s,d$ channels and the third peak with the $^3P\epsilon p$ channel.

8. Asymmetry Parameters

The angular distribution of detached photoelectrons is completely characterized by the asymmetry parameter, β, a quantity which measures the deviation of the emission pattern from isotropy ($\beta = 0$). Positive (negative) valves of β correspond to preferential electron emission parallel (perpendicular) to the direction of the electric field vector of the plane-polarized radiation. The asymmetry parameter is restricted to the range, $-1 \leq \beta \leq 2$. For plane-polarized radiation in the electric-dipole approximation, the photoelectron angular distribution for the single photon-single electron detachment process takes the form, $f(\theta) = 1 + \beta P_2(\cos \theta)$. Here it is assumed that the ion is randomly polarized prior to photoabsorption. The angle θ represents the angle between the polarization vector of the radiation and the direction of emission of the photoelectrons. A recent study by Thompson et al.,[17] of the spectral dependence of the angular distribution of photoelectrons detached in the process: $h\nu + He^-(1s2s2p\,^4P) \rightarrow He(1s2p\,^3P) + e^-(\epsilon p)$ is used to illustrate how the presence of an excited negative ion state may perturb the value of β. In the independent electron model, the outgoing electron in this process should be a p wave and be represented by $\beta = 2$ for all photon energies. The experiment studied β over the range of photon energies from ~ 1.8 - 2.4 eV. At the highest photon energy used in the experiment, there exists agreement with the independent electron model i.e. $\beta = 2$. As the photon energy is reduced toward threshold, however, the value of β is seen to decrease quite sharply. Correlation of some type apparently causes β to deviate from the value predicted by the independent electron model. One possible explanation is that the angular distribution of this channel is being influenced by the $(1s2p^2)\,^4P$ state that manifests itself as a shape resonance lying ~11 meV above the 2^3P threshold. In this case, the photodetachment process would proceed via an intermediate step: $h\nu + He^-(1s2s2p\,^4P) \rightarrow He^-(1s2p^2\,^4P) \rightarrow He(1s2p\,^3P) + e^-(\epsilon p)$. This resonance has been studied experimentally by Peterson et al.[11] and theoretically by Hazi and Reed[18] and Watanabe.[19] A recent MCHF calculation by Saha and Compton[20] seems to support the conclusions of Thompson et al[17] i.e. that the strong shape resonance may have an influence on the photoelectron angular distribution even though the peak lies some one hundred resonance halfwidths below the range of measurements. A comparison of the experimental

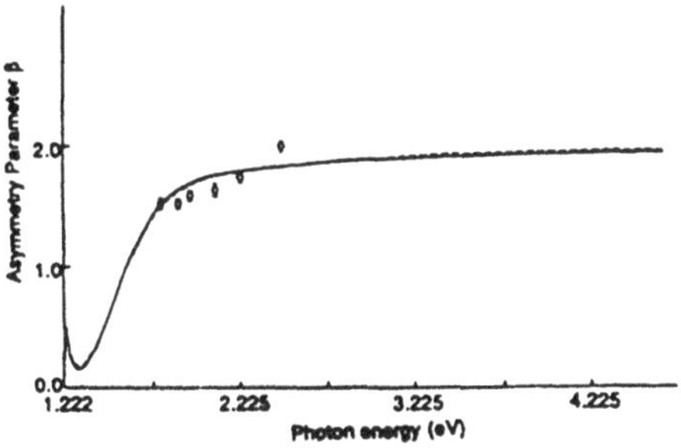

Fig. 3. Spectral dependence of the asymmetry parameter for the $^3P\epsilon p$ channel in the photodetachment of the metastable He^- ion. Solid lines are from a MCHF calculation (Ref. 20) and circles are experimental data (Ref. 17).

and theoretical results is shown in Figure 3.

9. Cross Sections

Calculations of photodetachment cross sections depend critically on the bound and continuum wave functions used to represent the initial and final states. To illustrate the role played by correlation, comparisons of experimental and theoretical cross sections are made for single photon-single electron detachment of He^- and Li^- ions. The partial cross sections for leaving the He atom in the 2^3S and 2^3P states have been measured over the spectral range 1.8 - 2.1 eV by Pegg et al.[21] The results are in good agreement with the MCHF calculations of Saha.[22] The total cross section, the sum of the partial cross sections, has been calculated by Hazi and Reed[18] and Saha and Compton.[20] Again, there is good agreement between experiment and theory. Measurements of the cross sections for photodetaching an electron from the Li^- ion are currently in progress.[23] In this experiment, the ratio of cross sections for photodetaching the Li^- ion and the D^- reference ion is measured. In order to eliminate the dependence of this ratio on the unknown efficiencies for the collection and detection of electrons detached from each beam, the ion beam energies are chosen to kinematically shift the different electron peaks to the same energy in the laboratory frames. The energy-aligned spectra are shown in Figure 4. The cross section ratio is corrected for the differences in ion beam densities and solid-angle transformation factors for the Li^- and D^- beams. These quantities depend on the ion beam velocities which can be accurately measured from a comparison of the photoelectron spectra of Li^- and D^- with that of He^- taken at the same ion beam energy. At a wavelength of 662.7 nm, for example, the measured ratio is $\sigma(Li^-)/\sigma(D^-) = 1.98 \pm 0.15$. A scale for this measurement is established by the use of a theoretical value for $\sigma(D^-)$. The value, arbitrarily chosen, is that of Stewart[24] which has an estimated accuracy of ~3%. The normalized ratio yields a cross section of $\sigma(Li^-) = 73.5 \pm 6.0$ Mb. The result is in good agreement with the theoretical predictions of Moores and Norcross[13] and the recent work of Moccia and Spizzo.[25]

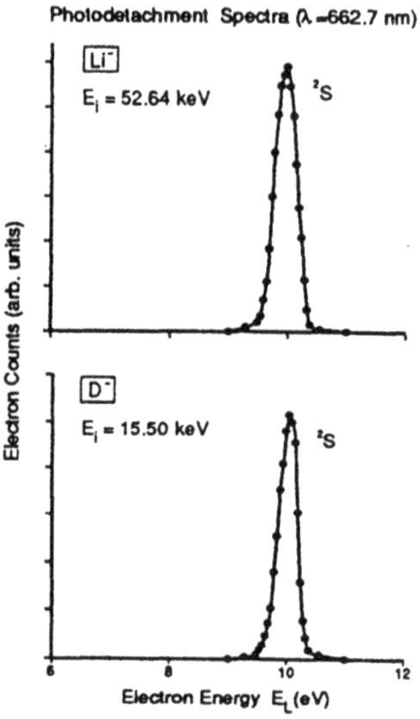

Fig. 4. Spectra of photodetached electrons from Li^- (top) and D^- (bottom) beams. The ion beam energies were chosen to kinematically shift the peaks to the same energy in the laboratory frame.

Research is supported by the U.S. Department of Energy, Office of Basic Energy Sciences, Division of Chemical Sciences through the University of Tennessee and Oak Ridge National Laboratory. ORNL is managed by Martin Marietta Energy Systems, Inc. under contract DE-AC05-840R21400 with the U.S. Department of Energy. The author wishes to acknowledge his current collaborators: J. Dellwo, Y. Liu and G. D. Alton.

References

1. H. Hotop and W. C. Lineberger, *J. Chem. Phys. Ref. Data* **14** (1985) 731.
2. A. V. Bunge and C. F. Bunge, *Phys. Rev.* **A30** (1984) 2179.
3. J. S. Sims, S. A. Hagstrom, D. Munch and C. F. Bunge, *Phys. Rev.* **A13** (1976) 560.
4. C. F. Bunge, M. Galan, R. Jauregui and A. V. Bunge, *Nuclear Instr. and Methods* **202** (1982) 299; A. V. Bunge, *Phys. Rev.* **A33** (1986) 82.
5. Y. Bae and J. R. Peterson, *Phys. Rev.* **A30** (1984) 2145.
6. T. J. Kvale, G. D. Alton, R. N. Compton, D. J. Pegg and J. S. Thompson, *Phys. Rev. Lett.* **55** (1985) 484.
7. D. Hanstorp, P. Devynck, W. G. Graham and J. R. Peterson, *Phys. Rev. Lett.* **63** (1989) 368.
8. D. J. Pegg, J. S. Thompson, R. N. Compton and G. D. Alton, *Phys. Rev. Lett.* **59** (1987) 2267.
9. J. B. Donahue, P. A. Gram, M. V. Hynes, R. W. Hamm, C. A. Frost, H. C. Bryant, K. B. Butterfield, D. A. Clark and W. W. Smith, *Phys. Rev. Lett.* **48** 1538 (1982); Y. K. Bae, M. J. Coggiola and J. R. Peterson, *Phys. Rev.* **A28** (1983) 3378; Y. K. Bae and J. R. Peterson, *Phys. Rev.* **A37** (1988) 3254.
10. For example, H. C. Bryant, K. B. Butterfield, D. A. Clark, C. A. Frost, J. B. Donahue, P. A. Gram, M. E. Hamm, R. W. Hamm and W. W. Smith, *Atomic Physics 7*, eds D. Kleppner and F. M. Pipkin (Plenum Press, New York, 1981); S. Cohen et al., *Phys. Rev.* **A36** (1989) 4728; P. A. Gram et al., *Phys. Rev. Lett.* **40** (1978) 107; M. E. Hamm et al., *Phys. Rev. Lett.* **43**, (1979) 1715; D. W. MacArthur et al., *Phys. Rev. Lett.* **56**, (1986) 282; P. G. Harris et al, *Phys. Rev. Lett.* **65** (1990) 309.
11. J. R. Peterson, M. J. Coggiola and Y. K. Bae, *Phys. Rev. Lett.* **50**, (1983) 664.
12. T. A. Patterson, H. Hotop, A. Kasdan, D. W. Norcross and W. C. Lineberger, *Phys. Rev. Lett.* **32** (1974) 189; J. Slater, F. H. Read, S. E. Novick and W. C. Lineberger, *Phys. Rev.* **A17** (1978) 201.
13. D. L. Moores and D. W. Norcross, *Phys. Rev.* **A10** (1974) 1646.
14. D. J. Pegg, in *The Physics of Electronic and Atomic Collisions*, eds A. Dalgarno, R. Freund, P. Koch, M. Lubell and T. Lucatorto (AIP, New York, 1990) p. 233.
15. C. Fischer, J. Lagowski and S. H. Vosko, *Phys. Rev.* **59** (1987) 2263.

16. For example, W. R. Johnson, J. Sapirstein and S. A. Blundell, *J. Phys.* **B22** (1989) 2341; L. Kim and C. H. Greene, *J. Phys.* **B22** (1989) L175; S. H. Vosko, J. Lagowski and I. L. Mayer, *Phys. Rev.* **A39** (1989) 446; P. Fuentealba, A. Savin, H. Stoll and H. Preuss, *Phys. Rev.* **A40** (1989) 2163; P. Krylstedt, W. Elander and E. Brändas, *J. Phys.* **B22** (1989) 1623 and G. F. Gribakin, B. V. Gul'tsev, V. K. Ivanov and M. Yu. Kuchiev, *J. Phys.* **B23**, (1990) 4505.
17. J. S. Thompson, D. J. Pegg, R. N. Compton, and G. D. Alton, *J. Phys.* **B23** (1990) L15.
18. A. U. Hazi and K. Reed, *Phys. Rev.* **A24** (1981) 2269.
19. S. Watanabe, *Phys. Rev.* **A25** (1982) 2093.
20. H. P. Saha and R. N. Compton, *Phys. Rev. Lett.* **64** (1990) 1510.
21. D. J. Pegg, J. S. Thompson, J. Dellwo, R. N. Compton and G. D. Alton, *Phys. Rev. Lett.* **64** (1990) 278.
22. H. P. Saha, private communication.
23. D. J. Pegg, J. Dellwo, Y. Liu and G. D. Alton, private communication.
24. A. L. Stewart, *J. Phys.* **B11** (1978) 3851.
25. R. Moccia and P. Spizzo, *J. Phys.* **B23** (1990) 3557.

LONG-RANGE DIPOLE FIELD EFFECTS ON H^- DETACHMENT PROCESSES ABOVE THE H(n=2) THRESHOLD

ANTHONY F. STARACE
Department of Physics and Astronomy
The University of Nebraska
Lincoln, NE 68588-0111

ABSTRACT

The long-range dipole field interaction between a low energy electron and the hydrogen atom in the n = 2 state is shown to be the key for understanding significant features of a number of H^- detachment processes. Collisional, single photon, and two-photon detachment processes are examined, each of which is shown to be sensitive to different aspects of low-energy H(n=2)–e^- interactions. Specifically, these detachment processes are best suited respectively for exhibiting the finite threshold cross sections for certain $^1S^e$, $^1P^o$, and $^1D^e$ final-state channels, the shape resonance in one of the $^1P^o$ final state channels, and the near–threshold Galitis–Damburg oscillations in certain $^1S^e$ and $^1D^e$ final state channels. The latter oscillations have yet to be observed experimentally in any process.

1. Introduction

This paper is focused on processes involving the detachment of the H^- ion and the simultaneous excitation of the resulting H atom to its n=2 state. The processes we shall discuss are quite different from one another and involve different experimental technologies. They are high energy collisional detachment, photodetachment, and two-photon detachment, i.e.,

$$H^- + T \rightarrow H(n=2) + e^- + T^* \tag{1}$$

$$H^- + \gamma \rightarrow H(n=2) + e^- \tag{2}$$

$$H^- + 2\gamma \rightarrow H(n=2) + e^- \quad . \tag{3}$$

In Eq. 1, T denotes an atomic target atom and the asterisk on the right hand side of Eq. 1 denotes that this target atom may be in either its initial state, an excited bound state, or an ionized state as a result of the collision. (Typically, the final state of the target in collisional detachment processes is not measured experimentally.)

The aim of this paper is to show that key features of the very different detachment processes in Eqs. 1–3 may be understood from a unified point of view. This point of view has as its focus the low energy states of the fundamental H(n=2) – e^- three-body system. It will be shown that each of the processes in Eqs. 1–3 is ideally suited for exhibiting different aspects of these low energy states.

In what follows, we discuss first the low-energy states of the H(n=2) – e^- three-body system, using an adiabatic hyperspherical representation.[1] We then examine how these states affect each of the detachment processes in Eqs. 1–3 in turn. Readers interested in further details of the theoretical work (i.e., on collisional detachment[2–4] and on one- and two-photon detachment[5] processes) upon which the present paper is based are referred to recent publications of the author and his collaborators.[2–5] These publications should also be consulted for more complete references to related experimental and theoretical work by others.

2. Low-Energy States of the H(n=2) – e^- System

In our calculations we describe the H(n=2) – e^- three-body system in an adiabatic hyperspherical representation[1,6,7] since this is known to describe fairly accurately the key dynamical features of this system[8] and since this representation is known to diagonalize asymptotically the long-range dipole interaction for this system.[1,8]

In the hyperspherical approach the exact two-electron wave function $\psi(\mathbf{r}_1,\mathbf{r}_2)$ is expanded in a complete set of adiabatic eigenfunctions $\phi_\mu(R,\alpha,\hat{\mathbf{r}}_1,\hat{\mathbf{r}}_2)$, which depend parametrically on a hyperradius $R \equiv (r_1^2+r_2^2)^{1/2}$ and are functions of the five angular variables $\alpha \equiv \tan^{-1}(r_2/r_1)$, $\hat{\mathbf{r}}_1$ and $\hat{\mathbf{r}}_2$.

$$\psi(R,\alpha,\hat{\mathbf{r}}_1,\hat{\mathbf{r}}_2) \equiv (R^{5/2} \sin\alpha \cos\alpha)^{-1} \sum_\mu F_\mu(R)\, \phi_\mu(R,\alpha,\hat{\mathbf{r}}_1,\hat{\mathbf{r}}_2)\ . \qquad (4)$$

The ϕ_μ satisfy an angular equation[1,6,7] having eigenvalue $U_\mu(R)$. The F_μ satisfy a set of coupled radial equations[1,6,7]; however, in the adiabatic approximation all but the diagonal coupling matrix elements are dropped so that each $F_\mu(R)$ satisfies a one-dimensional radial Schrödinger equation,

$$\left[\frac{d^2}{dR^2} - V_\mu(R) + k^2\right] F_{\mu k}(R) = 0 . \tag{5}$$

In Eq. 5 the effective radial potential $V_\mu(R)$, which characterizes the dynamical features of a particular hyperspherical channel μ converging to the nth level of the H atom is defined by

$$- V_\mu(R) \equiv \frac{U_\mu(R) + \frac{1}{4}}{R^2} + \left(\phi_\mu, \frac{d^2\phi_\mu}{dR^2}\right) - \frac{1}{n^2} , \tag{6}$$

where $(\phi_\mu, d^2\phi_\mu/dR^2)$ is the R-dependent diagonal coupling matrix element for the μth channel. Since the long-range dipole interaction due to the degeneracy of the H(n=2) states[9] is diagonal in the hyperspherical representation,[1,8] the asymptotic form of the effective radial potential is

$$V_\mu(R) \underset{R \to \infty}{\sim} \lambda_\mu(\lambda_\mu + 1)/R^2 . \tag{7}$$

In Eq. 7 λ_μ is an effective orbital angular momentum, which may be real or complex depending on the channel μ. For channels μ in which the long-range dipole interaction[9] is repulsive at asymptotic distances, λ_μ is real. Hence at threshold the cross section for any excitation to the channel μ is zero since it depends on $| k^{\lambda_\mu + 1/2}|^2$, which is zero for $k \to 0$. On the other hand, for channels μ in which the long-range dipole interaction[9] is attractive at asymptotic distances, one may write quite generally,[10]

$$\lambda_\mu = -\frac{1}{2} + i\alpha_\mu \tag{8}$$

As a consequence,the threshold value of the cross section for any excitation to the channel μ is finite[11] since it depends on $|k^{\lambda_\mu+1/2}|^2 = 1$. In addition, as noted by Gailitis and Damburg,[11] the transition matrix elements for channels having complex λ_μ are influenced above threshold by the term $k^{\lambda_\mu+1/2} = k^{i\alpha_\mu}$ (cf. Eq. 8), which, when re-written as exp $(i\alpha_\mu \ln k)$, may be seen to oscillate as a function of ln k.

The effective potentials $V_\mu(R)$ which converge asymptotically to the H(n=2) threshold and which are most relevant for the detachment processes in Eqs. 1–3 are shown in Fig. 1. All of the $^1S^e$ and $^1P^o$ potential curves are shown as well as the most important $^1D^e$ potential curve. Since the total orbital and spin angular momenta are insufficient to specify the potential curves uniquely, additional specification is necessary. In Fig. 1 we have employed abbreviated labels corresponding to Lin's classification of doubly excited states.[12]

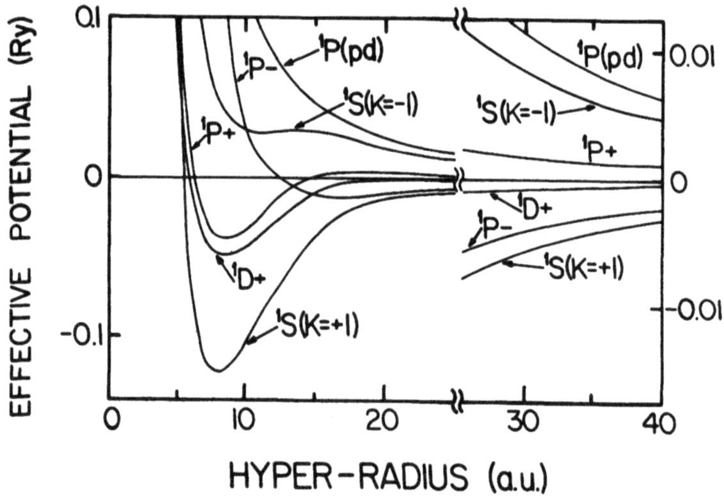

FIG. 1. Effective radial hyperspherical potentials V_μ in Ry plotted vs the hyperradius R for six channels converging to H(n=2): $^1S(K=\pm 1)$, $^1P\pm$, $^1P(pd)$, and $^1D+$. Note that the zero of energy is chosen to be the H(n=2) threshold and that near R = 25 the vertical energy scale is changed.

Note finally that the adiabatic $^1P+$ and $^1P-$ curves have a sharply avoided crossing near $R \approx 13.3$ a.u., implying a very strong interaction between them at this value of R. Following others,[1,8,12] we take this strong interaction into account by replacing the adiabatic potentials by the corresponding diabatic potentials (which cross near $R \approx 13.3$ a.u.) and ignore any residual interaction between the potentials. We emphasize that this use of the diabatic approximation is limited to these two potentials. All of the other potentials (as well as the $^1P+$ and $^1P-$ potentials outside the $R \approx 13.3$ a.u. region) are calculated by solving the adiabatic hyperspherical equations.

The key features of the interactions within the H(n=2) – e^- system are clearly exhibited in the effective potential curves shown in Fig. 1. These are, first, that the $^1P+$ potential is attractive at short distances and weakly repulsive at large distances thereby giving rise to a shape resonance (which is seen experimentally at about 18 meV above threshold).[8] This shape resonance feature dominates the cross section of any process which populates the $^1P^o$ final state channels above the H(n=2) threshold.

Second, because of their long-range repulsive behavior, the $^1P+$, $^1P(pd)$, and $^1S(K = -1)$ potentials all have zero cross sections at threshold.

Third, the three potentials corresponding to the $^1S(K{=}+1)$, $^1P-$, and $^1D+$ channels are attractive at asymptotic distances. As discussed above, they therefore have complex effective angular momenta. Hence the excitation cross section for each of these channels is finite at threshold (within the center-of-mass frame of the H(n=2) – e^- system). Futhermore, the transition amplitudes for excitations to these three channels having complex effective angular momenta oscillate on a ln k scale above threshold.[11]

In what follows we shall examine each of the processes in Eqs. 1–3 in turn in order to show how these key features of the interactions within the H(n=2) – e^- system are manifested both experimentally and theoretically.

3. Collisional Detachment of H⁻

The cross sections for high-energy H^- detachment collisions, particularly those differential in the energy and angle of the detached electron, depend sensitively on the low-energy states of the fundamental H–e⁻ three-body system. These low-energy states in the H^- frame are related kinematically to the small–angle (i.e., forward-direction) behavior of the detachment cross section measured in the laboratory. The closer to threshold in the H^- frame that one wishes to probe, the closer in angle to the forward direction one must measure. These same kinematic relations, furthermore, magnify the energy scale over which the fundamental dynamics of the H–e^- system may be observed: e.g., for 0.5–MeV H^- detachment collisions, features in the H^- frame measured on a scale of tens of meV appear in the laboratory frame on a scale of tens of eV.

It is useful to examine how each of the final state channels for the H(n=2) – e^- system contribute individually to the doubly differential cross section (DDCS) for the following collisional detachment process:

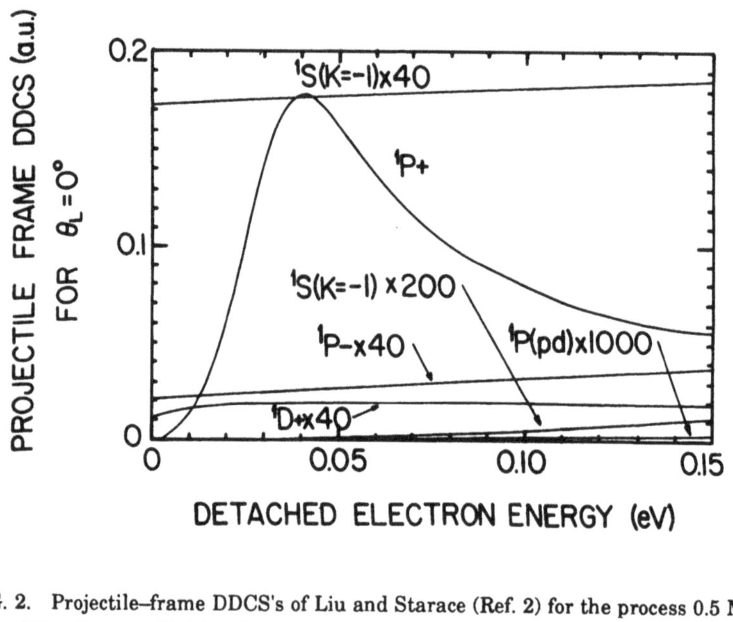

FIG. 2. Projectile–frame DDCS's of Liu and Starace (Ref. 2) for the process 0.5 MeV H^- + He → H(n=2) + e^- + He* for electrons detached at $\theta_P = 0°$, for each of the six hyperspherical channels $^1S(K = \pm 1)$, $^1P\pm$, $^1P(pd)$, and $^1D+$ taken separately, plotted vs electron energy.

$$0.5\text{–MeV } H^- + He \rightarrow H(n{=}2) + e^- + He^* . \tag{9}$$

The asterisk on the right–hand side of Eq. 9 indicates that the He atom may be either in an excited state, bound or continuum, or in the unexcited ground state. The projectile frame DDCS's for electrons detached in the forward direction ($\theta_P = 0°$) are shown in Fig. 2 in the theoretical approximation that each final state channel is treated separately (i.e., incoherently rather than coherently). One sees from Fig. 2 that the $^1P^+$ shape resonance channel dominates the collisional detachment process near threshold except for the region right at threshold. At threshold the cross sections for channels having asymptotic repulsive potentials are all zero. In contrast, the cross sections for the three channels having asymptotic attractive potentials are finite at threshold. Fig. 2 shows DDCS results calculated for a He target, but it is typical of results one obtains for any rare gas target. We examine now how the features shown in the projectile frame in Fig. 2 appear in the laboratory frame in actual experimental measurements.

In Fig. 3 we compare our laboratory frame DDCS's with experimental data of Duncan and Menendez[13] for the process,

$$0.5\text{–MeV } H^- + Ar \rightarrow H^* + e^- + Ar^*, \tag{10}$$

for three laboratory–frame detachment angles. The experimental measurements did not determine the final state of the H atom, which fact is indicated by the asterisk on the right hand side of Eq. 10. Our theoretical calculations[4] were carried out for both H(n=1) and H(n=2) final states. Clearly the large DDCS for the H(n=2) final state is a major feature of the experimental data. In our calculations this feature is shown to be due to the $^1P^+$ shape resonance. For small angles θ_L it is clear that the shape resonance must be taken into account in order to achieve good agreement with the experimental data. For larger angles θ_L, the kinematics of the transformation from the projectile frame to the laboratory frame results in progressively less contribution of the $^1P^+$ shape resonance to the H(n=2) cross section for the process in Eq. 10.

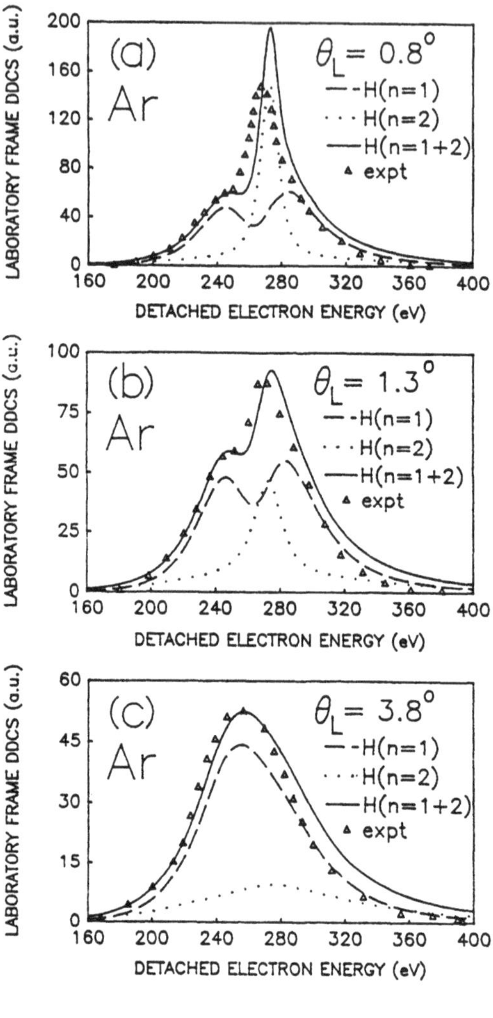

FIG. 3. Laboratory–frame DDCS's for the process 0.5 MeV H^-+ Ar → H* + e^- + Ar* for electron–detachment angles of (a) θ_L = 0.8°, (b) θ_L = 1.3°, and (c) θ_L = 3.8°. Theoretical results of Liu and Starace (Ref. 4) are indicated as follows: Dashed curves, results for H(n=1) final states; dotted curves, results for H(n=2) final states; solid curves, sum of H(n=1) and H(n=2) results. Experimental results of Duncan and Menendez (Ref. 13) are indicated by triangles; they are normalized to the solid curves at the position of the lower-energy peak for θ_L = 0.8° and 1.3° and at the position of the single peak for θ_L = 3.8°.

In order to exhibit the fact that three channels of the H(n=2) – e^- system have finite cross sections at threshold in the projectile frame (cf. Fig. 2), one must measure the detached electrons in the laboratory at or very close to $\theta_L = 0°$. One experiment which has done this is that of Anderson, Bangsgaard, and Sørensen[14] for the process,

$$0.1\text{–MeV}\ H^- + Xe \rightarrow H(n{=}2) + e^- + Xe^* \ . \tag{11}$$

Their results are shown in Fig. 4 in comparison with our theoretical results[4] for this process. The sharp cusp feature in both the experimental and theoretical results is the earmark of the finite threshold cross sections in the projectile frame. The cusp arises because in transforming the finite projectile frame DDCS's to the laboratory frame one must divide the DDCS's by k_P, the detached electron's momentum in the projectile frame.[15] At and near threshold, $k_P \approx 0$, which gives rise to the cusp behaviour seen in the laboratory frame results in Fig. 4. The experimental results[14] were measured at 0° in the laboratory with an angular acceptance of 0.506°. As the theoretical results show, the cusp feature is disappearing rapidly as θ_L increases from $\theta_L = 0°$. Thus a convolution of the theoretical results over a finite angular range is necessary for a precise comparison with the experimental results. However, qualitatively, the agreement of theory and experiment regarding this cusp feature is quite good.

The shoulder features on either side of the cusp in both theoretical and experimental results is another earmark of the low–energy states of the H(n=2) – e^- system. These shoulders are due to the $^1P+$ shape resonance shown in Fig. 2 for the related detachment process given in Eq. 9. For angles $\theta_L \lesssim 0.4°$ the kinematic transformation from the projectile to the laboratory frame results in a sampling of the $^1P+$ shape resonace feature twice, once on either side of the so–called equal velocity energy of the detached electron (in which the detached electron travels with the same velocity as the incident H^- projectile). The asymmetry seen experimentally in the relative magnitude of the $^1P+$ shoulder features is not reproduced well by the theoretical calculations and remains a question for future investigations.

Exhibition of Gailitis–Damburg oscillations[11] by studying collisional detachment processes is difficult. In collisional detachment all the final

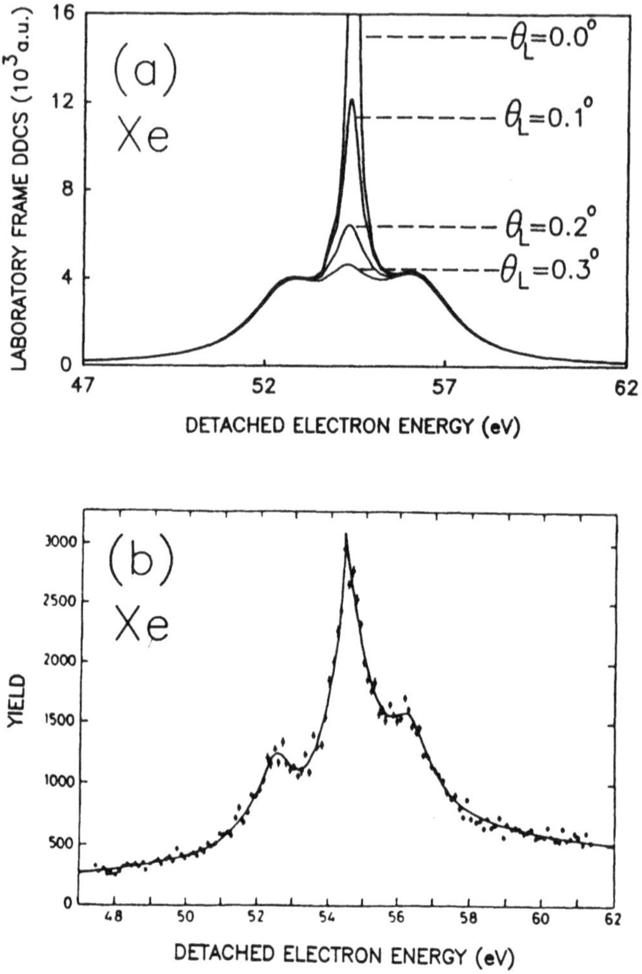

FIG. 4. Laboratory–frame DDCS's for the process 0.1 MeV $H^- + Xe \rightarrow H(n{=}2) + e^- + Xe^*$. (a) Theoretical results of Liu and Starace (Ref. 4) for electron detachment angles $\theta_L = 0.0°$, 0.1°, 0.2°, and 0.3°. (b) Experimental results of Andersen, Bangsgaard, and Sørensen (from Ref. 14) for $\theta_L = 0.0°$. The experimental angular acceptance is 0.506°.

state channels are populated. Hence the strong $^1P+$ shape resonance feature dominates the cross sections except for the very narrow energy region near threshold in which the $^1P+$ channel cross section is smaller than those for the channels which have finite threshold cross sections. But in this region, the laboratory frame DDCS has a cusp. Thus another process must be used to make the Gailitis–Damburg oscillations manifest.

4. Photodetachment of H^-

Because of electric dipole selection rules, the photodetachment process in Eq. 2 is a much more selective process than collisional detachment for studying $H(n=2) - e^-$ states. Since the initial state of H^- has ^1S symmetry, photodetachment accompanied by excitation of H(n=2) populates only the three $^1P^o$ final state channels whose effective potentials are shown in Fig. 1, i.e., the $^1P+$, $^1P-$ and $^1P(pd)$ channels.

As discussed in detail by Lin,[8] the $^1P+$ potential is attractive at short distances and weakly repulsive at large distances thereby giving rise to a shape resonance (seen experimentally at about 18 meV above threshold). The $^1P(pd)$ potential is strongly repulsive at all distances. Finally, the $^1P-$ potential is repulsive only at short distances, but is attractive at large distances. This channel has a complex effective angular momentum, λ_μ, and, as discussed above, gives rise to a small but finite cross section at threshold. However, due to the shape resonance near threshold, Gailitis–Damburg oscillations[11] near threshold are obscured. In addition, Gailitis–Damburg oscillations are extremely weak for the $^1P-$ channel.[16] For these reasons, we do not examine the near threshold behavior of the cross section in detail for this process.

The dominant feature of the near threshold photodetachment cross section of H^- is the $^1P+$ shape resonance feature. Our results for the total n = 2 cross section, i.e., $\sigma_{2s} + \sigma_{2p}$, are shown in Fig. 5 in comparison with the relative experimental data of Butterfield.[17] As pointed out by Lin,[8] the hyperspherical potential $^1P+$ predicts a shape resonance about 18.9 meV too high. In order to compare our n = 2 cross sections with experiment, we have therefore shifted our curves 18.9 meV lower in energy *for this figure only*. The experimental data in Fig. 23 of Ref. 17 have a non–zero back ground below threshold; we have subtracted this background from the data above threshold. Furthermore, we have normalized the experimental data to our theoretical prediction at the peak of the shape resonance. As shown in Fig. 5, our theoretically predicted n = 2 cross section is somewhat wider in energy than measured experimentally. Nevertheless, the agreement is quite reasonable considering that our final state hyperspherical potentials are uncoupled. Indeed, the adiabatic hyperspherical results appear to be intermediate in accuracy between the three–channel close–coupling results of Hyman, Jacobs, and Burke[18] and the very detailed results for the total cross section of Broad and Reinhardt,[19] as discussed in detail elsewhere.[5]

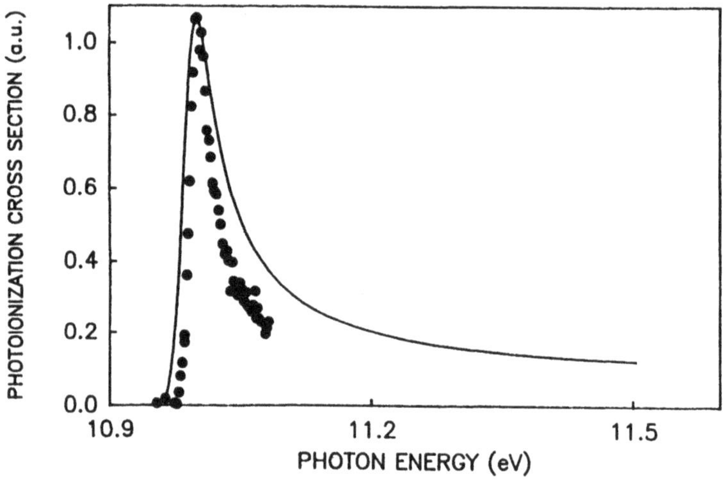

FIG. 5. Photodetachment cross section for the process $\gamma + H^- \rightarrow H(n{=}2) + e^-$. Solid line: (dipole length) adiabatic hyperspherical results of Liu, Du, and Starace (Ref. 5) shifted to the experimental peak position. Solid circles: relative experimental data of Butterfield (Ref. 17) normalized to the theoretically predicted peak height.

5. Two–Photon Detachment of H^-

The two–photon detachment process in Eq. 3 is a very favorable one for observing Gailitis–Damburg oscillations[11] above the H(n=2) threshold.[5] This is so for two reasons. First, electric dipole selection rules do not permit population of $^1P^o$ final state channels. Hence the strong shape resonance in the $^1P+$ final state channel about 18 meV above threshold cannot obscure these near–threshold oscillations. Second, the two–photon process does populate $^1S^e$ and $^1D^e$ final state channels, one of which, the $^1D+$ channel, is the only one with significant, undamped oscillations above threshold.[5]

Before demonstrating these Gailitis–Damburg oscillations[11] for this process, we must ask how we can be sure that the wiggles our calculations give for the two–photon detachment plus excitation cross sections of H^- are really due to long–range dipole field affects and are not due to some other cause. The answer is that the generalized quantum defect theory (QDT) of Greene, Fano, and Strinati[20] for a long range dipole field enables us to disentangle dipole–field effects from our numerical results analytically. In this way we are able to state with assurance which features of our cross section results are truly the Gailitis–Damburg oscillations[11] and which features are energy–dependent wiggles arising from other causes.

Through use of the QDT for long–range dipole fields,[20] one may show that our adiabatic hyperspherical radial functions, defined by Eq.(5), tend asymptotically to

$$F_{\mu k}(R) \underset{R\to\infty}{\sim} (2/\pi k)^{1/2} \sin(kR + \xi_\mu + \eta_\mu), \tag{12}$$

where η_μ is the phase shift in the μth channel and ξ_μ is an analytic phase dependent on the effective angular momentum λ_μ characterizing the long-range dipole interaction of the H(n=2) – e^- system.[20] For real values of λ_μ,

$$\xi_\mu \equiv -\frac{1}{2}\pi\lambda_\mu \quad , \tag{13}$$

while for complex values of λ_μ,[20]

$$\xi_\mu \equiv -\frac{1}{4}\pi + \theta_\mu \quad , \tag{14}$$

where (cf. Eq. 8)

$$\theta_\mu \equiv -\tan^{-1}\frac{\tan[\alpha_\mu \ln(k/2)+x_\mu]}{\tanh(\pi\alpha_\mu/2)} \tag{15}$$

and

$$x_\mu \equiv \arg\Gamma(1-i\alpha_\mu) \ . \tag{16}$$

The generalized QDT may also be used to extract the long–range dipole–field–induced energy dependence of $F_{\mu k}(R)$ by representing our adiabatic hyperspherical radial wave functions as[21]

$$F_{\mu k}(R) = N_\mu(k)\, \mathring{F}_{\mu k}(R) \tag{17}$$

where $N_\mu(k)$ is an effective normalization factor an which determines essentially all of the energy dependence of the radial wave function near $R \approx 0$, and where $\mathring{F}_{\mu k}(R)$ is a more smoothly varying function of k. The oscillatory, energy–dependent normalization factor $N_\mu(k)$ is an analytically known function of ln k.[5,21]

There are two ways in which an attractive dipole field introduces oscillations in measured cross sections on a ln k energy scale. The first is due to the rapid variation of the analytically determined dipole phase θ_μ (cf. Eqs. 14 and 15) for those hyperspherical channels μ having complex values of the effective angular momentum λ_μ. This analytically determined phase θ_μ (through ξ_μ) appears explicitly in the phase factor included in the two–photon transition amplitudes.[5] Interference effects between different amplitudes, such as occur commonly in calculating the angular distributions for the detached electrons, generally lead to sizable, undamped oscillations in the corresponding cross sections due to the rapid decrease of the analytically determined phases θ_μ with increasing ln k. This analytic behavior is shown in Fig. 6 for all three channels having complex λ_μ above the H(n=2) threshold.

The second way the long–range dipole field introduces oscillations in the cross sections is through the effective normalization $N_\mu(k)$ introduced in Eq. 17. Its behavior is shown in Fig. 7 for each of the three channels above the H(n=2) threshold having complex λ_μ. One sees clearly that whereas the long–range dipole–field–induced oscillations of $N_\mu(k)$ for the $^1S(K = +1)$ and $^1P-$ channels are strongly damped, those for the $^1D+$ channel are quite sizable.[5]

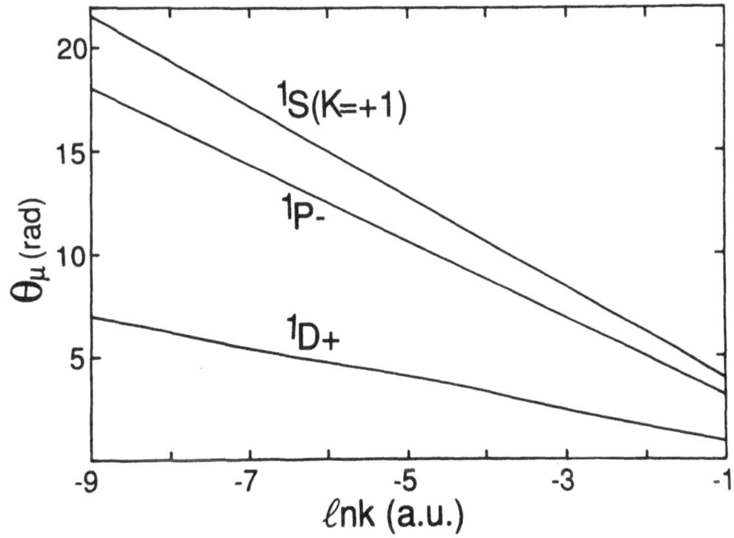

FIG. 6. Analytic phase θ_μ [defined in Eq. 15] vs lnk, where k (a.u.) is the detached–electron momentum, for the three adiabatic hyperspherical channels $\mu = {}^1S(K = +1)$, $^1P-$, and $^1D+$.

The total cross sections for the two–photon detachment of H^- with excitation of H(n=2) (cf. Eq. 3) are given for the cases of linearly (L) and circularly (C) polarized light in Figs. 8(a) and 8(b) respectively. In Figs. 8(c) and 8(d) we have used the generalized QDT[20,21] to extract analytically the energy–dependence arising from the long–range dipole field in order to give renormalized cross sections.[5] One sees clearly that for electron momenta such that ln k $\lesssim$ – 3.0, the oscillations of the cross sections are due to the long–range dipole field. Now, for ln k $\lesssim$ – 6.0, the assumed degeneracy of H(2s) and H(2p) breaks down due to spin–orbit and Lamb shift effects. Thus, for – 6.0 ≤ ln k ≤ – 3.0 or, alternatively, for detached electron kinetic energies from ≈ 0.1 meV to ≈ 34 meV, the energy dependence of the cross sections may be ascribed to Gailitis–Damburg oscillations.[11] As shown in Figs. 8(a) and 8(b), this energy region corresponds to a half–cycle of such oscillation over which the L cross section increases by ≈ 30% and the C cross section increases by ≈ 50%.

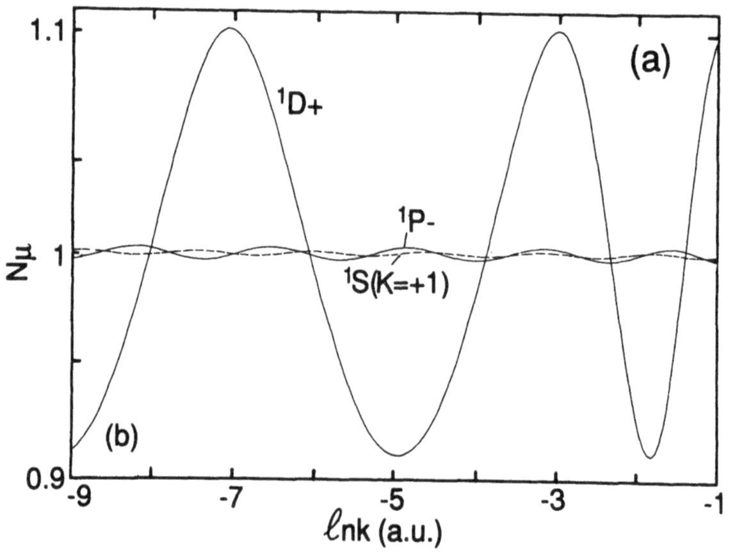

FIG. 7. Normalization factors $N_\mu(k)$ [cf. Eq. 17] for the three adiabatic hyperspherical channels $\mu = {}^1S(K = +1)$, ${}^1P-$, and ${}^1D+$ vs ln k, where k is the detached–electron momentum.

In Fig. 9 we present our results[5] for the total n = 2 differential cross section, which is the sum of the differential cross sections for the 2s and 2p states. Results are given for the six angles, $\theta_k = 0°, 18°, 36°, 54.7°, 72°$, and $90°$. [Note that the results labelled $\theta_k = 54.7°$ are actually calculated for $\theta_k = 54.7356°$, at which $P_2(\theta_k) = 0$.] One sees from this figure that the energy dependence of the differential cross section in the region $-6 \lesssim \ln k \lesssim -3$ (over which long–range dipole field effects play the major role) is highly dependent on the angle θ_k at which the photoelectron is detected. This energy dependence may be enhanced by use of linearly polarized light and small angles of detection, θ_k.

Note that the energy–dependences of the total two–photon detachment cross sections presented in Fig.8 are governed primarily by the long range dipole field normalization factors $N_\mu(k)$ (Cf. Fig. 7). The differential cross sections in Fig. 9 are strongly influenced in addition by the rapidly decreasing analytic phases θ_μ (Cf. Fig. 6). Indeed the energy–dependence of the

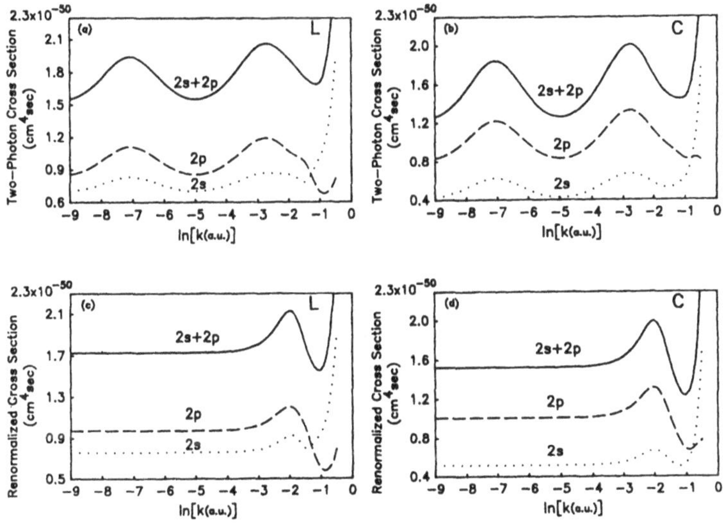

FIG. 8. Generalized two photon cross sections for the cases of linearly (L) and circularly (C) polarized light for the process $2\gamma + H^- \rightarrow H(2s,2p) + e^-$ plotted vs. ln k, where k(a.u.) is the photoelectron momentum. (a) L results; (b) C results; (c) renormalized L results; (d) renormalized C results.

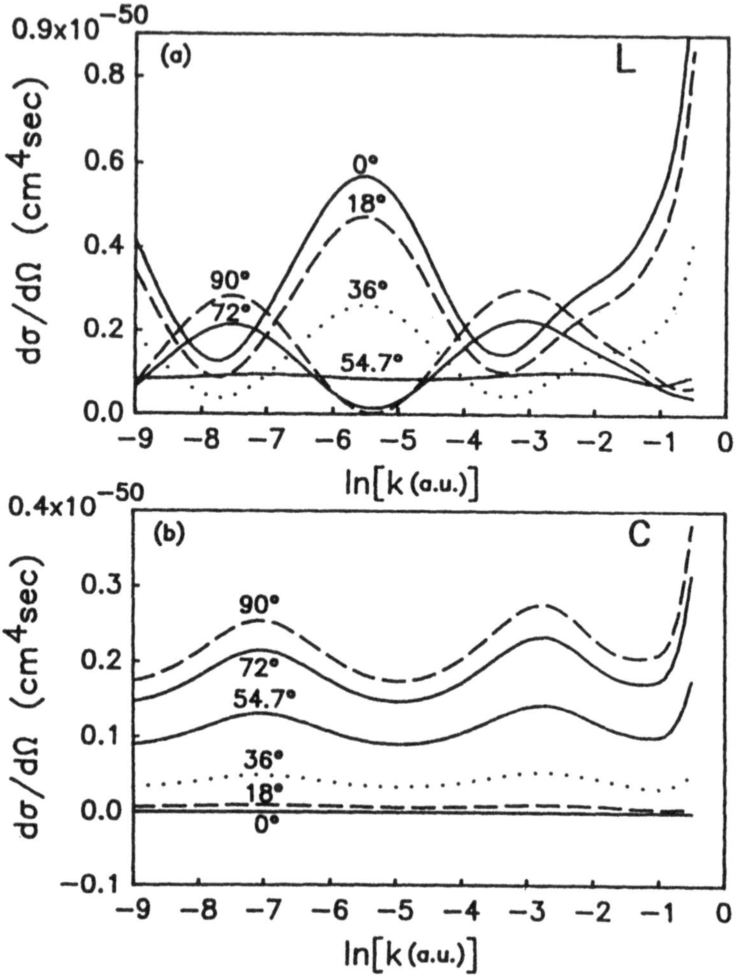

FIG. 9. Differential cross sections for the cases of linearly L and circularly (C) polarized light for the process $2\gamma + H^- \rightarrow H(n{=}2) + e^-$ plotted vs. ln k, where k(a.u.) is the photoelectron momentum, for the detached electron angles $\theta_k = 0°, 18°, 36°, 54.7356°, 72°$, and 90°. (a) L results; (b) C results.

asymmetry parameters for the two-photon process (Cf. Ref. 5) is primarily governed by these analytic phases. Thus, the long-range dipole field effects due primarily to $N_\mu(k)$ can be found by measuring the total cross sections, while those due primarily to θ_μ can be found by measuring the angular distribution asymmetry parameters.[5]

6. Conclusions

In this paper we have shown how the low-energy states of the fundamental H(n=2) – e^- three-body system influence three very different processes for detaching the H^- ion. We have also shown that these three different detachment processes are each best suited to make manifest a different one of the properties of the long-range dipole field interactions within the H(n=2) – e^- system. Thus, collisional detachment of H^- with measurement of the detached electron in the forward direction is best suited for demonstrating the finite threshold cross sections of those channels which have an attractive dipole-field interaction at asymptotic distances. This is so because the kinematic transformations from the projectile frame to the laboratory frame lead to a cusp behavior in the laboratory frame DDCS's which is possible only because of the finite threshold cross sections in the projectile frame.

Single photon detachment of H^- with excitation of H(n=2) is ideal for observing the shape resonance feature in one of the $^1P^o$ final state channels above the H(n=2) threshold. Electric dipole selection rules result in only final state channels having $^1P^o$ symmetry. The $^1P+$ channel, in which the shape resonance occurs, has a cross section so much larger than those for other allowed $^1P^o$ channels that the shape resonance feature can be observed very clearly.

Finally, the two-photon detachment of H^- with excitation of H(n=2) is the most likely process in which to observe the long-range dipole-field-induced oscillations above threshold that were predicted by Gailitis and Damburg[11] but which have yet to be observed experimentally in any process. On the one hand, the $^1P^o$ shape resonance feature is not populated in this process, and hence cannot obscure these oscillations. On the other hand, the $^1D+$ final-state channel which is populated has the largest amplitude for such oscillations of any final state channel. Theoretical calculations[5] predict a half cycle of such oscillation over the energy region from 0.1 meV to about 34 meV above threshold.

7. Acknowledgment

This work was supported in part by the U.S. Department of Energy, Office of Basic Energy Sciences, Division of Chemical Sciences, under Grant No. DE–FG02–88ER13955.

8. References

1. J. Macek, J. Phys. B **1** (1968) 831.
2. C.R. Liu and A.F. Starace, Phys. Rev. Lett. **62** (1989) 407.
3. C.R. Liu and A.F. Starace, Phys. Rev. A **40** (1989) 4926.
4. C.R. Liu and A.F. Starace, Phys. Rev. A **42** (1990) 2684.
5. C.R. Liu, N.Y. Du, and A.F. Starace, Phys. Rev. A **43** (1991), in press.
6. U. Fano, Rep. Prog. Phys. **46** (1983) 97.
7. A.F. Starace, in Fundamental Processes of Atomic Dynamics, edited by J.S. Briggs, H. Kleinpoppen, and H.O. Lutz (Plenum, New York, 1988), pp 235–258.
8. C.D. Lin, Phys. Rev. Lett. **35** (1975) 1150.
9. M.J. Seaton, Proc. Phys. Soc. (London) **77** (1961) 174.
10. C. Greene, U. Fano, and G. Strinati Phys. Rev. A **19** (1979) 1485.
11. M. Gailitis and R. Damburg, Zh. Eksp. Teor. Fiz. **44** (1963) 1644 [Sov. Phys. JETP **17** (1963) 1107], and Proc. Phys. Soc. London **82** (1963) 192; M. Gailitis, in *Atomic Physics 6*, edited by R. Damburg and O. Kukaine (Plenum, NewYork, 1978), pp. 249–266.
12. C.D. Lin, Phys. Rev. A **29** (1984) 1019; Adv. At. Mol. Phys. **22** (1986) 77. Specifically, our channels μ converging to the H(n=2) level have the following labels in Lin's $(K,T)^A$ notation: $^1S(K = \pm 1)$: $(\pm 1,0)^+$; $^1P+$: $(0,1)^+$; $^1P-$: $(1,0)^-$; $^1P(pd)$: $(-1,0)^0$; $^1D+$: $(1,0)^+$.
13. M.M. Duncan and M.G. Menendez, Phys. Rev. A **16** (1977) 1799.
14. L.H. Anderson, J.P. Bangsgaard, and J. Sørensen, Phys. Rev. Lett. **57** (1986) 1558.
15. See, e.g., Eq. (44) of Ref. 3.
16. C.H. Greene and A.R.P. Rau, Phys. Rev. A **32** (1985) 1352.
17. Kenneth B. Butterfield, The University of New Mexico, Ph.D. Thesis (Los Alamos National Laboratory Thesis No. LA–10149–T, 1984), Sec. IV. Values for the n = 2 cross section in Fig. 23 appear on pp. 124–5.
18. H.A. Hyman, V.L. Jacobs, and P.G. Burke, J. Phys. B **5** (1972) 2282.
19. J.T. Broad and W.P. Reinhardt, Phys. Rev. A **14** (1976) 2159.
20. C. Greene, U. Fano, and G. Strinati, Phys. Rev. A **19** (1979) 1485.
21. U. Fano and A.R.P. Rau, Atomic Collisions and Spectra (Academic, New York, 1986), Secs. 5.4 and 5.7.3.

BOUND-BOUND AND BOUND-FREE TRANSITION PROBABILITIES OF C, N^+ AND O

VLADIMIR ESCALANTE
Instituto de Astronomía, UNAM
Ap. Postal 70 – 264, México, DF 15900, MEXICO

ABSTRACT

A semiempirical model potential approximation has been developed for the accurate theoretical calculation of bound-bound and bound-free transition probabilities involving Rydberg states of one electron systems outside open shell cores. The model potential includes a core polarization term, and uses two parameters that are adjusted so that observed energies of a given Rydberg series are reproduced. In many cases comparison with other published results shows significant differences, especially when core polarization has not been included in frozen core calculations. The results have been used in the calculation of the spectra produced in cosmic plasmas by fluorescence by excitation of resonant lines, and by the radiative recombination of electrons in the continuum to excited states of atoms or ions, and subsequent decay via one-electron bound-bound transitions.

1. Introduction

The interpretation of observations of astrophysical and atmospheric plasmas often requires knowledge of transition probabilities of excited states of non-hydrogenic atoms and ions. Observations of low density plasmas often show emission lines produced by permitted transitions from excited states. The excitation of these lines is due in many cases to radiative and/or dielectronic recombination of ions and electrons to excited states, and decay of the excited state X^* to another bound state by a radiative transition:

$$X^+ + e^- \to X^* + \text{photon} \to X^{*\prime} + \text{photon} . \tag{1}$$

An alternative excitation mechanism that has often been suggested to produce the line emission is fluorescence by photoexcitation from the ground state:

$$X + \text{photon} \to X^* \to X^{*\prime} + \text{photon} . \tag{2}$$

In order to predict the emission spectrum of a low density recombining plasma, it is necessary to consider free-bound and bound-bound transition probabilities to many

Rydberg states. Experimental values are available for only low excitation states, and many *ab initio* calculation methods consume an excessive amount of time to give accurate probabilities.

A model potential method can be used to calculate in an economic and accurate way probabilities of dipole-allowed transitions. One-valence electron model potentials have been used to study alkali-like atomic and molecular systems, but the theory can be extended to the case of two-valence electrons outside a closed shell.[1] In low density plasmas doubly excited states play a limited role in the radiative recombination mechanism. In most cases the electron core of the atom or ion is in its ground configuration, and the calculations can be simplified by using a one-electron model potential. In the following section the model potential method is described, and, in section 3, some results are compared with other theoretical and experimental values.

2. The model potential.

The interaction of a valence electron with a core of m electrons and a nuclear charge Z can be described asymptotically by the potential:

$$V(r \to \infty) = -\frac{Z-m}{r} - \frac{\alpha_D}{2r^4} - \frac{\alpha_Q}{2r^6} + \frac{3\beta}{r^6} , \tag{3}$$

where α_D, and α_Q are the dipole and quadrupole polarizabilities of the core, and β is the first order non-adiabatic retardation correction.[2] Rydberg states that do not strongly penetrate the core can be described by a one electron Hamiltonian that uses the above potential with some corrections. If the valence electron wave function ϕ is separated from the core electrons wave function, $\psi(\mathbf{r_1},\ldots,\mathbf{r_m}|\mathbf{r}) = \chi(\mathbf{r_1},\ldots,\mathbf{r_m}|\mathbf{r})\phi_{\mathbf{nl},\gamma}(\mathbf{r})$, and it is assumed that the valence electron moves independently of the core electrons, $\phi_{nl,\gamma}(\mathbf{r})$ can be found from

$$\left[-\frac{1}{2}\nabla^2_{\mathbf{r}} + V(r) - E_{nl,\gamma}\right]\phi_{nl,\gamma}(\mathbf{r}) = 0 , \tag{4}$$

where $V(r)$ is a model potential that must be chosen so that the eigenvalues $E_{nl,\gamma}$ reproduce the experimental values. Atomic units are used throughout. The subscript γ refers to the total symmetry of the wave function described by appropriate quantum numbers.

If the excitation energies of the valence electron are much smaller than the excitation energies of the core, the potential is approximately independent of the energy, and will depend only on the total symmetry being considered nl, γ. The model potential that we employ approximates V(r) as[1,3]

$$V(r) = -\frac{Z}{r} + V_c(r) + V_1(r) + V_2(r) \ , \tag{5}$$

where

$$V_c(r) = \sum_{i=1}^{m} \langle \chi | \frac{1}{|\mathbf{r} - \mathbf{r}_i|} | \chi \rangle \tag{6}$$

is the repulsive potential due to the charge distribution of the core electrons,

$$V_1(r) = a_\gamma \exp(-k_\gamma r) \tag{7}$$

represents a short range local approximation to the exchange interaction of the valence electron with the core and other short range correlation effects, and

$$V_2(r) = -\frac{\alpha_D}{2r^4} W_6(r/r_0) - \frac{\lambda}{2r^6} W_8(r/r_1) \tag{8}$$

accounts for the dipole, and possible quadrupole components of the core. The polarization α_D can be obtained from theoretical or experimental estimations. Cutoff functions $W_n(x) = 1 - \exp(-x^n)$ suggested by Caves and Dalgarno[4] must be introduced to obtain the correct behavior for small r. Since the interaction of the valence electron with the core is approximately described by the polarization terms, the core wave functions $\chi(\mathbf{r}_1, \ldots, \mathbf{r}_m | r)$ used in the calculation of Eq. 6 can be substituted by the frozen core wave function $\chi(\mathbf{r}_1, \ldots, \mathbf{r}_m | \infty)$. The core potential can be obtained in this way from analytic antisymmetrized Hartree-Fock orbitals[5] for the ground state of the ion with charge $Z - m$. In many cases the model potential term V_1 can effectively compensate the inaccuracies introduced in the calculation of α_D or V_c.

The model potential parameters a_γ, k_γ, r_0, and r_1 must be adjusted so that the calculated energies $E_{nl,\gamma}$ reproduce experimental values for the symmetry being considered. More exponential terms can be added in V_1 if higher accuracy is needed, but this is only possible if more experimental energy values are known. For many atoms and ions, the number of experimental energies is limited for each Rydberg series, and the number of adjustable parameters must be reduced. The cutoff radius r_0 can be chosen as the radius that contains 99% of the core charge.[6,7] Several Cooper-Fano minima have been correctly predicted with this choice. We have not considered quadrupole terms in the model potential. Theoretical calculations for an $np\ ^2P^o$ core indicate that quadrupole terms introduce perturbations in the energy levels that are smaller than the accuracy of our calculated energies.[8]

In open shell systems, the valence electron is equivalent to the core electrons for ground configurations, and the electron-electron interaction perturbs the energy of the state strongly. Thus the simple form of $V_1(r)$ cannot reproduce the observed energy of the ground configuration and many terms in the Rydberg series with the same set of values $(a_\gamma,\ k_\gamma)$, and large errors can be expected in calculated probabilities involving the ground state. One way to solve the problem is to find a special set of values to be used in the calculation of the ground state wave function, $(a'_\gamma,\ k'_\gamma)$, by fitting the ground energy and a few excited state energies. Alternatively published theoretical values calculated with more sophisticated techniques or experimental transition probabilities involving the ground configuration can be used in the computation of the recombination spectrum.

Eq. 4 separates, and the wave function can be written as

$$\phi_{nl,\gamma}(\mathbf{r}) = \frac{1}{r} R_{nl,\gamma} Y_l^m(\hat{r}) \ . \tag{9}$$

The least squares method used here to find the model potential parameters describes the wave function $R_{nl,\gamma}(r)$ with a basis set of Slater-type orbitals. The number of orbitals needed increases with the number of energies in the Rydberg series that must be fit. Typically 20 Slater functions and 8 exponents were used in the basis set. When only two free parameters are being adjusted in the model potential, at least three energies must be included in the least squares fit. However, it is better to include as many energies as possible in the fit because this introduces more constraints in the range of possible values for the parameters. The fit to the parameters a_γ, and k_γ for given values of α_D and r_0 is not unique for a fixed degree of accuracy in reproducing the observed energies. Different choices of basis set can produce large variations in the parameters that, nevertheless, give equivalent forms of the potential. Care must be exercised to avoid linear dependence, and to find a model potential that reproduces the asymptotic behavior of the energies in the Rydberg series being considered. When the fit is done correctly, the calculated energies show a smooth asymptotic behavior for the quantum defect, and adding observed energies of higher levels into the fit does not alter drastically the model potential parameters.

Fitting as many observed energies as possible is important because wave functions for free states can be calculated using the same model potential parameters for the corresponding symmetries of bound states. For this approximation to be valid, the parameters must have been calculated using energies of excited states that have attained the asymptotic behavior of the Rydberg series. Correctly fitted parameters give a continuous dipole matrix element across the ionization threshold (Fig. 7).

In order to check the accuracy and correctness of the fit, Eq. 4 is solved by direct numerical integration on a grid for $R_{nl,\gamma}(r)$. Numerical solutions in a variable interval grid are more accurate than variational solutions and can be used to obtain accurate dipole matrix elements for transitions between bound states and between bound and free states. Since excited states essentially do not penetrate the core and the model potential is constructed so that the correct wave function nodal structure is obtained, there is no need to orthogonalize the resulting wave function with the core wave function, and only normalization to unity is necessary.

2. Results

2.1. Bound-bound transitions

The oscillator strength for a dipole-allowed transition from state α to state α' with energies in atomic units E_α and $E_{\alpha'}$ is given by

$$f_{\alpha\alpha'} = \frac{2}{3}(E'_\alpha - E_\alpha)\frac{S_{\alpha\alpha'}}{g_\alpha} , \tag{10}$$

where g_α is the statistical weight of the initial state, and $S_{\alpha\alpha'}$ is the line strength. Although calculated energies with the model potential reproduce observed energies with accuracy better than 1%, it is better to use the observed energy differences of the transitions in Eq. 10 because large errors are introduced by using calculated energy differences in long wavelength transitions. Agreement with experimental f values generally improves in this cases.

The dipole moment is reduced by the core polarization in Eq. 5. When first order corrections are used in the core wave function, the corrected dipole moment element is[7,9]

$$\mathbf{M}_{\alpha'\alpha} = \langle\phi_{\alpha'}| - \mathbf{r}\left(1 - 2\frac{\alpha_D}{r^3}W_n(r/r_0)\right)|\phi_\alpha\rangle , \tag{11}$$

where the cutoff function $W_n(x) = 1 - \exp(-x^n)$ is used to give the correct behavior at small r. The value of n is somewhat arbitrary. Weisheit and Dalgarno[10] recommend $n = 3$, or 4. In any case the oscillator strengths were found to vary by no more than 3% for values of n between 2 and 6, except in cases in which there is severe cancellation in the dipole moment, and the f value is small and very sensitive to the small r behavior of the wave functions. The value of the cutoff radius r_0 was taken equal to the the value used in the model potential cutoff function for the dipole polarizability term. This choice seems to predict many Cooper-Fano minima[7,10,11] (see for instance Fig. 1). The line strength is then given in terms of reduced radial wave functions by

$$S_{\alpha\alpha'} = C(\alpha,\alpha')l_{max}\left(\int_0^\infty R_{\alpha'}r[1-2\frac{\alpha_D}{r^3}W_3(r/r_0)]R_\alpha dr\right)^2 , \qquad (12)$$

where $C(\alpha,\alpha')$ is the well known angular factor for transtions between the terms of a multiplet.

Extensive tables of f model potential values have been presented for C and N^+ for principal quantum numbers $n \leq 12$ and orbital angular momenta $l \leq 2$[11,12] Figures 1 to 6 show comparisons of model potential calculations and experimental values of the absorption f values for some C and O Rydberg series.[11,13] Hofsaess[14,15] has calculated f values for C and O using a Thomas-Fermi-Hartree-Fock method. Cohen and McEachran[16] calculated f values for C using a frozen core Hartree-Fock method without taking into account the core polarization. McEachran and Cohen[17] calculated f values for N^+ and gave an improved set of values for C using the frozen core Hartree-Fock method with an l-dependent core polarization potential.

In most cases all calculations and experimental values agree in the first value of the Rydberg series because the cancellation in the dipole element is minimal. For higher terms in the series, discrepancies among different calculations increase substantially. There is better agreement between the model potential values and the Hartree-Fock values of McEachran and Cohen[17] that include core polarization effects in C. The f values of Cohen and McEachran[16] and Hofsaess[14,15] show large systematic differences with the model potential values. Discrepancies are much smaller for transitions involving higher angular momentum states because the non-hydrogenic effects are less important in those cases (Fig. 4). The differences can be understood by realizing that Hartree-Fock calculations that assume a frozen core do not properly treat the interaction between the valence electron and the core, and a correction term, like the dipole polarization potential, must be introduced to account for this interaction. The effect of the polarization term seems to be more important in the modification of the potential, and the derived wave function, rather than in the modification of the dipole operator. If α_D is set to zero in Eq. 12, but wave functions obtained with the polarization potential are used, the f values change by less than one percent, except in cases in which severe cancellation occurs ($f \lesssim 10^{-5}$).

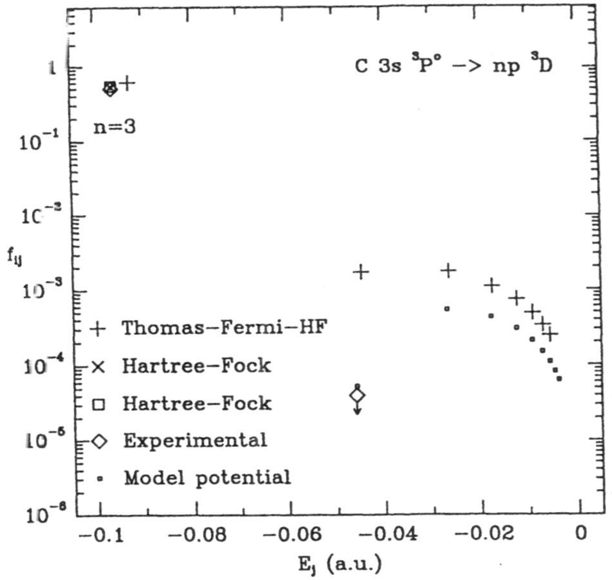

Figure 1.- Absorption f value for C $3s\ ^3P^o$ – $np\ ^3D$ for n = 3 to 12 versus the upper state energy in atomic units (1 a.u.= 27.21 eV). The Thomas-Fermi-Hartree-Fock calculation is from Ref. 14, Hartree Fock calculations are from Ref. 17 (×, includes core polarization) and Ref. 16 (squares). Model potential calculations are from Ref. 11. Experimental values are from Ref. 26 for $n = 3$ and Ref. 27 for $n = 4$ (upper limit).

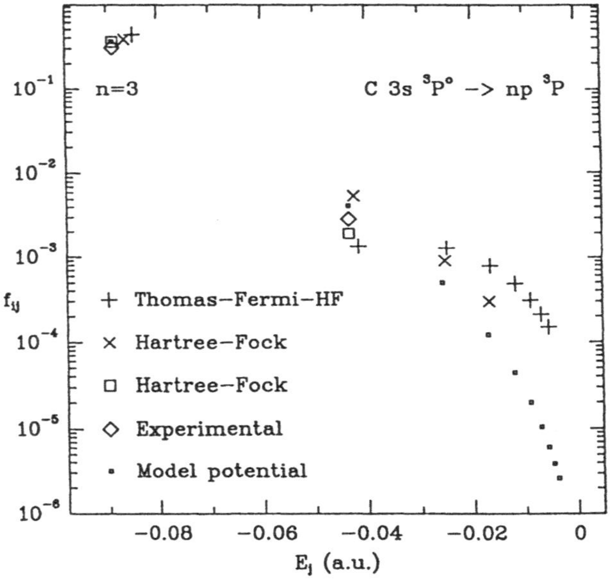

Figure 2.- Same as Fig. 1, for C $3s\ ^3P^o$ – $np\ ^3P$.

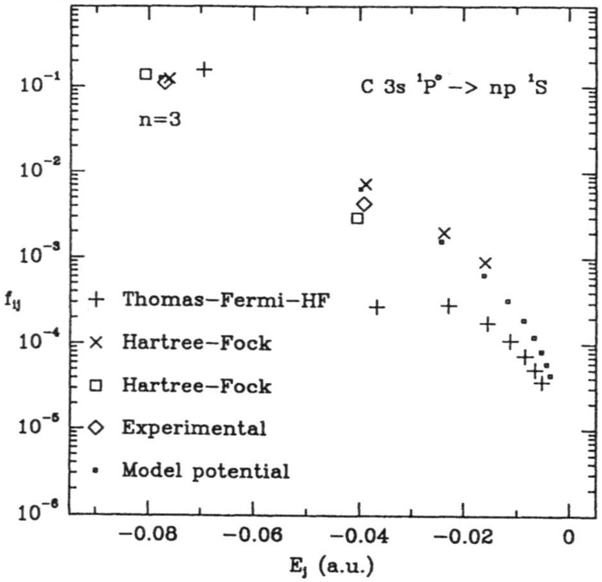

Figure 3.– Same as Fig. 1, for C $3s\ ^1P^o\ -\ np\ ^1S$.

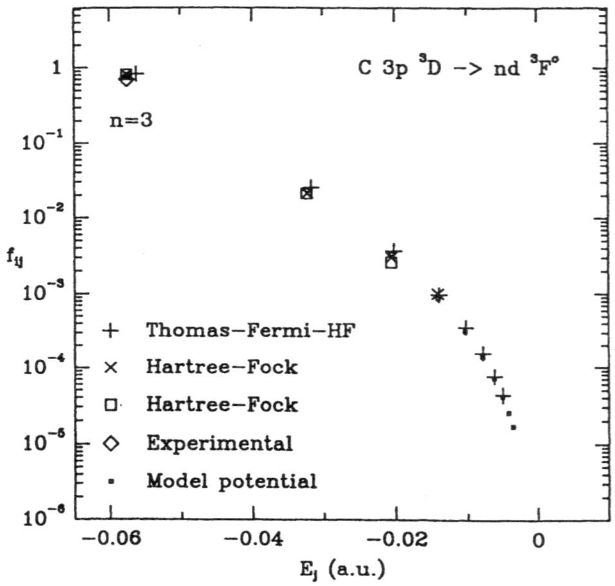

Figure 4.– Same as Fig. 1, for C $3p\ ^3D - nd\ ^3F^o$. The experimental value for $n = 3$ is from Ref. 26.

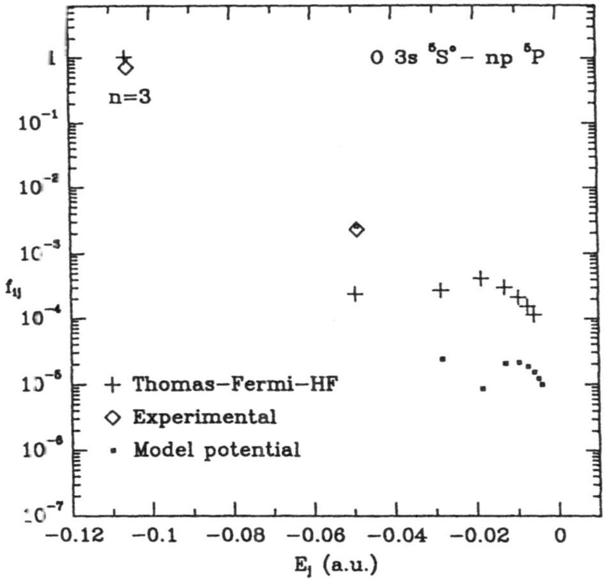

Figure 5.– Same as Fig. 1, for O $3s\ ^5S^o\ -\ np\ ^5P$. The Thomas-Fermi-Hartree-Fock values are from Ref. 15. Experimental values are from Ref. 28 for $n = 3$ and Ref. 26 for $n = 4$.

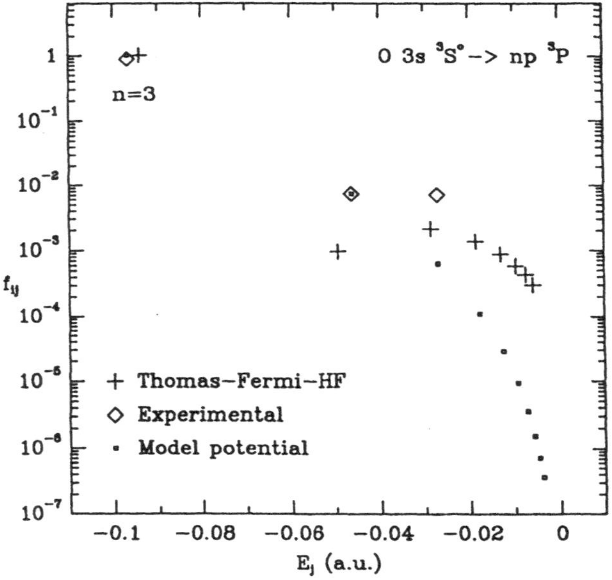

Figure 6.– Same as Fig. 1, for O $3s\ ^3S^o\ -\ np\ ^3P$. Experimental values are from Ref. 26 for $n = 3$, Ref. 28 for $n = 4$, and Ref. 29 for $n = 5$.

2.2 Bound-free transitions.

The calculations of photoionization cross sections can be done using the same model potential for the bound wave functions. The cross section for photoionization of a bound state with energy $E_b < 0$ and statistical weight g_b to a free state f of definite symmetry as a function of the free electron energy E_f is given in cm^2 by

$$\sigma^p_{b\to f}(E_f) = \frac{4\pi^2 a_0^2 \alpha}{3}(E_f - E_b)\frac{C(b,f)}{g_b} l_{max}\left(\int_0^\infty R_f r(1 - 2\frac{\alpha_D}{r^3})W_3(r/r_0)R_b\, dr\right)^2 , \tag{13}$$

where R_b, and R_f are the bound and free reduced wave functions respectively, and $C(b,f)$ is the same angular factor given for the bound-bound case in Eq. 12. The total photoionization cross section $\sigma^p_b(E_f)$ is obtained by summing over all symmetries of angular momenta of the continuum state for which the transition $b \to f$ is allowed. The recombination cross section is found from the detailed balance relation

$$\sigma^r_b(E_f) = \frac{g_b}{g_f}\sum_f \frac{mv^2}{h\nu}\sigma^p_{b\to f}(E_f) , \tag{14}$$

where $g_f = (2L_p + 1)(2S_p + 1)$ is the statistical weight of the recombining ion.

Tables of model potential total photoionization cross sections and comparisons with other calculations will be given.[18] There have been few theoretical calculations of photoionization cross sections for non-hydrogenic excited states. Hofsaess[19] used the scaled Thomas-Fermi method extrapolated to the continuum to find the free electron wave functions of C and O. Peach[20] gave a revised general formula for the cross section based on the quantum defect method.[21] Large discrepancies exist in the lowest angular momentum state cross sections between the different calculations.[18] In many cases there is severe cancellation in the dipole element near threshold. For instance, the Cooper-Fano minimum found in the discrete spectrum of the C $3s\ ^3P^o - np\ ^3D$ series (Fig. 1) decreases significantly the total photoionization cross section of the $3s\ ^3P^o$ term near threshold, and is responsible for the large difference with other calculations. A similar case is shown in Fig. 7 for the O $3s\ ^5S^o$ cross section. However, large differences with other calculations have been found even in cases where there is no significant cancellation in the dipole element.[18]

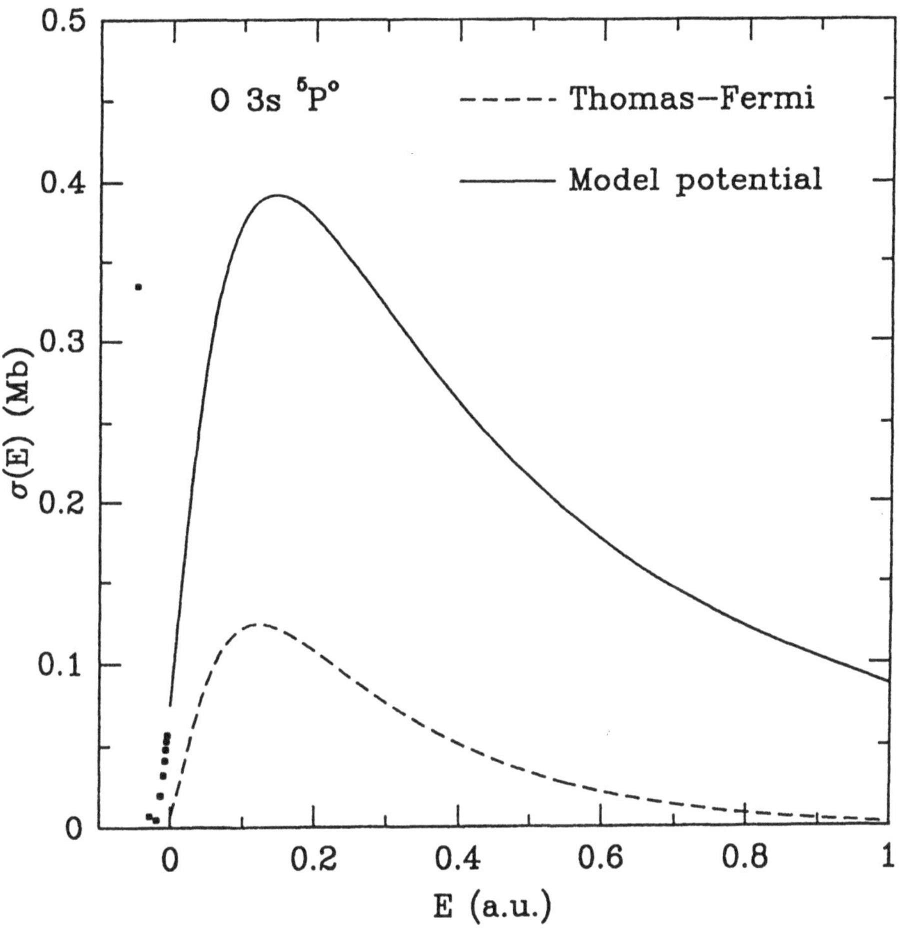

Figure 7.- Comparison of photoionization cross sections (1 Mb= 10^{-18} cm^{-2}) for O $3s$ $^5S^o$ as a function of the free electron energy in atomic units. The scaled Thomas-Fermi calculation is from Ref. 19. The dots are the continuation of the absorption cross section across threshold for the series $3s$ $^5S^o$ – np ^5P for $n = 4$ to 12. The point for $n = 3$ is out of the figure at 39.3 Mb and $E = -0.10576$ a.u. (Ref. 13).

3. Applications and Conclusions

In order to calculate the emission spectrum produced by recombination and/or fluorescence at low densities, the statistical equilibrium equations for the energy levels of the atom or ion must be solved. In a low density plasma irradiated by a low intensity radiation field, population and depopulation of moderately excited states ($n < 40$) takes place primarily by electron-ion recombination and dipole allowed transitions. Stimulated emission terms in atoms can be neglected, and absorption of radiation, occurs only from the ground and perhaps metastable levels. For a radiation field of intensity $J(\nu)$ at frequency ν, the statistical equilibrium equation for the number density of atoms or ions in energy level j, n_j, is

$$n_j A_j = \sum_{k>j}^{\infty} n_k A_{k,j} + n_e n_z \alpha_j(T) + \sum_m n_m J(\nu_{m,j}) B_{m,j} \ , \tag{15}$$

where $A_{k,j}$ and $B_{m,j}$ are the Einstein coefficients for radiative decay and absorption, and $A_j = \sum_{i<j} A_{j,i}$ is the total decay rate. n_m is the density corresponding to the ground or metastable levels. n_e and n_z are the free electron and recombining ion number densities, respectively, and $\alpha_j(T)$ is the bound state recombination coefficient as a function of electron temperature defined as the average over the Maxwellian velocity distribution of the electron flux and the recombination cross section given by Eq. 14:

$$\alpha_j(T) = \langle v \sigma_j^r(E_f) \rangle \ . \tag{16}$$

Eq. 15 has a solution of the form

$$n_j A_j = n_e n_z \alpha_j^{eff} + \sum_m n_m \beta_{m,j}^{eff} \ , \tag{17}$$

where

$$\alpha_j^{eff} = \sum_{k \geq j}^{\infty} \alpha_k C_{k,j} \tag{18}$$

is the effective recombination coefficient for level j, and

$$\beta_{m,j}^{eff} = \sum_{k\geq j}^{\infty} J(\nu_{m,k}) B_{m,k} C_{k,j} \tag{19}$$

is the effective fluorescence coefficient for photoexcitations from the ground or metastable level m and cascade to level j. The "cascade matrix" $C_{k,j}$ is the probability that a recombination or a photoexcitation to level k produces state j after cascading through all possible intermediate states that can be reached by consecutive radiative decays.[22] In a plasma where self-absorption effects can be neglected the emissivity of lines excited by recombination or fluorescence is thus given by

$$4\pi j_{j,i} = h\nu_{j,i} \frac{A_{j,i}}{A_j} (\alpha_j^{eff} n_e n_z + \sum_m n_m \beta_{m,j}^{eff}) \ . \tag{20}$$

At low densities the number of states to be considered is very large because the series in Eq. 18 converges very slowly. In a non-hydrogenic system described in the LS coupling scheme, each l symmetry of a Rydberg state produces several states with different total angular momentum L and total spin S. Thus we typically have to consider a few hundred states and several thousand transitions in the statistical equations that describe the population of the states. Tables of Eq. 18 using model potential data for C and N^+ have been presented by Escalante and Victor.[23]

Cosmic plasmas offer unique opportunities to observe atoms and ions in conditions not attainable in the laboratory, but reliable diagnostic methods to elucidate the quantum processes taking place are necessary. Generally a spectrum produced by recombination is dominated by many lines from high angular momentum states while a fluorescence spectrum enhances only certain emission lines that can be excited after a few cascades following the initial absorption. Comparison with observations of emission lines of N^+ in nova ejecta[24] and ionized nebulae[25] show that the excitation mechanism of the lines in those plasmas is radiative recombination.

The need for more theoretical calculations and experimental measurements of transition probabilities for excited atoms and ions is evidenced by comparing theoretical calculations carried out with different methods. Significant differences can sometimes be found even in cases where there is no severe cancellation in the dipole moment element. The model potential method that we have discussed provides a way to calculate accurately and economically transition probabilities of excited states that in many cases compare well with results from calculations of similar or higher sophistication.

4. References

1. C. Laughlin and G. A. Victor, in *Atomic Physics* **3**, ed. S.J. Smith and G.K. Walters, (1973) p. 247.
2. A. Dalgarno, G. W. F. Drake, and G. A. Victor, *Phys. Rev.* **176** (1968) 194.
3. C. Bottcher and A. Dalgarno, *Proc. Royal Soc. London A.* **340** (1974) 187.
4. T. C. Caves and A. Dalgarno, *J. Quant. Spectrosc. Radiat. Transfer* **12** (1980) 1539.
5. E. Clementi and C. Roetti, *Atomic Data and Nuclear Data Tables* **14** (1975) 177.
6. C. Laughlin, E. R. Constantinides, and G. A. Victor, *J. Phys. B* **11** (1978) 2243.
7. J. C. Weisheit, *Phys. Rev. A* **5** (1972) 1621.
8. E. S. Chang, (1982) private communication.
9. M. Cohen and A. Dalgarno, *Proc. Roy. Soc.* **A293** (1966) 359.
10. J. C. Weisheit and A. Dalgarno, *Chem. Phys. Lett.* **9** (1971) 517.
11. G. A. Victor and V. Escalante, *Atomic Data and Nuclear Data Tables* **40** (1988) 203.
12. G. A. Victor and V. Escalante, *Atomic Data and Nuclear Data Tables* **40** (1988) 227.
13. G. A. Victor and V. Escalante, in preparation.
14. D. Hofsaess, *J. Quant. Spectrosc. Radiat. Transfer* **28** (1982) 131.
15. D. Hofsaess, *J. Quant. Spectrosc. Radiat. Transfer* **42** (1989) 45.
16. M. Cohen and R. P. McEachran, *J. Quant. Spectrosc. Radiat. Transfer* **20** (1978) 295.
17. R. P. McEachran and M. Cohen, *J. Quant. Spectrosc. Radiat. Transfer* **27** (1982) 119.
18. V. Escalante and G. A. Victor, submitted.
19. D. Hofsaess, *Atomic Data and Nuclear Data Tables* **24** (1979) 285.
20. G. Peach, *Monthly Notices of the Royal Ast. Soc.* **71** (1967) 13.
21. A. Burgess and M. J. Seaton, *Monthly Notices of the Royal Ast. Soc.* **120** (1960) 121.
22. M. J. Seaton, *Monthly Notices of the Royal Ast. Soc.* **119** (1959) 70.
23. V. Escalante and G. A. Victor, *Astrophysical Journal Supplement Series* **73** (1990) 513.
24. V. Escalante and A. Dalgarno, *Astrophysical Journal* **369** (1991) 213.
25. V. Escalante, A. Sternberg, and A. Dalgarno, to appear in *Astrophysical Journal* (1991).
26 . W. L. Wiese, M. W. Smith and B. M. Glennon, *Atomic Transition Probabilities NBS 4* (Washington, DC, 1966).
27 . D. W. Jones and W. L. Wiese, *Phys. Rev.* **129** (1984) 2597.
28 . A. Goly, *Acta Phys. Pol.* **36** (1968) 33.
29 . J. Bromander, N. Duric, P. Erman and M. Larsson, *Phys. Scripta* **77** (1978) 119.

FLUORESCENCE STUDIES OF $He^+(n=2)$ PRODUCED BY PHOTOIONIZATION

J. JIMENEZ-MIER,† C. D. CALDWELL, M. G. FLEMMING
Department of Physics, University of Central Florida
Orlando, FL 32816

and

D. L. EDERER
National Institute of Standards and Technology
Gaithersburg, MD 20899

ABSTRACT

A partition of the helium photoionization cross section $\sigma(n=2)$ into the partial-wave cross sections σ_{2skp}, σ_{2pks} and σ_{2pkd} is discussed in this work. This partition was carried out by measuring the angular distribution and the enhancement, due to an external electric field, of the decay fluorescence at 304 Å. Threshold and resonance effects in these partial cross sections are discussed. Very good agreement is found with the results of a many-body perturbation theory (MBPT) calculation.

1. Introduction

Helium is the symplest atomic system in which the effect of electron-electron correlations can be investigated. These correlation effects are readily apparent in the shake-up process

$$He(1s^2) + h\nu \rightarrow He^+(n=2) + e$$

a phenomenon that cannot be explained within the context of the single particle approximation. Recently the attention focused on the photon energy region between the n=2 and n=3 thresholds at 65.4 and 73.0 eV respectively. This region is particularly interesting because of the presence of the autoionizing resonances converging to the n=3 state of the ion.

†Permanent address: Instituto de Ciencias Nucleares, Universidad Nacional Autónoma de México, 04510 México D.F., México

Experiments using synchrotron radiation sources have been carried out to determine the cross section branching ratio[1,2] $\sigma(n=2)/\sigma(n=1)$, the n=2 total cross section,[3] the 2p to 2s branching ratio $R = \sigma_{2p}/\sigma_{2s}$ at selected energies,[4] the photoelectron angular distribution parameter β_{eff},[3,5-7] and the alignment A_c^0 of the 2p ions near the n=2 ionization threshold.[8] Theoretical results for some of these parameters are given in an R-matrix approach,[9] in a many-body perturbation theory (MBPT) calculation,[10] and in a configuration interaction in the complex plane representation calculation.[11]

In this work we discuss the results that can be obtained from a detailed study of the decay fluorescence from $He^+(n=2)$ produced by photoionization.[12] A low resolution measurement of $\sigma(n=2)$, of the 2p to 2s branching ratio R, and of the alignment A_0^c was carried out in the photon energy region between the n=2 and n=3 thresholds. These three measurements allow a partition of $\sigma(n=2)$ into the partial-wave cross sections σ_{2skp}, σ_{2pks} and σ_{2pkd} as follows.[12]

There are three open channels leading to production of n=2 ions, namely, $2skp$, $2pks$, and $2pkd$. Therefore, five independent real parameters are needed to completely determine the density matrix of the system (the partial-wave cross sections σ_{2skp}, σ_{2pks}, and σ_{2pkd}, and two relative phases δ_{12} and δ_{23}).

Since the 2s state in He^+ is metastable, the 2s ions do not participate in the decay proces

$$He^+(n=2) \;\rightarrow\; He^+(1s) \;+\; h\nu'(304\text{Å})$$

On the other hand, the polarization and spatial distribution of the 2p decay fluorescence are given by[13]

$$I(\theta) \;=\; I_0\{1 \;-\; \frac{1}{2}h^{(2)}A_0^c P_2(\cos\theta) \;+\; \frac{3}{4}h^{(2)}A_0^c \sin^2\theta \cos 2\chi \cos 2\beta\} \qquad (1)$$

where we are assuming that linearly polarized radiation was used to ionize the helium atoms, $h^{(2)}$ is a ratio of 6-j symbols that include the effect of the ground state of the ion in the decay process, θ is the angle between the electric field vector of the ionizing radiation and the detection direction, P_2 is the second order Legendre polynomial, χ is the angle between the polarization selected by the analyzer and the electric field vector, in the collision frame the fluorescence polarization vector is written as $\hat{\epsilon} = (\cos\beta, i\sin\beta, 0)$, and A_0^c is the alignment of the ion

$$A_c^0 \;=\; \frac{\sum_{m_j}(3m_j^2 - j(j+1))\sigma(m_j)}{j(j+1)\sum_{m_j}\sigma(m_j)} \qquad (2)$$

In this equation $\sigma(m_j)$ stands for the relative cross section for the production of the m_j state of the ion.

In the case of helium a measurement of the fluorescence polarization is more difficult than a measurement of its angular distribution. Therefore, if no selection of polarization is made, one has to take the average of Eq. 1 over the polarization directions, this makes the third term cancell out, leaving an angular distribution of the form

$$I(\theta) = I_0\{1 - \frac{1}{2}h^{(2)}A_0^c P_2(\cos\theta)\} \tag{3}$$

The fluorescence signal at the magic angle $\theta_m = \frac{1}{2}\cos^{-1}(-\frac{1}{3})$ is independent of the ion alignment, and is thus proportional to the 2p cross section $\sigma_{2p} = \sigma_{2pks} + \sigma_{2pkd}$. Furthermore, in the presence of an external electric field, the 2s and 2p states of the ion are mixed and the fraction of ions left in the 2s state will decay. Therefore, the 2p to 2s branching ratio R can be obtained from a comparison of the fluorescence signal with and without an electric field.

Finally, the alignment A_0^c is related to the 2p partial-wave ratio $\varsigma = \sigma_{2pkd}/\sigma_{2p}$ through[8,12]

$$A_0^c(j = \frac{3}{2}) = -\frac{4}{5} + \frac{18}{25}\varsigma$$
$$A_0^c(j = \frac{1}{2}) = 0 \tag{4}$$

We therefore have three sets of measurements from which the three partial-wave cross sections σ_{2skp}, σ_{2pks} and σ_{2pkd} can be obtained.

2.Experiment

The experiment took place on beam line 1 of the storage ring SURF-II of the National Institute of Standards and Technology. Monochromatized synchrotron radiation was used to ionize a pseudo beam of helium atoms. The angular distribution of the fluorescence in a plane perpendicular to the direction of incidence of the ionizing beam was measured by rotating a detector about an axis lying along the direction of the synchrotron beam. The angle θ was measured with respect to the major axis of the synchrotron radiation polarization ellipse. The toroidal grating monochromator operating at grazing incidence of beam line 1 dispersed the synchrotron light, producing an image of the storage ring beam at the entrance of a capillary that separated our experimental chamber from a differential pumping chamber. The main limitation of this apparatus was its resolution, which was limited by aberrations in the toroidal grating. The best resolution we could get, estimated from the width of the 1_3 autoionizing resonance of helium, was 0.5 eV at a photon energy of 70 eV. The fluorescence signal was detected with a gas-flow proportional counter[14] separated from the interaction region by a VYNS thin film window.

The angular distribution of the fluorescence is given by Eq. 3 only in the case of 100 % linearly polarized ionizing radiation. If the synchrotron radiation is partially polarized, with polarization p Eq. 3 becomes

$$I(\theta) = I_0\{1 + \frac{5}{48}A_c^0(\frac{3}{2})[1 + 3p\cos(2\theta)]\} \quad (5)$$

To determine this polarization p we used a single gold mirror reflecting polarimeter.[15] A gold mirror making an angle of 45° with respect to the direction of incidence of the radiation beam is used to reflect it onto a tantalum photodiode The radiation intensities reflected when the plane of incidence is parallel and perpendicular to the major axis of the polarization ellipse are measured, and, by combining these data with the optical constant of gold[16] one can obtain the polarization of the monochromatized radiation.[17]

The relative throughput of the monochromator was also obtained from the signal reflected by the gold mirror with the plane of incidence parallel and perpendicular to the major axis of the polarization ellipse.

The 2s ions are metastable, and at low pressures they do not participate in the decay process. However, at the helium pressures used in this work ($2 - 4 \times 10^{-4}$ $Torr$), collisions of neutral helium atoms with 2s ions produce a mixing of the 2s and 2p states, and there is a finite contribution from these ions to the fluorescence signal.[18] To determine this 2s contribution, the $He^+(n = 2)$ fluorescence signal with and without an external electric field and at the magic angle was recorded at two helium pressures. We found that the signal due to collisional quenching of the 2s ions was about 11 % of the total fluorescence.

The angular distribution of the fluorescence was measured at selected photon energies. A separate measurement of the alignment was carried out by recording the fluorescence signal at 90° and 0° with respect to the electric field vector of the synchrotron radiation at a finer energy mesh. Both sets of data gave results for the alignment that were in agreement with each other, within the experimental error.

3. Results

The signal recorded at the magic angle and with an external electric field is shown in Fig. 1. These data were fully corrected for background, for changes in the storage ring beam current, and for the relative throughput of the monochromator. These results were further normalized to the absolute cross sections given by Lindle *et al*[3] across the energy region of overlap of both sets of data. Also shown are the results of Lindle *et al*,[3] a many-body perturbation theory (MBPT) calculation by Salomonson *et al*[10] of $\sigma(n = 2)$, and the result of a convolution of these theoretical data with a triangular slit function of 0.5 eV FWHM. Taking into account our low resolution, we can say that there is good agreement with the higher resolution results of Lindle *et al*[3] and with the theoretical results across the 1_3 and 1_4 resonances. However, in the vicinity of the n=2 threshold our results indicate that $\sigma(n = 2)$ might be as small as 50 % of the theoretical prediction.

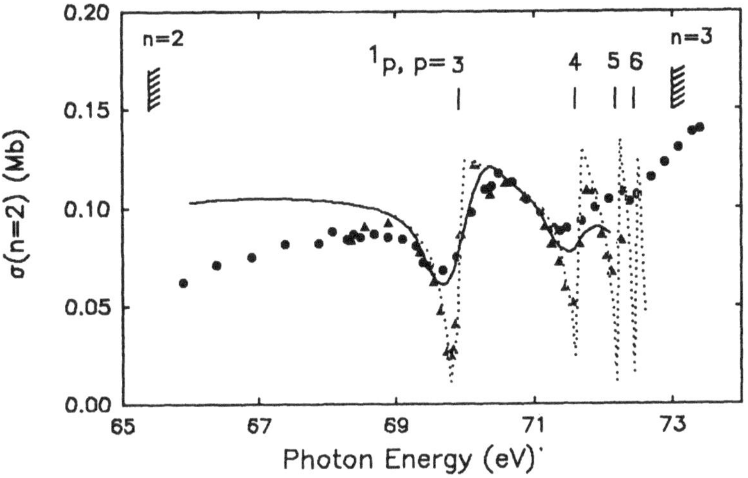

Fig. 1 Partial photoionization cross section of helium leaving the ion in the n=2 state. • this work, △ experimental results of Lindle *et al*, Ref. 3, - - - - MBPT calculation, — result of convolution of MBPT data with a triangular slit function of 0.5 eV FWHM. The position of the $n = 2$ and $n = 3$ thresholds and of the 1_p autoionizing resonances is also indicated.

From the fluorescence recorded at the magic angle with and without an external electric field that completely quenched the 2s ions we obtained the branching ratio R. Combining the results for R with $\sigma(n = 2)$ we were able to determine the partial cross sections σ_{2skp} and σ_{2p}. The 2p partial-wave ratio $\varsigma = \sigma_{2pkd}/\sigma_{2p}$ was obtained from the alignment results, and combining the values of ς with the results for σ_{2p} the 2p partial-wave cross sections were determined.

The partition of $\sigma(n = 2)$ into the partial-wave cross sections σ_{2skp}, σ_{2pks} and σ_{2pkd} is shown in Fig. 2. The results of the MBPT calculation of Salomonson *et al*[10] are also presented here. Once again, for purposes of comparison, these theoretical data were convoluted with a triangular slit function of 0.5 eV FWHM. Near the n=2 threshold the three experimental partial-wave cross sections are again smaller than the corresponding theoretical predictions, with the dominant contribution coming from the 2*pks* channel. The 2*skp* and 2*pkd* cross sections slowly rise with approximately the same slope. The behavior of the cross section across the 1_3 resonance is well reproduced by theory in the case of σ_{2skp} and σ_{2pks}, however, the experimental results indicate that the minimum in σ_{2pkd} should appear after the minimum in σ_{2pks}, and the MBPT calculation gives the opposite behavior.

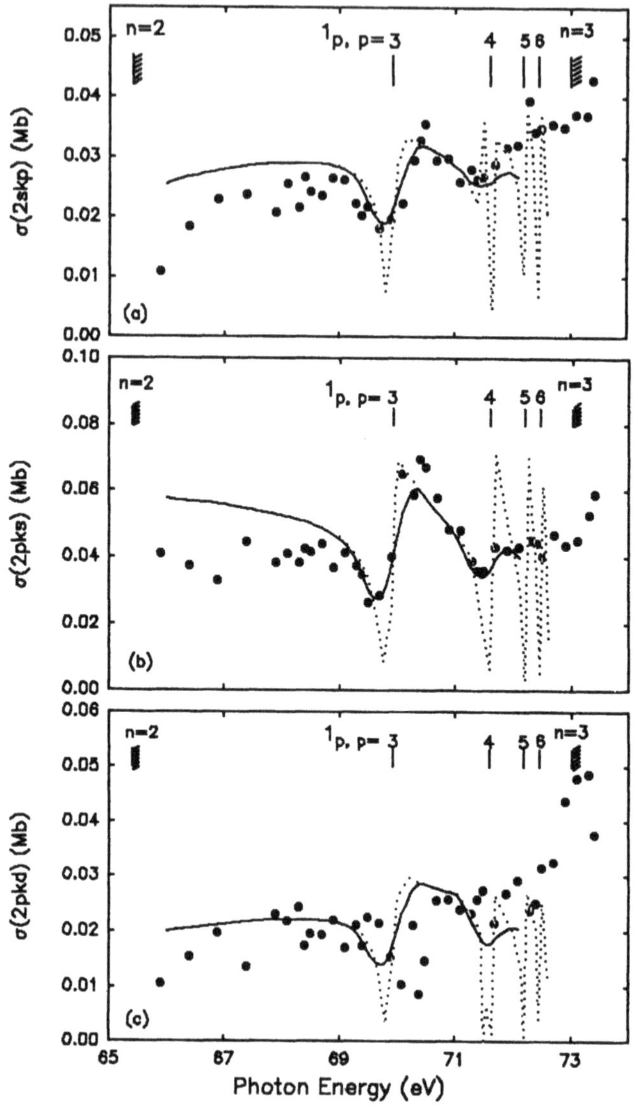

Fig. 2. Helium photoionization partial-wave cross sections. a) σ_{2skp}, b) σ_{2pks}, c) σ_{2pkd}. In all cases: • Experimental results, - - - - MBPT calculation, — result of convolution of MBPT data with a triangular slit function of 0.5 eV FWHM. The position of the $n = 2$ and $n = 3$ thresholds and of the 1_p autoionizing resonances is also indicated.

4. Future work

From these results one can see that a higher resolution study of the 304Å fluorescence should provide excellent means of comparison with theory for the behavior of the paritial-wave cross sections across the region of autoionizing resonances. In particular, one could determine the Starace parameters[19] of the broad 1_p resonances, and, with high enough resolution, one could study the effect of the first few narrow 2_q resonances.

5. Acknowledgements

This work was supported through the National Science foundation under grant PHY-8701193. JJM whishes to thank the Department of Physics of University of Central Florida for their hospitality. JJM, CDC and MGF wish to thank the SURF-II staff for their assistance.

6. References

[1] P.R. Woodruff and J.A.R. Samson, *Phys. Rev. Lett.* **45**, 110 (1980).
[2] F. Wuilleumier, M.Y. Adam, N. Sandner, and V. Schmidt, *J. Phys.-Lettres* **41**, L-373 (1980).
[3] D.W. Lindle, T.A. Ferret, U. Becker, P.H. Kobrin, C.M. Truesdale, H.G. Kerkhoff, and D. A. Shirley, *Phys. Rev. A* **94**, 714
[4] P.R. Woodruff and J.A.R. Samson, *Phys. Rev. A* **25**, 848 (1982).
[5] J. M. Bizau, F. Wuilleumier, P. Dhez, D. L. Ederer, T. N. Chang, S. Krummacher, and V. Schmidt, *Phys. Rev. Lett* **48**, 588 (1982).
[6] V. Schmidt, H. Derenbach and R. Malutzki, *J. Phys. B* **15**, L523 (1982).
[7] P. Morin, M.Y. Adam, I. Nenner, J. Delwiche, M.J. Hubin-Franskin, and P. Lablanquie, *Nucl. Instr. and Meth.* **208**, 761 (1983). (1985).
[8] J. Jiménez-Mier, C.D. Caldwell and D.L. Ederer, *Phys. Rev. Lett.* **57**, 2260 (1986).
[9] M.A. Hayes and M.P. Scott, *J. Phys. B* **21**, 1499 (1988).
[10] S. Salomonson, S.L. Carter and H.P. Kelly, *Phys. Rev. A* **39**, 5111 (1989).
[11] S.M. Burkov, N.A. Letyaev, S.I. Strakhova and T.M. Zajac, *J. Phys. B* **21**, 1185 (1988).
[12] J. Jiménez-Mier, C. D. Caldwell, M. G. Flemming and D. L. Ederer, to be published.
[13] U. Fano and J. H. Macek, *Rev. Mod. Phys.* **45**, 553 (1973).
[14] Gas Flow Proportional Counter, J. E. Manson Co., P.O. Box 1288, Concord, MA 01742.

[15] K. Rabinovitch, L. R. Canfield and R. P. Madden, *Appl. Opt.* **4**, 1005 (1965).
[16] D. W. Lynch and W. R. Hunter, "Comments on the Optical Constants of Metals and an Introduction to Data for Several Metals" in *Handbook of Optical Constants of Solids*, E.D. Palik, ed., Academic Press, Orlando, 1985, pp. 275-367.
[17] J. A. R. Samson, *Nucl. Instr. and Meth.* **152**, 225 (1978).
[18] M. H. Prior and E. C. Wang, *Phys. Rev. A* **9**, 2383 (1974).
[19] A. F. Starace, *Phys. Rev. A* **16**, 231 (1977).

IX. HIGHLY CHARGED ION COLLISIONS

LOW-ENERGY COLLISIONS OF MULTIPLY CHARGED IONS WITH ELECTRONS, ATOMS AND SURFACES

R. A. Phaneuf, F. W. Meyer, D. C. Gregory, C. C. Havener
P. A. Zeijlmans van Emmichoven, S. H. Overbury, and D. M. Zehner
Oak Ridge National Laboratory
Oak Ridge, TN 37831, U.S.A

G. H. Dunn, J. S. Thompson, and E.K. Wåhlin*
Joint Institute for Laboratory Astrophysics
Boulder, CO 80309, U.S.A.

A.C.H. Smith
University College, London, WCIE 6BT, U.K.

ABSTRACT

Results are presented from recent experiments at the ORNL-ECR Multicharged Ion Research Facility which have addressed low-energy interactions of multiply charged ions with electrons, hydrogen atoms and metal surfaces. The specific processes for which results are reported are electron-impact excitation, electron capture and ion neutralization. The objective of these studies is a more detailed understanding of the mechanisms which govern the interactions of multicharged ions at low (or near-threshold) collision energies.

1. Introduction

Low-energy interactions of multiply charged ions are generally dominated by enhanced Coulomb or polarization forces, and/or by the appreciable neutralization energy that such ions carry into the interaction. This neutralization energy is equal to the sum of the binding energies of the q electrons that have been removed from an atom to produce an ion of net positive charge q. This may be thought of as a reservoir of stored electronic potential energy for certain types of interactions. In experimental studies with low-energy ion beams, these q electrons are removed from the atom in an ion source prior to acceleration. The development of advanced sources of multiply charged ions such as the electron-cyclotron-resonance (ECR) and electron-beam ion source (EBIS) now makes possible experimental studies with low-energy ions having charges of 20 or more and carrying hundreds to even thousands of electron volts of neutralization energy. During the last few years, the increasing availability of such ion sources for atomic physics research has spawned considerable activity in this area.

The large Coulomb forces between highly charged ions and other charged particles dominate long-range interactions, modifying particle trajectories, and producing large cross sections for some processes. The large neutralization energy of the ion can dominate kinetic effects in slow collisions. For multiply charged ions, the reduced screening of the nucleus due to missing electrons increases the binding energy of the outer valence electrons, but has a lesser effect on inner-shell electrons. Thus processes involving inner-shell electrons may become relatively more important as the ionic charge increases. For example, in electron-impact ionization of multiply charged ions, inner-shell excitation-autoionization can dominate direct outer-shell ionization by an order of magnitude or more in some cases[1]. Polarization forces due to ion-induced dipole interactions are also enhanced in multicharged ion interactions with neutral particles, since the attractive potential is proportional to the square of the ionic charge:

$$V(r) = -\alpha q^2/2r^4 \qquad [1]$$

where α is the dipole polarizability of the neutral particle, q is the ionic charge, and r is the inter-particle distance. This interaction is relatively weak and short-range compared to the Coulomb interaction, but can play a role in multicharged ion-neutral collisions at electron-volt energies and below.

In this paper, three recent experiments conducted at the ORNL-ECR Multicharged Ion Research Facility will be described which address these properties of multicharged ion interactions. Specifically, results will be presented for low-energy interactions of multicharged ions with electrons, atoms and metal surfaces.

2. Merged-Beams Electron-Energy-Loss Experiment for Electron-Impact Excitation of Ions

In addition to being a fundamental process of intrinsic interest, electron-impact excitation of multiply charged ions is critical to radiative power loss and diagnostics of magnetic fusion plasmas. High-resolution cross-section measurements in the near-threshold region are needed to benchmark detailed close-coupling calculations, and to elucidate the role of resonances. All existing absolute excitation cross-section measurements for multiply charged ions have been based on the crossed-beams approach and on absolute intensity measurements of the radiation emitted as the excited states decay[1]. Such measurements are severely limited by low photon intensities and detection efficiencies ($\sim 10^{-4}$), difficulties with absolute optical calibration, as well as low electron energy resolution (~2 eV). Cascade contributions from excitation of higher levels may also contribute to the measured photon emission, complicating such measurements.

A different experimental approach[2] which circumvents these limitations has been developed at the Joint Institute for Laboratory Astrophysics in Boulder, Colorado for use in conjunction with the ORNL-ECR multicharged ion source. The technique, illustrated in Fig. 1, involves merging fast beams of electrons and multiply charged ions in a uniform axial magnetic field, and detecting electron-impact excitation events via electron energy-loss spectroscopy. Trochoidal (**E** x **B**) analyzers are used to merge and demerge the beams. The magnetic field ensures complete collection of inelastically scattered electrons, which are directed onto a position-sensitive detector. Detailed electron-trajectory modelling calculations are performed in order to predict the locations where both elastically and inelastically scattered electrons will strike the position-sensitive detector. A fast two-beam modulation scheme is used to separate detector counts due to scattered electrons produced in beam-beam events from those due to interactions of either beam with residual gas or surfaces. To minimize such background events, a vacuum of at least 1×10^{-10} torr is maintained with beams present.

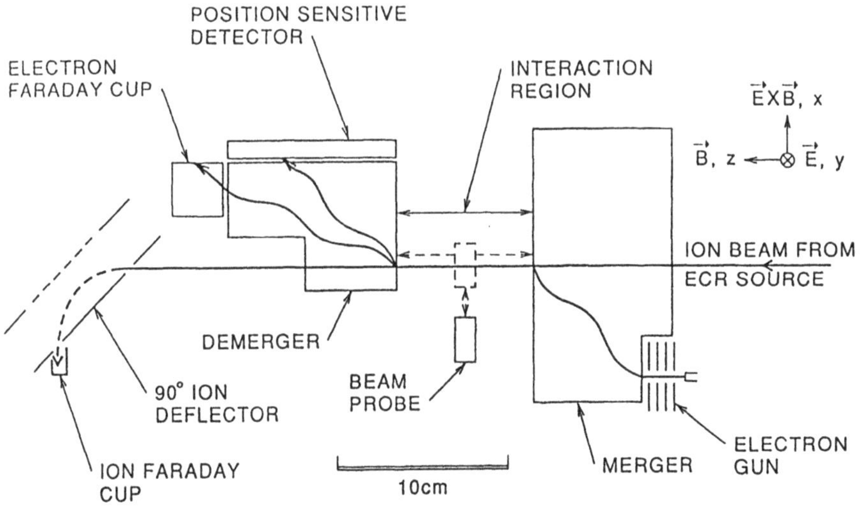

Figure 1. Schematic of JILA merged-beams electron-energy-loss apparatus for the measurement of cross sections for electron-impact excitation of ions.

A movable beam probe consisting of a phosphor screen, micro-channel plate and charge-injection-device (CID) camera is inserted into the interaction volume to quantify the three-dimensional spatial intensity distributions and overlap of the beams. Real-time spatial images of the electron and ion beams are displayed on an oscilloscope, facilitating their tuning and alignment.

This apparatus was moved from JILA and installed at the ORNL-ECR facility during 1990. The first measurements on a multiply charged ion with this apparatus were reported recently[2] for 3s-3p excitation of Si^{3+}:

$$e + Si^{3+}(3s\ ^2S_{1/2}) \rightarrow e + Si^{3+}(3p\ ^2P_{1/2,3/2}) - 8.88\ \text{eV} \qquad [2]$$

This system was chosen for the first measurements because cross-section calculations for Na-like ions are expected to be of high reliability, and thus a comparison of experiment and theory serves to test the validity of both. The absolute experimental cross-section measurements[2] are compared in Fig. 2 to benchmark 7-state close-coupling calculations by Badnell, Pindzola, and Griffin[3]. The cross section for electron-impact excitation of an ion is finite at threshold, and usually reaches its largest value there. Thus an abrupt jump from zero is expected at the 8.9 eV 3s-3p threshold. For the sake of comparison, the theoretical cross section in Fig. 2 has

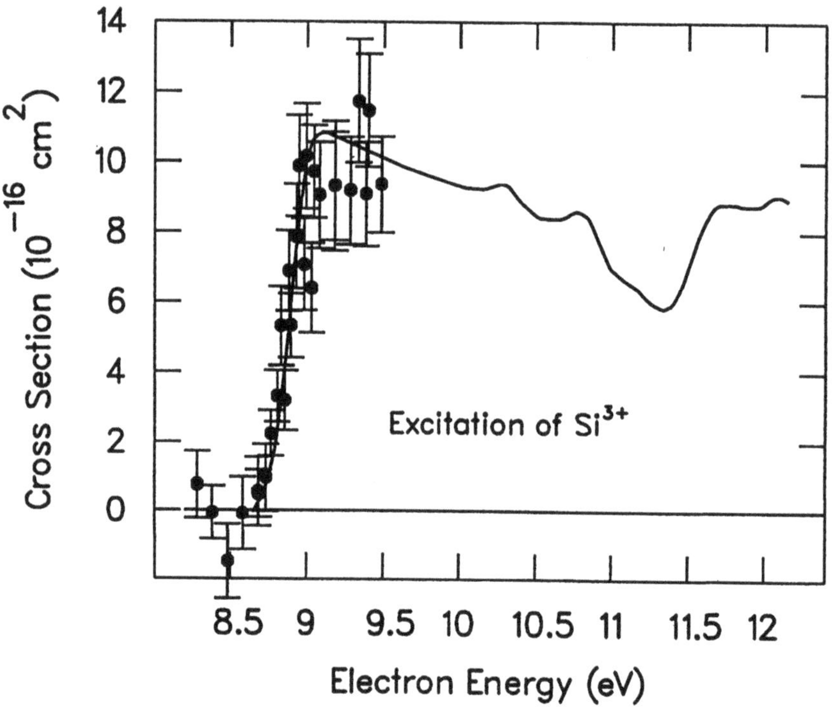

Figure 2. Comparison of absolute cross-section measurements for 3s-3p excitation of Si^{3+} with 7-state close-coupling calculations by Badnell, Pindzola, and Griffin. The theoretical cross section has been convoluted with a 0.2 eV FWHM Gaussian distribution to simulate the energy resolution of the merged-beams experiment.

been convoluted with a Gaussian electron-energy distribution of 0.2 eV (FWHM), which was selected empirically to simulate the energy resolution of the experiment. In fact, there are actually two fine-structure components in the 3s-3p excitation ($^2P_{1/2}$, $^2P_{3/2}$), and thus two expected steps in the cross section which are separated by 0.06 eV. Although the experimental energy resolution is insufficient to resolve these two steps, it may be slightly better than 0.2 eV. In any case, this represents an order-of-magnitude improvement over previous ion excitation measurements based on detection of photon emission[4]. The absolute agreement in cross-section magnitude between experiment and theory in the critical threshold region is seen to be very satisfactory.

Trajectory modelling of the primary electron beam, and of both elastically and inelastically scattered electrons, indicates that the merged-beams electron-energy-loss apparatus is, in principle, capable of covering almost the entire energy range presented in Fig. 2. This could not be realized in these initial measurements due to high background levels encountered as the relative energy was further increased above the excitation threshold. It is believed that this elevated background was due to components in the primary electron beam with increased transverse energy, whose trajectories either impacted the position-sensitive detector directly, or interacted with nearby surfaces. Such effects become increasingly severe as the relative energy is increased above the threshold, since the difference in energy between the primary and scattered electron beams decreases, and thus their spatial separation in the demerger is reduced.

Improvements to the electron gun and merger are in progress, based on more detailed computer trajectory-modelling calculations. It is expected that these will permit the energy range of the technique to be extended further above the excitation threshold, where strong resonance effects are predicted to occur for many ions. The structure in the theoretical cross section shown in Fig. 2 is due to such resonances. Their apparent effect has been reduced somewhat because of the convolution of the theoretical cross section with the simulated experimental electron energy distribution. These first data serve to establish the proof of principle of the merged-beams electron-energy-loss technique for excitation of multiply charged ions, and the threshold measurement demonstrates that the experimental resolution should be sufficient to uncover such resonance structures. The energy range accessible to the technique is constrained in principle only by the decreasing dispersion of elastically and inelastically scattered electrons as the energy is raised above the inelastic threshold. Future experiments will focus on fusion-relevant ion species, and will be selected in order to provide the most sensitive tests of theory.

3. Electron Capture by Multiply Charged Ions from Hydrogen Atoms Energies Below 1 eV

Low-energy electron-capture collisions involving highly ionized ions play an important role in non-equilibrium plasmas which are found both in nature and in the laboratory. Reactions of the type:

$$X^{+q} + H \rightarrow X^{+q-1} + H^{+} + \Delta E \qquad [3]$$

are characterized by large cross sections that depend strongly both on the electronic energy-level structure of the colliding system and on the interaction mechanism. The dominant reaction channels are those which, depending on q, are exoergic by several to tens of eV, and lead to population of excited states of the X^{+q-1} product.

At collision energies in the eV range, the interaction potential energy becomes an appreciable fraction of the total available energy. Electrons of the transient quasi-molecule formed during such collisions have sufficient time to adjust to the changing interatomic field as the nuclei approach and separate. Theoretical calculations in this energy regime require sophisticated coupled-states approaches[5], which often disagree with each other in their predictions of capture cross sections. Parallel quantitative theoretical and experimental determinations of cross sections for such processes can therefore yield important new information about quasimolecular structure and collision dynamics. For example, the cross section for a particular reaction channel may be enhanced at very low energies by the weak ion-induced-dipole attraction between the reactants (Eq. 1), which can permit the collision to sample potential-curve crossings which are smaller than the impact parameter[6]. Since this attractive force is weak, it will be significant only at very low collision velocities. This is the basis for the so-called Langevin-orbiting model[7], which predicts classical orbiting trajectories for a certain critical impact parameter, and a resulting reaction cross section which varies inversely with velocity. This model successfully explains the behavior of thermal rate coefficients for many ion-molecule reactions. It cannot, however, be arbitrarily applied to electron-capture collisions[8], since an active potential-curve crossing must exist at an appropriate internuclear separation in order for the capture cross section to be enhanced by such a "trajectory effect." Since the polarization potential varies as q^2, its influence on the collision may be felt at higher velocities as the ion charge is increased.

To address such issues, an ion-atom merged-beams apparatus has been developed at Oak Ridge National Laboratory and applied to experimental studies of electron-capture collisions of multicharged ions with hydrogen atoms at the lowest attainable energies. The technique[9] is illustrated in Fig. 3. A reactant beam of ground-state H atoms is prepared by intra-cavity laser photodetachment of a 6-9 keV beam of H^- ions. A beam of multicharged ions of similar velocity from the ORNL-ECR ion

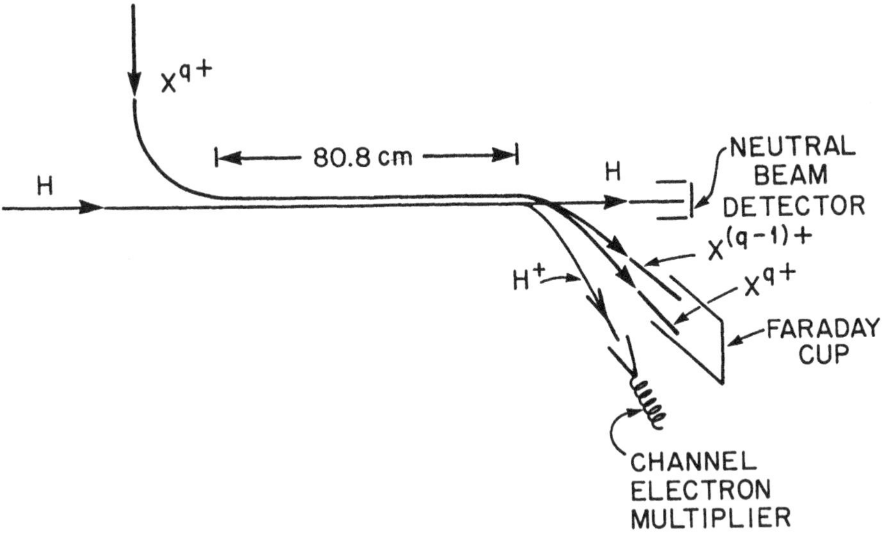

Figure 3. Simplified schematic of ion-atom merged-beams technique for measurement of cross sections for electron capture by multicharged ions from H atoms.

source is merged electrostatically with this neutral beam in a region of ultrahigh vacuum ($P < 1 \times 10^{-10}$ T). A rotating-wire beam scanner and two scanning-slit beam probes quantify the spatial intensity distributions and overlap of the beams. An analyzing magnet demerges the beams and separates the product H^+ ions from the reactants. The H^+ products are further dispersed electrostatically and counted by a channel electron multiplier; the reactant ion and neutral beams are collected in a Faraday cup and on a secondary-emission detector, respectively. A standard fast two-beam gating scheme is used to separate beam-beam signals from backgrounds produced by collisions with residual gas.

A number of major improvements were made during the past year to the original apparatus to increase the angular acceptance for reaction products and to improve the signal-to-background ratio. This redesign included the addition of differential and active pumping in the merge path, shortening the merge path from 80 to 50 cm, and enlarging the apertures of the signal-collection lenses and the demerging magnet chamber. The signal-to-background ratio has been increased by a factor of three, permitting measurements to be made at lower energies, where the reaction rate in the apparatus becomes vanishingly small. A deflector was added just before the intersection of the H^- beam with the photodetachment laser in order to remove H atoms from the reactant beam which have collisionally detached on residual gas. This and the vacuum improvements give a reactant H beam with a ground-state purity of 0.9998, compared to 0.995 for the previous arrangement. This is critical to obtaining meaningful measurements, because electron capture by multicharged ions from excited H atoms has much larger cross sections, particularly at very low collision energies. With all of these improvements, the effective dynamic energy range of the merged-beams apparatus has been extended downward by an order of magnitude to 0.1 eV/amu, which is the practical limit due to the finite divergences of the reactant beams.

On the basis of recent calculations[10] of differential cross sections for the O^{5+} + H(D) system, Andersson and co-workers have suggested that a disagreement of earlier merged-beams data with theory resulted from incomplete collection of the $H^{+}(D^{+})$ products in the 2 to 20 eV/amu energy range. This discrepancy has now been resolved by further measurements with the improved apparatus, which also permitted the energy range to be extended down to 0.14 eV/amu. These new measurements are compared with previous data and with theory in Fig. 4. The newer data are consistent within the experimental uncertainty with theory over the entire energy range, and clearly show the predicted increase at the lowest energies due to the onset of capture into the 4p and 4d states. The 10 to 30% reduction in the

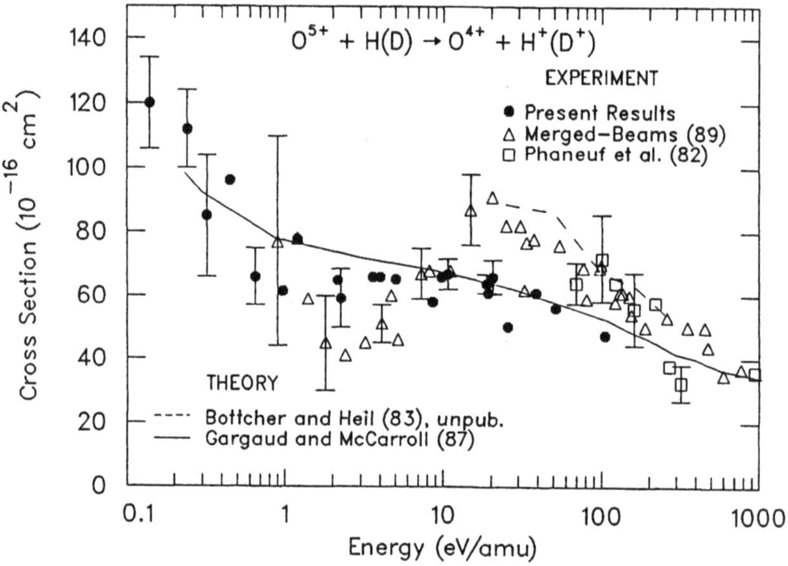

Figure 4. Comparison of merged-beams cross-section measurements for electron-capture in O^{5+} + H(D) collisions using improved apparatus (solid points) with previous measurements and theoretical calculations.

measured cross section compared to the previous data in the 10-100 eV/amu energy range is due to the improvement in ground-state purity of the H(D) reactant beam. Previous data for O^{3+} + H(D) have also been extended with the new apparatus down to 0.17 eV/amu, clearly showing the predicted contribution of capture into the 3p state at the lowest energies[11].

An upgrade of the ORNL-ECR ion source is in progress, which will increase its performance with respect to charge state and beam intensity. This will open up more highly charged collision systems for study with the merged-beams technique. Due to the enhanced polarization attraction between the reactants, such systems may exhibit "trajectory effects" or even orbiting resonances at the lowest energies accessible to the merged-beams technique. It should also be possible to study electron-capture collisions of bare ions such as C^{6+} or O^{8+} with hydrogen atoms. Since only one electron and two nuclei are involved, these are true three-body systems, and will provide the most fundamental tests of theory. Plans also exist to extend the merged-beams technique to the study of collisions of multicharged ions with excited hydrogen, for which cross sections are expected to be extremely large.

4. Neutralization of Multicharged Ions in Grazing Interactions with Metal Surfaces

The interaction of highly charged ions with metal surfaces has become a topic of considerable experimental investigation[12-14]. One remarkable and unexpected result of recent work has been the observation of extremely rapid neutralization of highly ionized projectiles at metal surfaces. Our understanding of such processes follows from the pioneering work of Hagstrum[15], who first proposed the so-called resonance-neutralization model nearly 40 years ago, and from work in the 1970's by Arifov, Parilis and co-workers[16]. The process is illustrated in Fig. 5. As a highly charged bare (or nearly bare) ion approaches a metal surface, the potential barrier between the ion and the surface is lowered, and a negative image charge is induced at the surface. At some critical distance, electrons flow from the valence band of the metal into empty excited levels of the ion, forming so-called "hollow" or "super-excited" atoms. Classically, over some range of ionic charge q, this distance is estimated to be q+7 atomic units[17]. Of course, quantum-mechanical tunnelling through the barrier is possible at somewhat larger distances. As this "hollow" atom de-excites near the surface by Auger electron emission, electrons continue to flow from the metal to keep the ion neutralized. The emission of K X-rays[12], or of K-Auger electrons[13,14,18] may be used as a "clock" to provide information about the formation and decay of this exotic atom.

The emphasis of recent ORNL research in this area has been on elucidation of the time scales over which projectile ion neutralization occurs during grazing interactions with metal surfaces. The experimental approach is to measure the energy spectrum of projectile K-Auger electron emission as a function of the angle

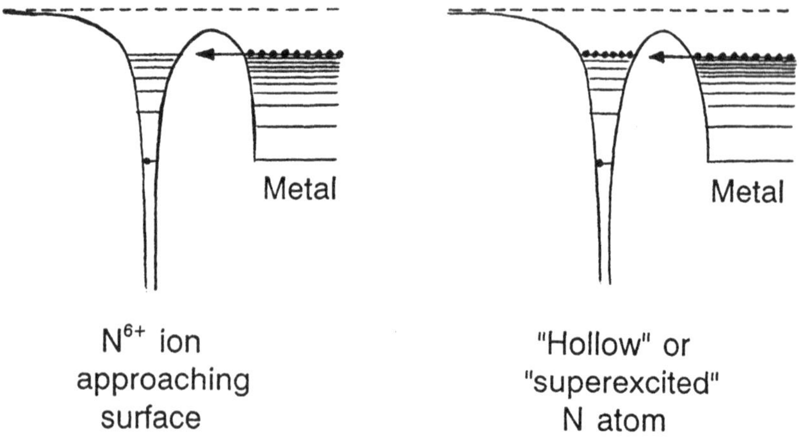

Figure 5. Formation of a "hollow" or "super-excited" N atom by the resonance neutralization of a N^{6+} ion at a metal surface.

of incidence of the ion on the surface, and therefore as a function of the time the projectile spends in the vicinity of the surface.

Figure 6 shows the variation of the measured yield[18] of N KLL Auger electrons with angle of incidence for 60-kev N^{6+} projectiles on Au (110). In order to try to account for the measured K-Auger electron yields, a computer simulation[14] of the autoionization cascade was made using calculated atomic Auger transition rates. Radiative decay rates are negligible compared to Auger rates in this case. The cascade is assumed to start when the ion is 30 a.u. above the surface with 6 electrons resonantly captured in the n=7 shell of the N^{6+} projectile, which is nearly resonant with the Fermi energy of the metal. Re-neutralization into n=7 is assumed subsequent to each autoionization step. The result of this simulation is shown in Fig. 7 for 24 keV N^{6+} incident on a metal surface at an angle of 10°, and indicates that the autoionization cascade is much too slow (~100 times!) to explain the observed K-Auger yields.

It is unlikely that this discrepancy is due to the neglect of additional de-excitation processes which become possible when the ion is close to the surface, or to possible enhancements of the atomic Auger rates in the proximity of the surface. Rather, the large observed K-Auger yields are more likely due to the contribution of sub-surface processes. This conclusion has been reached on the basis of extensive Monte-Carlo simulations of ion trajectories in the solid[18]. Once the ion enters the solid, the L shell may be filled rapidly either by direct capture to the M shell from the Au 5p and/or valence band, followed by rapid LMM Auger decay, or by capture directly into the

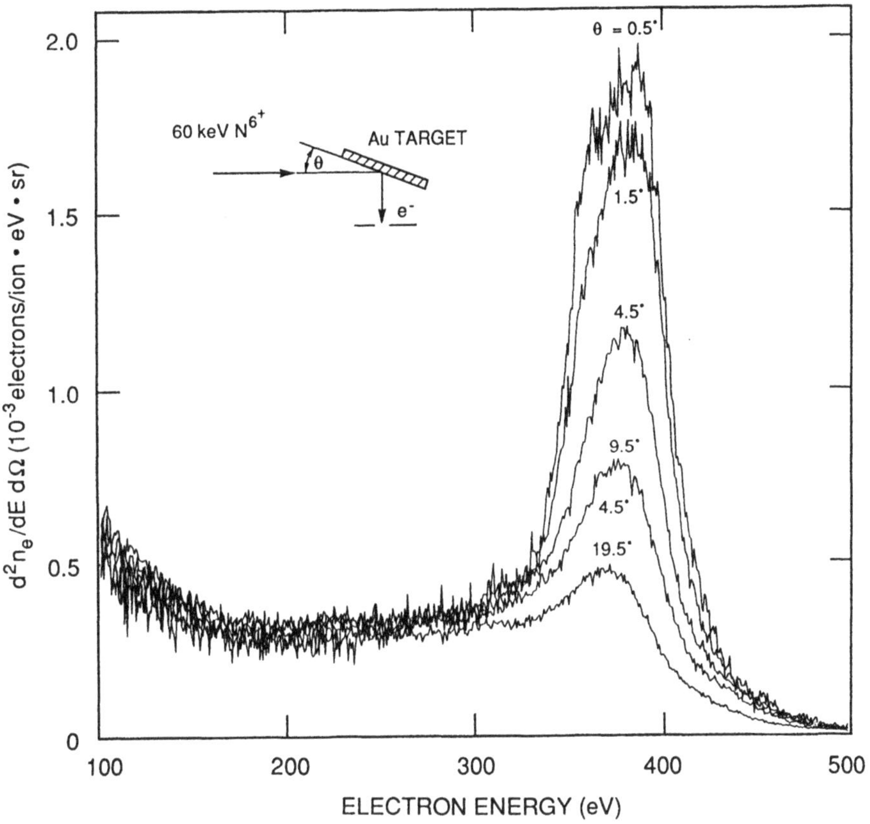

Figure 6. Absolute measured projectile K-Auger electron yields for 60-keV N^{6+} ions incident on an Au (110) surface at five different angles.

L shell. This makes KLL emission possible with minimum delay. The simulations show that, due to the small inelastic mean free paths of the KLL electrons in the solid, the observed KLL emission originates from only the first few atomic layers of the bulk, prior to appreciable angular scattering of the projectile ion. As a result, this sub-surface KLL emission has the same Doppler signature as electron emission on the incident projectile ion trajectory. Also, when allowance is made for the different escape probabilities for LMM and KLL electrons emitted from below the surface, the anomalously low measured ratio[13,14,18] of L to K Auger electron yields is at least partially explained.

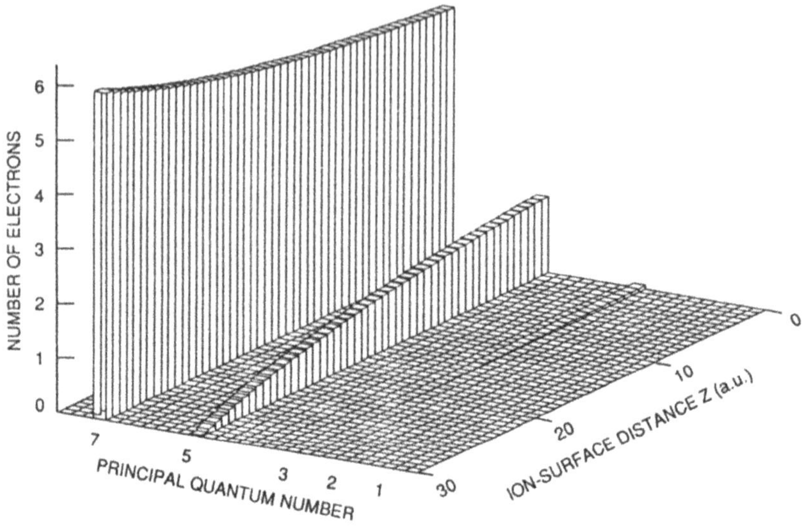

Figure 7. Calculated distribution of electrons over the principal quantum shells as a function of ion-surface distance. Atomic rates were used in the atomic ladder model for 24 keV N^{6+} incident at 10° on a metal surface. The single K-shell electron carried by the N^{6+} ion into the interaction is not indicated.

Examining the electron yield measurements in Fig. 6 in more detail, one observes that the shape of the K-Auger peak on the low-energy side is different at the two smallest angles, suggesting the appearance of an additional feature. This feature is emphasized in Fig. 8, in which the shape of the peak at 19.5° has been subtracted from the normalized spectrum at each angle, and is attributed to above-surface neutralization. Its time evolution (i.e., dependence on ion incidence angle) is consistent with the Auger cascade model described above. Only at the smallest angles of incidence is there sufficient time available for above-surface de-excitation to contribute appreciably to the total yield of projectile K-Auger electrons. The manifestation of "hollow atoms" in multicharged ion-surface interactions is thus perhaps more subtle than was previously thought.

The upgrade of the ORNL ECR ion source will enable studies to be made of ion-surface interactions with more highly charged ions, for which the neutralization energy is substantially increased, thus providing more severe tests of evolving models for neutralization. Plans also call for construction of a new apparatus with provision for ion deceleration prior to interaction with the surface. This will allow measurements to be made for different ion incidence angles, but the same ion

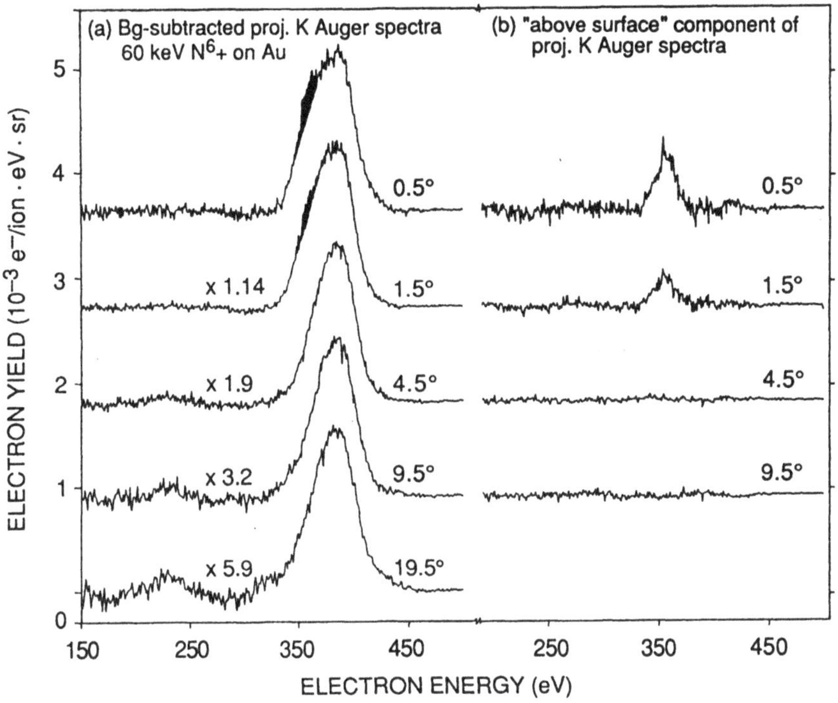

Figure 8. (a) Background-subtracted K-Auger spectra from Fig. 6 scaled to the peak of the 0.5° spectrum. (b) Above-surface components obtained by subtracting the 19.5° spectrum from the spectra in (a) at other angles.

velocity perpendicular to the surface (and thus interaction time). It is possible that the above- and below-surface effects would then be more clearly resolvable. Interactions with semiconductors, where a gap exists between the valence and conduction bands, is another intriguing possibility.

5. Summary

A novel merged-beams electron-energy-loss apparatus developed at JILA has been implemented at ORNL for studies of electron-impact excitation of multiply charged ions, with the first absolute cross-section measurements obtained for 3s-3p excitation of Na-like Si^{3+}. The near-threshold measurements are in excellent agreement with theory, and indicate an energy resolution of 0.2 eV for the technique.

Recent improvements to the ORNL ion-atom merged-beams apparatus have permitted measurements of absolute electron-capture cross sections for collisions of multiply charged oxygen atoms with deuterium atoms down to energies as low as 0.14 eV/amu. The data provide stringent tests of theory, and verify the predicted increase of the cross section at energies below 1 eV/amu due to the contributions of capture into the 3p and 3d levels.

In angle- and energy-resolved measurements of electrons ejected in grazing interactions of multicharged nitrogen ions with gold and copper single crystals, components have been identified in the spectrum due to both above- and below-surface emission by the ion. These data and their interpretation via computer simulations resolve some outstanding discrepancies concerning the time scale for neutralization of highly charged ions at solid surfaces.

An upgrade of the ORNL-ECR multicharged ion source is in progress which will substantially increase the capabilities and scope of studies such as those described in this paper.

6. Acknowledgments

This research was sponsored by the U.S. Department of Energy in part under Contract No. DE-AC05-84OR21400 with Martin Marietta Energy Systems, Inc., and in part under Contract No. DE-AI101-76PR16010 with the National Institute of Standards and Technology.

A.C.H.S. participated in the electron collision experiment as a Visiting Fellow at JILA and a Guest Scientist at ORNL during 1989-90. The authors are grateful to D. C. Griffin, M. S. Pindzola, and N. Badnell for permission to quote theoretical results prior to publication, and to C. Bottcher for insightful comments on this manuscript.

7. References

* Present address: MS-E526, Los Alamos National Laboratory, Los Alamos, NM 87545.

1. R. A. Phaneuf, "Experiments on Electron-Impact Excitation and Ionization of Ions," p.117-156 in *Atomic Processes in Electron-Ion and Ion-Ion Collisions*, ed. F. Brouillard, NATO Advanced Science Institutes Series B, Vol. 145 (Plenum, New York and London, 1986).
2. E. K. Wåhlin, J. S. Thompson, G. H. Dunn, R. A. Phaneuf, D. C. Gregory and A. C. H. Smith, *Phys. Rev. Lett.* **66** (1991) 157.

3. N. Badnell, M. S. Pindzola and D. C. Griffin, *Phys. Rev. A* (to be published).
4. P. O. Taylor, D. Gregory, G. H. Dunn, R. A. Phaneuf and D. H. Crandall, *Phys. Rev. Lett.* **39** (1977) 1256.
5. M. Kimura and N. F. Lane, in *Advances in Atomic and Molecular Physics*, ed. D. Bates and B. Bederson, Vol. 26, Academic Press (1990), pp.79-160 .
6. C. C. Havener, M. S. Huq, F. W. Meyer and R. A. Phaneuf, *J. de Physique Coll.* C1 (1989) C1-7.
7. G. Gioumousis and D. P. Stevenson, *J. Chem. Phys.* **29** (1958) 294.
8. M. Henchman, in *Ion-Molecular Reactions*, J. L. Franklin, ed., Vol. 1, Plenum Publishing Co. (1972) pp. 101-259.
9. C. C. Havener, M. S. Huq, H. F. Krause, P. A. Schulz and R. A. Phaneuf, *Phys. Rev. A* **39** (1989) 1725.
10. L. R. Andersson, M. Gargaud, and R. McCarroll, in *Electronic and Atomic Collisions*, XVI International Conference on the Physics of Electronic and Atomic Collisions, Abstracts of Contributed Papers, New York (1989), p. 566.
11. M. Gargaud, R. McCarroll, and L. Opradolce, *Astron. Astrophys.* **208** (1989) 251.
12. J. P. Briand, L. de Billy, P. Charles, S. Essaba, P. Briand, R. Geller, J. P. Desclaux, S. Bliman and C. Ristori, *Phys. Rev. Lett.* **65** (1990) 159; and presentation at this workshop.
13. L. Folkerts and R. Morgenstern, *Europhys. Lett.* **13** (1990) 377.
14. P. A. Zeijlmans van Emmichoven, C. C. Havener, and F. W. Meyer, *Phys. Rev. A* **43** (1991) 1405.
15. H. D. Hagstrum, *Phys. Rev.* **96** (1954) 336; H. D. Hagstrum and G. E. Becker, *Phys. Rev. B* **8** (1973) 107.
16. U. A. Arifov, L. M. Kishinevskii, E. S. Mukhamadiev and E. S. Parilis, *Zh. Tekh. Fiz.* **43** (1973) 181 [*Sov. Phys. Tech. Phys.* **18** (1973) 118].
17. K. J. Snowdon, *Nucl. Instrum. Methods Phys. Res. B* **34** (1988) 309.
18. F. W. Meyer, S. H. Overbury, C. C. Havener, P. A. Zeijlmans van Emmichoven and D. M. Zehner, "Evidence for Above-Surface Neutralization during Interactions of Highly Charged Ions with a Metal Surface" (submitted to *Physical Review Letters*, 1991).

ELECTRON CAPTURE PROCESSES FOR HIGHLY-CHARGED, HEAVY PROJECTILES

P.H. MOKLER
Gesellschaft für Schwerionenforschung,
P.O. Box 11 05 52, D-6100 Darmstadt, Germany

Abstract

Electron capture processes into swift few-electron projectiles are reviewed. Special emphasis is given to hydrogenic ions with atomic numbers around and beyond 30. The projectile velocities discussed are smaller or comparable to the orbital velocity of the 1s-projectile electron. Targets with the lowest atomic numbers are preferred; there the pure Coulomb or kinematical capture can be neglected to the capture channels with quasefree electrons as radiative electron capture (REC) and dielectronic capture (resonant transfer and excitation, RTE). In the considered cases all the capture processes yield to a characteristic x-ray emission and the emitted x rays give information on the processes. Using coincidence techniques, two examples how to use this for structure investigations of heavy few-electron projectiles are shown. On the one hand the REC radiation is used to determine the K-binding energy precisely. On the other hand via RTE the 1s2s $^1S_0$ state is populated and its two-photon decay is studied. An outlook on the heaviest projectiles will be given.

1. Introduction

In heavy ion-atom collisions inner-shell vacancy production has been studied over years. Nowadays, it is in general a well-established atomic reaction mechanism predominantly investigated via its x-ray emission. In these investigations the target ionization has been considered first for light projectiles colliding with heavy target atoms, $Z_1 << Z_2$. Lateron, in more symmetric collisions, $Z_1 \approx Z_2$, the inner-shell vacancy production has been studied for both the collision partners. However, the field of very heavy projectiles colliding with light target atoms, $Z_1 >> Z_2$ was not easily accessible due to the lack of appropriate heavy ions. With the advent of new heavy ion accelerator facilities this situation changed now. Even for the heaviest ion species (Pb ... U), few-electron projectile beams can now be provided for experiments.

Following the way of Madison and Merzbacher [1], we introduce two parameters to characterize the collision system:

(i) asymmetry parameter, $\alpha = Z_1/Z_2$

(ii) adiabaticity parameter, $\eta = (v/u)^2$

where v and u are the collision velocity and the orbital velocity for the electron

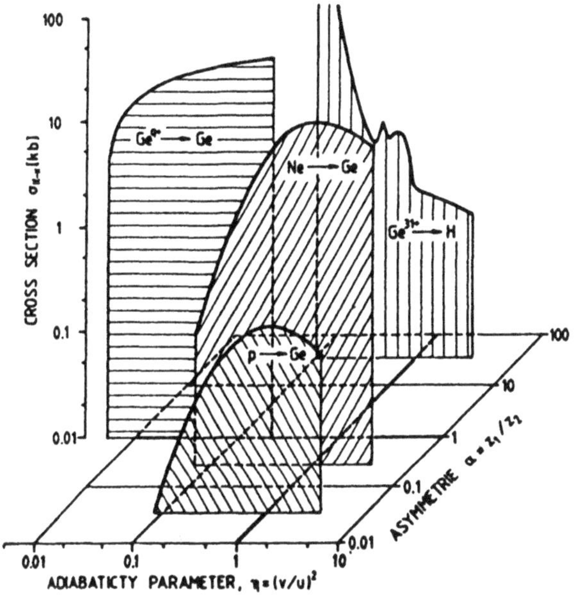

Figure 1: General cross section dependencies for various collision systems with Ge as one partner. The K-vacancy production or decay is shown.

(or the vacancy !) of concern. If we define η as a scaled collision energy — $\eta = (E_{coll} \cdot m)/(M \cdot E_B)$, with E_{coll} the collision energy, m/M the electron to projectile mass ratio, and E_B the binding energy of the electron of concern — the term is also valid for high projectile energies, In Fig. 1 within the parameter field η, α estimates for excitation functions $\sigma(\eta)$ at different α are given. All the cases refer to the K-shell of Ge (Z = 32). For $\alpha < 1$ ($\alpha = 0.03$ for p → Ge) we are in the field of direct ionization: For $\eta << 1$ we have asymptotically $\sigma_K \sim \eta^4$; σ_K has its maximum around $\eta \approx 1$ and decreases with η^{-1} beyond. Increasing α, the cross section increases with Z_1^2 which is indicated by the Ne → Ge case ($\alpha = 0.3$). For higher α we leave the domain of direct ioniation and enter for small η ($\eta << 1$) the region of quasi-molecular excitation. Here, during collision the partners form a quasimolecule and vacancies can be transfered along quasimolecular levels via couplings into inner shells (see e.g. Refs. [2] and [3]). Large inner-shell excitation cross sections at low scaled collision energies are typical, see the Ge^{q+} → Ge excitation function (α =1) in Fig. 1. Magnitude and shape of the excitation function depend on incoming charge state (i.e. vacancies) and dynamics of the involved couplings. In the shown case it is the characteristics of the $2p\pi - 2p\sigma$ rotational coupling.

Increasing α beyond 1, i.e. going to an inverted collision system ($Z_1 > Z_2$), we can consider both, the target ionization or the projectile inner-shell vacancy formation.

For the target ionization we are in the region of direct, multiple Coulomb ionization. This field of highly ionized slow recoil ions is not treated in this contribution, see e.g. Ref. [4]. We concentrate on the projectile. In Fig. 1 the "excitation" function for projectile K x-ray emission is given for hydrogenic Ge-projectiles colliding with H_2. At low energies electron capture into "excited" projectile levels produces the inner-shell vacancies and determines the steeply decreasing excitation function with the structures. Only at the highest energies, η, the true excitation — i.e. projectile K to L excitation — contributes to the excitation function.

It is the aim of this paper to give insight into the various capture processes of heavy few-electron projectiles at intermediate small scaled energies. As these capture processes populate well defined levels the associated x-ray emission will give specific information on the structure of heavy few-electron projectiles. Examples how to use the capture processes for atomic structure investigations will be given.

In the following chapter 2 the capture processes will be described. In chapter 3 measurements for radiative electron capture (REC) will be shown and precise information on the 1s binding energy in the projectile will be given. Chapter 4 deals with the resonant capture process (resonant transfer and excitation, RTE). Here, special emphasis is given to the two-photon decay of the 1s2s $^1S_0$ state into the groundstate $1s^2$ $^1S_0$ of the He-like ejectile. In the last chapter an outlook to the heaviest projectiles is given.

2. The Electron Capture Processes

During capture of an electron to a swift, heavy projectile conservation of energy and momentum has to be considered. Hence, a free electron cannot be captured without involving a third body. This means free electrons can only be captured either by emitting a photon — this is the radiative recombination [5], or by exciting another projectile resonantly — this is the dielectronic recombination [6], or if the electron density is high enough by three-body recombination [7] — i.e. involving a further electron to conserve energy and momentum.

In collisions with bound target electrons the target atom residue takes the part of the third body and we have the normal charge exchange process, the kinematical or Coulomb capture [8]. Here, the initial binding of the electron to the target atom is essential. Consequently, the higher bound target electrons will dominantly be captured. This is especially true for heavier target atoms in the intermediate velocity region. As the overlap of initial and final state wave functions decrease with projectile energy, the kinematical capture cross section decreases with about $\eta^{-5.5}$ (around $\eta \approx 1$). For heavy projectiles and light target atoms the Coulomb or kinematical capture may only predominate at small η ($\eta \leq 0.5$). Due to the overlap of states the capture populates preferentially higher projectile shells, yielding to projectile x-ray emission during cascading.

For heavy projectiles target electrons bound to light atoms can be considered as quasifree; i.e. the binding to the target can be neglected, only the momentum distribution of the bound electrons around the target, the Compton profile, has to be taken into account. For capture of quasifree electrons we have to consider similar processes as for free electron capture. Here, they are named radiative electron capture, REC, and resonant transfer and excitation, RTE, instead of radiative recombination and dielectronic capture. The equivalent to the three-body recombination which may be effective for multi-electron targets is normally not taken into account.

In Fig. 2 the kinematics for the dominant quasifree electron capture processes are schematically shown. For REC the energy gained by capture is converted into a photon. The capture to the ground state and to an excited state, e.g. capture to the L shell, is shown. In the second case, the ground state may be reached by an additional x-ray emission (e.g. for hydrogenic projectiles). For RTE the energy gained by capture is transfered to excite another projectile electron. This is a resonance process as a strong energy condition has to be fulfilled ! The produced doubly excited state stabilizes radiatively, for hydrogenic projectiles in the end with two K-x-rays. At resonance energy the REC and RTE processes yield to same x-ray cascade emission with the same sum energy and same intermediate states (e.g. L-REC and KLL-RTE). Here, both processes are non-distinguishable and may interfer.

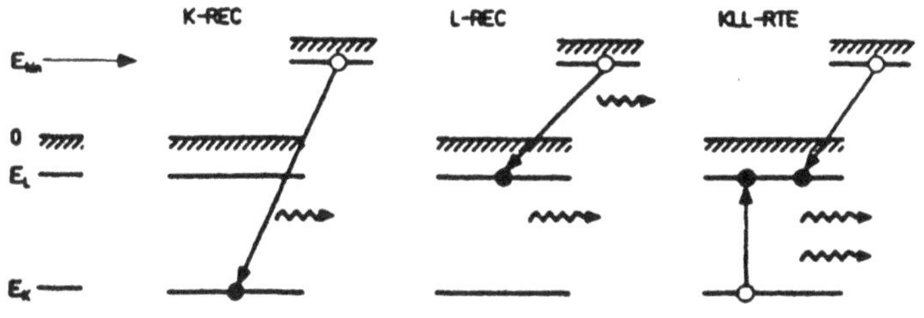

Figure 2: Capture of quasifree electrons into a heavy projectile; REC: radiative electron capture and RTE: resonant transfer and excitation.

Nevertheless, both processes are basically different: Whereas for REC the projectile (sreened nucleus) interacts with the quasifree electron, we have to consider the electron-electron interaction for RTE. Common to both processes is that they are time reversed processes to well-known fundamental atomic effects: The REC can be treated in analogy to the time reversed photoeffect and the RTE is analogous to the time reversed Auger effect followed by a radiative stabilization. Using the principle of detailed balance valid for time reversal and assuming validity of the impulse approximation — i.e. assuming the target electrons as quasifree — the capture cross sections can be calculated from the time inverted fundamental atomic processes. Along this way the cross sections have been calculated [9]. The corresponding excitation function are reproduced in Fig. 3.

In Fig. 3 the production cross sections for the various L x-ray emitting processes are given for $Ge^{31+} \rightarrow H_2$ collisions. Additional to REC and RTE, the Coulomb capture and pure projectile excitation are included into the graph. The REC drops with η^{-1}, less strongly than the Coulomb capture yielding an equivalent cross section around $\eta \approx 0.5$. Only the excitation process stays about constant around $\eta \approx 1$. For high energies it is the dominant L-x ray producing process. RTE shows quite prominent resonance structures at $\eta = 0.5$ and around $\eta \geq 0.6$ for the KLL and KLn with n > 2, respectively. (For the resonance assignment we use the Auger notation.) Due to the binding energy scaling for inner shells in hydrogenic atoms the RTE resonances are fixed at the same η for all projectiles ! In order convert the reaction cross sections into x-ray emission cross sections one has to multiply the curves in Fig. 3 by appropriate factors correcting for cascade multiplication, for metastable state reduction a.s.o. The factors vary between 0.5 and 1.5 (reasonable values are EXC: 0.5, CEC: 0.8, RTE: 1.5, L-REC: 0.8 to 1.5 at $\eta = 0.5$). With these values the measured x-ray cross sections can well be described. In Fig. 3 the measured L x-ray emission cross sections [9, 10] are also shown for $0.4 \leq \eta \leq 0.7$. The RTE resonances are clearly visible.

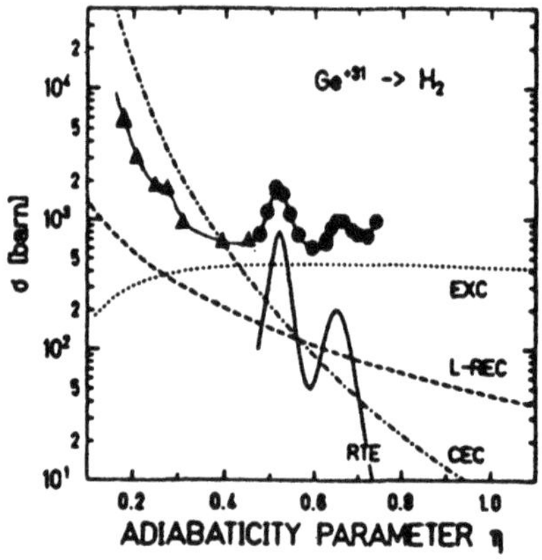

Figure 3: Excitation functions for inner-shell processes in $_{32}Ge^{31+} \rightarrow H_2$ collisions; the experimental cross sections are from [9] and [10].

3. Radiative Electron Capture

X-Ray spectra for $Ge^{31+} \rightarrow H_2$ collisions are shown in Fig. 4 at low scaled velocities $\eta = 0.20$ and 0.27 [10]. The spectra are measured in coincidence with single electron capture ejectiles to reduce background and pure excitation contributions. One clearly observes (at 90° to the beam direction) the characteristic projectile radiation, Ge – K_α, – K_β, – L..; the K-, L-, and M-REC contributions clearly shift with projectile energy.

From the energy conservation law we can deduce for the REC centroid positions in the projectile frame.

$$E_{REC} = E_P - E_T \cdot \gamma + m_0c^2(\gamma - 1)$$

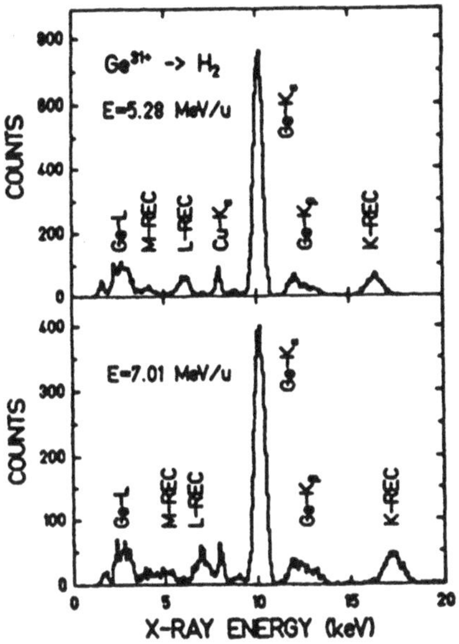

Figure 4: X-ray spectra for $Ge^{31+} \rightarrow H_2$ collision at a scaled energy of $\eta = 0.20$ and 0.27. A particle coincidence condition was set to one-electron capture events.

where $\gamma = (1 - v^2/c^2)^{-1/2}$ is the relativistic factor, E_P and E_T are the electron binding energies in the projectile and the target atom, respectively, and $m_0c^2(\gamma - 1)$ is the initial kinetic energy of the target electron in the projectile frame.

The initial momentum distribution of the target electrons determines the width of the REC-lines and increases about linearly with projectile momentum or with $\eta^{1/2}$. As the (K-)photoeffect shows a dipole emission characteristics, REC does the same. As the retardation cancels the Doppler effect, a dipole emission pattern is observed in the target rest frame.

In Fig. 5 the centroid energies (in the lab.-frame) for the K-, L-, M-, and N-REC radiation are plotted as a function of projectile energy [9, 10]. (The quadratic Doppler effect has not been included into Fig. 5.) A linear dependence of the different centroid energies with projectile energy is observed. The intercept extrapolated to zero velocity gives the binding energy difference. As E_T for H_2 is only 15.4 eV within the accuracy of the drawing the intersept gives directly the projectile binding energies: For the K-shell it is the 1s binding energy in the He-like system, for the higher shells one has to average over the involved levels appropriately.

A detailed evaluation of the K-data gives an accurate spectroscopic information on the binding energy of the 1s level. We finally result with a 1s binding energy of 13531 ± 20 eV [11]. This value compares nicely with Drake's prediction of 13557.44 eV [12]. The accuracy of the experimental value is already sensitive to relativistic

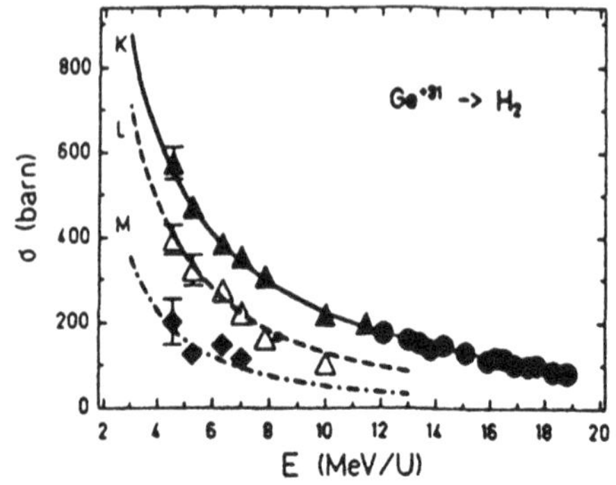

Figure 6: REC cross sections for $Ge^{31+} \to H_2$ collisions [10]. The predictions according to Ref. [5] are normalized by a factor of 0.7.

the η representation with Z^{-1}; i.e., the heavier the projectile is the better resolved are the adjacent shell resonances.

For $Ge^{31+} \to H_2$ collisions the KLL resonance is clearly separated from higher KLn resonances with $n > 2$, see Fig. 3. This can be seen more clearly from x-ray / x-ray coincidence measurements. In this case pure excitation or kinematical capture are excluded as these processes do not yield to two emitted K-x rays. However, in the KLL-resonance RTE and L-REC cannot be distinguished from one another. On the other hand, the L-REC is small in cross section compared to the RTE values in the resonance. This can also be read from the deep minimum between the KLL and KLn resonances in the K-x-ray / K-x-ray coincidence data shown in Fig. 7 [9, 13].

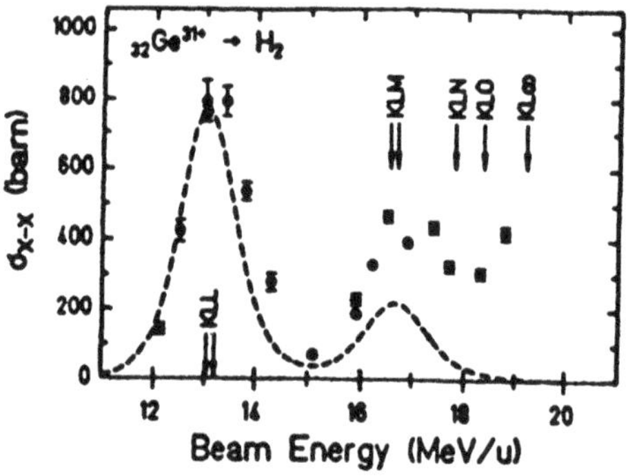

Figure 7: RTE excitation function for $Ge^{31+} \to H_2$ collisions measured by K-ray / K-ray coincidences [9, 13].

To the KLL resonance about 10 resonances contribute. One of these resonances leads to the 2s2p $^1P_1$ intermediate state which decays instantaneously to the 1s2s $^1S_0$ level. As the ground state is also a $^1S_0$ level ($1s^2\ ^1S_0$) this state can only decay via

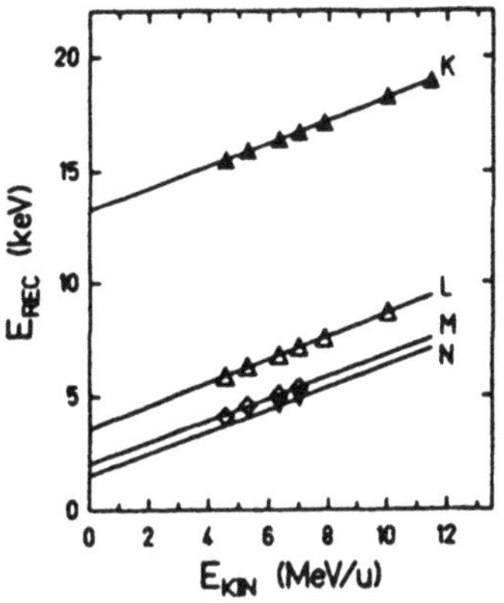

Figure 5: Centroid positions for the K-, L-, M-, and N-REC radiation as a function of the projectile energy ($Ge^{31+} \rightarrow H_2$) [10, 11].

effects and to the Breit term. It is limited in the moment by the uncertainty in beam energy determination ($\Delta E/E \sim 10^{-3}$) and to a smaller extent by the statistics, i.e. the measuring time of about 15 minutes per spectrum. The accuracy can easily be increased by a factor of 10, or maybe by 100 and will in the end compete with the accuracy gained by high resolution x-ray spectroscopy.

Another information which can be extracted from the particle-coincident x-ray spectra shown in Fig. 4 are the REC-cross sections. Assuming a dipole emission characteristics for all shells, the extracted cross section values are plotted in Fig. 6 for K-, L-, and M-REC. Additionally, theoretical predictions using the formula given by Bethe and Salpeter [5] are shwon. The theoretical curves are multiplied by a factor of 0.7 to fit the data. This factor is compatible with the experimental normalization error. Hence, a good accordance between theory and experiment can be stated ! Nevertheless, a distinction between the contributions from various subshells cannot be extracted from the measurements shown here. In order to do so one has to observe the cascading to the ground state which can be done by x-ray / x-ray coincidence techniques, in particular by x-ray / x-ray / charge-changed particle coincidences. Those measurements will be shown in the next chapter.

4. The Resonant Transfer and Excitation

RTE can be the dominant x-ray producing process, see Fig. 3, and the dominant capture process at the resonance energy. Using a hydrogenic scaling for the binding energies ($E_n = Z^2 / n^2 \cdot Ry$) one finds from the energy conservation law the KLL resonance at $\eta = 0.5$ and the KLn resonances extent for $n \rightarrow \infty$ to $\eta = 0.75$. As in the case of REC, the width for the RTE resonance is determined by the velocity distribution of the target electrons and is proportional to the projectile momentum. As the resonance stays at the same η, the resonance width decreases in two photon emission to the ground state. Hence, we have a three photon emission problem: First, a K_α^H hypersatellite is emitted, then two photons with a sum energy corresponding to the K_α line, $E_1 + E_2 = K_\alpha$. Restricting the coincident x-ray emission to the condition $E_1 + E_2 = K_\alpha$, one observes the population and decay of the 1s2s $^1S_0$ level. As this level at the KLL region is dominantly (90 %) fed by RTE via the 2s2p $^1P_1$ resonance one can deduce directly the 2s2p $^1P_1$ cross section [13, 14]. The result is given in Fig. 8. There, also the population mechanism is elucidated. Once more, the deep minimum between KLL and KLn (n > 2) shows that the direct population of the 1s2s $^1S_0$ via REC is small. For the KLL resonances the accordance between theory and experiment is excellent.

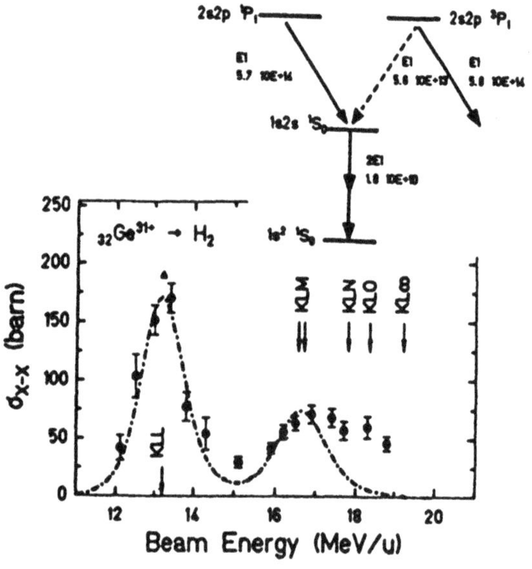

Figure 8: The 1s2s $^1S_0$ population cross section for $Ge^{31+} \rightarrow H_2$ collisions measured by the two-photon decay of this state [9, 14]. In the insert the population mechanism in the KLL resonance is sketched.

Considering Fig. 8 it seems clear that RTE is the dominant population mechanism for the 1s2s $^1S_0$ state; and one can study via this mechanism the 2E1 decay, in particular the shape of the two-photon emission spectrum. The 2E1 energy spectrum is determined by the overlap of all bound and continuum P states with our metastable S state. Due to the change in overlap with Z the spectral shape of the 2E1 changes with Z. In Fig. 9 normalized spectral distributions for He-like cases (1s2s $^1S_0$) are given. All spectra are normalized to the spectrum for hydrogen (2s $^2S_{1/2}$ case) [15]. Due to relativistic and correlation effects the Z-dependences are inverted for H-like and He-like cases. It is emphasized that until today there are no full relativistic ab initio calculations for the He-like sequence, and still triplet contributions are not considered. Hence, it is desirable to study the spectral shape for heavy, He-like projectiles. Such a study is much more sensitive than lifetime measurements which are an integration over the spectrum.

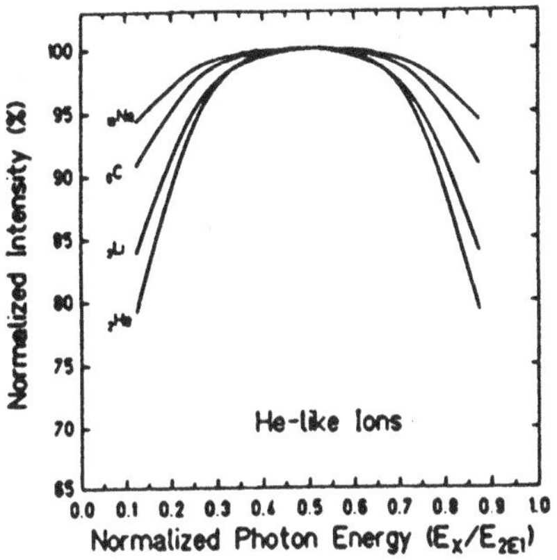

Figure 9: Two-photon energy spectra for the decay of the 1s2s $^1S_0$ state in He-like systems normalized to the distribution in hydrogen (2s $^2S_{1/2}$ state) [15].

In order to get a first information on the spectral shape we measured the 2E1 decay in He-like Kr at the KLL resonance in $Kr^{35+} \rightarrow H_2$ collisions using x-ray / x-ray / one-electron capture-event coincidences. In Fig. 10 an event plot is shown [9, 10]. Beside the dominant K_α / K_α peak, three ridges are observed K_α / E_1, E_2 / K_α, E_1 / E_2 with $E_1 + E_2 = K_\alpha$. These three ridges are caused by the cascade 2s2p $^1P_1$ - (K_α) $\rightarrow$ 1s2s $^1S_0$ - (E_1, E_2) $\rightarrow$ 1s^2 $^1S_0$. The additional structures can be explained by REC and the following cascades. For information some decay possibilities are shown in the graph on the right hand side. A detailed description of the different features will not be given here. However, from the cascades subshell differential REC cross sections can be deduced [10].

In Fig. 11 the experimental 2E1 spectrum along the diagonal, i.e. using the condition $E_1 + E_2 = K_\alpha$, is shown [9, 10]. On both ends of the spectrum, structures are superimposed on the symmetric 2E1 continuum. These structures are also caused by REC: e.g. the M-REC + the L x-ray from the cascade via the L shell fulfills at the KLL resonance also the energy condition M-REC + L-X = K_α. Corresponding conditions exist for RTE to higher shells. Depending whether we observe in the one detector the characteristic radiation (e.g. L-lines) or the REC (e.g. to the M-shell) we get narrow or broadened lines. The various contributions are indicated in Fig. 11. Moreover, Drake's prediction [16] for the 2E1 spectrum normalized and multiplied by the detection efficiency is shown in Fig. 11. On the level of accuracy a reasonable accordance between experiment and theory for the spectral distribution of the 2E1 decay is found. A detailed analysis is under way.

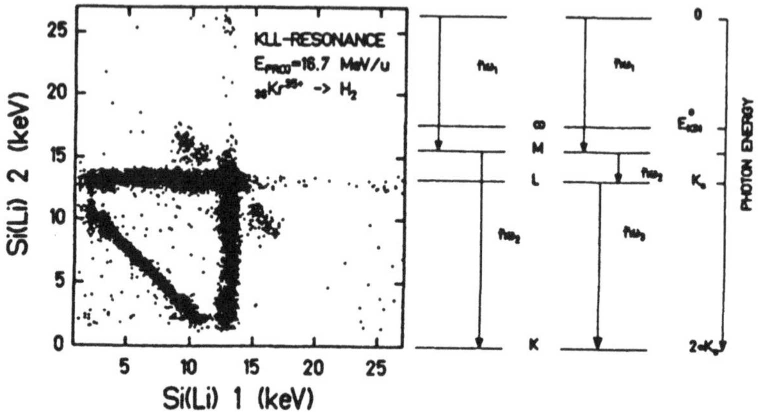

Figure 10: X-ray / x-ray event plot for $Kr^{35+} \rightarrow H_2$ collisions measured for one-electron capture conditions. Various REC events and the following x-ray cascades are sketched on the right side [9].

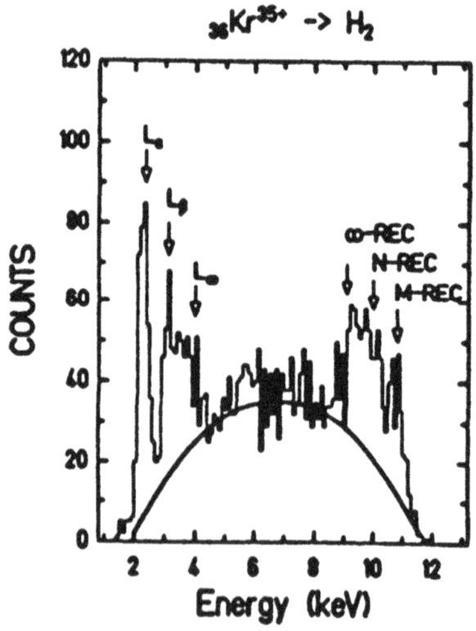

Figure 11: Two-photon coincidence spectrum measured along the diagonal $E_1 + E_2 = K_\alpha$ (see Fig. 10). REC cascade contributions are indicated on top of the efficiency corrected 2E1 distribution [10, 16].

5. Application to the Heaviest Projectiles

In the previous chapters the electron capture processes into swift, heavy, few-electron projectiles have been reviewed. Emphasis was given to the example of $_{32}Ge^{31+} \rightarrow H_2$ collisions at scaled energies of $0.2 \leq \eta \leq 1$. Single electron capture yields to singly (e.g. via REC) or doubly (e.g. via RTE) excited He-like ions $Ge^{30+*(*)}$. The emitted x-rays do not only give detailed information on the capture processes themselves, but reveal also subtleties of the structure of the He-like ion. On the one hand the REC and RTE processes could be determined in great detail. On the other hand it was examplified that from the K-REC the 1s binding energy can be extracted with high accuracy and from the RTE we can extract the spectral distribution of the 2E1 decay.

In order to extract this information, clean experimental conditions have to be guaranteed: By particle / x-ray coincidences one has to select the one electron capture events. Using x-ray / x-ray (/particle) coincidences one restricts the result to the wanted cascade events. The gas target has to be well localized and to be thin enough to ensure single collision conditions. Moreover, the ion beam has to be prepared very carefully and has to be charge-state selected and clean. In particular, for REC measurements at low scaled energies η one has to apply the sophisticated ACEL/DECEL technique [17]. Additionally, there the ion beam energy has to be determined to a very high precision.

Those stringent conditions can nowadays be provided even for the heaviest ions at the new storage ring facilities. Hence, in the following the extrapolation of the capture processes to higher Z projectiles is investigated. Within the picture of scaled energies, η, this is relatively simple.

In Fig. 12 we show for $\eta = 0.5$, i.e. at the KLL-RTE resonance, the cross section dependences for the various reactions induced by $Z^{1e} \rightarrow H_2$ collisions [9]. Except for REC which shows no Z dependence at fixed η all the reaction cross sections decrease in the high Z limit at $\eta = 0.5$. The collisional capture decreases with Z^{-5}, the excitation with Z^{-4} and RTE in the limit with Z^{-3}. However, the last dependence will slightly be reduced by relativistic effects. Considering capture alone RTE and REC are the only dominant processes yielding to the same competetitive cross sections. Hence, for very heavy projectiles interferences between both processes may be observable at $\eta = 0.5$ which is interesting in itself.

For the structure investigations we are interested in, it might be better to go to even lower η applying the ACEL/DECEL technique. There, the increase in REC cross section may compensate the loss in RTE contributions to populate e.g. the 1s 2s $^1S_0$ level. Also the collisional capture for instance at $\eta = 1/4$, is still two orders of magnitude smaller than the L-REC. Moreover, the REC contributions will move out of the diagonal in the x-ray / x-ray event plot (cf. Fig. 10) so that the 2E1 spectrum is expected to be clean. Hence, valuable information on the 2E1 decay in He-like very heavy projectiles can be expected.

Also for determining the 1s binding energy low η will increase the accuracy. There is the hope to reach an accuracy for the heaviest systems which is comparable

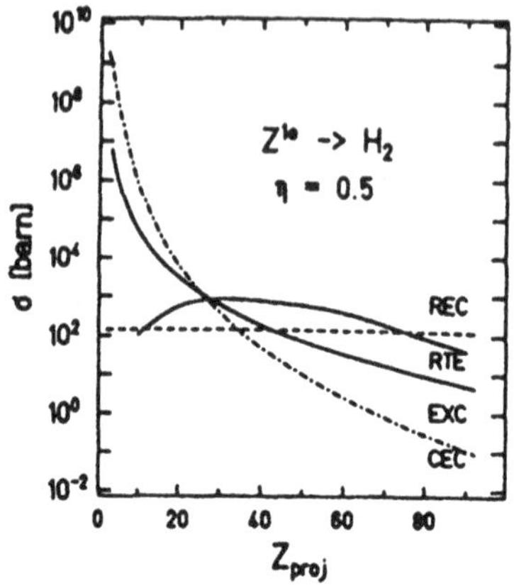

Figure 12: Cross section extrapolations for inner-shell projectile processes in $Z^{1s} \rightarrow H_2$ collisions [9].

to those yielded by high resolution x-ray spectroscopy. However, here the binding energy is measured directly.

References

[1] D.H. Madison, E. Merzbacher, in "Atomic Inner-Shell Processes" (ed. B. Crasemann; Academic Press, 1975) p. 1

[2] P.H. Mokler, D. Liesen, in "Progress in Atomic Spectroscopy", Part C (eds: H.J. Beyer, H. Kleinpoppen; Plenum, 1983) 321

[3] C.D. Lin, P. Richard, in "Advances in Atomic and Molecular Physics, Vol. 17 (eds. D. Bates, B. Bederson; Academic Press, 1981) 275

[4] H. Schmidt-Böcking, C.L. Cocke, S. Kelbch, R. Mann, P. Richard, J. Ullrich, in "High-Energy Ion-Atom Collisions" (eds.: D. Berényi, G. Hock; Académiai Kiado, Budapest, 1985) 155

[5] H.A. Bethe and E. Salpeter, Handbuch der Physik 35 (Springer, 1957) 88

[6] A. Burgess, Astrophys. J. 139 (1964) 776

[7] J. Stevefelt, J. Boulmer, J.-F. Delpech, Phys. Rev. A12 (1975) 1246

[8] H.-D. Betz, in "Atomic Physics", Methods of Experimental Physics, Vol. 17 (ed.: P. Richard; Academic Press, 1980) 73

[9] P.H. Mokler, Th. Stöhlker, C. Kozhuharov, J. Ullrich, S. Reusch, Z. Stachura, A. Warczak, A. Müller, R. Schuch, E.A. Livingston, M. Schulz, Y. Awaya, T. Kambara, in "X-Ray and Inner-Shell Processes" AIP Conf. Proc. 215 (eds.: T.A. Carson, M.O. Krause, S.T. Manson; AIP, New York, 1990) 335

[10] Th. Stöhlker, thesis (University of Gießen) and GSI (report, 1991) to be published

[11] P.H. Mokler, Th. Stöhlker, C. Kozhuharov, Z. Stachura, A. Warczak, Z. Phys. D (1991) to be published

[12] G.W. Drake, Can. J. Phys. 66 (1988) 586

[13] S. Reusch, thesis (Universität Gießen, 1988) and report GSI-88-19 (1988, ISSN 0171-4546)

[14] P.H. Mokler, S. Reusch, A. Warczak, Z. Stachura, T. Kambara, A. Müller, R. Schuch, M. Schulz, Phys. Rev. Lett. 65 (1990) 3108

[15] G.W.F. Drake, Phys. Rev. A34 (1986) 2871

[16] G.W.F. Drake, private communication (1989)

[17] P.H. Mokler, D.H.H. Hoffmann, W.A. Schönfeldt, D. Maor, W.E. Meyerhof, Z. Stachura, Nucl. Instr. Meth. B4 (1984) 34

X. SYNCHROTON RADIATION

ELECTRON CORRELATION EXPLORED THROUGH ELECTRON SPECTROMETRY USING SYNCHROTRON RADIATION

C. D. CALDWELL, S. B. WHITFIELD, AND M. G. FLEMMING
Department of Physics, University of Central Florida
Orlando, FL, 32816, USA

M. O. KRAUSE
Oak Ridge National Laboratory, Oak Ridge, TN, 37831, USA

ABSTRACT

The development of synchrotron radiation facilities as a research tool has made possible experiments which provide new insights into the role which correlation plays in electron dynamics and atomic and molecular structure. Features such as autoionizing resonances, normal and resonant Auger decay modes, and ionization threshold structure have become visible in a wealth of new detail. Some aspects of this information drawn from recent experiments on the alkaline earth metals and the rare gases are presented. The potential for increased flux and resolution inherent in insertion device-based facilities like the Advanced Light Source should advance this understanding even further, and some future directions are suggested.

1. Resonant Auger Decay

For photons which are energetic enough to allow two electrons to be removed from an atom, double ionization will ultimately occur. This double ionization can take place via a number of routes. The most common is the normal Auger effect, in which the first electron creates a hole in an inner shell, producing a highly excited ion which decays by emission of a second electron. Direct double ionization of two outer electrons can also occur, but this is generally low in intensity and therefore more difficult to observe. Of recent interest is the shake off[1] of electrons following excitation of an inner-shell electron to a Rydberg orbital which lies above the double-ionization threshold in energy and which can decay by emitting two electrons. These shake-off processes can be detected by looking at electrons which leave the excited atom with zero kinetic energy; a second electron carries all the additional energy of the photon.

Within the context of experiments which we have been conducting recently on the decay of inner shell excitations, we have begun an examination of the role of shake-off as a possible decay channel. We focus on shake-off as the limit of a shake-up process in which the final ionic state is the doubly charged ion, rather than a highly excited state of the singly charged ion. The technique we use is an extension of the Constant Ionic State (CIS) approach which has been very effective in determining the excitation paths which feed decay channels leading to specific final states of an ion. The normal single photoionization which occurs through an intermediate excited state may be represented according to:

$$A + h\nu \rightarrow A^{*} \rightarrow A^{+}(E_b) + e^{-}(\epsilon) \qquad (1)$$

The energy E_b is equal to the first ionization potential plus the energy of the excited state above the ground state. In the CIS technique both the photon energy $h\nu$ and the electron kinetic energy ϵ are scanned simultaneouly; thus, the electrons which are observed correspond to the production of an ion with energy E_b.

The functioning of the CIS scheme applied to shake-off studies is based on an extension of the above process to include the production of two electrons:

$$A + h\nu \rightarrow A^{*} \rightarrow A^{2+}(E_t) + e^{-}(\epsilon_1) + e^{-}(\epsilon_2) \qquad (2)$$

In this extended version, the energy E_t is the threshold for the production of a given final state of the ion. Once the final state is selected by a choice of E_t and ϵ_1 for the CIS scan, with $E_t+\epsilon_1=h\nu$, then the energy ϵ_2 must be zero (within the resolution of the electron detector). Thus, we are not only able to determine that an electron of zero kinetic energy has been observed; we are also able to specify the final state of the ion produced when that electron emerges. In a sense this experiment is a type of coincidence experiment in which only those electron pairs are detected in which one electron has zero kinetic energy. In principle and practice, this continuum method (CCIS) can of course be further extended to define an energy $\epsilon_2>0$ for one of the electrons.

The system we have examined is the production of the doubly ionized xenon ion following excitation of one of the 4d electrons to an np orbital.[2] For n≥6 this orbital lies above the $5s^2 5p^4(^3P_2)$ ground state of Xe^{++}, and shake-off becomes a possibility. Using the continuum CIS approach, we have measured the production of this state throughout the energy region beginning with n=6. The spectrum is given in Fig.1.[2] Appearance of electrons corresponding to ions at energy E_t is apparent at all the excitations which could be clearly resolved.

While it is tempting to regard the spectrum in Fig. 1 as arising solely from shake-off, the interpretation for the case of xenon is not as simple as that. The p^4 configuration in Xe^{++} gives rise to five terms, and there are Rydberg series ending on each of these terms. Thus, there is the possibility of an overlap in energy between the ionization limit of one of the doubly charged ionic states and a Rydberg orbital corresponding to a series ending on a higher limit. This is illustrated in the energy level diagram in Fig. 2. If this occurs, then an electron would appear as a consequence of a shake-up process which would produce the singly charged ion in that excited state. This would subsequently autoionize into the corresponding continuum, producing a near zero-energy electron. It is impossible to distinguish between these two processes by using only one (ϵ_1,ϵ_2) pair through either the continuum CIS method or through the detection of zero-energy electrons.

Although some measurements and calculations exist,[3] the locations of the higher-lying members of the Rydberg series ending on the Xe^{++} limits are not known. However, they can be calculated based on a quantum defect analysis. Even including possible inaccuracies in the calculated energy levels, there are numerous Rydberg levels which lie in close proximity to the Xe^{++} threshold which we are measuring. As we have observed in other experiments which will be discussed below, there is a strong probability for shake-up into these orbitals. Thus, we are inclined toward the interpretation that the structure shown in Fig. 1 is primarily due to shake-up, followed by autoionization, rather than to shake-off. This belief is supported by preliminary measurements which we have since conducted on the $^1S_0$ threshold, where essentially no structure is to be observed in the CCIS spectrum and where the $^1S_0$ threshold lies high enough in energy to be completely free from any overlapping shakeup lines. (See Fig. 3.)

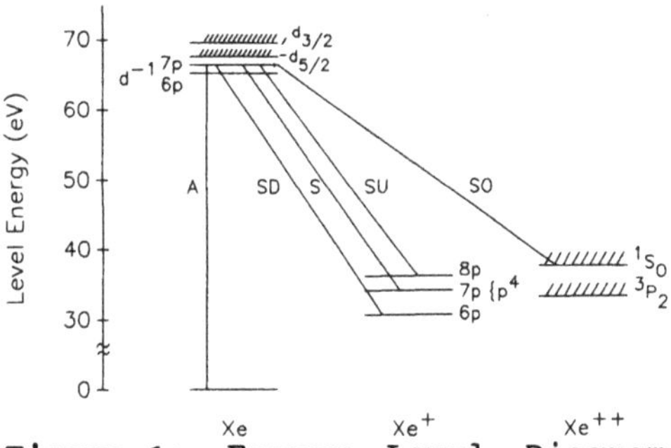

Figure 1: Energy Level Diagram for Xe

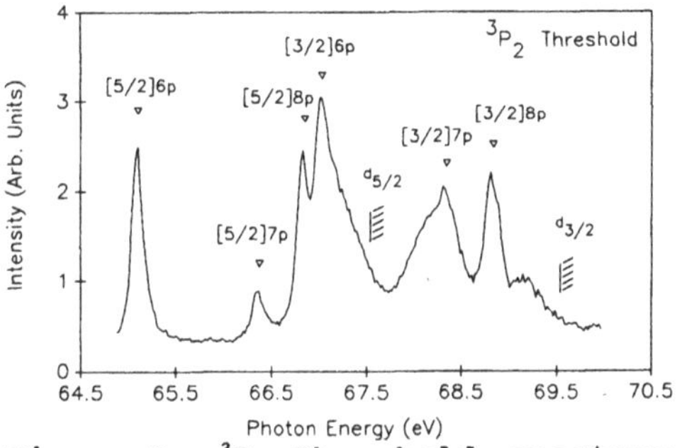

Figure 2: $^3P_2$ Threshold spectrum in Xe.

2. Excited-State Population Enhancement in the Alkaline Earths

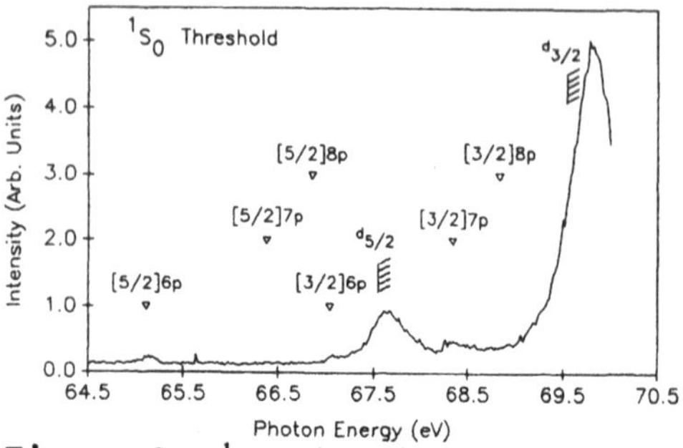

Figure 3: $^1S_0$ Threshold spectrum in Xe

Decay of Rydberg levels resulting from inner-shell excitations has been analyzed through the simple approach of shake theory.[4] Briefly summarized, within this description the decay is presumed to occur primarily through the break-up of an inner-shell orbital, and the electron which was originally excited has little role to play. This electron either remains in the orbital to which it was initially promoted, i.e. retains the same n value, and becomes a "spectator," or it is shaken up or down, i.e. the n value increases or decreases by one or more units, respectively, in response to the change in the effective Z value of the decaying core. The most obvious consequence of the correlation effects which this theory seeks to explain is the frequent appearance of enhancements in the population of excited states of the resulting ion over those created by the direct production as satellites in the case of non-resonant excitation. In many cases production of these excited states is not only enhanced, but becomes the predominant mode of decay. We note parenthetically that for higher n's, the ground state is rarely produced.

2.1 Beryllium

In an experiment which we completed some years ago[5] we observed a most startling result which occurred when the $1s2s^22p$ level of beryllium was observed to decay by electron emission: rather than produce the ground state of the ion, the decay proceeded in such a fashion as to create 95% of the ions in the $1s^22p$ excited state. Indeed, rather than form the ground state of the ion, the decay even preferred to shake up to create the $1s^23p$ state. A series of scans toward higher-lying $1s2s^23p$ and $1s2s^24p$ excitations demonstrated a general pattern of spectator and shake-up processes always predominating over ground state production and shake-down processes. An example of the PES spectrum following both the 3p and the 4p excitations[5] at low resolution which illustrates this behavior is given in Fig. 4. Notice that, while the 3p excitation produces the ion primarily in the 3p state, the 4p excitation generates not only the 4p state but an equal number of ions in the 5p state.

2.2 Magnesium

In the work on beryllium information emerged which indicated that the results discovered for this species might be a general property of all beryllium-like ions.[6] An intriguing question then becomes, to what extent this behavior might be a property of the more complicated members of the alkaline earth series. To explore this question we recently conducted a series of measurements on the decay of excitations of one of the 2p electrons in magnesium.[7] While more complicated than beryllium, magnesium is still rather simple compared to the other alkaline earth elements. In particular, the energy region of the $Mg^{++}(1s^2 2s^2 2p^6)$ ground state is not overlapped by ionization limits of series of two-electron excitations as is the case for calcium and strontium. Decay of the 2p hole might also be expected to occur through a break-up of the $3s^2$ orbital in the same fashion as decay of the 1s hole in beryllium occurred through a break-up of the $2s^2$ orbital.

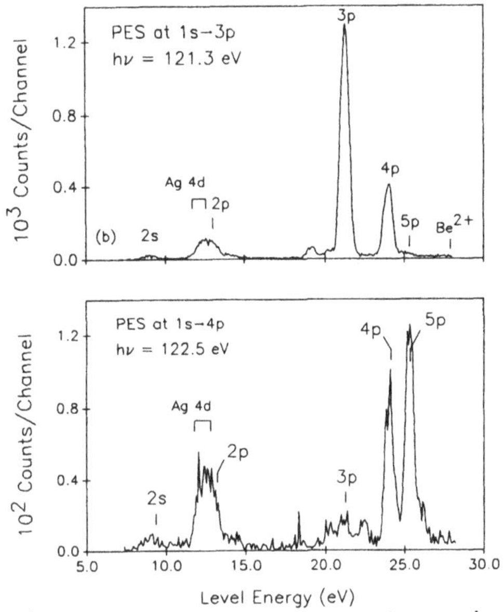

Figure 4: Resonant Auger decay in Be

Excitation from the 2p orbital in magnesium is very different in one regard from that of the 1s in beryllium in that there are two possible series, $2p^6 3s^2 \rightarrow 2p^5 3s^2 ns$ and $2p^6 3s^2 \rightarrow 2p^5 3s^2 nd$. Two examples of the results of the magnesium experiment are given in Fig. 5.[7] These particular spectra are chosen because they illustrate two important features of the behavior of the decay of these excited states. In panel (a) is shown the photoelectron spectrum resulting from the decay of the $2p_{3/2}6d$ and the $2p_{1/2}5d$ states. (These could not be resolved with the monochromator we used.) The appearance of the 5d and 6d spectators is clear, along with the 7d shake-up line. A spectrum of this type is rather typical for all the nd excitations we observed except for the $2p^{1/2,3/2}3d$ orbital and the $2p_{1/2}6d$ orbital. In the case of 3d excitation decay results primarily in the production of the ground state, in complete discordance with the predictions of shake theory. The spectrum in panel (a) is also rather typical for excitations into all ns orbitals which we observed. The behavior which is most surprising is illustrated in panel (b) of the figure, the $2p_{1/2}6d$ excitation.

In panel (a) two small features appear to the right of the spectrum. These are L_2-M_1M_1 and L_3-M_1M_1 Auger electrons which are produced by second order radiation from the diffraction grating. These are observed in all decay spectra which we measured and always appear with rather weak intensity, as expected from the small fraction of second order radiation which we expect. However, the same is not true for the excitation depicted in panel (b). Rather than appear as a simple weak feature due to second order radiation, the L_2-M_1M_1 Auger line appears as the dominant decay mode for this excitation. While this is certainly allowed energetically, it can only correspond to a shake-up transition in which the ion is left in the 11d final state. That this should occur is completely out of step with the general trend of the other decay modes. This normal pattern predicts that the 6d or 7d level of the ion should have appeared with the maximum intensity, not the 11d as observed. Also unclear is why only the L_2-M_1M_1 line should appear and not the L_3-M_1M_1 line. That this is true has been borne out by an independent series of measurements on the production of these two Auger lines following excitation to the resonances indicated above. While we are convinced that this is a real phenomenon, quantitative calculations are urgently needed if we are to understand this behavior.

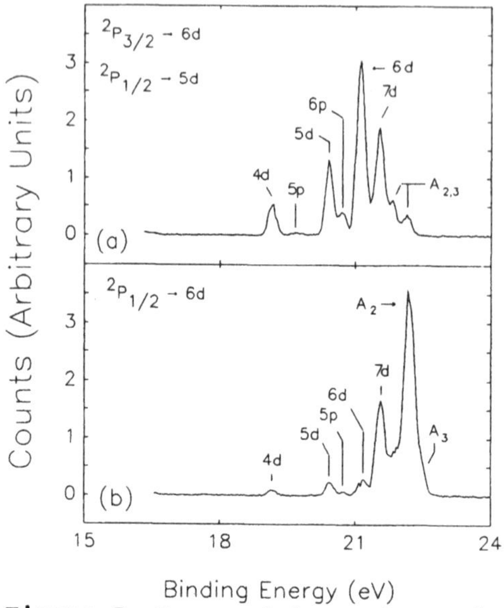

Figure 5: Resonant Auger Decay in Mg

2.3 *Strontium*

In order to investigate the possible trends which exist

for the decay of excitations in the alkaline earth elements, we have recently made a series of measurements on the resonance enhancement of excited state ion production in strontium as well. Our data for this atom are still quite preliminary, but the type of behavior we are seeing can be illustrated by two simple spectra. In Fig. 6 are shown CIS scans focused on the 5s (dotted line) and 4d (solid line) states of the strontium ion. The very broad feature near 25 eV is the principal $4p \rightarrow 4d$ excitation. This feature appears in CIS scans fixed on both final states, albeit in quite different forms. This difference can be attributed to the number of different states which arise from the different couplings within the complex. Also visible at higher energies on the right are two weak features, one of which is only apparent in the 4d channel. Corresponding photoelectron spectra taken at these two features are given in Fig. 7. These spectra are not normalized to each other in absolute value; rather the contributions to the 4d production have been set arbitrarily equal to illustrate the observation regarding enhancement. In particular, one notices that there is virtually no ground state 5s production at feature 1. This is behavior which we have observed in other experiments as well, namely, that features which appear very weak in a normal absorption spectrum can be made readily visible through exploitation of the decay enhancement which they create. Although our analysis is still far from complete, we believe that the information which we acquire through these enhancement measurements can help us not only to identify excitations but to understand in more detail the dynamics of the decay process.

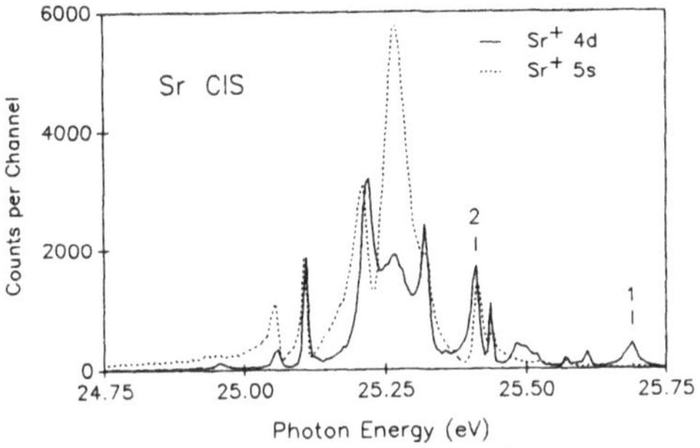

Figure 6: CIS spectrum for Sr

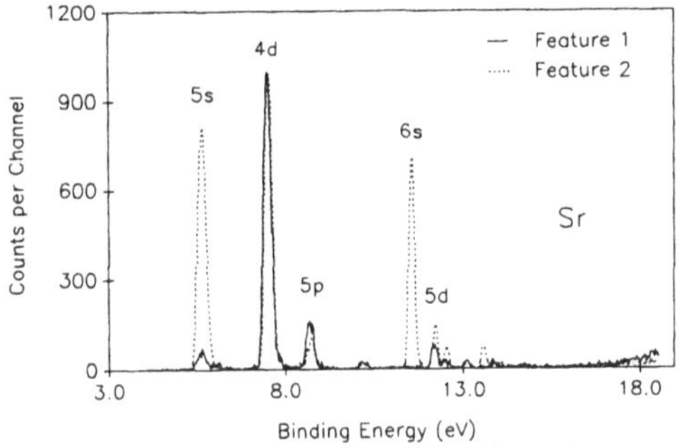

Figure 7: PES Spectrum for Sr

3. Gas-Phase Experiments at the Advanced Light Source

Although experiments on atoms and molecules using synchrotron radiation have helped create a revolution in our understanding of the details of electron dynamics and atomic and molecular structure, there are still limitations on the information we can acquire. These limitations are imposed by the levels of flux and resolution which are available in traditional radiation sources based on bending magnets. The next breakthroughs in the study of gas phase processes using synchrotron radiation are going to demand both higher flux and increased resolution, in other words, the third generation of synchrotron radiation sources. One such instrument is the Advanced Light Source (ALS) currently being constructed at the Lawrence Berkeley Laboratory in California. Based on insertion devices as radiation sources, this facility will have an extremely high brightness, contributing both to improved flux and improved resolution.

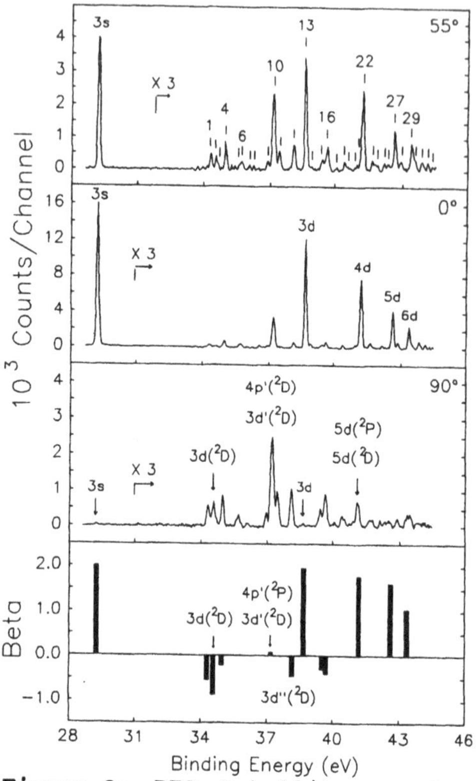

Figure 8: PES Satellite spectra of Ar

From a series of proposals which were submitted to the ALS Scientific Review Panel last year, two groups of scientists were selected as Principal Research Teams (PRTs) for cooperation with the ALS staff in determining the characteristics of the first two beamlines which the ALS will have in operation. One of the lines, called U-8 in consequence of its derivation from an 8-meter undulator source, will be devoted to gas phase processes in atoms and molecules. It will cover an energy range from 20-300 eV, with resolution up to 10,000 over part of the spectrum. Although the list of experiments which are proposed for use on this line is quite long, a considerable portion of the activity will be devoted to electron spectrometry. We anticipate that this new beamline will make measurements such as coincidence experiments, alignment determinations, spin determinations, and high resolution electron spectra routine.

As a last item in this presentation is shown in Fig. 8 an electron spectrum of the 3s state in argon,[8] together with the 3s,3p correlation satellites. This particular measurement was performed at the Synchrotron Radiation Center in Wisconsin on the undulator line which was in operation there for a time and provided a glimpse of what the future will bring. It is simply not possible to acquire the detail which is available in this spectrum through use of even the best bending magnet sources. The high degree of polarization of undulator radiation (P~100%) makes even the measurement of angular distributions for these features a much simpler task. Other measurements which we performed using this undulator source provide the same type of tantalizing results, leading us to expect that the marriage of electron spectrometry with synchrotron radiation will continue to create new insight into the details of electron correlation and dynamics.

4. Acknowledgements

This research was supported by the National Science Foundation under grant NSF-PHY-8907286. MOK acknowledges support by the Division of Chemical Sciences, U.S. Department of Energy under contract DE-AC05-84OR21400 with Martin Marietta Energy Systems, Inc.

5. References

1. P.A. Heimann, D.W. Lindle, T.A. Ferrett, S.H. Liu, L.J. Medhurst, M.N. Piancastelli, D.A. Shirley, U. Becker, H.G. Kerkhoff, B. Langer, D. Szostak, and R. Wehlitz, *J. Phys.* **B20** (1987) 5005.
2. C. D. Caldwell, (To appear in Nuclear Instruments and Methods, 1991).
3. J.E. Hansen and W. Persson, *Phys. Scrip.* **36** (1987) 602.
4. K.G. Dyall and F.P. Larkins, *J. Phys.* **B15** (1983) 3339.
5. C.D. Caldwell, M.G. Flemming, M.O. Krause, P. v.d. Meulen, C. Pan, and A.F. Starace, *Phys. Rev.* **A41** (1990) 542.
6. D. Petrini, *J. Phys.* **B14** (1981) 3839.
7. S.B. Whitfield, C.D. Caldwell, and M.O. Krause, *Phys. Rev.* **A43** (1991) 2338.
8. M.O. Krause, S.B. Whitfield, J.-Z. Wu, C.D. Caldwell, P. v.d. Meulen, and R.W.C. Hansen, (To be published).

STORED IONS AS TARGETS FOR SYNCHROTRON RADIATION

D. A. CHURCH
Physics Department, Texas A&M University
College Station, Texas 77843, USA

Abstract

Atoms, and multi-charged ions stored in a Penning ion trap, have been recently used as targets for focussed broad-band synchrotron X-radiation. For example, the K-shell photoionization of Ar and Ar^{2+} each result in distributions of Ar charge states which critically depend on the details of the vacancy cascade which follows the inner electron excision; details such as relative Auger and fluorescence rates, and electron shake-off probabilities. Confinement of the ions from these resulting charge state distributions in the Penning trap permits a simple measurement of the relative number of ions found in each charge state. However, electron capture collisions by the stored charges can modify the distributions. Corrections can be made using measured electron transfer rate coefficients and the target and residual gas densities. Potential studies of photoionization using improved methods are discussed, with reference to the already-completed measurements on Ar and Ar^{2+}. Possibilities for the use of undulator radiation from next-generation synchrotrons are also mentioned.

1. Introduction

The use of stored ions as a target for photoionization studies[1,2] presents both advantages and disadvantages. A distinct advantage is the relative simplicity and cost effectiveness of this form of ion target, compared to a beam of multi-charged ions from a sophisticated ion source. The apparatus is relatively small and self-contained, and can be mounted on a wheeled cart for introduction into the synchrotron radiation (SR) beam. An ion target with a particular mass-to-charge ratio m/q can be selectively stored. Product ions from the photoionization are also stored in the trap, and can be accumulated as a simple form of signal integration. There is a disadvantage related to this procedure, however, since the confined ions can capture electrons in collisions with residual gas, thus modifying the measured charge distribution from that originally created. The original distribution can be reconstructed from the data, provided that the electron transfer rate coefficients are known, and that the partial pressures of target gases have been measured.[1] Excellent vacuum technique can minimize the electron capture.

The magnetic field associated with Penning ion confinement inhibets the use of particle detectors. Analog ion detection can be used, but with lower sensitivity, requiring longer data collection times. In an ion trap, absolute target ion densities, and hence absolute cross sections, are difficult to determine, but with care relative cross section measurements at a useful level of accuracy and precision appear feasible. With brighter SR beams becoming available in the near future from the third-generation SR facilities, it will be feasible to study SR photoionization along isoelectronic sequences, or as a function of level of ionization, or to determine the dependence on atomic number Z, all as a function of photon energy.

Ion traps are also useful for collision and spectroscopy studies on the low energy multi-charged ions resulting from SR removal of inner shell electrons.[3] The stored ions in the distribution can have temperatures as low as 400 K, corresponding (for argon ions) to relative Doppler-limited transition widths $\frac{\Delta\nu}{\nu} = 10^{-6}$, or less than 1 GHz in the visible spectral region. This is better than can be obtained with many multiply-charged ion sources using a collinear laser-beam ion-beam geometry. The measurement of rate coefficients for electron transfer collisions is another important application of stored multiply-charged ions.[4] These collisions are important in laboratory and astrophysical plasmas, and significant theoretical effort has been devoted to the development of calculational techniques for these collisions, in recent years.

2. Technical Considerations

A Penning ion trap employs a static electric quadrupole potential

$$\phi(r,z) = \frac{V_0}{4{z_0}^2}(r^2 - 2z^2) \tag{1}$$

and a uniform axial magnetic field $\vec{B} = B_0\hat{z}$ to confine ions in a limited region of space.[5] Ions with low charge states q are often produced in the trap by electron impact ionization of target gas atoms or molecules. Ion densities to $10^7/q\,cm^{-3}$ have been produced in this way. Although such densities are comparable to gas pressures only in the 10^{-10} Torr range, they can exceed neutral densities in an ultra-high vacuum (uhv) chamber, and are comparable to ion densities in beams from multi-charged ion sources such as the Electron Cyclotron Resonance Ion Source. Since the confined ions move in the trap with well-defined frequencies which depend on the ion mass-to-charge ratio m/q, a particular charge state can be selectively confined by removal of the other charge states using radio-frequency excitation. Thus, stored ions may be used as a target for interaction with suitable projectiles. Neutral beams, such as photons, are particularly appropriate, and extensive work with laser excitation of singly-charged positive ions,[6] and photodetachment of negative ion targets,[7] has been completed. Inner shell photoionization occurs with orders-of-magnitude smaller cross sections however, and the available monochromatized photon fluxes are considerably less than those available at longer wavelength from lasers. Nevertheless, the lack of experimental information on the important topic of ion photoionization encourages the development of new techniques.

The product ions of the photoionization interaction have very low kinetic energies, since little net momentum is transferred in the photoelectron ejection or Auger cascade.[3] The product ions could be extracted from the trap along the magnetic field lines, through a transition region to lower magnetic field to a channeltron or microchannel-plate detector. With care, time-of-flight (TOF) m/q analysis would probably be effective. Despite higher energies, the electrons would also be confined radially by the axial magnetic field, and could similarly be extracted. Energy definition would be lost for the electrons, since the electron energies would be distributed both parallel and perpendicular to the magnetic field lines.

Employment of analog detection of the ions in the trap is a simple alternative to these particle counting schemes. The axial oscillation of the ions at angular frequency

$$\omega_z = \left(\frac{qV_0}{m{z_0}^2}\right)^{1/2} \tag{2}$$

is excited by ramping the trap potential V_0 to bring ω_z into resonance with an excited tuned circuit, composed of the trap capacitance and an external inductance. The driven coherent motion of the ions induces a radio-frequency potential difference superposed on the exciting rf on the tuned circuit. This results in an absorption peak which is proportional to the ion number N and charge state q. As N is increased, this signal tends eventually to saturate in amplitude and to broaden, limiting resolution.

The presence of more than a single ion m/q ratio leads to signal shifts due to space charge, and also due to coherent coupling between the driven motion of different m/q ratios. Despite these deficiencies, the sensitivity of this detection method, and its ease of application, often make it the method of choice.

The brightest, quasi-monochromatic SR currently available is produced by undulators. In the third-generation synchrotron facilities now under construction, such as the Advanced Photon Source (APS) at Argonne National Laboratory, undulator radiation will produce a flux as high as 10^{16} photons/s in a < 1 mm^2 spot 30 m from the accelerator, with a bandwidth of about 10% of the photon energy.[8] Photon energies in the 4 keV to 20 keV range may be chosen. Since the photon energy is tunable, this source will potentially be very effective in inner-shell photoionization studies.

In the meantime, focussed broadband bending magnet radiation from the x-ray ring of the National Synchrotron Light Source (NSLS) at Brookhaven National Laboratory has been used. Two 250 mil thick water-cooled beryllium windows were used to isolate the vacuum of the ion trap, the beam transport line, and the electron storage ring. These windows served to strongly attenuate the photon flux just below the argon K edge near 3.2 keV, so K-shell photoionization dominated. A 1:1 cylindrical grazing incidence x-ray mirror was mounted ten meters from the source to focus the radiation through the ion trap. The platinum overcoat effectively reflected x-rays to energies near 15 keV. Consequently, the radiation used in the measurements was broadband, although the flux dropped by a factor near sixty over the energy band from 3 to 15 keV, and the rapid decrease of the photoionization cross section with energy above threshold concentrated the most effective photon bandwidth for photoionization near the K edge. Using a mean current in the x-ray ring of about 140 mA during the actual measurements, the flux distribution with energy F(E)dE integrated over the cross section for K-shell photoionization σ(E) produced a result $G = \int_{E_K}^{\infty} \sigma(E)F(E)dE = 2.8 \times 10^{-4}$ cm^2/s. This number may be compared with a comparable number $G' \equiv \int_{E_K}^{\infty} \sigma(E)F'(E)dE$ for an undulator planned for the APS. Just above threshold, with a mean ring current near 70 mA of stored positrons (based on a maximum ring current of 100 mA), $G' \cong 5 \times 10^{-3}$ cm^2/s, more than a factor of ten higher than feasible with the focussed bending magnet radiation, and with significantly improved resolution.

An ion target can be easily produced in the trap by electron impact ionization of target gas. An electron current parallel to the magnetic field lines from an electron gun mounted behind one end electrode of the Penning trap produces ions with an initial distribution closely centered on the trap axis. The initial ion distribution is defined radially by the apertures that define the electron beam, but under the action of space charge fields, the ions gradually drift radially outward in the trap, consistent with conservation of canonical angular momentum. The ions assume an axial distribution characterized by a mean axial energy about one tenth of the axial potential well energy.

In the present measurements,[1] the synchrotron radiation was focussed through the trap in the radial plane, perpendicular to the magnetic field. The original measurements with this system were carried out before the focussing mirror was available. Radial introduction of the unfocussed radiation maximized the flux through the ion trap.[3] If the focussed SR is used to produce target ions by photoionizing a target gas, the ion distribution in the radial plane peaks at the trap center, but is spread throughout the radial plane by the $\vec{E} \times \vec{B}$ drift of the ion cloud about the symmetry axis of the trap at the magnetron frequency $\omega_- = \omega_c/2 - \omega_0$, where

$$\omega_0 = (\omega_c{}^2/4 - \omega_Z{}^2/2)^{1/2} \text{ and } \omega_C = qB/m. \tag{3}$$

Thus, the effective ion density in the focussed SR beam is reduced, compared to that for a target produced by electron impact. A superior geometry would have the focussed SR incident axially through the trap, producing an ion target with properties similar to those of the electron target described above. The subsequent interaction between radiation and target ions would then be maximized.

Since the target ion density is effectively in the 10^{-10} Torr range, it is important to reduce the density of similar atoms to well below this value. An excellent possibility for an atom target, from which to produce the ion target by photoionization, would be a beam of metal atoms passing through the trap, effusing from a differentially-pumped oven. The beam could be interrupted by a mechanically-actuated shutter when the presence of the atoms was undesired. Due to the high sticking probability of metal atoms on metal surfaces, the partial pressure of these atoms in the target chamber would be extremely low, when the beam was blocked. On the other hand, the effective density of the unblocked beam could easily be in the high 10^{-9} Torr range, permitting rapid target ion formation and accumulation. Special attention to ultra-high vacuum technique could reduce the partial pressures of typical residual gases to below the 10^{-11} Torr range. Since these gases have low SR ionization cross sections at keV energies, their primary influence would be in causing charge-changing collisions with the stored target and product ions.

In the current measurements, a pulsed argon source was used.[1] Argon atoms entered an intermediate chamber when a pulsed valve was briefly opened. This chamber was evacuated by a large cryopump, and was loosely coupled to the ion trap vacuum system by the orifices of a microchannel plate fragment. The transient pressure increase in the intermediate chamber produced a corresponding pressure pulse (of much smaller magnitude) in the ion trap. This increased atom density was ionized by the electron pulse, and was then pumped away with a time constant near 100 ms by cryopumps on each end of the tube containing the trap, leaving the stored ion target behind. It is thought that the relatively high pressure of H_2 observed in the trap region (near 10^{-8} Torr) arose from effusion from the intermediate chamber.

3. Data Collection and Analysis

The measurement sequence consisted of the brief electron pulse synchronized with the argon gas pulse, to produce the Ar^{2+} target in the trap. The trapping magnetic field was set sufficiently low to make Ar^{+} unstable, so that it could not build up in the trap either by direct ionization or by electron capture to Ar^{2+}. After the ion target was formed, it was held for two seconds while the gas pulse was pumped away. Then a two second SR pulse was gated on, by rotating a mechanical shutter in the beam line, which blocked the SR beam during the target formation interval. Following the target irradiation, the ions in the trap were detected. These ions consisted of the target Ar^{2+} ions, plus other ions produced from residual gas by the electron pulse, photoions formed by ionizing the residual target gas, photoions arising from ionization of the Ar^{2+} target, and any ions formed by electron capture (primarily from residual H_2) to these stored ions. To clearly identify the ions resulting solely from photoionization of Ar^{2+}, two background ion signals were acquired and subtracted from the ions detected in the signal cycle (denoted as "A" in Figure 1.)

To determine the ions formed by the electron pulse only, a cycle (denoted B) which lacked only the SR pulse was employed. The ions detected in this cycle consisted of the Ar^{2+} target, plus other ions produced by the electrons, and any product ions from electron transfer that were stored. The ions produced by photoionizing neutral argon were measured with cycle C, which differed from cycle A only by the absence of the electron pulse, and hence the Ar^{2+} ion target.

The base level for the true ion signals in Figure 1 was verified by subtracting either background B or C from A; with double background subtraction the zero offset of the signals becomes meaningless. When both backgrounds were subtracted,

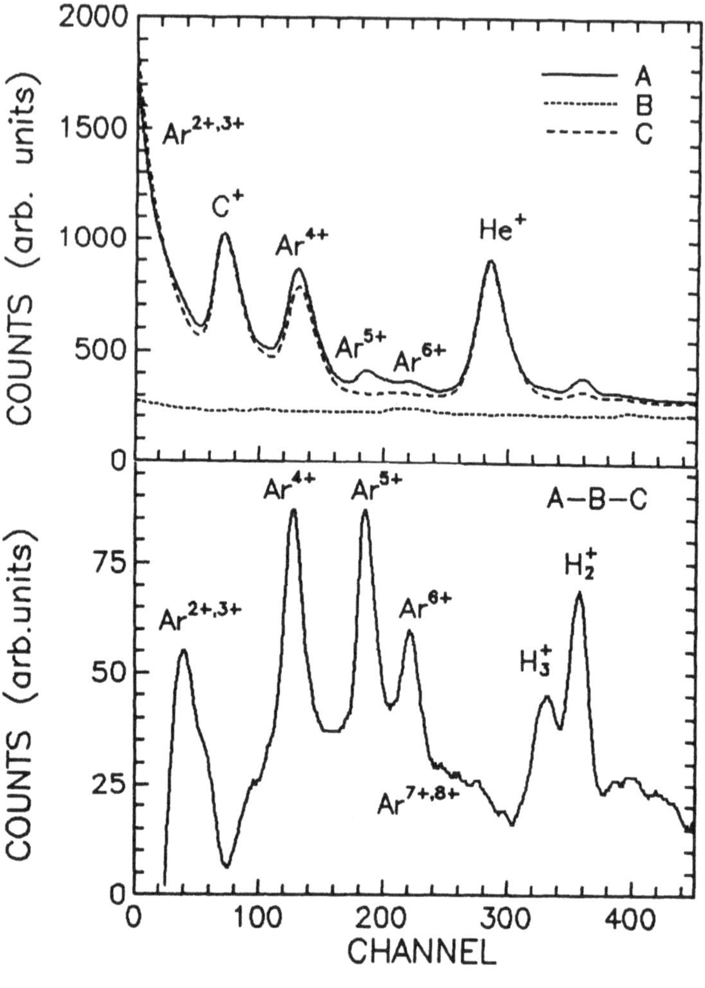

Figure 1 (a) The observed signal (A) and background (B and C) traces for 1000 cycles of integration. (b) The signal minus background spectrum resulting from photoionization of stored Ar^{2+}. The hydrogen peaks are a consequence of electron capture collisions of the photons with residual H_2 (from ref. 1, with permission).

the true ion signal due to photoionization of Ar^{2+} remained, plus ions formed by electron capture to these photoions. Charge states from Ar^{3+} to Ar^{8+} were observed. It was expected that little Ar^{3+} would be formed in the vacancy cascade, since at least one Auger electron was emitted along with the K-shell photoelectron. Consequently the presence of Ar^{3+} is due primarily to electron capture collisions of Ar^{4+}. Single capture was dominant by far, since the primary residual gas was H_2. In fact, the only observed product ions from collisions observed after background subtraction were H_2^+ and H_3^+. The H_2^+ ions were formed directly by electron capture collisions with the argon ions, but the H_3^+ ions were tertiary ions formed by the reaction of H_2^+ with H_2.

The mean of the distribution of the measured photoion charge states resulting from the photoionization of Ar^{2+} was significantly higher than that for the corresponding photoionization of neutral argon, as expected. However, due to electron capture during the ion storage, the mean charge was lowered compared to its value upon ion production. Using the measured target gas densities and electron capture rate coefficients for the different charge states, plus a system of coupled differential equations which described both the photoion production and electron capture rates of formation and loss for each charge state, an analytical procedure was used to generate the initial photoion charge distribution from the measured photoion charge distribution.[1] This distribution had Ar^{6+} as a most probable charge state, and a mean charge $\bar{q} = 5.84 \pm 0.25$. Of course, to the extent that the residual gas pressure could be reduced by better vacuum technique, the correction could be minimized in future measurements. There are theoretical predictions for mean charges, and for charge distributions, which are formed by vacancy cascades following inner shell photoionization of certain ion charge states, but unfortunately none as yet for Ar^{2+}. These calculations typically do not include electron shakeoff, so a comparison between theory and experiment would provide significant input about the role, and Z and charge dependences, of this important physical process.

4. Summary and Conclusion

Photoionization of a stored ion target in a Penning ion trap has been successfully observed and studied, using focussed bending magnet radiation from the x-ray ring of the NSLS. The signal was the number of ions in each photoion charge state resulting from the vacancy cascade following K-shell photoionization of Ar^{2+}. This photoion distribution provides useful information on the relative fluorescence and

Auger rates following vacancy production, as well as on the effects of electron shake-off. A broad distribution of photon energies was used in the measurement, since monochromatization would have reduced the photon flux to a level at which signal detection would have been difficult. However, the future employment of an undulator, which has a relatively narrow bandwidth, tunability, and high brightness, should enable more sophisticated measurements in the near future.

Analog detection of the product ions is the simplest procedure when using a Penning trap, due to the magnetic field required for ion confinement. This leads to lower signal sensitivity than could be obtained with particle counting methods, and makes the determination of an absolute photoionization cross section difficult. However, relative cross sections, measured as a function of photon energy or target charge state, could be obtained and normalized to an absolute measurement made by other means. On the other hand, it appears that with effort, a particle counting system to detect ions or electrons could also be installed.

With the current apparatus, the ion target is produced by electron impact ionization of target gas. If the photon beam were directed along the symmetry axis of the trap, a multi-charged ion target could be generated by photoionization, permitting subsequent efficient interaction between the photons and the ion target, and the generation of higher target charge states. A particular target ion charge state could be selected and accumulated, and with a base pressure near 10^{-9} Torr, the target storage time would be sufficiently long to permit the study of quite high ion charge states. If a gated metal atom beam were used instead of a gas pulse, the interval between target preparation and photoionization could be greatly reduced, speeding the data collection. It thus appears that ion targets stored in traps will have sufficient utility to justify further development of the technique.

5. Acknowledgments

It is a pleasure to acknowledge the important contributions of all collaborators in the research discussed here, which was carried out on the X-26C beam line of the x-ray ring of the NSLS. The collaborators were my Ph. D. student at Texas A&M, Scott Kravis; Brant Johnson, Mati Meron, and Keith Jones of Brookhaven National Laboratory; John Levin and Ivan Sellin of the University of Tennessee and Oak Ridge National Laboratory; Yoshiro Azuma, Nora Berrah-Mansour, and Gordon Berry of Argonne National laboratory, and Michel Druetta of the University de St. Etienne, St. Etienne, France. The Texas A&M portion of the collaboration was funded by the National Science Foundation, and significant start-up funds were also provided by Texas A&M University.

6. References

1. S. D. Kravis, D. A. Church, B. M. Johnson, M. Meron, K. W. Jones, J. Levin, I. A. Sellin, Y. Azuma, N. Berrah-Mansour, H. G. Berry, and M. Druetta, (to be published).
2. S. D. Kravis, D. A. Church, B. M. Johnson, J. C. Levin, Y. Azuma, I. A. Sellin, M. Meron, K. W. Jones, M. Druetta, N. Mansour, H. G. Berry, and R. T. Short, *Nuc. Instrum. Methods* **B** (in press).
3. D. A. Church, S. D. Kravis, I. A. Sellin, C. S. O, J. C. Levin, R. T. Short, M. Meron, B. M. Johnson, and K. W. Jones, *Phys. Rev.* **A36** (1987) 2487.
4. D. A. Church, *Studies of Ion Collisions in Ion Traps*, *Physica Scripta* **T22** (1988) 164.
5. H. G. Dehmelt, *Adv. At. Mol. Phys.* **3** (1967) 53 and **5** (1969) 109.
6. D. J. Wineland, W. M. Itano, and R. S. Van Dyke, Jr., *Adv. At. Mol. Phys.* **19** (1983) 135.
7. D. J. Larson, N. Kwon, P. S. Armstrong, T. Olsson, and R. Trainham, *Nuc. Instrum. Meth.* **B40/41** (1989) 248.
8. G. K. Shenoy, P. J. Viccaro, and D. M. Mills, *Characteristics of the 7-GeV Advanced Photon Source: A Guide for Users*, Argonne National Laboratory Report ANL-88-9 (1988).

THE ADVANCED LIGHT SOURCE: A NEW TOOL FOR RESEARCH IN ATOMIC AND MOLECULAR PHYSICS

Fred Schlachter and Art Robinson

Advanced Light Source, Accelerator and Fusion Research Division, Lawrence Berkeley Laboratory, University of California, Berkeley, CA 94720, USA

ABSTRACT

The Advanced Light Source at the Lawrence Berkeley Laboratory will be the world's brightest synchrotron radiation source in the extreme ultraviolet and soft x-ray regions of the spectrum when it begins operation in 1993. It will be available as a national user facility to researchers in a broad range of disciplines, including materials science, atomic and molecular physics, chemistry, biology, imaging, and technology. The high brightness of the ALS will be particularly well suited to high-resolution studies of tenuous targets, such as excited atoms, ions, and clusters.

1. Introduction

Bright, wavelength-tunable light sources make life easier for researchers interested in spectroscopy and scattering experiments. For those working in the ultraviolet or x-ray regions of the spectrum, synchrotron radiation is usually the tool of choice. In fact, the availability of intense, tunable, collimated, polarized radiation has driven the evolutionary development of dedicated facilities optimized for the generation of synchrotron radiation. By 1993, so-called third-generation synchrotron sources will be coming on line in the United States, Europe, and Asia. Next-generation synchrotron light facilities of the type now under construction specialize in producing either soft (long-wavelength) x and ultraviolet radiation (collectively, the XUV) or hard (short-wavelength) x rays by means of insertion devices (undulators and wigglers).

The prospect of exploiting high spectral brightness (flux per unit area of the source, per unit solid angle of the radiation cone, and per unit bandwidth) is the main selling point of the new third-generation synchrotron facilities. The storage rings at these facilities are designed to optimize brightness in two ways. First, the rings contain lengthy straight sections to accommodate long undulators (typically 5 meters in length) with up to 100 or so periods. In general, both photon flux and undulator brightness increase linearly with the number of periods, thereby putting a premium on long undulators with many periods. Second, the rings are designed to have electron (or positron in some cases) beams with a small cross section and a small angular divergence; that is, the storage rings have a small emittance. Since the size and divergence of the electron beam set lower bounds for the radiation-source size and divergence, a low-emittance electron beam translates directly into a high-brightness undulator photon beam. The combination of a very-low-emittance storage ring with optimized undulators makes possible the generation of radiation with a spectral brightness that is a factor of 20 or more greater than that of existing, second-generation sources, depending on the spectral range. In the past, order-of-magnitude increases in brightness have led to qualitatively new developments in spectroscopic and structural studies of both gas-phase and condensed matter.

Most directly benefiting from high brightness are researchers in both the life and physical sciences who hope to achieve enhanced spatial resolution down to distance scales of about 100 Å in x-ray microscopy and spatially resolved XUV spectroscopy. An example of the latter arises from the study of solid surfaces, which are mostly heterogeneous, making interpretation of spectroscopic data obtained from illuminating the entire surface difficult. With spatial resolution, spectral features could be directly associated with specific surface areas and structures. With existing synchrotron sources, there are a growing number of tantalizing results in x-ray microscopy (including holography) and in spatially resolved x-ray absorption and photoelectron spectroscopy. Moreover, the imaging technology is advancing in step with that of undulators.

For the spectroscopist, the benefits of brightness are substantial but more indirect. Brightness leads to the ability to achieve high spectral resolution without the usual penalty of reduction in signal and increase in measuring time. Experiments with once impracticably long measuring times, perhaps because of inherently weak signals, now become reasonable to contemplate.

Finally, brightness and the naturally pulsed nature of synchrotron radiation bring the ability to observe short-lived or transient systems by means of time-resolved spectroscopic, scattering, and imaging experiments. For example, life scientists hope that, if radiation damage due to the x-ray exposure can be avoided or minimized, it may be possible to image changes in functioning biological cells and cellular structures in near-natural environments. The ultimate time resolution, made possible because there are enough photons in a single pulse of bright synchrotron light to generate a useful signal, would be to follow events in real time on the sub-nanosecond time scale.

Undulators are not invariably the preferred source of synchrotron radiation. For example, the possibility of time-resolved experiments derives more from high photon flux than high brightness. As it happens, undulators tend to generate higher photon fluxes in their spectral regions than do other sources, although any relative advantage is considerably less for photon flux than for brightness. If the feasibility of a contemplated experiment, time resolved or otherwise, is dominated by flux considerations, a wiggler may be the better source of synchrotron light.

2. The Advanced Light Source

The Advanced Light Source, a next-generation synchrotron source of XUV radiation, is under construction in Berkeley, California, at the Lawrence Berkeley Laboratory (LBL) of the University of California. The ALS is a U.S. Department of Energy-funded construction project with a total estimated cost of $99.5 million. Congress has already appropriated 93 percent of this amount, and the remainder is in the President's fiscal year 1992 budget. The project is scheduled to be completed in April 1993. As a national user facility, the ALS will be available to visiting and in-house researchers from university, industrial, and federal laboratories.

As the highest priority project at LBL, the ALS is poised for success. LBL has a long history and extensive base of experience building and operating state-of-the-art accelerator facilities. An exceptionally strong and experienced team of scientists and engineers is dedicated to make the ALS succeed. LBL also has a demonstrated commitment and considerable experience operating national user facilities, such as the National Center for Electron Microscopy, which offers a broad array of imaging and analytical capabilities down to atomic length scales to a user community spanning many disciplines. In short, all the elements are in place for success.

The ALS facility consists of an accelerator complex, a complement of insertion devices, beamlines, and associated experimental apparatus, and a building to house this equipment. The state-of-the-art (low-emittance) storage ring has 10 straight sections available for insertion devices (undulators and wigglers). Initially, 24 high-quality bend-magnet ports will also be available. The undulators will generate high-brightness radiation at photon energies from below 10 eV to above 2 keV in the XUV region of the spectrum. The wigglers will access the hard x-ray region by generating broad-band radiation up to 20 keV. Bend magnets will be useful below 10 keV. Infrared radiation will also be available from the bend magnets. In the normal operating mode, the time structure of the radiation will comprise pulses with a full-width-half-maximum of about 30 ps and separation of 2 ns.

These capabilities will support an extensive research program in a broad spectrum of scientific and technological areas in which x-ray and ultraviolet radiation is used to study and manipulate matter in all its varied gaseous, liquid, and solid forms. New opportunities for research in materials, interface, and surface science, in atomic and molecular physics, in chemistry, in the life sciences, in earth science, and in x-ray optics will be available. The ALS will also serve those interested in developing the fabrication technology for micro- and nanostructures, as well as for characterizing them. In sum, the high brightness of the ALS will open new areas of research while the facility continues to serve as a reliable source for experimental techniques that have proven their worth in solving problems in basic research and applied technology.

3. Accelerator Overview

The ALS accelerator complex is described in considerable detail in a conceptual design report. An overall layout of the facility's accelerator complex, which consists of a 50-MeV linac, a 1-Hz, 1.5-GeV booster synchrotron, and an electron storage ring, is shown in Fig. 1. Table 1 lists the major parameters of the storage ring. Although the energy of the storage ring will range from 1 to 1.9 GeV, its performance is optimized at 1.5 GeV. Performance characteristics of the ALS are determined primarily by the design of the storage-ring magnet lattice. The ALS lattice has 12 long straight sections that are joined by 12 achromatic arcs, each containing three bend magnets. This structure is referred to as a triple-bend achromat. Each of the 12 identical segments (superperiods) of the ALS lattice also contains six quadrupole focusing magnets and four sextupole magnets. The ALS lattice is optimized for the use of insertion devices. Of the 12 straight sections in the storage ring, one is used for injection and one is occupied by rf cavities, leaving 10 full straight sections available for undulators and wigglers up to 4.5 m in length. Each arc of the storage ring is fitted with four bend-magnet ports (Fig. 2). Of the maximum of 48 ports, 24 are so-called prime ports with superior properties and will be developed first.

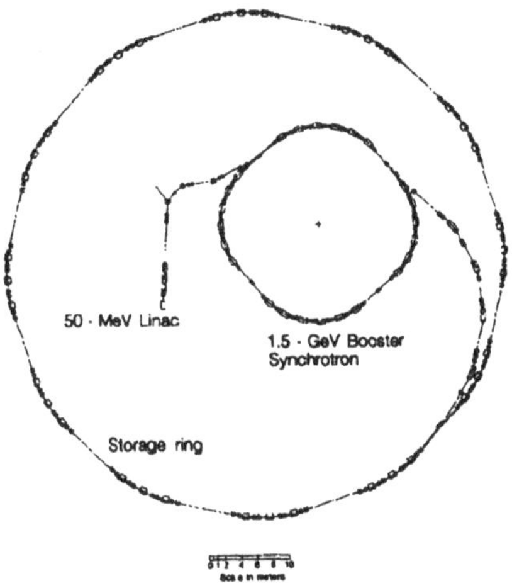

Fig. 1. Layout of the ALS accelerator complex showing the placement of the 50-MeV electron linear accelerator, the 1.5-GeV booster synchrotron, and the storage ring.

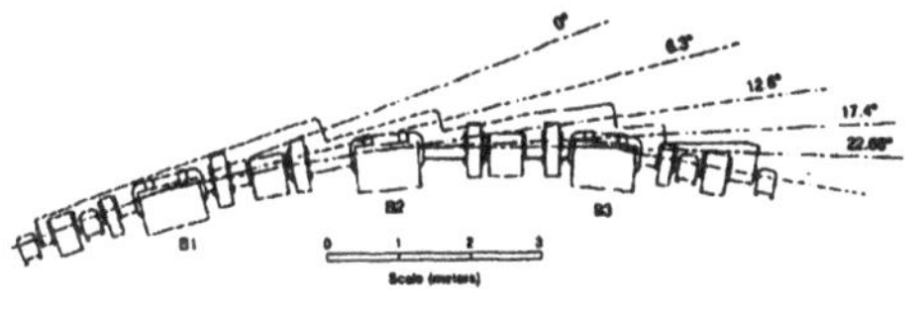

Fig. 2. Arrangement of photon-beam ports in one superperiod of the ALS storage-ring lattice showing an insertion-device port and four bend-magnet ports.

The ALS produces electron beams that are bunched rather than continuous. The storage-ring rf system has a frequency of 500 MHz, so the spatial separation between bunches is 0.6 m and the temporal separation is 2 ns. The storage-ring lattice, the rf system, and the

Table 1. Main Parameters of the ALS Storage Ring

Beam energy [GeV]	
Nominal	1.5
Minimum	1.0
Maximum	1.9
Circumference [m]	196.8
Beam current [mA]	
Multibunch	400
Single bunch	7.6
Beam emittance, rms [m·rad	
Horizontal	$<10^{-8}$
Vertical	$<10^{-9}$
Relative rms momentum spread	
Multibunch	8.0×10^{-4}
Single bunch	13.0×10^{-4}
Radiation loss per turn [keV]	91.5
Beam lifetime [h]	8
Nominal bunch duration, FWHM [ps]	30-50
Number of straight sections	12
Number available for insertion devices	10
Length for insertion devices [m]	4.5

impedance of the vacuum-chamber hardware determine the length (spatial and temporal) of the bunches. For the ALS at the nominal current of 400 mA, the predicted full-width-at-half-maximum (FWHM) value of the bunch length is 35 ps. Fig. 3 shows the electron-bunch structure. To avoid trapping positive ions in the potential well of the negatively charged electron beam, the multibunch mode with a 400-mA current will have 250 contiguous bunches, followed by a gap of 78 empty buckets. For particular experiments—for example, those involving time-of-flight measurements—it can be advantageous to have only one or a few circulating electron bunches in the storage ring. In the few-bunch mode, the nominal current per bunch will be 7.6 mA and the bunch length (FWHM) is predicted to be 55 ps, although still larger bunch currents may be tolerated (with additional bunch lengthening) before the beam becomes unstable.

4. Insertion Device Characteristics

The magnetic structure of an insertion device consists of a linear array of north-south dipoles of alternating polarity. The normal vertical orientation of the dipoles causes relativistic electrons of energy E to undergo a nearly sinusoidal electron trajectory of period λ_u in the horizontal plane, causing the emission of synchrotron radiation, as shown in Fig. 4. The peak magnetic field, B_o, on the undulator axis depends exponentially on the ratio R of the gap between the dipole north and south pole faces, g, to the period of the undulator, λ_u. For so-called hybrid devices made with the permanent magnet material neodymium-iron-boron, B_o, is approximately

$$B_o[\text{Tesla}] = 3.44\ e^{-R(5.47\ -\ 1.8\ R)}$$

for $0.07 < R < 0.7$.

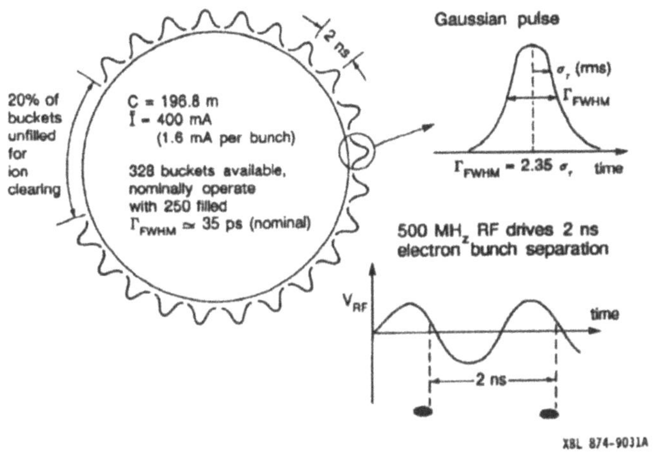

Fig. 3. Schematic illustration of the electron bunch structure in the ALS storage ring during multibunch operation. As shown at the upper right, each bunch has a full width at half maximum of about 35 ps. The spacing between bunches, dictated by the rf frequency, is 2 ns.

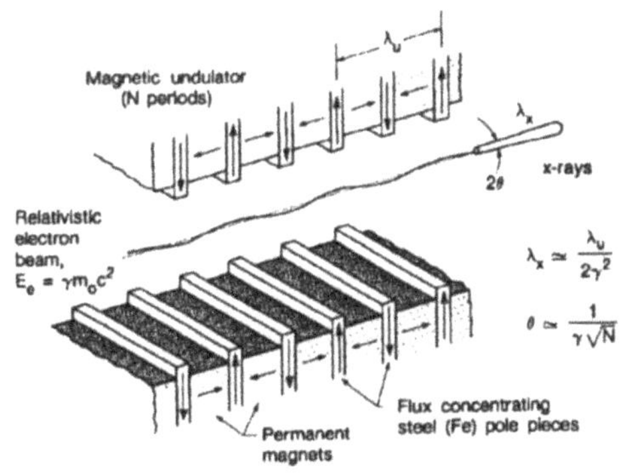

Fig. 4. Diagram of the magnetic structure of an undulator shows the main features of the hybrid technology that uses permanent-magnet material to drive steel pole pieces. North-south dipoles of alternating polarity bend electrons moving at nearly the speed of light into a sinusoidal trajectory, thereby causing the emission of a narrow cone of XUV radiation of high spectral brightness at a wavelength (λ_x) determined primarily by the electron-beam energy (E_e) and the undulator period (λ_u).

The deflection parameter K is the ratio of the maximum angular deviation of the electron trajectory from the insertion-device axis to the natural opening angle of the synchrotron-radiation cone. The cone has a natural opening angle that is inversely proportional to the electron-beam energy. Expressed in terms of undulator parameters, K is approximately

$$K = 0.934\ B_O[\text{Tesla}]\ \lambda_u[\text{cm}].$$

Purists would say $K = 1$ defines the break point between an undulator and a wiggler. However, an insertion device retains significant undulator properties at higher K values, and most of the undulators planned for the new synchrotron sources operate in this intermediate range with $K > 1$. When $K >> 1$, the structure is called a wiggler.

In the undulator regime, an important effect comes into play that gives insertion devices special properties. Because of the rather gentle perturbation of the electron trajectory in an undulator (several micrometers amplitude), the electron trajectory during a pass through the device lies within one radiation-cone opening angle. This condition means that the radiation emitted from successive undulator periods add coherently. The interference behavior due to the coherence gives rise to a sharply peaked radiation spectrum consisting of a fundamental and several harmonics. The photon energy, ε_n, of the nth harmonic is

$$\varepsilon_n[\text{keV}] = \frac{0.949\ n E^2[\text{GeV}]}{\lambda_u[\text{cm}]}\left(\frac{1}{1 + K^2/2 + \gamma^2\theta^2}\right),$$

where γ is ratio of the electron relativistic and rest masses and θ is the angle of emission relative to the undulator axis. By contrast, the lack of interference in a wiggler means that its synchrotron radiation spectrum is like the broad, continuous spectrum from the dipole or bend magnets in the curved sections of the storage ring.

The angular dependence of the undulator photon energy give rise to a ring-like spatial pattern (Fig. 5), which has been vividly confirmed in photographs of rainbow rings at visible wavelengths produced by undulators in low-energy electron storage rings. As the angle increases, eventually the on-axis photon energy of, for example, the fundamental will reappear as the second harmonic of the off-axis fundamental. The angle for the mth ring of the same photon energy as the on-axis harmonic n is

$$\theta_{n,m} = \frac{1}{\gamma}\sqrt{\frac{m}{n}\left(1 + \frac{K^2}{2}\right)},$$

where $n+m$ is the ring harmonic.

It is important to the experimenter to be able to select the photon energy of the undulator light—that is, the light should be tunable. Providing this tunability is a challenge to undulator designers. In principle, the photon energy is tuned from high to low values primarily by decreasing the undulator gap g from a maximum to a minimum distance, thereby

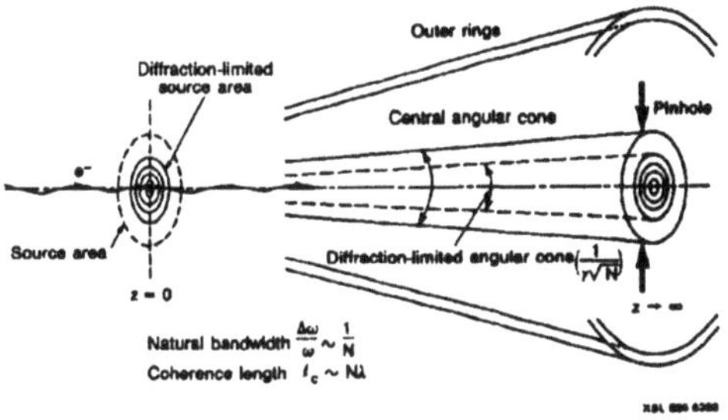

Fig. 5. Schematic illustration of the ringed radiation pattern from an undulator. A pinhole or other aperture can be used to select only the central cone of undulator radiation, thereby passing a sharply peaked spectrum centered around the on-axis fundamental and its harmonics.

increasing the field B_0 and the value of K. Both the minimum and the maximum are arbitrarily set by the drop off of the photon flux at low and high gap values, but are also subject to constraints such as the vertical diameter of the storage ring vacuum chamber. At the ALS, use of the third and fifth harmonics of the undulators is planned to extend their spectral range to higher photon energies than can be reached with the fundamental alone.

While tunable, undulators retain their most desirable properties in a comparatively small photon-energy range. For $K << 1$, the only significant photon flux is in the fundamental. As K is increased, the number of harmonics with measurable photon flux grows rapidly. It becomes useful to define a critical harmonic n_c above which half the total radiated power occurs. The critical harmonic is given by

$$n_c = 0.75K\left(1 + \frac{K^2}{2}\right).$$

The cubic dependence of n_c on K means that most of the undulator radiation is in the harmonics when tuning the undulator to lower photon energies (Fig. 6). These harmonics are spectrally broadened by electron-beam emittance and by the magnet errors discussed below and are not useful as high-brightness sources. But they are intense enough to require that substantial liquid cooling be used to avoid distortion or destruction of the first optical elements in the beamlines that guide the radiation from the storage ring to the experiment. Essentially, increasing K too far turns the undulator into a wiggler, as the higher harmonics blend together into a broad spectrum.

Many undulator properties improve with the number of periods N. For example, the intrinsic, on-axis spectral width (Full-Width-Half-Maximum) of the nth undulator harmonic is approximately

$$\Delta\varepsilon_n/\varepsilon_n = 0.9/nN\ .$$

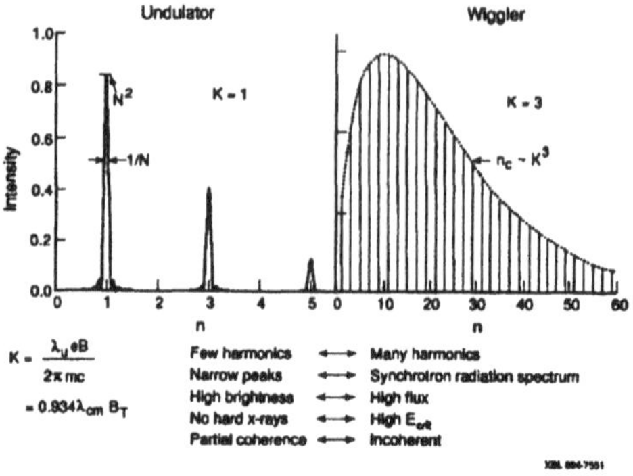

Fig. 6. The photon energies of the fundamental and harmonics of an undulator are conveniently tuned by varying the so-called deflection parameter, K, *which is proportional to the peak magnetic field,* B, *and the undulator period,* λ_u. *But, as the magnetic field of the undulator increases, the device becomes more and more like a wiggler. The figure compares the properties of a undulator to those of a wiggler and shows how rapidly an undulator takes on wiggler properties.*

In practice, the lower bound of the spectral width is limited by the angular divergence of the electron beam. Because electrons in the beam are not perfectly collimated, radiation observed on axis is actually emitted over a range of angles with a consequent spread in photon energies. When the intrinsic width drops below this spread, increasing the number of periods no longer improves the spectral resolution.

5. ALS Insertion Devices

Operating at 1.5 GeV, the ALS is optimized for insertion-device operation in the XUV spectral regions. Between them, the undulators will be able to excite the K shell of elements through silicon and the L shell of elements up to krypton, while the wiggler will be able to excite the L shell of nearly every element in the periodic table.

Table 2 lists the properties of three undulators (U8.0, U5.0, and U3.9, where the number refers to the period length in centimeters) that span the XUV spectral region when the ALS operates at 1.5 GeV. In addition, undulator U10.0, also included in Table 2, has been requested by a user group specifically to provide low-energy photons in the range 6-25 eV. The wiggler in Table 2 generates a broad continuous spectrum characterized by a critical photon energy ε_c, defined as the photon energy above which half the total power is radiated. At the high end of the broad wiggler spectrum, the flux drops rapidly but is still one-tenth of its maximum value at photon energies near $4\varepsilon_c$.. With a wiggler ε_c of 3.1 keV, the ALS spectral range extends into the hard x-ray region above 10 keV. High-quality synchrotron radiation will be available from the 24 prime bend-magnet ports as well. The critical photon energy of the bend magnets is 1.56 keV.

Table 2. Parameters for initial ALS insertion devices

Name	Period [cm]	No. of periods	Photon energy range [eV] [a]	Critical energy [keV]
Undulators				
U10.0	10.0	45	— [b]	—
U8.0	8.0	55	5.4–220 [c] [16.2–660] [27–1100]	—
U5.0	5.0	89	52–380 [156–1140] [260–1900]	—
U3.9	3.9	115	169–500 [507–1500] [845–2500]	—
Wiggler				
W16	16	16	—	3.1

[a] The photon energy range of the fundamental and of the third and fifth harmonics (shown in brackets) as the deflection parameter K decreases from its maximum value to approximately 0.5, when the electron-beam energy is 1.5 GeV.
[b] Only the range from 6 to 25 eV in the fundamental will be used.
[c] Below about 8 eV in the fundamental, the peak field in undulator U8.0 exceeds the bend-magnet field and may adversely affect storage-ring operation.

Engineering designs are well under way for the initial complement of insertion devices. The design philosophy is to create a generic design, as shown in Fig. 7, and thereby reduce engineering and fabrication costs and enhance maintainability. The goals of very high brightness and useful fifth-harmonic output impose unusually tight tolerances on the magnetic-field quality and thus on the mechanical structure of the undulators. The ideal undulator is designed to be as nearly transparent to the electron beam as possible; that is, the beam emerging from the downstream end should be the same as if there were no undulator at all. Transparency is never fully achieved, and real undulators may have bending, focusing, and other effects on the electron beam. At the ALS, however, accelerator physicists have made detailed numerical simulations of the effects of undulators and wigglers on the stored beam that demonstrate that, while there is a significant effect, it is only slightly greater that those expected from random magnetic-field errors in the storage-ring magnets.

To minimize effects on the stored electron beam due to undulators, as well as wigglers, it is necessary to put severe limits on the allowed net dipole, quadrupole, and other components of the magnetic field over the length of an undulator. These limits in turn place equally severe restrictions on random magnetic field errors. The cumulative effects of these

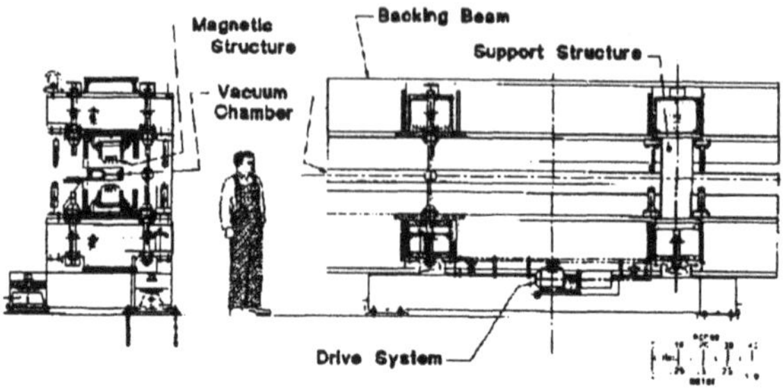

Fig. 7. Drawing of a generic insertion device for the straight sections of the ALS storage ring showing the main structural features that all undulators and wigglers will have in common. Undulators will be about 4.5 meters in length. The figure of the engineer give a sense of scale for the structure needed to precisely maintain and manipulate the undulator's magnetic structure in the face of large magnetic forces.

random errors can also significantly degrade the brightness of undulator radiation. Random field errors cause a loss of phase coherence that spectrally broadens the undulator harmonics, so that the brightness in a given spectral interval (the spectral brightness) drops. Moreover, systematic errors, such as a continuous change in the amplitude of the sinusoidal magnetic field over the length of the undulator, reduce the brightness because the spectrum of the undulator radiation shifts from one part of the device to the next (the spectrum chirps).

In general, the use of high harmonics from undulators with many periods in a low-energy ring obliges the engineer to take the most care to minimize field errors. The ALS fits into all three categories. Taking all error sources into account, physicists have arrived at a magnetic-field rms error budget that allows for a net error of 0.25% for undulator U5.0, which is just at the state of the art. The error budget for U8.0 is slightly more relaxed.

6. Spectral Brightness and Coherence

In a storage ring with electron current I, the total power P_{tot} radiated by an undulator of length L over all harmonics and angles is

$$P_{tot}\,[\mathrm{kW}] = 0.633 E^2[\mathrm{GeV}] B_0^2[\mathrm{Tesla}] I[\mathrm{Amp}] L[\mathrm{m}].$$

For many purposes, it is only the central cone of undulator radiation that experimenters use. The spectral flux F_n or flux per unit bandwidth, BW, for the nth harmonic integrated over the cone is

$$F_n[\mathrm{photons/sec/0.1\%\ BW}] = 1.43 \times 10^{14}\, I[\mathrm{Amp}] N Q_n(K),$$

where Q_n is a function of value less than or equal to 1 that increases with K and decreases with n. The flux increases linearly with N

The spectral brightness is the spectral flux per unit area of the source and unit solid angle of the radiation cone. In the jargon of statistical mechanics, the brightness is the density of flux in phase space. The average spectral brightness B_n in the central cone is approximately

$$B_n[\text{photons/sec/mm}^2/\text{mrad}^2/0.1\%\text{BW}] = \frac{F_n}{(2\pi)^2\Sigma_x\Sigma_{x'}\Sigma_y\Sigma_{y'}} = \frac{3.62 \times 10^{12}\ \text{I[Amp]}NQ_n(K)}{\Sigma_x\Sigma_{x'}\Sigma_y\Sigma_{y'}},$$

where Σ_x and Σ_y are the rms source sizes in x and y coordinates for a beam with a Gaussian density distribution, and $\Sigma_{x'}$ and $\Sigma_{y'}$ are the rms cone opening angles (angular divergence). The source size is obtained by adding in quadrature the size of the electron beam and the source size of a diffraction-limited photon beam. The angular divergence is obtained in an analogous manner. In most cases, the brightness varies linearly with N. Brightness is a conserved quantity that cannot be improved by an optical system; it therefore represents a true source characteristic. Absorption, scattering, and poor design can, of course, degrade the brightness downstream from the synchrotron source.

The critical harmonic of a low-K undulator lies at lower photon energies than the critical energy of the same device operated as a high-K wiggler. But, when the radiation is concentrated in a few harmonics, these become much brighter (up to several hundred times) than slices of the same spectral width from a wiggler. As a bonus, in so-called planar undulators, the radiation in the plane of the electron orbit turns out to be highly linearly polarized with the electric field vector in the orbit plane. Spectral brightness is shown in Fig. 8 for the undulators and wiggler described in Table 2, the ALS bend magnets, and for representative sources at other facilities. The spectral brightness of ALS undulators is shown in more detail in Fig. 9 for the fundamental and the third and fifth harmonics of the undulators.

A significant fraction of the radiation from the ALS undulators is spatially coherent. The criterion for spatial coherence is that the product of the area of the light source and the solid angle into which it emits be no larger than the square of the wavelength of the light. Since this is the diffraction condition, spatially coherent light is also said to be diffraction-limited. In accordance with the diffraction condition, the electron-beam emittance e sets the minimum wavelength at which all the radiation can be diffraction limited, according to the relation

$$\varepsilon = \lambda_{\min}/4\pi.$$

Even at wavelengths below the minimum, part of the radiation remains diffraction limited, the fraction decreasing as the square of the wavelength. In addition, for practical optical systems, it is possible to use several radiation modes without significantly losing resolution (Fig. 10), thereby increasing the effective fraction of coherent radiation.

Although phase-sensitive techniques, such as holography, most naturally come to mind when thinking about coherent radiation, a more general virtue is the ability to focus. For example, a Fresnel zone plate can focus a coherent beam of soft x-rays to a spot whose radius is approximately 1.2 times the width of the outermost zone. With state-of-the-art microfabrication techniques, such as electron-beam lithography, it is possible to make zone plates with outer zone widths of about 300 Å. This capability can be exploited in scanning systems in which the focused x-ray beam sweeps across a sample to generate imaging or spatially resolved spectroscopic information with a comparable resolution.

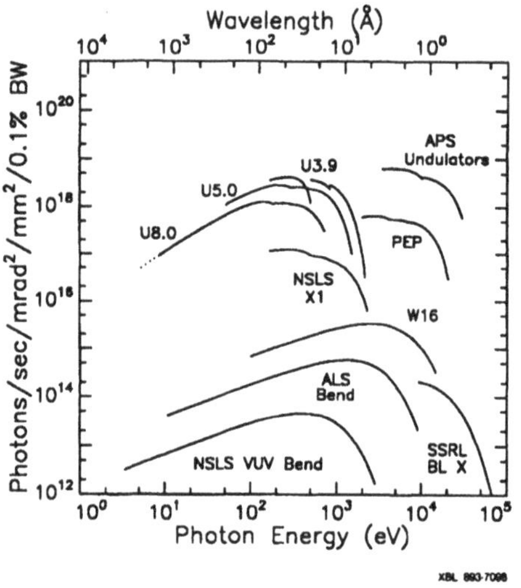

Fig. 8. With the use of more than one undulator harmonic it is possible to cover a wide spectral range. The spectral brightness envelopes are shown for the three ALS undulators and wiggler described in Table 2, the ALS bend magnets, undulators planned for the Advanced Photon Source (APS), and representative insertion-device and bend-magnet sources at the National Synchrotron Light Source (NSLS) and the Stanford Synchrotron Radiation Center (SSRL). X-1 is a recently installed undulator beamline at NSLS. The discontinuities in the undulator curves represent a shift from the envelope of the fundamental to that of the third harmonic, etc. The spectral brightness of the undulators greatly exceeds that of other sources shown in the drawing.

7. Insertion-Device Beamlines

Low-emittance storage rings and insertion devices have created new challenges for designers of XUV optics. First, the source size and divergence have become smaller. For ALS undulators at the high-photon-energy end of the spectral range, the rms size is typically 330 μm horizontal by 65 μm vertical and the rms divergence is typically 40 μrad horizontal by 30 μrad vertical. The smaller source size requires tighter tolerances for relay optics and monochromator components in both optical figure and finish to avoid loss of light (e.g., rms surface roughness ≈ 0.5 nm and tangential slope error <1 μrad for a condensing mirror). The attainment of higher resolution by use of smaller slits also becomes practical (the spectral-resolution goal of monochromators in undulator beamlines is $\Delta E/E \approx 10^{-4}$). Monochromator components therefore need tighter tolerances to avoid loss of resolution. Finally, the photon-beam power has increased to several kw/cm^2. The requirement that thermal distortions and stress be controlled complicates the design [for example, water cooling in a UHV (≈ 1 ntorr) environment is required] and limits the choice of materials.

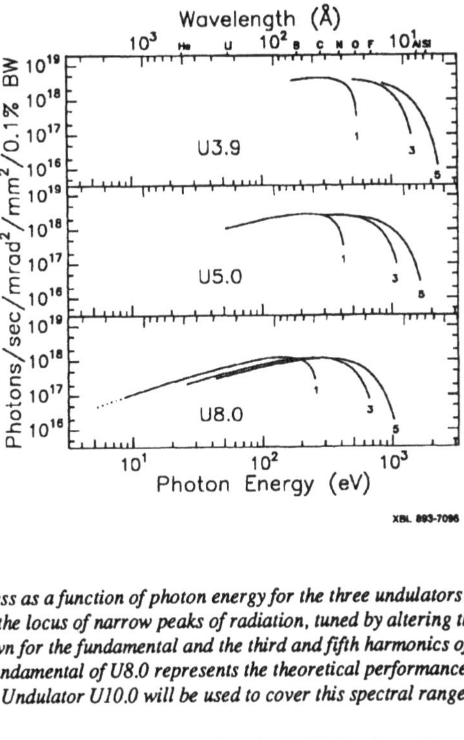

Fig. 9. Spectral brightness as a function of photon energy for the three undulators described in Table 2. Each undulator curve is the locus of narrow peaks of radiation, tuned by altering the undulator gap. Separate curves are shown for the fundamental and the third and fifth harmonics of each undulator. The dotted extension of the fundamental of U8.0 represents the theoretical performance at large values of the deflection parameter K. *Undulator U10.0 will be used to cover this spectral range.*

The most serious limitation is that of optical fabrication tolerances. It is difficult to fabricate aspheric optical surfaces (paraboloids, ellipsoids, toroids, etc.) sufficiently accurately in mirrors, monochromators, and other optical elements to take full advantage of the undulator source. One way to address this problem is to avoid the use of aspheric surfaces and to build beamlines entirely with plane and spherical surfaces. Although studies at LBL and elsewhere have confirmed that, when a large heat load is present, thermal considerations dominate the design and cost of the beamline, undulators usually generate lower total power than wigglers. For the ALS undulator beamlines, the power problem is significant but within the state of the art.

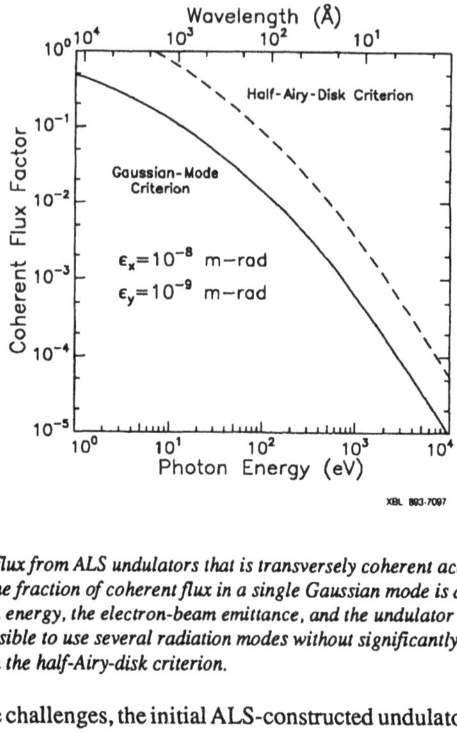

Fig. 10. Fraction of the flux from ALS undulators that is transversely coherent according to two criteria. From statistical optics, the fraction of coherent flux in a single Gaussian mode is a dimensionless ratio involving only the photon energy, the electron-beam emittance, and the undulator length. For practical optical systems, its is possible to use several radiation modes without significantly losing spectral resolution, as reflected in the half-Airy-disk criterion.

To meet these challenges, the initial ALS-constructed undulator beamlines are based on the spherical-grating monochromator system with water-cooling. The strategy is to use spherical surfaces for all mirrors and gratings and to actively water-cool the optics. Depending on the beamline requirements, the condensing system chosen for ALS undulator beamlines consists of one or two spherical mirrors. (Users are free to choose other optical designs for beamlines that they construct, subject to approval by the ALS.) Because of the low emittance of the ALS storage ring, the monochromator can accept the entire undulator beam in most cases, even at a slit-width of 10 μm.

Engineering design of the ALS-constructed beamlines is now under way, with numerous changes occurring as individual problems are encountered and solved. The undulator beamlines, comprising a front end inside the storage-ring shielding wall and a branchline outside the wall, that are actually constructed may differ in some ways from the following descriptions.

Figure 11 shows the beamline layout for the U5.0 beamline front end (top) and branchline (bottom). The front end components, with the exception of the photon beam-position monitor, are primarily for protection of the beamline equipment, the storage-ring vacuum, and, most important, personnel. The photon shutter is water cooled. The beam-position monitors provide information to a feedback system that will locate and stabilize the position and angle of the electron beam at the center of the undulator to 10 percent of its rms size and divergence.

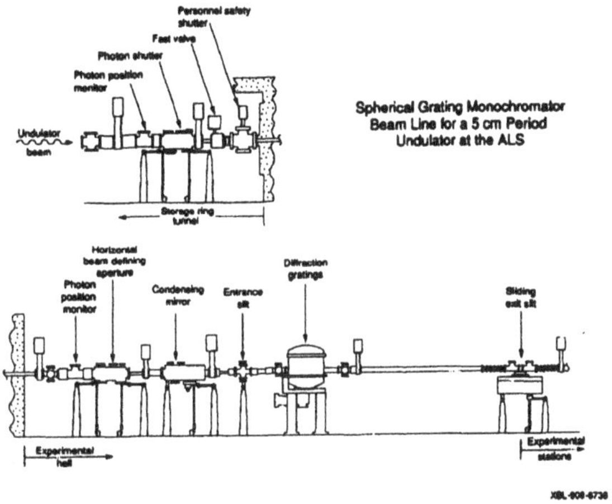

Fig. 11. Beamline layout for the U5.0 beamline. The front end inside the storage-ring shield wall is shown on the top; the branchline outside the storage-ring shield wall is shown on the bottom. User-supplied optics beyond the monochromator and the experimental stations are not shown.

The branchline includes a spherical-grating monochromator with three water-cooled diffraction gratings that can cover the range of photon energies from 65 eV to 1500 eV. When the undulator gap is at a minimum (maximum K), as discussed earlier, photons will be generated in large numbers from harmonics higher than the fifth and at photon energies higher than 1500 eV. The primary branch is designed to deliver photons from the fundamental and the third and fifth harmonics with minimum losses and high spectral resolution. The water-cooled beam-defining aperture passes the entire central cone of undulator radiation. The condensing mirror, also water-cooled, focuses a vertical image of the source, demagnified by a factor of about 15, onto the monochromator entrance slit. The grating images the entrance slit onto an exit slit that moves in such a way that the image remains in focus as the grating rotates. The Rowland circle condition is satisfied at only two wavelengths, but third-order aberrations are small enough so that the resolution remains determined by the size of the entrance slit. To focus the source image onto the sample, refocusing optics after the monochromator exit slit are required and depend on the specific application for which the beamline is intended. The pressure in the beamline must be less than 0.5 ntorr.

Figure 12 shows the beamline layout for the U8.0 beamline front end (top) and branchline (bottom). The branchline includes a spherical-grating monochromator with three water-cooled diffraction gratings that can cover the range of photon energies from 20 eV to 300 eV. Apart from the photon-energy ranges covered, there are two main differences between the U5.0 and U8.0 branchlines. The first is that, in addition to the primary U8.0 branchline, deflection mirrors could be introduced to generate secondary branches that would bypass the monochromator. This opportunity arises from the larger angles of incidence possible for mirrors at lower photon energies. The second difference is that a horizontal focusing mirror is the first optic in the U8.0 branchline. Horizontal focusing before the monochromator is required when operating the undulator at large values of K because the divergence of the undulator beam increases in the horizontal direction under these conditions. Moreover, the beam power is highest at large K and the horizontal mirror can absorb power as the first optic without affecting beamline performance.

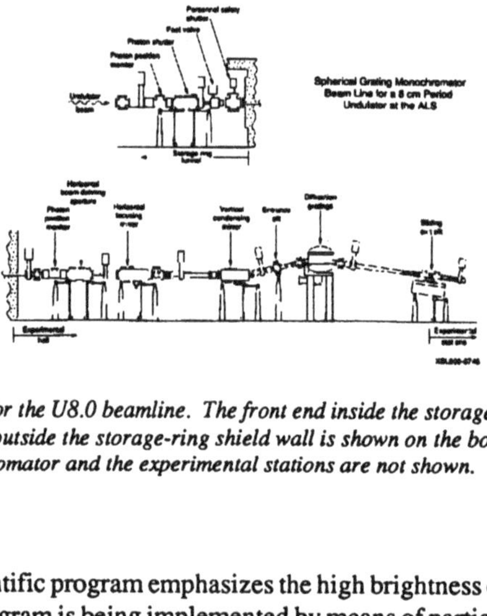

Fig. 12. Beamline layout for the U8.0 beamline. The front end inside the storage-ring shield wall is shown on the top; the branchline outside the storage-ring shield wall is shown on the bottom. User-supplied optics beyond the monochromator and the experimental stations are not shown.

8. Scientific Program

The initial scientific program emphasizes the high brightness of XUV light available from the ALS. The program is being implemented by means of participating research teams (PRTs) consisting of investigators with related research interests from one or more institutions. The primary responsibility for experimental apparatus rests with the PRTs; the responsibility for the beamlines and, where appropriate, insertion devices will be shared between the ALS and the PRTs. In return for its commitment, each PRT receives a guaranteed fraction of the ALS operating time at its beamline. A substantial fraction of the ALS running time at each beamline will also be made available to independent investigators not affiliated with a PRT by means of a proposal process.

PRTs working with undulator or wiggler beamlines are called insertion-device teams. Other PRTs work with bend-magnet beamlines and are known as bend-magnet teams. A Call for Proposals issued in the spring of 1989 resulted in 18 proposals from both insertion-device and bend-magnet teams. Nine insertion-device teams, which subsequently coalesced into the eight that are listed in Table 3, were approved in December 1989. Figure 13 shows a layout of the ALS building with possible locations of insertion devices and associated beamlines. Four bend-magnet teams that were approved in June 1990 are listed in Table 4. Including three additional ports allocated to insertion-device teams, seven of the 24 available prime bend-magnet ports have been spoken for. More than 200 persons, mostly from universities and federal laboratories, participated in the first round of PRT proposals.

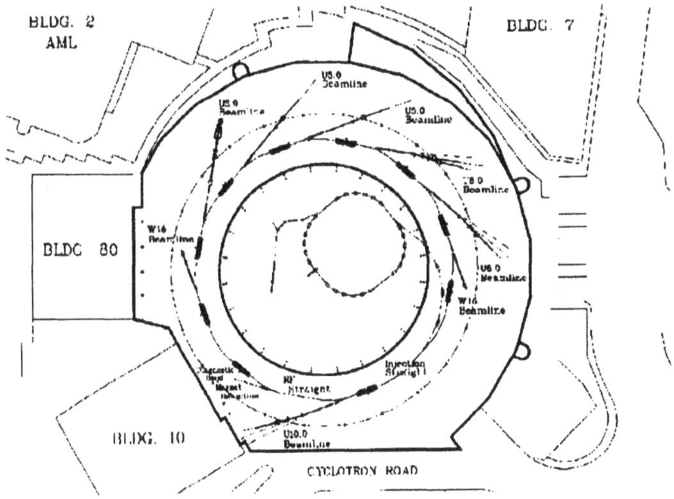

Fig. 13. Floor plan showing a possible placement of insertion devices and locations of associated beamlines in the ALS building. The maximum beamline length is about 35 m. The prime bend-magnet ports to be developed initially are located in the center bend magnet of each storage-ring arc. At two ports per magnet, there is a total of 24 ports.

Some of these groups are expected to use the high brightness of the ALS undulators to open new areas of research in the materials sciences, such as spatially resolved photon and electron spectroscopy (spectromicroscopy). Biological applications will include x-ray microscopy with element-specific sensitivity in the water window of the spectrum (23-44 Å) where water is much more transparent than protein, thereby allowing soft x-rays to penetrate the natural aqueous environment of these systems. The ALS will also be an excellent research tool for atomic physics and chemistry because the high flux will allow measurements to be made with tenuous gas-phase targets. The short pulse width (30-50 ps) will facilitate time-resolved experiments. A future option is the construction of special devices to generate radiation with a controlled elliptical polarization.

Table 3. Insertion-Device Participating Research Teams (PRTs)

Insertion Device	Scientific Focus	Spokesperson
U10.0	Chemical dynamics	Tomas Baer U. of North Carolina
U8.0	Atoms, molecules, ions	Denise Caldwell U. of Central Florida
U8.0	Pump-probe, timing, dynamics experiments	R. Stanley Williams U. of California, Los Angeles
U5.0	Surfaces and interfaces	Brian Tonner U. of Wisconsin-Milwaukee
U5.0	Surfaces and interfaces	Joachim Stöhr IBM Almaden Research Center
U3.9	X-ray imaging and optics for the life and physical sciences	Stephen Rothman U. of California, San Francisco
W16	Atomic, molecular, optical physics; materials science	Bernd Crasemann, U. of Oregon Philip Ross, LBL
W16	Life sciences	Stephen Cramer, U. of California, Davis

Major research areas proposed for ALS undulator beamlines include: (1) soft x-ray microscopy of materials, surfaces, and biological systems, (2) spatially resolved spectroscopy (spectromicroscopy) of materials, surfaces, and biological systems, (3) high-resolution soft x-ray spectroscopy of materials and surfaces, (4) soft x-ray gas-phase spectroscopy of atoms and molecules, (5) molecular spectroscopy and dynamics with synchrotron radiation/laser pump-probe methods (6) spin-polarized photoemission spectroscopy. and (7) polarization-dependent experiments, such as circular dichroism of biological systems.

Wiggler-based x-ray studies will include spectroscopy of atoms in both the gas phase and in condensed matter, spatially-resolved elemental analysis with an x-ray microprobe, grazing-incidence x-ray scattering from surfaces, and x-ray diffraction of large biological molecules (protein crystallography). Bend-magnet research will include studies of physical and biological systems with polarized radiation and infrared absorption spectroscopy of solids, surfaces and gases.

Table 4. Bend-Magnet Participating Research Teams (PRTs)

Scientific Focus	Spokesperson
Soft x-ray spectroscopy and time-resolved studies of materials and surfaces; x-ray optics	James H. Underwood Lawrence Berkeley Laboratory
EUV spectroscopy and time-resolved studies of materials, surfaces, and gases; x-ray optics	James H. Underwood Lawrence Berkeley Laboratory
Infrared spectroscopy and time-resolved studies of surfaces, solids, and gases; fast IR detectors	Gwyn P. Williams Brookhaven National Laboratory
Polarized photon studies: accelerator and beam diagnostics, spin-dependent interactions of photons with matter, chiral biological molecules	Mervyn Wong Lawrence Berkeley Laboratory

9. Atomic and Molecular Physics

An outstanding feature of the ALS will be its high brightness, which will result in increased intensity on the target relative to less bright sources for two kinds of experiments:

- Experiments with tenuous targets which occupy a small volume (e.g., when the target is a focused beam of ions), or experiments where only those interactions which take place in a small volume are of interest, (e.g., focusing the x rays into the entrance volume of an electron analyzer);
- High-resolution experiments, where the high brightness serves to allow focusing all of the x-ray beam onto the entrance slit of a monochromator.

The ALS will therefore be useful for a variety of experiments in atomic and molecular physics and materials science, where the tenuous target might be an atomic vapor, laser-excited atoms or molecules, an ion beam, ions in a trap, clusters, or actinides (which might be too radioactive or too rare to permit a large target). The high resolution will be useful for experiments with atoms or molecules (e.g., looking at doubly excited states of atoms or vibrational structure of molecules).

Another class of applications of high brightness consists of coincidence experiments, where more than one reaction product must be detected simultaneously; a very intense x-ray beam is required to generate a large enough signal to compensate for the limited solid angle or low efficiency for detecting one or both products.

The short pulse length of the ALS will be useful for a variety of experiments in chemistry, in which two or more photons are used to study a system (often called two color or multi-color experiments, or pump-probe experiments). As applied at the ALS, one photon will typically be a VUV photon or x ray from an undulator, and the second photon could be from an adjacent bend magnet or from an auxiliary laser.

An active research program in most of these areas is planned; please contact the authors for additional information regarding access to the ALS.

RECENT DEVELOPMENTS IN ATOMIC PHOTOIONIZATION USING SYNCHROTRON RADIATION: EXCITED ATOMS AND IONS

François J. Wuilleumier, D. Cubaynes and J.M. Bizau
Laboratoire de Spectroscopie Atomique et Ionique and LURE, Université Paris-Sud, Bât. 350, 91405 ORSAY Cedex, FRANCE

ABSTRACT

After the pionnering studies of atomic photoionization on closed shell atoms, mostly the rare gases, during the 70's, synchrotron radiation is now widely used in many places for systematic measurements in open shell atoms as well as for a fine analysis of the photoionization processes in the rare gases. New ways have also been successfully explored. The first experiment combining the use of laser and synchrotron radiation for studies on excited atoms took place ten years ago; several cases have been studied since, involving first resonant photoionization, then direct photoionization into the continuum in some alkali-and alcaline earth atoms. The present status of these experiments is described. Most recently, photoelectron spectrometry has been successfully used, for the first time, to study photoionization of a singly charged ion beam with undulator radiation. These latest results are presented and a program devoted to the study of photoionization experiments on multiply charged ions over the coming years is described

I. INTRODUCTION

Thirty years have passed since the first use of synchrotron radiation to investigate atomic photoabsorption in the ultraviolet region.[1] Since this early time, continuous improvements have been achieved in the production and in the use of synchrotron radiation. Pioneering experiments[2] which were at the ultimate edge of the technical possibilities have become routine experiments which are systematically performed in many places to go deeper and deeper into the detailed knowledge of atomic properties and of the dynamics of excitation and deexcitation in photoionization processes. Other atomic systems, more difficult to produce and to study, have become amenable to exploration.

During the first ten years, experiments in atomic physics were mostly confined to total photoabsorption or reflectivity studies. In a simple photoabsorption experiment, the excited states of the neutral atoms show up as resonance transition series and the total photoabsorption cross section can be measured. Thus, information can be gained only on the sum of all partial cross sections for the various processes contributing to the attenuation of the photon beam. These processes are single ionization of an electron into the continuum, simultaneous excitation and ionization of the atom-leaving the residual positive ion into various excited ionic states - double, eventually triple photoionization. If hν is the energy of the monochromatic photon beam, all atomic electrons whose binding energy is lower than hν, are involved in these processes. Thus, various techniques of increasing sophistication have been developed to explore the various ionization channels : energy resolved, angular resolved and spin resolved electron spectroscopies, ion spectroscopy, fluorescence spectroscopy. More recently, coincidence techniques have been introduced to allow for more selective experiments to be made, such as electron-ion, photon-ion, photoelectron-Auger electron, photon-photon coincidences. Nowadays, it is possible to measure the partial photoionization cross sections, the angular distribution of photoelectrons and Auger electrons, the spin polarization parameters. These parameters are related to the wavefunctions of the initial and final states of the atomic system and to an operator which describes the interaction. In this context, the growing use of synchrotron radiation has provided with a large amount of data showing evidence of phenomena that could not be explained in the one-electron model, but required the introduction of electron correlations, and also relativistic formalism, into the theoretical models. Thus, theory has also taken a great benefit from the development of synchrotron radiation since the experimental results have inspired the theoreticians to develop more and more accurate approximations to describe better the atomic structure and the dynamics of the photon-atom interaction.

From the experimental point of view, a much more intense and well collimated photon beam is required for selective spectroscopies than for photoabsorption experiments, because the pressure in the interaction zone of a differential analyzer has to be generally lower by at least four orders of magnitude and the length of the effective source volume is typically hundred times lower. In addition to these requirements, no window can be used to contain the sample when electrons or ions have to be detected.

This explains why this type of experiments was extremely difficult to achieve before the advent of new storage rings, exclusively built and optimized during the 80's for the production of synchrotron radiation, and of the improved instrumentation developed to make the best use of it, such as the toroïdal grating monochromator. With the better conditions offered by the synchrotron radiation centers, many more scientists, in atomic physics and molecular physics, made use of this light and an impressive growth of the experimental activity in atomic physics occured within a few year period. Nowadays, such experiments are performed at most of the storage rings.[3]

Several opportunities were open to new (and early) investigators: to improve the results previously obtained on the rare gases ; to develop new experiments in the widely unexplored fields of the metallic vapors, especially in the open shell atoms ; to pave the way to more challenging goals such as the photoionization of atoms not anymore in the ground state, but brought into excited or ionic states. This paper is devoted to a review of the recent experimental studies on excited atoms and ions. After having briefly summarized the interest to study these atomic species and the main results obtained in the works published before 1989, we will describe in more details the recent measurements that occured in these fields since 1989.

II - PHOTOIONISATION OF ATOMS IN EXCITED STATES

The development of this field has started with the advent of tunable dye lasers. The first experiments have combined electron excitation of atoms in the ground state with laser ionization of the excited electron.[4] In this way, only outer shell photoionization in excited atoms have been tested over a few electron volts above the first ionization threshold. Later on, pulsed dye lasers, tuned to the excitation energy of some atomic transitions have been used, in synchronization with a pulsed source of continuum radiation in the VUV (BRV source, laser produced plasmas) to start inner-shell photoabsorption experiments in alkali and alkaline earth atoms and in manganese.[5-12] Since the successful combination of a cw dye laser with synchrotron radiation[13-16] in 1981, photoexcitation and photoionization of inner-electrons have been probed by electron spectrometry over a large photon energy range extending from about 15 electron volts up to several hundred electron volts. Here we will concentrate on photoionization studies of free atoms excited in two step experiments where one photon excites a valence electron and a second one probes the excited atom in inner- or outer shells.

1°) Interest

For the atomic physicist, excited atoms provide with a wealth of new information. The wavefunction of the excited outer electron extends usually over a radius which is much larger than for an atom in the ground state. Therefore, in the photoionization of an excited atom, the atomic potential is probed over a larger spacial range. Moreover, photoexcitation and photoionization from excited atomic states provide access to states of parity opposite to that of the ground state. The excited state can be prepared with a well defined set of quantum numbers and the photoionization dynamics of these precisely defined states can be studied as a function of photon energy. In addition, the angular distribution of the electrons photoionized from a specifically prepared excited state will reflect the geometry and dynamical properties of that state and yield phase shift information about the continuum wavefunctions which cannot be obtained from total cross section measurements. The measurement of the spin polarization parameters is another mean of providing the missing link for a determination of the continuum wave function phase shifts. Also, several theoretical calculations predict the occurence of several new feature for some photoionization parameters[17-20] but have not yet been tested. The total potential of an atom may be very sensitive to small changes in the electron orbit. The promotion (or ionization) of a valence electron to an excited state (or to a continuum state) modifies sometimes dramatically the effective potential of the core electrons, producing eventually a radical redistribution of the oscillator strength density.[9]

There are also many applications that require information about oscillator strengths and cross sections from these excited states. Some of these applications are resonance ionization spectroscopy, X-UV lasers where highly excited metastable atoms (and ions) may form the basis for obtaining a population inversion, radiative transfers in laboratory plasmas and in stellar atmospheres.

2°) Experimental conditions

We will refer to previous reviews[21-23] for a detailed discussion of the problems raised by the combination of two different sources of photons having almost opposite characteristic properties. In short, synchrotron radiation (SR) photon beams are characterized by a broad energy range available- typically 15-350 eV in the experiments discussed here - a relatively low energy resolution ($\Delta E/E \sim 10^{-3}$, at most 10^{-4}) and a low photon flux (10^{12} to 10^{13} photons in a 1% band width), while a cw dye laser has a very limited tunability in a narrow energy range, but an extremely narrow spectral width (10^{-8} to 10^{-9}) and a very high photon flux (at least 10^{18} photons/sec). In addition, the time structure of SR is such that it delivers nanosecond pulses with a repetition rate of 10^7 to 10^8. Thus, the intensity of SR in one pulse is extremely low (10^5 to 10^6 photons/pulse) while the repetition rate of a pulsed laser is very low (~ 10 Hz). This explains why the time structure of SR has not yet been used in connection with pulsed lasers, even though the use of lasers mode locked to the synchrotron radiation or the availability of new SR sources with higher photon flux may open new opportunities to pulsed experiments. Let us mention also that SR is elliptically polarized and the laser either linearly or circularly polarized.

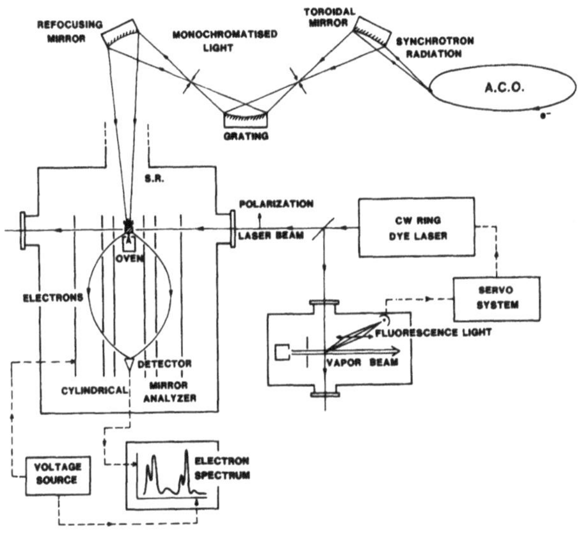

FIGURE 1
Layout of the experimental setup for photoionization studies on laser-excited atoms

We show in Fig.1 a schematic diagram of the experimental setup used in Orsay since more than ten years[13-16,24] first at ACO, then at Super-ACO[25] storage rings. A toroïdal grating monochromator delivers a monochromatized photon beam (10^{12} photons/sec in a 1% band width) for photon energies between 15 eV and 150 eV. This flux is focused with a toroidal mirror on the source volume of a cylindrical mirror analyzer (CMA) electron spectrometer. The CMA is usually operated with a kinetic energy resolution $\Delta E/E = 0.9$ %. An oven, mounted on the axis of the CMA and of the VUV beam emits a weakly collimated beam of metal atoms (ground state density up to 10^{13} atoms/cm^3).

The laser beam irradiates the atomic beam perpendicularly to the CMA axis to minimize the Doppler effect. The beam waist of the laser is adjusted to fit the size of the source volume of the spectrometer which constitutes the interaction region of the sample (2x4 mm^2). The laser beam is produced by an argon (or krypton) ion pumped dye laser which has an output power of about 500 mW, and is linearly polarized in the horizontal plane. It is locked to the transition between the ground state and the first excited state of the atoms under investigation by observing the fluorescence emitted from the excited atoms in an auxilliary beam. Various dyes have been used to study laser excited sodium (rhodamine 6G), potassium (rhodamine 700) and barium (rhodamine 110) atoms. In some experiments, a second ring dye laser was used to promote the excited atom into higher excited states via a cascade two photon excitation.[34-36]

A similar experimental set up was used at DESY for the first experiments achieved by Sonntag and his collaborators in the same field.[26] It has been more recently improved[27] to allow for a rotation of the CMA in order to measure the angular distribution of the electrons emitted in the autoionization of resonantly produced core-excited states in laser excited lithium, using an undulator station of BESSY.

The maximum percentage of atoms which can be brought into the first excited state depends on the polarization of the laser. In our experimental conditions, this maximum is 31% for sodium. But the key point to obtain high enough densities of excited atoms to allow for accurate quantitative measurements to be made is the initial density of atoms in the ground state. For low densities, such as 10^{10} to 10^{11} atoms/cm^3, one does not succed to bring more than a few percent of the atoms in the excited state. But, if one increases this ground state atomic density, the trapping of the radiation into the atomic beam helps to increase the density of atoms brought into the excited state via reabsorption of the fluorescence radiation emitted in the decay of this excited state. At ground state densities of 10^{12} to 10^{13} atoms/cm3, we easily succeded to get 25% to 30% of sodium or potassium atoms in the resonantly excited 3p $^2P_{3/2}$ and 4p $^2P_{3/2}$ states, respectively, and almost 100% of Ba atoms in the 5d $^{1,3}D$ metastable states. On the other hand, the alignment or orientation of the excited atoms produced by the laser is lost at high density. It has been shown[28,29] that, above 10^{12} atoms/cm^3, the polarization degree of the fluorescence emitted in the decay of the $^2P_{3/2}$ states of Na and Li atoms is zero, and that experiments aiming to make use of the alignement or orientation of the excited atoms induced by the laser should not be carried out at ground state atomic densities higher than 10^{11} atoms/cm^3. Thus, a critical choice of this density has to be made, depending on the type of experiments that is persued.

In Orsay, we are presently interested in studying direct photoionization into the continuum. Working at high densities, we have an unpolarized target and the geometry we have chosen is such that the measured differential cross sections are proportional to the partial photoionization cross sections.[30] When the influence of the laser polarization on the relative intensity of autoionization lines is to be studied, other choices have to be made.

3°) Earlier measurements

Because of the poor tunability range of the cw dye lasers, alkali-and alkaline earth atoms were the most obvious candidates for SR experiments on laser-excited atoms. The feasibility of such experiments involving electron spectrometry was demonstrated in 1981 for sodium atoms[13-16] and in 1982 for barium atoms[16,31] at LURE. In these experiments, direct photoionization into continuum states as well as resonant excitation to autoionizing states were observed. Similar investigations were initiated in Germany,[26] at DORIS, in 1985.

Since the direct photoionization cross sections are lower by orders of magnitude than the resonant photoionization cross sections, quantitative studies on autoionization following two electron-two photon transitions were first achieved, in atomic sodium. The transitions between a laser excited $2p^63p$ $^2P_{3/2}$ state and autoionizing levels with $2p^53s3p$ configuration were systematically investigated.[24,32] Values of the oscillator strengh associated to the excitation of different L.S states were determined. Shakeup processes occuring[23] in the core - excitation of $2p^63p$ $^2P_{3/2}$ state to $2p^53s4p$ $^2D_{5/2,\ 3/2}$ autoionization states were also measured. Similar experiments were performed later for the $2p^63p$ $^2P_{1/2}$ state.[29,33] Two lasers beam were also combined with synchrotron radiation to produce autoionizing states with $2p^53s5s$, $2p^53s4d$, $2p^53s4p$ and $2p^53s4s$ configurations: the energy of these states and the relative oscillator strenghs of the inner-shell transitions have been determined.[34-36] Studies of the 5p excitation resonances in laser excited $5p^66s5d$ $^{1,3}D$ barium atoms were also undertaken.[22]

The first measurements of direct photoionization cross sections for outer and inner-shell ionization in an excited atom have been determined in the barium case,[37,38] between 20 eV and 140 eV photon energy : the energy dependence of the 5d, 5p and 5s photoionizaton cross sections were shown to be strongly enhanced via intershell correlation with the 4d subshell, like in the case of xenon[39] and barium[40] atoms in the ground state.

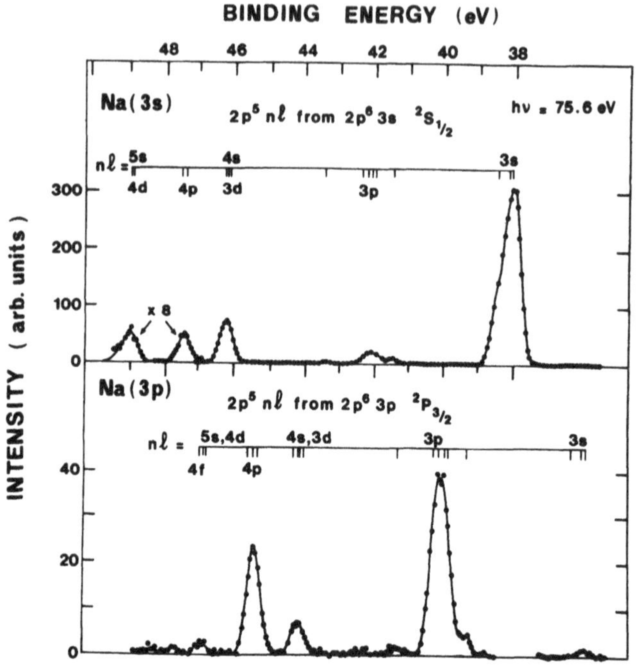

Fig.2
Photoelectron spectra following ionization of sodium atoms in the ground state (Top) and in the 3p-excited state (Bottom).

sodium atoms (energy difference of 2.1 eV). In the upper part of Fig. 2, we show the photoelectron spectrum emitted from sodium atoms in the ground state. With increased photon flux, it was possible to observe some previously undetected satellite lines, corresponding namely to $2p^5 4p$ and $2p^5 5s$ or to $2p^5 4d$ final ionic state electronic configuration. In the lower part of the figure is shown the corresponding photoelectron spectrum following photoionization in the 2p-subshell of 3p-excited sodium atoms. Here, the 2p main line is shifted to higher binding energies (40.1 eV against 38.0 eV in the ground state), reflecting the change in the outer screening, as well as the conjugate shakeup satellite lines ($2p^5 4s$ and $2p^5 3d$ final ionic states, $\Delta l = \pm 1$ for the shaken electron) while the shakeup satellite lines ($2p^5 4p$ final ionic states, $\Delta l = 0, +2$) are shifted in the other direction. In addition, one observes also shake down satellites corresponding to $2p^5 3s$ final ionic states.

New experimental capabilities were offered by the use of new storage rings especially built for the production of synchrotron radiation, such as BESSY in Berlin and Super ACO in Orsay, and by the possibility of using insertion devices. With the higher photon flux and brightness expected from these new sources, it was predicted[22] to be able to extend the tunability range of CW dye lasers in compensating the loss of laser power with a higher SR flux, to work with lower atomic densities- i.e. to study alignment and orientation effects, in resonant photoionization first - and with an improved resolution in order to better discriminate between photoelectron lines from the ground state and from the excited state. Two of these goals were already reached in 1987 with the observation of resonant photoionization in laser excited 4p-calcium atoms [41] and with the first measurements of the influence of the laser polarization on the relative intensities of various autoinization lines in laser-excited lithium [42] and sodium [29,33] atoms.

Results obtained until 1986 have been reviewed in a chapter of Advances in Atomic and Molecular Physics [21]. Later on, two shorter reviews have been devoted to the works achieved in 1987 and in 1988, respectively [22,23]. Since 1989, two major progress have been made : a) the use of the Super ACO storage ring made possible the first accurate measurements of inner-shell direct photoionization cross sections in excited sodium [25] and potassium [43] atoms; b) the use of a BESSY high flux wiggler/undulator station allowed to determine, for the first time, the angular distribution of the electrons ejected upon autoionization of resonantly core-excited states in laser-excited lithium atoms.[27] The results of experiments b) have not yet been published in details, but a summary of them can be found in a recent review paper on synchrotron radiation [27]. In the following, we will concentrate on the new results a) obtained with Super ACO, refering the reader to the previous reviews for more information about earlier works.

4°) Recent experimental results

a) 2p - photoionization cross section in 3p - excited sodium atoms [25]

The photoelectron spectrum following ionization in the 2p-subshell of sodium atoms in the ground state exibit the so-called main line, corresponding to single photoionization of the 2p electrons, and several so-called satellite lines at higher binding energies, due to simultaneous excitation of the 3s electron and ionization of one 2p electron. These satellites are due to electronic correlations and the corresponding final ionic states have either $2p^5 3p$ (conjugate shakeup) or $2p^5 4s$, 3d (shakeup states) electronic configuration. This spectrum has been studied in details with ACO. With an increase in the photon flux of about a factor 3 with 300 μm slits, as compared to ACO, a better resolution of the monochromator could be used with Super ACO. The energy range from 50 eV to 110 eV was used with a nearly constant bandwith (0.30 eV to 0.35 eV). In these conditions, it was possible to accurately determine the integrated area under the photoelectron lines due to 2p-photoionization in the ground state and in the 3p - excited

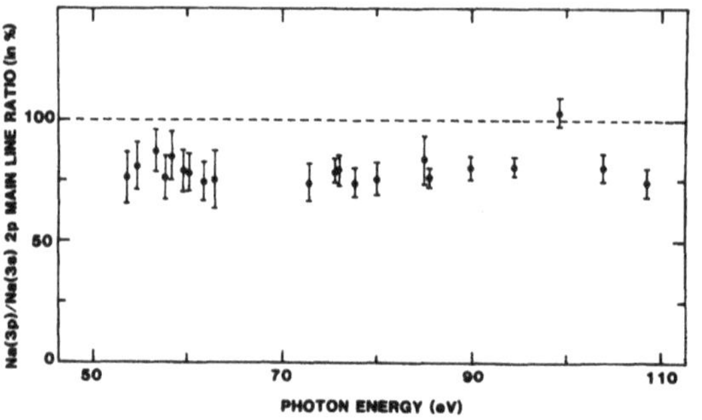

Fig.3
Ratio between the 2p-main lines in sodium atoms in excited and ground states.

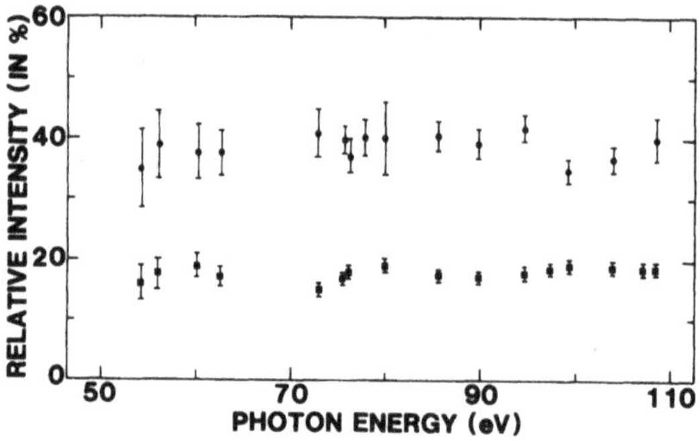

Fig.4
Relative intensity of the shakeup satellites in excited and in ground state sodium atoms.

The two main results [25] obtained from these measurements are : i) the ratio between the single 2p photoionization cross sections (main lines) in the excited state and in the ground state, shown in Fig. 3, is equal to 0.78 ± 0.05, which demonstrates that the single 2p-photoionization cross section in the excited state is about 20% lower than in the ground state ; this result was, at first sight, in contradiction with the results of a many body calculation [44] predicting that the magnitude and the photon energy dependence of the 2p-photoionization cross section should be independent of the atomic orbit that the valence electron occupies ; ii) the relative intensity of the conjugate shakeup satellites is about the same in the ground state and in the excited state while, at all photon energies, the relative intensity of the shakeup satellites, shown in Fig. 4, is strongly enhanced in the excited state, by a factor 2 (40%, against 20% in the ground state). When one adds the contribution from the satellite lines to the 2p-main lines, the 2p photoionization cross section is now about the same (ratio of 0.92 ± 0.07) in the excited state and in the ground state, which brings the theory in agreement with experiment, although this enhancement of the shakeup satellite intensity has not yet been reproduced by theory. The contribution of the direct double photoionization cross section, involving the simultaneous ejection of a 2p electron and of a 3s (or 3p) electron is still missing, but would still increase the cross section ratio ; however, recent measurements[45] seem to indicate that this double photoionization cross section is quite small for sodium atoms in the ground state.

The conclusion of similar experiments were found to be the same in sodium atoms for photoionization in the 2s-subshell [46]. This similarity in the behavior of two inner subshell photoionization cross sections for sodium was a motivation to check whether a similar effect would also occur for the first inner-p subshells of other alkalis. In the next paragraph, we show the first results obtained in the case of potassium atoms.

b) 3p-photoionization cross section in 4p-excited potassium atoms.

The photoelectron spectrum due to 3p-photoionization of potassium atoms in the ground state is strongly influenced by the existence of the empty 3d-orbitals. The $3p^5 3d$ ionic states are very close in energy to the $3p^5 4s$ states and the corresponding photoelectron lines overlap partly. Thus, the fine structure of the 3p-main line is more complicated than in the case of the 2p-main line in sodium, and is also quite different in the excited state.

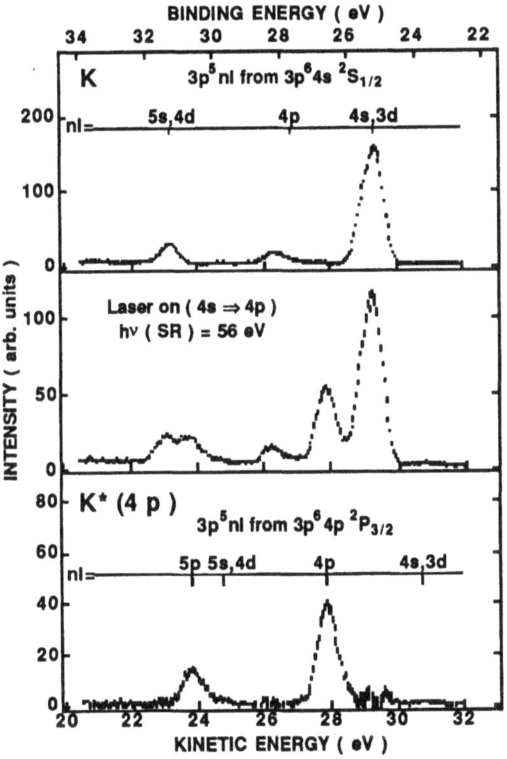

Fig.5
Photoelectron spectra following ionization of potassium atoms in the ground state(Top) and in the 3p-excited state(Bottom).

We show, in Fig. 5, a typical set of photoelectron spectra taken at 56 eV photon energy. They have been corrected for the transmission of the CMA. The upper panel shows the photoelectron spectrum of potassium atoms in the ground state after 3p-photoionization. To get enough counting statistics and, thus, a good accuracy on the ratio of the 3p cross sections, the resolution chosen for the photon energy was only 0.45 eV. With such a resolution, the fine structure of the 3p-main line is not observable. Conjugate shakeup satellites ($3p^5 4p$ final ionic state electronic configuration) and shakeup satellites ($3p^5 5s$, $3p^5 4d$) are far in binding energy and their relative intensity has been measured with accuracy. The photoelectron spectrum with the laser on is shown in the middle panel. The potassium beam is a mixture of potassium atoms in the $3p^64s$ $^2S_{1/2}$ ground state and in the $3p^64p$ $^2P_{3/2}$ excited state. The 3p-main line in the excited state is shifted to higher binding energy (1.5 eV) and has also a complicated fine structure observable at higher resolution. The shakeup and conjugate shakeup satellites partly overlap with the corresponding satellites from the ground state. The pure photoelectron spectrum following photoionization in the 3p-subshell of 4p-excited potassium atoms is shown in the lower panel. It has been obtained from substraction of the middle and upper spectra

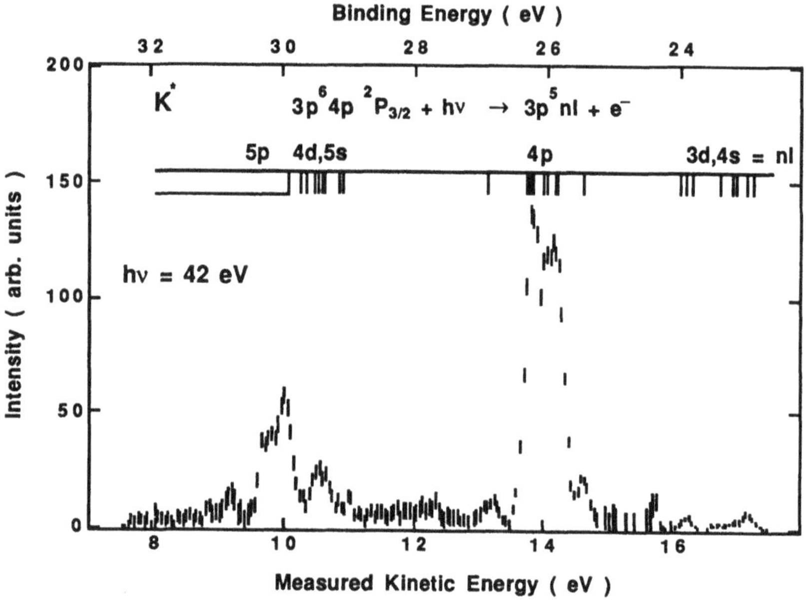

Fig.6
High resolution photoelectron spectrum following photoionization of 4p-excited potassium atoms.

after proper normalization. The percentage of potassium atoms laser-excited to the 4p-state is about 30% here. In this spectrum, the shakeup satellite lines, corresponding to the $3p^55p$ electronic configuration, and the conjugate shakeup satellites ($3p^55s$ and $3p^54d$ electronic configuration) are too close in binding energy to be well splitted with the moderate resolution chosen here. However, it is possible to resolve these different final ionic states if one accepts to loose in intensity in reducing the width of the monochromator slits. An example of such a spectrum, taken with higher resolution at 42

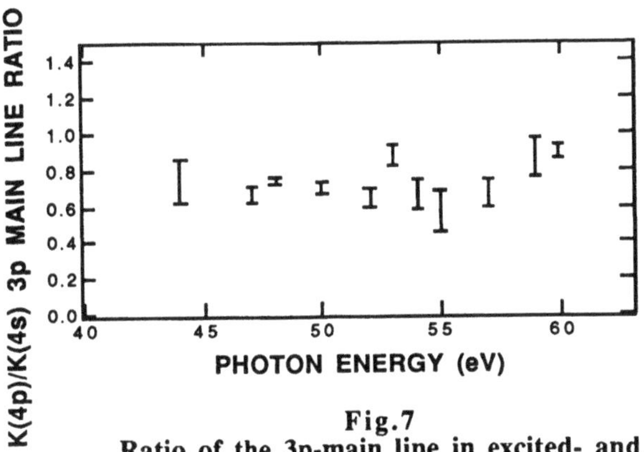

Fig.7
Ratio of the 3p-main line in excited- and groud state potassium atoms

eV photon energy, is shown in Fig. 6 for 4p-excited potassium atoms. Here the two groups of satellites are well splitted, around 30 eV and 29.5 eV binding energy, and the fine structure of the main line is partly observable.In addition, shakedown satellites, corresponding to $3p^53d$ and $3p^54s$ electronic configuration for the final ionic state, are also visible. Detailed comments on these photoelectron spectra will be presented elsewhere.[43] Here, we would like only to point out the main result: the ratio between the integrated area under the 3p - main lines in the 4p-excited state and in the 4s -ground states, respectively, is again lower than 1. The effect is even more pronounced than in the case of sodium, as it can be seen in Fig.7. This ratio does not show any strong photon energy dependence and its averaged value is equal to 0.70 ± 0.07. It is difficult, from these first spectra, to extract the relative intensity of the individual satellites over an extended photon energy range. However, the overall intensity of all satellites, relative to the intensity of the main line, can be determined fairly accurately to be 0.45 ± 0.05 in the excited state and 0.25 ± 0.03 in the ground state. Including these contributions from the satellite lines to estimate the ratio between the total 3p-photoionization cross sections in the excited state and in the ground state, one obtains a value of 0.81 ± 0.12. The missing contribution corresponding to direct double photoionization in the 3p and 4s (or 4p) subshells does not seem to be high enough, according to recent measurements, to bring the value of this ratio much closer to 1. In that case, there would remain a small difference between the experimental results ans the theoretical prediction. The complete results of this study will be published elsewhere [43].

To conclude with photoionization of excited atoms, it is clear that the advent of new sources being presently built, such as the Advanced Light Source in Berkeley, will still increase the range of feasible experiments, possibly up to the combination of synchrotron radiation with pulsed lasers, to study the dynamics of excited states.[22]

III - PHOTOIONIZATION OF ATOMIC IONS

1°) History

Previous experimental studies on the interaction of photons with positively charged ions were mostly dedicated to photoabsorption experiments. The first results were obtained more than twenty years ago using two 3-electrode vacuum sparks, one serving as the background source, the other one as the source of the absorbing metallic vapor.[47] Resonant absorption lines of beryllium and magnesium ions in various charge states, including autoionizing lines, were observed at that time.[48-51] Simultaneously, P. Jaeglé et al.[52] invented the dual laser plasma method (DLP) : two plasmas are generated, with a proper delay, by focusing, onto thick targets, two laser beams obtained by splitting the beam of a powerful pulsed laser (neodynium first, then also ruby lasers). The first plasma acts as the source of continuum radiation, the second plasma is the absorbing medium. The presence of a given charge state ion in the plasma is controlled in playing with the optical delay between the two laser beams. This DLP method was first introduced to study the production of population inversion in highly charged ion plasmas[53], then for the observation of the first X-UV laser emissions.[54] But the new experiments on singly and doubly charged ions were obtained with laser driven resonant ionization.[6-9] A dye laser tuned to an atomic resonance line of the atoms or singly charged ions to be studied produce the ionized medium : if the density of excited atoms is high enough, in the 10^{16} atoms/cm^3 range, full ionization of the vapor occurs rapidly, via a sequence of ionizing collisions, namely associative and Penning ionization, producing the seeding electrons which then serves to ionize the medium after superelastic collisions. A pulsed source of radiation, generating a VUV continuum, was used, in synchronization with the pulsed laser, to observe photoabsorption spectra of the ionic species. In the beginning, the continuum radiation was produced with a BRV discharge source. Later on, the continuum radiation emitted by a laser produced plasma created under impact of a powerful laser beam onto a thick target, like in the DLP method, replaced the BRV source.[11,56] This method has been mostly limited to singly, and sometimes doubly charged, alkali - and alkaline earth- ions, because of the limited tunability range of the dye lasers. Na^+, Li^+, Ca^+, Ba^+ Ba^{++}, Cs^+ and also Mn^+ ions have been investigated in that way. In the mean time the use of the DLP method was systematically extended, mostly by Carroll and Kennedy, in Northern Ireland, and by Tondello-Jannitti and Nicolosi in Italy. Li, Be, C, Al and Si ions in various stages of ionization have thus been studied and some photoabsorption cross sections have been determined.[57-68] Most recently the behaviour of La^{3+} ions was efficiently observed and interpreted.[69]

The first succesful photon-ion experiments [70] with synchrotron radiation was a merged beam experiment involving ion spectrometry. In the interaction of a singly charged beam of Ba^+ ions with a monochromatic beam of synchrotron radiation along a path of about 12 cm, some of the singly charged ions were photoionized into doubly charged ions. Singly and doubly charged ions were detected after magnetic separation. In these experiments, the density of the ionized medium was very low, about ten orders of magnitude lower than in the dense plasma studies. Photoionization cross sections have been measured, at the Daresbury storage ring, using the fact that these cross sections are resonantly enhanced in the photon energy region where autoionizing states can be produced by inner-shell excitation of the singly charged ions. A number of ions have been studied[71-74] in that way, in the 20 eV-30 eV photon energy range. This type of experiment is now stopped but will likely be reactivated when the Advanced Light Source in Berkeley comes into operation. Several attempts have also been made to measure photoionizations cross sections in trapped ions,[75,76] but this method has not yet proved to be very efficient.

2°) Feasibility experiments with electron spectrometry[77]

As explained in the introduction of this papers, photoelectron spectrometry is the most suitable technique to provide detailed insight into photoionization process. However, the low density of ions focussed in the small source volume of an electron analyzer (at most 10^8 ions/cm^3, i.e. lower by three to four orders of magnitude than the currently used density of the sample in similar investigations on neutral atoms) makes it extremely difficult to observe these electrons. In addition, the transmission of an electron spectrometer is about 1%, as compared to nearly 100% in the case of ion spectrometry, while the length of the interaction region in both cases is about 1 cm and 10 cm, respectively. This makes understandable why it was absolutely necessary to demonstrate the feasibility of such experiments before launching an extended program of research on ions and to have access to an undulator to get the highest flux of photons available from synchrotron radiation.

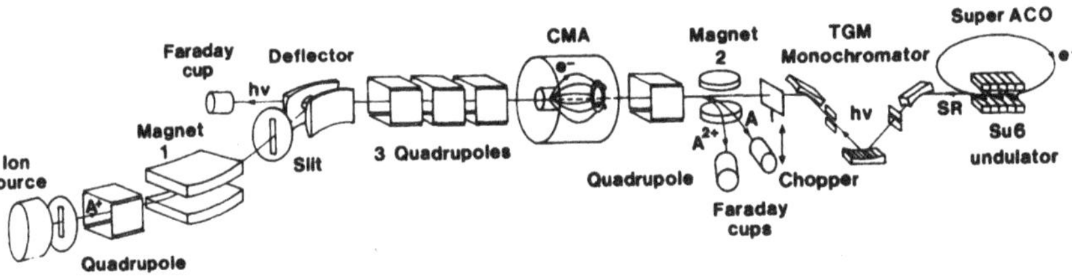

Fig.8
Layout of the PISA experimental setup.

The general scheme of the experimental setup is shown in Figure 8. In short, a Ca^+ ion beam of about 35 μA was focused, after magnetic selection and electrostatic deflection, into the source volume of a cylindrical mirror electron analyzer (CMA); the

The full photoionization process used in this feasibility experiment can be written as:

$$Ca^{+}\ 3p^{6}4s\ ^{2}S + h\nu \rightarrow Ca^{+*}\ 3p^{5}3d4s\ ^{2}P \rightarrow Ca^{++}\ 3p^{6}\ ^{1}S + e^{-}$$

Ca^{+} ions were selected because the energy and the cross section for photoionization of the $3p^{6}4s\ ^{2}S$ state via the $3p^{5}3d4s\ ^{2}P$ autoionizing state were known from previous photoabsorption[10] ($h\nu$ = 33.20 eV) and photoion[72] ($\sigma \sim 2200$ Mb) spectrometry measurements. Also the excitation energy of this resonance is in the tuning range of the undulator we have access to. The whole process can be described as the resonant photoionization of the 4s electron.

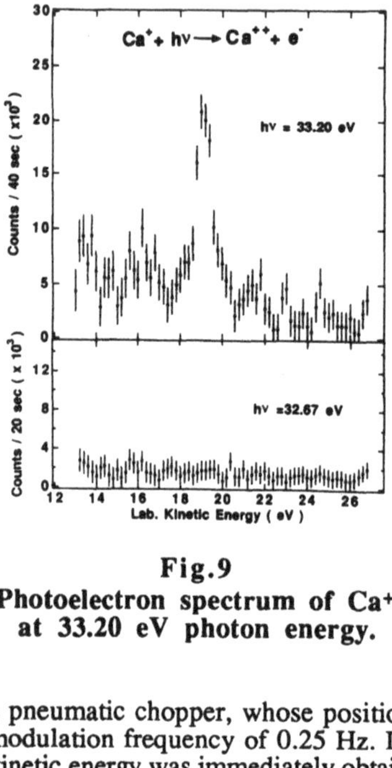

Fig.9
Photoelectron spectrum of Ca^{+} at 33.20 eV photon energy.

Taking into account the Doppler shift of the photons in the ion frame, the monochromator had to be set to 33.17 eV in the laboratory frame. The band pass was 0.3 eV.The photon flux of photons at 33.17 eV was optimized by tuning the gap of the undulator.

In a first series of tests, long term variations of the ion beam produced variations in the background of amplitude higher than the expected intensity of the photoelectron signal. Thus, even in reducing to 0.1 sec/channel the counting time, the substraction of the electron spectra measured with and without photons, respectively, in order to obtain the photon-induced spectrum, gave randomly scattered data. Then, we added on the path of the photon beam a pneumatic chopper, whose position was controled by the computer, with a typical modulation frequency of 0.25 Hz. In that way, the photon induced spectrum at each kinetic energy was immediately obtained by instantaneous substraction, in each channel, of the spectra measured with and without synchrotron radiation. The ionic current being constantly read by the computer, every time a discharge was detected the counting was repeated at the same analyzing energy in the spectrometer. This procedure revealed itself to be extremely efficient, minimizing the effect of ion beam fluctuations. A typical accumulation series was to count, in each of the 80 channels, the electron flux during typically 4 sec. with and without the photons in, to store them and to start again a new accumulation cycle. In that way, it was possible to accumulate a photoelectron spectrum during 40 to 50 sec/channel.

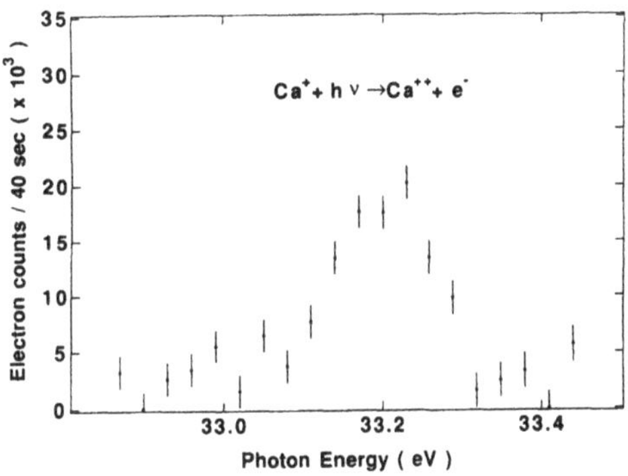

Fig.10
Constant-ionic-state spectrum over the Ca^+ resonance

Figure 9 shows two typical examples of the photoelectron spectra we have regularly obtained with this procedure, each time a beam of 400 mA was available at the injection in Super ACO. The ion beam current was measured to be about 35μA and the averaged beam current in Super ACO was 200 mA. In the upper part of the figure, one sees the photoelectron spectrum which is measured when the monochromator is tuned to the energy of the resonance. A very intense electron peak appears at about 19.2 eV kinetic energy. The ionization energy of Ca^+ ion being 11.87 eV, the kinetic energy of the autoionization electrons is equal to 21.30 eV when they are leaving the ion. However, two corrections must be applied to this value. The first one is due to the potential created by the ionic current itself. It has the effect to retard the electrons while they are leaving the ion beam. For a 35 μA ionic current, we calculated this retarding potential to be in the order of 5.5 Volts. From the shifted energy position of the Ne2p photoelectron line ejected by 33.20 eV photons with and without the ion beam going through the CMA, we measured this potential to be (5.6 ± 0.2 volts) for the spectrum of Figure 9, in excellent agreement with the calculated value. Thus, the electrons are ejected in the ion frame with an energy of (15.7 + 1.2) = 16.9 eV, + 1.2 eV being due to contact potentials in the CMA. The second correction is due to the velocity of the ions in the laboratory frame. 20 keV Ca^+ ions have a velocity of 3.0 x 10^7cm/sec. This velocity is equivalent to an accelerating potential in the laboratory frame. In the case of 16.9 eV electrons, this potential is equal to 2.4 V, rising the kinetic energy of the electrons in the laboratory frame to (19.3 ±0.2) eV, again in excellent agreement with the (19.2±0.1) eV measured value. On the left of the main electron line in Figure 9, one observes a low intensity structure which is due to the photoionization of the $1b_1$ state of H_2O present in the residual gas at a pressure of about 10^{-7} Torr. This is deduced from the measurement of the photoelectron spectrum emitted from the residual gas in the absence of ions. All main photoelectron lines of N_2 and H_2O are clearly observed in this spectrum. These lines feel the retarding potential induced by the ion beam, but not the accelerating potentiel due to the velocity of the ions. Thus, the $1b_1$ peak of H_20 in Figure 9 is shifted from its position in the spectrum without the ions by (-5.6 + 1.2) = -4.4eV, which brings its kinetic energy to 16.2 eV, in complete agreement with the measured value. In the lower part of Figure 9, we show the photoelectron spectrum which is observed when the photon energy is shifted by about 0.5 eV. The cross section for direct photoionization of the 4s electron being very weak (~1 Mb), the photoelectron line disappears, leaving a flat background, while the H_20 peak can still be detected.

monochromatized beam of synchrotron radiation emitted in the SU6 undulator of the super ACO positron storage ring was also focused into the source volume the CMA. The electrons emitted in the interaction were analyzed at an angle close to the magic angle.

The ions are produced in a plasma discharge ion source[78,79] which was developed at the Institut de Physique Nucléaire, in Orsay. The Ca^+ ions are formed by ionization of a Ca neutral vapor evaporated from an oven inside the source. The ions are extracted using an accelerating high voltage (20 keV here) applied between the source and the extraction electrode. After focusing by an electrostatic quadrupole, a first magnet (M_1) selects the ionic species of interest and focussed the desired q/m isotope on a collimating slit (2x30 mm^2). Then, the ions are deflected by a toroidal electrostatic deflector which makes the ion beam colinear to the axis of the CMA and to the photon beam. After focussing of the ion beam into the source volume of the CMA with the help of three electrostatic quadrupoles, the ions exit the CMA and are refocussed by another quadrupole at the entrance of a second magnet (M_2). This magnet deflects and separates the incident singly charged ions from the doubly charged ions produced either during the interaction with the photon beam or in collisional processes with the residual molecules, each ionic charge being counted with a Faraday cup.

The photon beam is produced in the SU6 undulator of Super ACO. The light is selected in wavelength by a toroidal grating monochromator and then focussed into the source volume of the CMA by a toroidal mirror. A gold foil, connected to an electrometer, could be introduced along the path of the photon beam to measure the variation of its intensity.

The CMA is basically the same electron analyzer already described in paragraph II-2 of this paper. However, it was slightly modified for this experiment. A colinear geometry was chosen to minimize the Doppler broadening of the photoelectron lines. Thus, to allow for the ions to go through the CMA, the electron detector was removed from the axis of the CMA. In addition, the diameter of the first inner cylinder was augmented to increase the maximum divergence angle possible for the ion beam. The averaged length of the interaction volume was thus about 17 mm along the axis of the CMA. Because of the high velocity of the ions, the electrons transmitted through the slits of the CMA are emitted at an angle higher than the magic angle (e.g. 60° for 20 eV electrons in the laboratory frame). Three high counting rate RTC channeltron placed in the focal plane of the photoelectrons at 120° from each other. One of the main problems of this kind of experiment is the high electronic background that comes with ion beams. It varies with the residual pressure in the experimental chamber. To minimize this background, a negatively polarized grid was placed in front of the channeltrons to protect them against the high rate of low energy electrons. The residual pressure was 1×10^{-7} Torr, after 150°C baking, with mostly, water and nitrogen molecules. The best overlap between the photon beam and the ion beam into the source volume of the CMA was achieved using a removable double side Faraday cup. The electron spectrum in the absence of photons, the photoelectron spectrum and the beam current were simultaneously recorded by a computer.

Finally, we show, in Figure 10, the spectrum recorded in the Constant Ionic State mode. In this mode, the pass electron energy in the CMA is kept at the top of the photoelectron line and the photon energy is scanned by regular intervall steps (0.03 eV). The resulting curve is the excitation function of the resonance which is proportional to the photoionization cross section. At variance with the case of a photoelectron spectrum, the width of the CIS function may be smaller than the value resulting from the convolution of the monochromator transmission curve with the natural shape of the resonance, which is the case here (0.18 eV). The CIS scan was very helpful to finely tuned the photon energy of the monochromator to the exact energy of the resonance, within 10 to 20 meV.

To conclude, we have demonstrated that it is already possible to observe and to measure photoelectron spectra resulting from the photoionization of a singly charged ion beam with undulator synchrotron radiation. In the present experimental conditions, the method is limited to the study of resonances with high cross sections. The next step is to build new experimental chambers with ultra high vacuum techniques, to improve the detection efficiency and, possibly, to increase the ion beam current focussed in the source volume of the CMA. With a reasonably expected gain of about a factor 4 in total, the counting rate at the top of a photoelectron line should reach about 1 count/Mb/sec. Having a residual gas pressure in the 10^{-10} Torr range, would drastically reduce the electron background due to photoionization of the residual gases present in the CMA. Thus, with the present ion source, it should be possible to study many resonant photoionization processes in singly charged ions, as well as Auger decay of core-excited ions using the white light emitted by the undulator. To go further, i.e. to study direct photoionization processes into the continuum and multiply charged ions, another ion source should be connected to an undulator beam line of Super ACO. This new source should be able to provide with high ion beam currents, at least in the 100 μA range, of low charge (l to 5) state ions in the source volume of the CMA. Presently, the best ion source seems to be an ECR ion source which would be optimized to deliver high currents of low charge state ions.

The project to connect an ECR ion source to an undulator beam line of Super ACO is presently being considered. Scientific justification for initiating new studies of the interaction processes between photons and multiply charged ions results from the fact that, except for a few measurements on mostly singly charged species, there is practically no quantitative data on ion targets, while they are of great importance for the understanding of astrophysical and laser-produced plasmas. From the theoretical point of view, there are calculations, mostly based on one-electron models, predicting the behaviour of photoionization parameters along various isoelectronic or isonuclear sequances, but these results need yet to be tested. For instance for highly charged ions, properties of the wave functions at small distances of the nucleus are expected to vary dramatically with the power of the atomic numbers. Relativistic and quantum electrodynamic effects are predicted to be of great importance in a few-electron highly charged ions.[80]

IV - CONCLUSION

The photoionization studies of atoms in excited states is about at the stage in which the studies on atoms in ground state were in 1980. For the ions, we are earlier, somewhere in the 70's. Thus, we expect that during the next ten years the pump-probe experiments (photoionization of excited atoms) and the experiments on atomic ions will know an increasing developement, similar to the one observed now in atoms and molecules. The advent of Infra-Red Free Electron Lasers and of mode loked lasers will extend considerably the field of excited atoms and molecules. The combination of an ECR source with synchrotron radiation, first (hopefully) at Super-ACO, later on the Advanced Light Source in Berkeley, the European Synchrotron Facility in Grenoble, the Advanced Photon Source in Argonne and possibly the future 2 GeV french storage ring, will widely open this new and exciting field, because stripped atoms vastly preponderate over neutral species in the universe.

V) ACKNOWLEDGEMENTS

The authors would like to thank Tom Morgan, Mathias Richter, S. Sorensen, E. Källne, F. Schlachter, B. Carré, J. Obert and J.C. Putaux for their efficient contribution to various parts of the works presented in this paper.

VI - REFERENCES

1. R.P. Madden and K. Codling, Phys. Rev. Lett 10, 516 (1963).
2. F.J. Wuilleumier, Atom. Phys. 7, 493 (1981).
3. F.J. Wuilleumier, in Workshop on "Photon -Ion Interactions", ESRF, Grenoble, 1989, p.3.
4. R.F. Stebbings, F.B. Dunning, F.K. Tittel, and R.D. Rundel, Phys. Rev. Lett. 30, 815 (1973).
5. T.B. Lucatorto and T.J. Mc Ilrath, Phys. Rev. Lett. 37, 428 (1976).
6. J. Sugar, T.B. Lucatorto, T.J. Mc Ilrath and A.W. Weiss, Opt. Lett. 4, 109 (1979).
7. T.J. Mc Ilrath and T.B. Lucatorto, Phys. Rev. Lett. 38, 1390 (1977).
8. T.B. Lucatorto and T.J. Mc Ilrath, App. Opt. 19, 3948 (1980).
9. T.B. Lucatorto, T.J. Mc Ilrath, J. Sugar and S. Younger, Phys. Rev. Lett. 47,1124(1981).
10. B.F. Sonntag, C.L. Cromer, J.M. Bridges, T.J. McIlrath and T.B. Lucatorto, in:"Short Wavelength Coherent Radiation Generation and Applications" (eds. D.T. Attwood and J. Bokor) p.412, Amer. Inst. of Phys. Conf. Proc. n°147, New York, 1986.
11. J.W. Cooper, C.W. Clark, C.L. Cromer, T.B. Lucatorto, B.F. Sonntag and F.S Tomkins, Phys. Rev. A35, 3970 (1987).
12. L.W.J. Kent, J.P. Connerade and A.P. Thorne, J. Phys. B20, 2947 (1987).
13. J.M. Bizau, J.M. LeGouët, D.L. Ederer, P. Koch, F.J. Wuilleumier, J.L. Picqué and P. Dhez, (1981), XI ICPEAC, Abstracts of post-deadline papers (S. Datz, ed.) p.1.
14. F.J. Wuilleumier, J. Physique, 43, C2-347(1982).

15. J.M. Bizau, F.J. Wuilleumier, P. Dhez, D.L. Ederer, J.L. Picqué, J.L. LeGouët and P. Koch, (1982), "Laser Techniques for the Extreme Ultraviolet Spectroscopy" (T.J. Mc Ilrath and R. Freeman, eds) American Institute of Physics Conference Proceeding n°90 (Am. Inst. of Phys. New York) p.331.
16. F.J. Wuilleumier, in: "X-ray and Atomic Inner-Shell Physics" (B. Crasemann, ed) AIP Conference Proc n°94, Am. Inst of Physics, New York, p.615.
17. A. Msezane and S.T. Manson, Phys. Rev. Lett. 35, 364 (1975).
18. A. Msezane and S.T. Manson, Phys. Rev. Lett. 48, 473 (1982).
19. J. Lahiri and S.T. Manson, Phys. Rev. Lett. 48, 614 (1982).
20. J. Lahiri and S.T. Manson, Phys. Rev. A37, 1047(1988).
21. F.J. Wuilleumier, D.L. Ederer and J.-L. Picqué, Adv. Atom. Molec. Phys. 23, 197(1988).
22. D. Cubaynes, J.M. Bizau, F.J. Wuilleumier, D.L. Ederer, J.L. Picqué, B. Carré, M. Ferray and F. Gounand, J. Physique 48, C9-453(1987).
23. D. Cubaynes, J.M. Bizau, B. Carré and F.J. Wuilleumier, in: "The Physics of Electronic and Atomic Collisions", XVI ICPEAC, AIP Conference Proceedings Series n°205,(A. Dalgarno, R.S. Freund, P.M.Koch, M.S. Lubbell and T.B. Lucatorto, eds.), 1990, Amer. Inst. Phys., New York, p. 224.
24. J.M. Bizau, F.J. Wuilleumier, D.L. Ederer, J.C. Keller, J.L. Le Gouët, J.L. Picqué, B. Carré and P.M. Koch, Phys. Rev. Lett. 55, 1281 (1985).
25. D. Cubaynes, J.M. Bizau, F.J. Wuillemier, B. Carré and F. Gounand, Phys. Rev. Lett. 63, 2460 (1989).
26. A. Nunnemann, T. Prescher, M. Richter, M. Schmidt, B. Sonntag, H.E. Wetzel, and P.J. Zimmermann, J. Phys. B18, L337 (1985).
27. B.F. Sonntag, Physica Scripta T34, 93(1991).
28. A. Fischer and I.V. Hertel, Z. Phys. A304, 103(1982).
29. M. Meyer, M. Pahler, Th. Prescher, E.V. Raven, M. Richter, B. Sonntag, S. Baier, W. Fiedler, B.R. Müller, M. Schulze and P. Zimmermann, Physica Scripta T31, 28 (1990).
30 - V. Schmidt, Phys. Lett. 45A, 63(1973).
31. J.M. Bizau, B. Carré, P. Dhez, D.L. Ederer, P. Gérard, J.C. Keller, P. Koch, J.L. Legouët, J.L. Picqué, G. Wendin and F.J. Wuilleumier, in: "Electronic and Atomic Collisions", XIII ICPEAC, eds. J. Leichler, W. Fitsch, I.V. Hertel, N. Stolterfoht and C. Wille, Contributed Papers, Berlin, 1983, p.27.
32. F.J. Wuillemier, in: "Laser Techniques in the Extreme Ultraviolet" (S.E. Harris and T.B. Lucatorto, eds.) AIP Conf. Proc. N° 119, 1984, Amer. Inst. of Physics, New York, p.220.
33. M. Meyer, B. Müller, A. Nunnemann, Th. Prescher, E.V. Raven, M. Richter, M. Schmidt, B. Sonntag and P. Zimmermann, J. Physique, 48, C9-551 (1987).
34. M. Ferray, F. Gounand, P. D'Oliveira, P.R. Fournier, D. Cubaynes, J.M. Bizau, T.J. Morgan and F.J. Wuilleumier, Phys. Rev. Lett. 55, 1281, (1987).
35. D. Cubaynes, J.M. Bizau, T.J. Morgan, F.J. Wuilleumier, M. Ferray, F. Gounand, P. D'Oliveira and P.R. Fournier J. Physique, 48, C9-5B (1987).
36. B. Carré, P. D'Oliveira, M. Ferray, P. Fournier, F. Gounand, D. Cubaynes, J.M. Bizau and F.J. Wuilleumier, Z. Phys. D15, 117 (1990).
37. J.M. Bizau, D. Cubaynes, P. Gérard, F.J. Wuilleumier, J.L. Picqué, D.L. Ederer, B. Carré and G. Wendin, Phys. Rev. Lett. 57, 306 (1986).

38. J.M. Bizau, D. Cubaynes, P. Gérard, F.J. Wuilleumier and J.L. Picqué, J. Physique 48, C9-491 (1987).
39. M.Y. Adam, F.J. Wuilleumier, N. Sandner, S. Krummacher, V. Schmidt and W. Mehlhorm, Japan J. Appl. Phys. 17-2, 43 (1978).
40. J.M. Bizau, D. Cubaynes, P. Gérard and F.J. Wuilleumier, Phys. Rev. A40, 3002, (1989).
41. M. Meyer, Th. Prescher, E.V. Raven, M. Richter, B. Sonntag, B.R. Müller, W Fiedler and P. Zimmermann, 48, C9-547 (1987).
42. M. Meyer, B. Müller, A. Nunnemann, T. Prescher, M.V. Raven, M. Richter, M. Schmidt, B. Sonntag and P. Zimmermann, Phys. Rev. Lett. 59, 2963(1987).
43. D. Cubaynes et al., to be published.
44. T.N. Chang and Y.S. Kim, J. Phys. B 15, L835 (1982).
45. J.M. Bizau et al., to be published.
46. M. Richter, J.M. Bizau, D. Cubaynes, T. Menzel, F.J. Wuilleumier and B. Carré, Euro Phys. Lett. 12, 35 (1990).
47. G. Mehlman-Balloffet and J.M. Esteva, Astrophys. J. 157, 945(1969).
48. J.M. Esteva, G. Mehlman-Balloffet and J. Romand 12, 1291 (1972).
49. G. Mehlman and J.M. Esteva, C. R. Acad. Sc. 276, 173(1973).
50. G. Mehlman and J.M. Esteva, Astrophys. J. 188, 191 (1974).
51. J.M. Esteva and G. Mehlman, Astrophys. J. 193, 747 (1974).
52. A. Carillon, P. Jaeglé and P. Dhez, Phys. Rev. Lett. 25, 140 (1970).
53. P. Jaeglé, G. Jamelot, A. Carillon, A. Sureau and P. Dhez, Phys. Rev. Lett. 33, 1070 (1974).
54. A. Carillon et al., submitted to Phys. Rev. Lett. (1991).
55. W. T. Hill, K.T. Cheng, W. R. Johnson, T.B. Lucatorto, T.J. Mc Ilrath and J. Sugar, Phys.Rev. Lett. 49, 1632 (1982).
56. J.W. Cooper, C.W. Clark, C.L. Cromer, T.B. Lucatorto, B.F. Sonntag, E.T. Kennedy and J. Costello, Phys. Rev. A39, 6074 (1989).
57. J.T. Costello, E.T. Kennedy, B.F. Sonntag and C.W. Clark, Phys. Rev. A43, 1441 (1991).
58. P.K. Carroll and E.T. Kennedy, Phys. Rev. Lett. 38, 1068(1977).
59. E. Jannitti, P. Nicolosi and G. Tondello, Physica. 124 C, 139 (1984).
60. E. Jannitti, M. Mazzoni, P. Nicolosi, G. Tondello and W. Yong-chang, J. Opt. Soc.Am. B2, 1078 (1985).
61. E. Jannitti, F. Pinghong and G. Tondello, Physica Scripta 33, 344 (1986).
62. E. Jannitti, P. Nicolosi and G. Tondello, Physica Scripta 36, 93 (1987).
63. E. Jannitti, P. Nicolosi and G. Tondello, Opt. Com. 63, 37 (1987).
64. E. Jannitti, P. Nicolosi and G. Tondello, Phys. Lett. A131, 186 (1988).
65. E. Jannitti, P. Nicolosi and G. Tondello, SPIE, vol. 911, 1988, p.11 and 157.
66. J. Brilly, E.T. Kennedy and J.P. Mosnier, J. Phys. B21, 368 (1988).
67. E. Jannitti, P. Nicolosi and G. Tondello, in:"Inner-shell Ionization and X-Ray Spectra", eds. S.T. Manson and M.O. Krause, AIP Conf. Proc. n° 215, 1990, Amer. Inst of Phys., New-York, P. 672.
68. P. Nicolosi, E. Jannitti and G. Tondello, J. Physique II, 1, CI-89 (1991).
69. J.E. Hansen, J. Brilly, E.T. Kennedy and G.O. Sullivan, Phys. Rev. Lett. 63, 1934 (1989).
70. I.C. Lyon, B. Peart, J.B. West and K. Dolder, J. Phys. B 19, 137 (1986).

71. I.C. Lyon, B. Peart and K. Dolder and J.B. West, J. Phys. B20, 1471 (1987).
72. I.C. Lyon, B. Peart and K. Dolder, J. Phys. B20, 1925 (1987).
73. B. Peart and I.C. Lyon, J. Phys. B20, 1673 (1987).
74. B. Peart, I.C. Lyon and K. Dolder, J. Phys. B20, 5403 (1987).
75. D.A. Church, S.D. Kravis, I.A. Sellin, J.C. Levin, R.T. Short, M. Meron, B.M. Johnson and K.W. Jones, Phys. Rev. A36, 2487 (1987).
76. S.D. Kravis, D.A. Church, B.M. Johnson, M. Meron, K.W. Jones, J. Levin, I.A. Sellin, Y. Azumah, N. Berrah Mansour, M.G. Berry and M. Druetta, Phys. Rev. Lett. 66, 2956 (1991).
77. J.M. Bizau, D. Cubaynes, M. Richter, F.J. Wuilleumier, J. Obert, J.C. Putaux, T. Morgan, E. Källne, S. Sorensen and A. Damany, Phys. Rev. Lett. 67, 576 (1991).
78. J.C. Putaux, J. Obert, L. Kotsila, B. Roussière, J. Sauvage-Letessier, C.S. Liang, A. Pehaire, P. Paris and J. Giroux, Nucl. Instr. Meth. 186, 321 (1981).
79. J.C. Putaux, J. Obert, J. Boissier and P. Paris, Nucl. Instr. Meth., 26 B, 213 (1987).
80. B. Crasemann, Invited paper presented at the XVII ICPEAC, Brisbane, 1991.

LIST OF PARTICIPANTS

ALVAREZ, IGNACIO
INSTITUTO DE FISICA, UNAM,
LABORATORIO DE CUERNAVACA
APDO. POSTAL 139–B
CUERNAVACA, MOR., 62190, MEXICO

AMAYA, ALEJANDRO
INSTITUTO DE FISICA, UNAM
LABORATORIO DE CUERNAVACA
APDO. POSTAL 139–B
CUERNAVACA, MOR., 62190, MEXICO

ANDERSEN, LARS H.
INST. PHYS., UNIV. AARHUS
DK–8000 AARHUS, DENMARK

BELTRAN, VIRGILIO
INSTITUTO DE CIENCIAS NUCLEARES
UNAM, CIRCUITO EXTERIOR, C.U.
MEXICO, D.F. 04510 MEXICO

BRIGGS, JOHN S
FAKULTAET FUER PHYSIK
UNIVERSITAET FRIEBURG
78 FREIBURG, GERMANY

BRYANT, HOWARD C.
DEPARTMENT OF PHYSICS & ASTRONOMY
THE UNIVERSITY OF NEW MEXICO
ALBURQUERQUE, N.M. 87131 USA

BUNGE, CARLOS F.
INSTITUTO DE FISICA, UNAM, MEXICO
P.O. BOX 20–364
MEXICO, D.F. 01000, MEXICO

CALDWELL, C.D.
DEPARTMENT OF PHYSICS
UNIVERSITY OF CENTRAL FLORIDA
ORLANDO, FL 32816 USA

CHAMPION, ROY L.
PHYS. DEPT., COLL. WILLIAM & MARY
WILLIAMSBURG, VA 231851, USA

CHURCH, D.
PHYSICS DEPARTMENT
TEXAS A & M UNIVERSITY

COLLEGE STATION, TX 77843 USA

CISNEROS, CARMEN
INSTITUTO DE FISICA, UNAM
LABORATORIO DE CUERNAVACA
APDO. POSTAL 139–B
CUERNAVACA, MOR., 62190 MEXICO

DUBOIS, ROBERT
BATTELLE PACIFIC NW LABS
P.O. BOX 999
RICHLAND, WA 99352 USA

ESCALANTE, V.
INSTITUTO DE ASTRONIMIA, UNAM
CIRCUITO EXTERIOR C.U.
MEXICO, D.F. 04510 MEXICO

FORD, A.L.
CENTER FOR THEORETICAL PHYSICS
TEXAS A & M UNIVERSITY
COLLEGE STATION, TX 77843

HAHN, YUKAP
DEPARTMENT OF PHYSICS
UNIVERSITY OF CONNECTICUT
STORRS, CT 06268 USA

HELM, HANSPETER
SRI INTL, MOLECULAR PHYSICS LAB
333 RAVENSWOOD AVE.
MENLO, CA 94025, USA

JAECKS, D.
PHYSICS DEPARTMENT
UNIVERSITY OF NEBRASKA
LINCOLN, NA 68588 USA

JAUREGUI,ROCIO
INSTITUTO DE FISICA, UNAM
P.O. BOX 20–364
MEXICO, D.F. 01000 MEXICO

JENSEN, R.
PHYSICS DEPARTMENT
YALE UNIVERSITY
NEW HAVEN, CT 06520 USA

JIMENEZ–MIER, J.
INSTITUTO DE CIENCIA NUCLEARES
UNAM, CIRCUITO EXTERIOR, C.V.

MEXICO, D.F. 04510 MEXICO

LEVENTHAL, JACOB J.
DEPARTMENT OF PHYSICS
UNIVERSITY OF MISSOURI
ST. LOUIS, MO 632121 USA

LEY–KOO EUGENIO
INSTITUTO DE FISICA, UNAM
P.O. BOX 20–364
MEXICO, D.F. 01000 MEXICO

LUBELL, M.
PHYSICS DEPARTMENT
CCNY
NEW YORK, NY 10031 USA

MACEK, J.
PHYSICS DEPARTMENT
UNIVERSITY OF NEBRASKA
LINCOLN, NA 68588 USA

MANSON, S.T.
DEPARTMENT OF PHYSICS & ASTRONOMY
GEORGIA STATE UNIVERSITY
ATLANTA, GA 30303 USA

MARTINEZ, HORACIO
INSTITUTO DE FISICA, UNAM
LABORATORIO DE CUERNAVACA
APDO. POSTAL 139–B
CUERNAVACA, MOR., 62190 MEXICO

McLAUGHLIN, BRENDAN M.
INSTITUTE FOR ATOMIC & MOLECULAR PHYSICS
HARVARD SMITHSONIAN, CENTER FOR ASTROPHYSICS
CAMBRIDGE, MA 02138 USA

MITCHELL, JBA
PHYSICS DEPARTMENT
UNIVERSITY OF WESTERN ONTARIO
LONDON, ONTARIO CANADA

MOKLER, PAUL H.
GSI, P.O. BOX 110552
D–6100 DARMSTAT GERMANY

MORGAN, THOMAS
PHYSICS DEPARTMENT
WESLEYAN UNIVERSITY
MIDDLETOWN, CT 06459–0155 USA

PEGG, D.
PHYSICS DEPARTMENT
UNIV. OF TENNESSEE
KNOXVILLE, TE 37916 USA

PETERSON, JAMES R.
MOLECULAR PHYSICS LAB.
PN 013, SRI INTL.
333 RAVENSWOOD AVE.
MENLO PARK, CA 94025 USA

PHANEUF, R.
PHYSICS DIVISION
OAK RIDGE NATIONAL LAB
OAK RIDGE, TN 37830 USA

RECAMIER, JOSE
INSTITUTO DE FISICA, UNAM
LABORATORIO DE CUERNAVACA
APDO–POSTAL 139–B
CUERNAVACA, MOR., 62190 MEXICO

SALAS, ALVARO
FACULTAD DE CIENCIAS, UNAM
CIUDAD UNIVERSITARIA
MEXICO, D.F., 04510 MEXICO

SALZBORN, E.
INSTITUTE F. KERNPHYSIK
UNIVERSITY OF GIESSEN
LEIHGESTERNER WEG 217
GISSEN, GERMANY 6300

SCHLACHTER, ALFRED S.
LAWRENCE BERKELEY LAB
4–230, UNIV. CALIFORNIA
BERKELEY, CA 94720 USA

STARACE, A.F.
PHYSICS DEPARTMENT
UNIVERSITY OF NEBRASKA
LINCOLN, NA 68588 USA

STAUFFER, ALAN D.
DEPARTMENT OF PHYSICS
YORK UNIVERSITY, TORONTO
CANADA M3J IP3

STOLTERFOHT, NICOLAUS
LABORATORIC DE SPECTROSCOPIE ATOMIQUE
ISMRa F14050

CAEN, CEDEX, FRANCE

SWENSON, JOSEPH K.
LAWRENCE LIVERMORE NATIONAL LABORATORY
LIVERMORE, CA 94550 USA

TANIS, JOHN A.
DEPT OF PHYSICS
WESTERN MICHIGAN UNIVERSITY
KALAMAZOO, MI 49008 USA

URQUIJO, JAIME DE
INSTITUTO DE FISICA, UNAM
LABORATORIO DE CUERNAVACA
APDO. POSTAL 139–B
CUERNAVACA, MOR., 62190 MEXICO

VAZQUEZ, GABRIEL
INSTITUTO DE FISICA, UNAM
LABORATORIO DE CUERNAVACA
APDO. POSTAL 139–B
CUERNAVACA, MOR., 62190 MEXICO

WUILLEUMIER, FRANCOIS
LURE–UNIVERSITY PARIS SUD
ORSAY, FRANCE 91405

www.ingramcontent.com/pod-product-compliance
Lightning Source LLC
LaVergne TN
LVHW020834261026
850861LV00047B/40